Weathering and Erosion Processes in the Natural Environment

Weathering and Erosion Processes in the Natural Environment

Edited by

Virendra Bahadur Singh
University of Delhi, New Delhi, India

Sughosh Madhav
Jamia Millia Islamia, New Delhi, India

Naresh Chandra Pant
University of Delhi, New Delhi, India

Ravi Shekhar
Centre for the Study of Regional Development, School of Social Sciences, Jawaharlal Nehru University, New Delhi, India

WILEY

This edition first published 2024

Registered Offices
John Wiley & Sons, Inc., 111 River Street, Hoboken, NJ 07030, USA
John Wiley & Sons Ltd, The Atrium, Southern Gate, Chichester, West Sussex, PO19 8SQ, UK

For details of our global editorial offices, customer services, and more information about Wiley products visit us at www.wiley.com.

Wiley also publishes its books in a variety of electronic formats and by print-on-demand. Some content that appears in standard print versions of this book may not be available in other formats.

Library of Congress Cataloging-in-Publication Data Applied for:
Hardback ISBN: 9781394157334

Cover Design: Wiley
Cover Image: Courtesy of Virendra Bahadur Singh

Set in 9.5/12.5pt STIXTwoText by Straive, Pondicherry, India
Printed and bound by CPI Group (UK) Ltd, Croydon, CR0 4YY

C9781394157334_201223

In the Loving memory of my beloved Mother, who had always been a source of inspiration for me.

Dr. Virendra Bahadur Singh

Contents

List of Contributors

Vikas Adlakha
Wadia Institute of Himalayan Geology
Dehradun, India

Amjad Al-Rashidi
Water Resources Development
Management Program
Water Research Center, Kuwait Institute
for Scientific Research, Kuwait

Bedour Alsabti
Water Resources Development
Management Program, Water Research
Center, Kuwait Institute for Scientific
Research, Kuwait

Uday Bhan
Department of Petroleum Engineering and
Earth Sciences, UPES
Dehradun, India

Roberto Dall'Agnol
Programa de Pós-graduação em Uso
Sustentável de Recursos Naturais em
Regiões Tropicais, ITV, Belém
Brazil; Instituto Tecnológico Vale (ITV)
Belém, PA, Brazil; and Programa de Pós-
graduação em Geologia e Geoquímica
Instituto de Geociências, Universidade
Federal do Pará, Belém, Brazil

Gessica da Silva
Programa de Pós-graduação em Uso
Sustentável de Recursos Naturais em
Regiões Tropicais, ITV
Belém, Brazil

Benidhar Deshmukh
Discipline of Geology, School of Sciences
Indira Gandhi National Open University
New Delhi, India

Rahul Devrani
Jindal School of Environment and
Sustainability, O.P. Jindal Global
University, Sonipat, Haryana, India

Prerna Diwan
Department of Microbiology, Ram Lal
Anand College, University of Delhi
New Delhi, India

Chandra S. Dubey
K.R. Manglam University
Gurugram, Haryana, India

Joystu Dutta
Department of Environment Science
Sant Gahira Guru Vishwavidyalaya Sarguja
Ambikapur, India; and IUCN Commission
on Ecosystem Management
Gland, Switzerland

Somenath Ganguly
Department of Petroleum Engineering and Earth Sciences, UPES,
Dehradun, India

Priyadarshan S. Ganvir
Department of Geology, M. G. Arts, Science and Late N. P. Commerce College
Armori, India

Lalit Goswami
Department of Chemical Engineering
Chungbuk National University
Cheongju, South Korea

Rajeeva Guhey
Department of Geology
Govt. N. P. G. College of Science
Raipur, India

Rakesh K. Gupta
Department of Microbiology
Ram Lal Anand College
University of Delhi
New Delhi, India

Sunila Hooda
Department of Microbiology
Ram Lal Anand College
University of Delhi
New Delhi, India

Sabu Joseph
Department of Environmental Sciences
University of Kerala
Thiruvananthapuram
Kerala, India

Aseem Kerketta
Department of Biotechnology
Sant Gahira Guru Vishwavidyalaya
Sarguja, Ambikapur, India

Suhas Damodar Khobragade
Hydrological Investigations Division
National Institute of Hydrology
Roorkee, India

Vamsi K. Kudapa
Department of Petroleum Engineering and Earth Sciences, UPES, Dehradun, India

Anil Kumar
Wadia Institute of Himalayan Geology
Dehradun, India

Ashish Kumar
Department of Biotechnology
Sant Gahira Guru Vishwavidyalaya
Sarguja, Ambikapur, India

Ashutosh Kumar
Department of Geology, Earth Science
Indira Gandhi National Tribal University
Amarkantak, Madhya Pradesh, India

Prem Kumar
Ram Lal Anand College
Department of Geology
University of Delhi, Delhi, India

Rohit Kumar
Discipline of Geology, School of Sciences
Indira Gandhi National Open University
New Delhi, India

Susheel Kumar
Centre for Petroleum Exploration
Mizoram University, Aizawl, India

Sushil Kumar
Department of Environmental Studies
Siksha Bhavana (Institute of Science)
Visva-Bharati, Santiniketan
West Bengal, India

Anupma Kumari
Department of Zoology
Patna University, Patna, India

Anamika Kushwaha
Department of Chemical Engineering
Chungbuk National University
Cheongju, South Korea

Ihsan U. Lone
Discipline of Geology, School of Sciences
Indira Gandhi National Open University
New Delhi, India

Nirlipta P. Nayak
Department of Petroleum Engineering and
Earth Sciences, UPES, Dehradun, India

Amrita K. Panda
Department of Biotechnology
Sant Gahira Guru Vishwavidyalaya Sarguja
Ambikapur, India

Neeraj Pant
Hydrological Investigations Division
National Institute of Hydrology, Roorkee
India; and School of Engineering
Design and Built Environment
Western Sydney University
Sydney, Australia

Mansi Podia
Department of Microbiology
Ram Lal Anand College, University of Delhi
New Delhi, India

Shailendra Pundir
Wadia Institute of Himalayan Geology
Dehradun, India; and Department
of Geology, Centre of Advanced Study
Kumaun University, Nainital, India

Leandro S. Quaresma
Programa de Pós-graduação em Uso
Sustentável de Recursos Naturais em
Regiões Tropicais, ITV
Belém, Brazil

Jaya Rai
Ram Lal Anand College
Department of Geology
University of Delhi, Delhi, India

Alagappan Ramanathan
School of Environmental Sciences
Jawaharlal Nehru University
New Delhi, India

Pooja Rani
Department of Vocational Education
Indira Gandhi National Tribal University
Amarkantak
Madhya Pradesh, India

Rakesh K. Ranjan
Department of Geology
Sikkim University, Sikkim, India

Chidambaram Sabarathinam
Water Resources Development Management
Program, Water Research Center
Kuwait Institute for Scientific Research
Kuwait

Prafulla K. Sahoo
Instituto Tecnológico Vale (ITV)
Belém, PA, Brazil; and Department of
Environmental Science and Technology
Central University of Punjab
Bathinda, India

Gabriel N. Salomão
Instituto Tecnológico Vale (ITV)
Rua Boaventura da Silva
Belém, PA, Brazil

Kiran Sathunuri
Discipline of Geology, School of Sciences
Indira Gandhi National Open University
New Delhi, India

Prabhat Semwal
Hydrological Investigations Division
National Institute of Hydrology, Roorkee
India

Singaraja Chelladurai
Department of Geology, Presidency
College, Chennai, India

Mohammed A. Sulaiman
Department of Zoology, Patna University
Patna, India

Shristy S. Swarnkar
Department of Biotechnology, Sant Gahira
Guru Vishwavidyalaya Sarguja
Ambikapur, India

Thilagavathi Rajendiran
Department of Geology
Anna University
Chennai, India

Jobin Thomas
Department of Geological and Mining
Engineering and Sciences
Michigan Technological University
Houghton, MI, USA

Thrivikramji Kythavilakom Pillai
Centre for Environment and
Development
Thiruvananthapuram
Kerala, India

Prerna Yadav
Department of Microbiology
Ram Lal Anand College
University of Delhi
New Delhi, India

Shailesh K. Yadav
Department of Geology
Sikkim University
Sikkim, India

Mohammad M. Zafar
Department of Zoology
Patna University
Patna, India

Preface

Weathering is a process in which the earth's surface breaks down, dissolves, or wears out under the influence of atmosphere, hydrosphere, cryosphere, biosphere, and nuclear radiation. Erosion is the movement of substances on or below the surface of earth by flowing water, ice, and wind. Weathering and erosion are essential geochemical processes responsible for the formation of soil, sediment, dissolved solute, landforms, etc. There are three types of weathering, namely physical, chemical, and biological weathering. Various factors such as the composition of rock minerals, texture, temperature, and rainfall affect weathering and erosion. Climate change also impacts the weathering and erosion processes. Various anthropogenic activities accelerate the weathering and erosion processes, while several management strategies are also applied for controlling the weathering and erosion processes. Weathering and erosion have great significance on different types of terrestrial and aquatic ecosystems. Weathering and erosion create various landforms on the earth. The water quality of aquatic ecosystems is also controlled by weathering and erosion processes. Chemical weathering in the aquatic environment plays an essential role in sequestration of CO_2 from the atmosphere that exerts substantial impact on the climate. Sediment flux and nutrient geochemistry of surface water body is governed by the weathering and erosion processes. Various toxic elements such as arsenic and fluoride are derived from weathering of hard rock minerals in the aquatic environment that causes multiple types of diseases in human beings. Weathering and erosion are among the most important natural phenomena with vital significance. Thus, there is an urgent need to know about the fundamentals and recent advancements in the weathering and erosion processes.

This book covers weathering and erosion processes in the terrestrial and aquatic environment. This book contains chapters related to physical and chemical weathering, role of microbes in weathering and erosion, rock–water interaction, textural and mineralogical signatures, atmospheric CO_2 consumption due to chemical weathering, climatic impact on weathering and erosion, sediment biogeochemistry, and causes of toxicity and pollution due to weathering in the aquifer. This current book deals with different geological and metrological factors that control weathering and erosion processes. The impact of climate change and anthropogenic activities on weathering and erosion processes has also been discussed in the current book. CO_2 sequestration induced with chemical weathering has also been discussed in this book. Sediment dynamics in the aquatic ecosystem and their relation with weathering will be part of this book. Metal fractionation and biogeochemistry in the aquatic ecosystem will also be part of this book. This book deals with the role of weathering in aquifer geochemistry and geogenic contamination in groundwater. It brings a holistic knowledge about the weathering and erosion processes, their measurement techniques, and significance to the natural environment. This book contains both practical and theoretical and latest and broad aspects of weathering and erosion.

This book will be useful for teachers, scientists, research scholars, and postgraduate students working in the field of geology, environmental sciences, weathering, erosion, geochemistry, hydrology, and climate change. We have tried to fill the gap in already published literatures by adding various aspects of weathering and erosion in this book. We have incorporated the book chapters of various authors having different backgrounds, which covers various aspects of weathering and erosion processes in the aquatic environment. We would like to express our gratitude to all the authors, reviewers, and publisher of this book for their dedicated work in publishing a remarkable and meaningful edited volume in the field of weathering and erosion.

1

Heavy Metals in the Sediment of River Ganga: A Review

Anupma Kumari, Mohammed A. Sulaiman, and Mohammad M. Zafar

Department of Zoology, Patna University, Patna, India

1.1 Introduction

Heavy metals are nonbiodegradable, persistent, and biomagnifying pollutant that can elucidate detrimental impact at a very low concentration, beyond which it can lead to several physiological malfunctions, edema of eyelids, nasopharynx congestion, cardiovascular disorders, and oxidative and genotoxic damages (Patel et al. 2021; Dustgeer et al. 2023; Hayder et al. 2023; Pandey and Kumari 2023). Many of such metals are significant constituents of several enzymes involved in the metabolic pathway of various redox reactions. However, their property to transit between different oxidative states and form organic complexes makes them potentially toxic to the biotic components of an aquatic ecosystem (Kumari et al. 2021). Exposure to certain heavy metals like arsenic (As), cadmium (Cd), chromium (Cr), lead (Pb), and mercury (Hg) is also reported by many studies to produce reactive oxygen species (ROS), causing a conformational change in DNA and even introducing double-strand DNA breaks, which further can lead to life-threatening cancerous diseases (Paul 2017; Abbas et al. 2019; Salama et al. 2019; Gökalp et al. 2022; Mishra et al. 2023; Singh et al. 2023). The route of heavy metals to the riverine system includes geogenic factors such as atmospheric deposition, metal corrosion, metal leaching, volcanic eruption, and rock-weathering processes and receives anthropogenic inputs from mining, smelting, and industrial, agricultural, and domestic effluents containing hazardous heavy metals (Matta et al. 2020; Liu et al. 2022; Xie et al. 2022). Once entered the aquatic environment, the availability of heavy metals cycles between the abiotic and biotic components of the ecosystem and is affected by temperature, pH, concentration and type of chelating ligands, the ionic state of the minerals, the grain size of soil, and microbial enzyme bioavailability (Pandey et al. 2015a; Jaiswal and Pandey 2018; Ehis-Eriakha and Akemu 2022). Moreover, river water is the transient host for the trace metal concentration. The bottom sediment of a riverine ecosystem serves as an absolute receptor for the toxic pollutants, which, when released into the overlying water column, exaggerate significant undesirable changes to the overall health of an aquatic environment (Verma and Pandey 2019; Aggarwal et al. 2022;

Weathering and Erosion Processes in the Natural Environment, First Edition. Edited by Virendra Bahadur Singh, Sughosh Madhav, Naresh Chandra Pant, and Ravi Shekhar.

Dhiman et al. 2022). A major portion of the transported toxic pollutants is diffused to the water system in sediment-associated form, which, after entering the river body, gets accumulated as inorganic complexes or gets bound to minerals in particulate form in the river sediments (Singh et al. 2017; Iordache et al. 2022; Zhang et al. 2022). Besides functioning as the source of a secondary pollutant, bottom sediment plays a crucial role in regulating the aquatic environment's nutrient cycle, food web, and trophic status. Unlike flowing water, the sedentary nature of sediment makes it a great asset for studying environmental conditions, pollution status, bio-concentration, and biomagnification pattern of the trophic level in a river ecosystem. Various statistical indices, including contamination factor (CF), enrichment factor (EF), geo-accumulation index (I_{geo}), and pollution load index (PLI), have used the bottom sediment as an indicator to assess the ecological health that marks the change in river water quality owing to some anthropogenic influences. The Ganga River Basin emerges from the Gangotri Glacier at Gomukh (30°36′N, 79°04′E) and extends to the Ganga Sagar (21°38′N, 88°04′E), sheltering more than 600 million people in India, Nepal, and Bangladesh and 446 vertebrate species of aquatic importance (NMCG 2022; Sati 2021) (Figure 1.1). It covers nearly 26.2% of the total geographical area of India, with a catchment area of 861,404 km^2. The length of the main stem of the River Ganga is 2525 km, which nourishes a population density of 482 person/km^2 in India. While passing through more than 100 cities of class I and class II categories and thousands of villages, different industrial, agricultural, and domestic effluents are discharged into the river ecosystem. According to the report of NMCG (2022), nearly 12,000 million liters per day (MLD) of sewage are

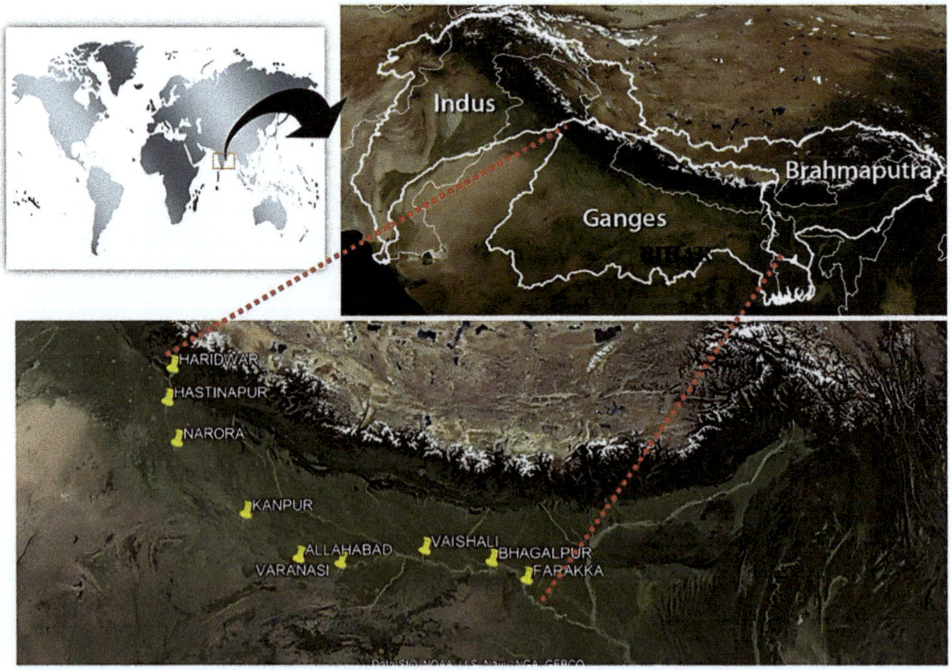

Figure 1.1 Map of the Ganga River Basin showing reviewed sampling locations of various researches since 1981. *Source:* Viste and Sorteberg (2015) / Copernicus Publications / CC BY 3.0.

generated, of which 3000 MLD of sewage are drained directly to the main stem of the River Ganga. Several studies have reported the presence of a considerable amount of heavy metals and other chemicals in the domestic sewage and industrial effluents drained into the Ganga River System (Pandey et al. 2015b). Dwivedi et al. (2018) have reported a ninefold increase in the untreated sewage discharge to the river since 1985. The pressure on various agricultural and industrial sectors to meet the demand of the rapidly increasing population has only contributed to worsening the conditions. Moreover, regular monitoring of river bed sediments by researchers at different sampling locations in the past 30 years has certainly kept track of the pollution status of River Ganga. Therefore, the chapter is aimed to compile and review various works done on the sediment of River Ganga and provides a descriptive analysis of the temporal variations in the heavy metal status in the sediment of River Ganga in the past three decades.

1.2 Source of Heavy Metals

1.2.1 Natural Sources

Heavy metals are distributed throughout the Earth's crust, and their presence in the water and sediments depends upon the regional geology, as rocks and soil are considered the chief sources of heavy metals in the environment (Aravinthasamy et al. 2021; Nivetha et al. 2021; Zimik et al. 2021). Both igneous and sedimentary rocks have been reported to store trace amounts of various heavy metals, but the role of the latter in soil formation is far greater than the former (Alloway and Jackson 1991; Okrusch and Frimmel 2020; Mukherjee et al. 2021). Natural emissions of heavy metals occur during volcanic eruptions, forest fires, and weathering of rocks (Bordoloi et al. 2020; Ali et al. 2021; Pandey and Tiwari 2021). The principal ores and heavy metals present are represented in Table 1.1.

1.2.2 Anthropogenic Sources

Anthropogenic sources have exceeded natural sources due to extensive human activities contributing to several heavy metals in the environment.

Lead is a bluish-green metal found naturally in Earth's crust, which has been designated as one of the most common industrial contaminants as it is found in trace amounts in food, water, and soil (Fu and Xi 2020; Zhang et al. 2019, Chen et al. 2021b). Several domestic and industrial applications contribute to the Pb level in the environment. These include mining, refining, smelting of ores, use of the lead acid battery, ceramics, leaded gasoline, ammunition, metal products (solder and pipes), stripping of leaded paints, and cable sheathings (Qader et al. 2021; Levin et al. 2021; Schindler et al. 2022). Cd is another important heavy metal of major concern worldwide (Chen et al. 2021a; Velusamy et al. 2021). The most common airborne Cd source is burning fossil fuels and domestic wastes. Industrial sources responsible for enhanced Cd levels in the environment include metal alloys, pigments, Ni—Cd batteries, electroplating other metals, coloring pigments, synthetic chemicals, and glass ceramics (Patel et al. 2021; Aggarwal et al. 2022). The use of phosphate fertilizers and manure in the agriculture sector further contributes to increased

Table 1.1 Heavy metals and their respective natural sources.

Sl. no.	Heavy metal	Ores	References
1	Pb	Galena (PbS), cerussite ($PbCO_3$), anglesite ($PbSO_4$), and minium (Pb_3O_4)	Hu et al. (2020); Xie et al. (2021)
2.	Cd	As impurity in sphalerite (ZnS)	Li et al. (2022); Sanz et al. (2022)
3.	Cr	Chromite ($FeCr_2O_4$)	Mohan et al. (2022); Rakhimov et al. (2022)
4.	Cu	Copper pyrite ($CuFeS_2$), tenorite (CuO), malachite ($Cu_2CO_3(OH)_2$), azurite ($C_2H_2Cu_3O_8$), chalcanthite ($CuSO_4{\cdot}5H_2O$), and brochantite ($Cu_4SO_4(OH)_6$)	Kang et al. (2021); Movahednia et al. (2022)
5.	Hg	Cinnabar (HgS) and calomel (Hg_2Cl_2)	Langcay et al. (2021); Petranich et al. (2022)
6.	Mn	Pyrolusite (MnO_2), manganite ($Mn_2O_3{\cdot}H_2O$), and braunite ($Mn^{2+}Mn^{3+}{}_6(SiO_4)O$)	Keshavarz et al. (2021); Luo et al. (2021)
7.	Zn	Sphalerite (ZnS), marmatite ((ZnFe)S), smithsonite ($ZnCO_3$), and zincite (ZnO)	Nikkhou et al. (2021); Saidi and Kadkhodayan (2022)
8	Ni	Pentlandite ($(Ni,Fe)_9S_8$), and garnierite ($(NiMg)_6Si_4O_{10}(OH)_8$)	Sibanda et al. (2019); Dzvinamurungu et al. (2022)
9.	Al	Bauxite ($Al_2O_3{\cdot}2H_2O$), corundum (Al_2O_3), cryolite (Na_3AlF_6), feldspar ($KAlSi_3O_8$), alunite ($KAl_3(SO_4)_2(OH)_6$), and kaolin (Al_2O_3 $2SiO_2{\cdot}2H_2O$)	Kuan et al. (2020); Onukwuli et al. (2021)

Cd level in the aquatic ecosystems (Haider et al. 2021; Cai and Li 2022). Chromium is a heavy metal whose elemental form, Cr does not occur naturally. It is found in various oxidation states with valency ranging from Cr (II) to Cr (VI). Chromium compounds have versatile uses in several industries and are a major anthropogenic contributor to the environment. Cr is extensively used in tanning, metal processing, electroplating, cement production, nonferrous alloy production, and paint manufacturing (Prasad et al. 2021; Younas et al. 2022). Emissions from Cr-based automotive catalytic converters and tobacco smoke also add to Cr level in the environment (Srivastava et al. 2021; Akpambang et al. 2022). Copper (Cu) is a naturally occurring heavy metal found in various ores. Cu is widely used due to its high thermal and electrical properties in various electrical cables, motors, circuit boards, and photovoltaic cells (Feary and Cullinan 2022). Major industries that discharge Cu include tanning, paper, pulp, and wood preservatives. Several fertilizers and fungicides also use Cu which may enter surface water bodies by runoff and groundwater by leaching (Chheang et al. 2021; Mir et al. 2021). Mercury is a naturally occurring volatile metal that is a shiny silver-white, odorless liquid in elemental form. Mercury is extensively used in various devices such as thermometers, hydrometers, pyrometers, and barometers (Zhao et al. 2021; Oteyola and Ola-Oladimeji 2022). It is also being used in dental preparations such as amalgams, nuclear reactors, pulp, and paper industries, as a component of batteries, and as a solvent for precious metals (Mensah et al. 2021). Arsenic is one of the most widely distributed heavy metals throughout the environment. Burning fossil fuels, smelting ores, and mining activities are major anthropogenic contributors to the environment (Kumari et al. 2021). Other sources include electrical wastes, pesticides, insecticides, and use of wood preservatives (Jang et al. 2016; Mehdi et al. 2021). Arsenic compounds have found application in medications for treating diseases in avian species, such as blackhead in turkeys, as well as in the treatment of African sleeping sickness (De Koning 2020; Nigam et al. 2021). Ni is a silvery-white metal having a slight golden tinge. The major anthropogenic sources of Ni contamination include silver refineries, mining, coal combustion, electroplating, zinc base castings, waste incineration, and various chemical fertilizers (Chow et al. 2021; Saidi and Kadkhodayan 2022; Kayode et al. 2022). The major anthropogenic sources of environmental Mn exposure are occupational, including wastewater discharges, sewage sludge, mining, mineral processing, emissions from alloy production (steel, silico-Mn alloy), combustion of fossil fuels, and use in fertilizers. Apart from these, Mn, to a lesser extent, is also contributed by the combustion of fuel additives (Gondek et al. 2021; Li et al. 2022). Zinc (Zn) is a lustrous, bluish-white metal that is one of the most widely distributed heavy metals in the Earth's crust (Hussain et al. 2022). Zinc is mostly discharged during several industrial activities such as industrial effluents from galvanizing industries, smelting of ores, wood preservatives, and polymer stabilizers. Mining, fossil fuel combustion, zinc-containing pesticides and fertilizers, and dispersants used in motor oil contribute to anthropogenic sources of Zn in the environment (Parmar and Thakur 2013). Aluminum is a versatile and the third most abundant element found in the Earth's crust (Gupta et al. 2009). Aluminum is central in several fields, such as transportation, food packaging, construction, and electronic hardware. The automobile and aircraft industries are the major contributors of Al followed by construction, utensil, and electric industries (WHO 2011; Alasfar and Isaifan 2021). The use of potash alum in treating drinking water is also another emerging source of contamination. The heavy metals are persistent for long durations, and their continuous addition from different sources contributes to elevated levels over time.

1.3 Effects on Human Health

Lead is considered a toxicant as its exposure could lead to several toxic effects on the body (Paithankar et al. 2021; Kianoush et al. 2012). Lead toxicity can adversely affect the functioning of the entire human body, but its effect is more impactful in children than adults (Wani et al. 2015). Hameed et al. (2019) have reported that Pb can easily cross the placental barrier and severely degrade the developing child's nervous system. Exposure to higher lead levels leads to an inflammatory response in various body organs and is associated with several neurological, urinary, and cardiovascular abnormalities (Wani et al. 2015; Balali-Mood et al. 2021; NIOSH). Cadmium is a potential human carcinogen whose exposure has been associated with lung, prostate, breast, pancreatic, nasopharynx, and urinary bladder cancer (Genchi et al. 2020). Cd exposure enhances the production of ROS, which results in oxidative stress promoting abnormal gene expression and the inhibition of DNA repair mechanism and apoptosis (Xu et al. 2019). Schaeffer et al. (2020) reported that the kidneys were most vulnerable to Cd leading to severe tubular dysfunction. Other complications have also been associated, such as osteomalacia, pulmonary edema, testicular function, and hemopoietic system damage. Chromium element is most stable as Cr (III) followed by Cr (VI), and its toxicity greatly depends upon the oxidation state of the metal. Cr (VI) compounds are powerful oxidizing agents compared to Cr (III) and have been characterized as Group I carcinogen by IARC (2013). Wise et al. (2019) have reported oxidative damage in Cr (VI)-exposed cells revealing its genotoxic nature. Inhalation by oral or dermal routes results in nasal irritation, nasal ulcer, asthma, and lung cancer. The kidney is the second most affected organ by Cr after the lung (Tsai et al. 2017). Dermal exposure to chromium (VI) may lead to skin ulcers and dermatitis (Hessel et al. 2021). Cu is an essential micronutrient that has been linked to Wilson's disease when exposed to high concentrations. Cases of jaundice and other liver-related complications have also been reported. Shah et al. (2019) have highlighted a large number of lung cancer cases associated with using copper-containing sprays. Mercury is a toxic heavy metal that occurs in elemental, organic, and inorganic forms. These three forms of mercury differ in their degree of toxicity and effects. Mold et al. (2019) reported kidney-related disparities, such as tubular necrosis, tubular dysfunction, membranous glomerulonephritis, and nephrotic-range proteinuria due to exposure to Hg. Hg ingestion directly affects the gastrointestinal system with symptoms such as indigestion, abdominal pain, ulcers, and bowel disease. Inhalation, ingestion, or dermal exposure to different mercury compounds alters the neurological response and leads to behavioral disorders. Hg compounds can easily cross the placental barrier and retard the fetal development. Arsenic is one of the most common environmental pollutants responsible for many deaths throughout the year (Xu et al. 2019). Arsenic affects almost every body organ system, including skin, cardiovascular, excretory, nervous, and respiratory. The International Agency for Research on Cancer has classified it as a class I human carcinogen. Skin, bladder, kidney, liver, prostate, and lung cancers are most common after arsenic exposure (Martinez et al. 2011). These degradation health effects occur chiefly through the induction of oxidative stress, histone modification, miRNA expression, and alterations to DNA methylation (Ray et al. 2014). Due to its versatile uses, nickel exposure is very common in humans (Genchi et al. 2020). Using Ni in jewelry results in constant contact with the skin, which may lead to dermatitis referred to as "Nickel itch" (Mahurpawar 2015). Soluble and insoluble Ni compounds are classified as Group 1 carcinogens, whereas Ni alloys are designated as Group 2B carcinogens (IARC 2013). Zambelli and Ciurli (2013) reported that chronic exposure to Ni might lead to lung cancer, lung fibrosis, and several other kidney and cardiovascular diseases. Zinc is an essential micronutrient, and consumption of small amounts of Zn is essential for survival (Laura et al. 2021). However, acute or chronic exposures can disrupt certain human body functions, leading to certain complications (Nriagu 2007). Inhalation of fumes containing ZnO leads to metal fume fever. The common symptoms include chest pain, cough, fever, and gastroenteritis (Plum et al. 2010). High doses of oral Zn uptake led to several gastrointestinal symptoms, such as nausea, vomiting, and abdominal cramps (Zhang et al. 2012). The acute toxicity of Al is very low but is known for its neurotoxic effects (Klotz et al. 2017). Wang et al. (2016) reported that Al exposure could lead to encephalopathy and dementia syndrome. Elevated Al^{3+} levels have been reported in Alzheimer's patients, but connections between the two are yet to be clarified (Ulusoy et al. 2015).

1.4 Status of Heavy Metal in the Sediment of River Ganga

Several studies on the heavy metal pollution in the sediment of River Ganga have been carried out in the past three decades (1990–2020). This section will present the findings of the spatial and temporal concentrations of heavy metals in sediment of River Ganga, source identification, fractionation of heavy metals, and the recently studied subject of assessing microbial enzyme activities on riverbed sediments and their relation with heavy metal bioconcentration in sediment. For a better understanding of the work done on heavy metal pollution in the sediment of River Ganga, this section is reviewed under two broad subheadings.

1.4.1 Heavy Metals in the Sediment of River Ganga Before 2010

Table 1.2 summarizes the research done on heavy metal pollution in the sediment of the Ganga River System before 2010. The entire River Ganga stretch has been assessed for heavy metal contamination. Ajmal et al. (1987) worked on nine different heavy metals in the 1100-km-long River Ganga and found a substantially increased level of heavy metal pollution. Kanpur was reported with a high concentration of Cd (3.10–3.48 mg/kg) and Co (13.00–14.4 mg/kg), while the sediment of the river system at Varanasi was found to be predominated by metal contamination of Cr (66.0–83.16 mg/kg), Mn (140.5–470.0 mg/kg), and Ni (24.7–28.80 mg/kg). The highest concentrations of Pb (20.0–21.8 mg/kg) and Zn (396.0–418.6 mg/kg) were observed at Monghyr and Fe at Norora. The study also compared the level of heavy metal concentration in different zones, where downstream sites were revealed to have a greater impact on metal contamination. Singh et al. (2003) carried out a similar study at 27 sampling locations, covering an area of 1700 km in the Ganga River System. The result has revealed that all the heavy metals have higher estimated values than their average shale concentration in the studied region of River Ganga except for Fe and Ni. It has comparatively discussed the metal profile in the sediment of the Upper Gangetic Region and the Lower Gangetic Region. The high concentrations of Mn, Fe, Co, Cu, and Cd in the upper segment of the studied area and Cr, Zn, Ni, and Pb in the lower stretch of River Ganga suggest the role of industrial and domestic contamination in the lower part of riverine system. Singh et al. (2002) performed a study to check the status of the heavy metal pollution in the Ganga River and its major tributaries. A massive increase in the cumulative EF was reported within the 15-km-long stretch of Yamuna River (a major tributary of River Ganga) at Delhi and Agra, which explains untreated discharges and accumulation of heavy metal in the sediment from the urban centers. Nearly every selected heavy metal shows EF > 1 in Gangetic Plain except Mn (0.87), Co (0.99), and Ni (0.74) at Kanpur and Ni (0.83) at Varanasi. The study further quantitatively categorizes the heavy metal pollution based on the geo-accumulation index (I_{geo}) and indicates very high pollution of Cd; moderate-to-high pollution of Cu and Zn; moderate pollution of Ni, Cr, and Pb; and least pollution of Fe, Mn, and Co in the Ganga River System. Kumari et al. (2021) and Purushothaman and Chakrapani (2007) also assessed the pollution status of heavy metals in the sediment of River Ganga using different indices and reported the effectiveness of statistical tools in assessing river water quality. Kumari et al. (2021) showed the abundance order of heavy metals as Cd > Cr > Pb at upstream and

Table 1.2 The concentration of heavy metals present in the sediment of River Ganga before 2010.

Study area	Sampling year	HMs concentration (in mg/kg)									References
		Fe	Mn	Zn	Pb	Cu	Cr	Ni	Co	Cd	
Norora—Monghyr	1981	2168.0–11,624.8	110.5–470.0	72.0–418.6	0.65–21.8	11.27–95.0	9.0–83.16	3.45–28.80	2.35–14.4	0.20–3.48	Ajmal and Khan (1987)
Hastinapur–Farakka	1993	34,100–46,000	1150–3070	87–181	18–353	44–69	121–200	35–63	14.7–25.3	0.41–1.31	Singh et al. (2003)
Delhi–Varanasi	1993	28,700–61,100	440–1750	90–1974	14–856	33–1204	115–817	35–538	11.7–29.0	0.14–114.8	Singh et al. (2002)
Varanasi	2001	–	–	–	3.3–181.73	–	13–112.07	–	–	0.2–6.5	Kumari et al. (2021)
Srinagar–Patna	2005	66,550–197,315	2982–2924, 365	301–878	190–1297	310–1004	208–4875	122–394	113–461	–	Purushothaman and Chakrapani (2007)
Allahabad	2005–2006	–	–	10.48–20.40	4.28–8.40	0.98–4.42	1.80–6.40	–	–	0.14–1.40	Gupta et al. (2009)
Bhagalpur	2006–2007	–	–	BDL-870	–	BDL-90	BDL-140	BDL-80	–	ND	Singh et al. (2012)

BDL = Below Detection Limit.

downstream sites, where downstream sites are marked by an increase in concentration from their reference sites (except Cd). High concentrations were noted in the pre-monsoon season, where the annual mean concentration does not show any significant variation in their sampling location and seasons. Geo-accumulation index (I_{geo}) has classified Cr as the least concerned heavy metal and Cd as the significant one, with a I_{geo} value of 2.51 upstream and 1.82 downstream sites. The study explains natural and anthropogenic factors such as synthetic use of agricultural discharge, urban-industrial release, and high traffic density as possible reasons for the heavy metal enrichment in the sediment of the riverine system. Purushothaman and Chakrapani (2007), working on 20 sampling locations from Srinagar to Patna, have inferred a high value of the geo-accumulation index, and the EF for all the heavy metals in the downstream region is attributed to the discharge of particle-laden industrial and urban effluents. They have also worked on fractionating heavy metals in Ganga River sediment. The result suggests Fe and Mn make up the bulk of the heavy metal fraction with the highest Fe concentration of 197,315 mg/kg at Varanasi in the main stem of Ganges and 231,244 mg/kg from Sone, a tributary of River Ganga. Mn has ranged from 2982 to 24,365 mg/kg at Bhagalpur. Other heavy metals that were noted high in the main stem of River Ganga include Co (461 mg/kg) at Bhagalpur, Cu (1004 mg/kg) and Pb (1297 mg/kg) at Srinagar, Zn (878 mg/kg) at Kanpur, Ni (394 mg/kg) at Varanasi, and Cr (4875 mg/kg) at Farakka. The heavy metal was analyzed for different fractions, including exchangeable ion and CO_3^{2-} fraction, Fe—Mn hydroxide fraction, associated with sulfide and organic fraction, and associated lithogenic mineral fractions. Lithogenic or residual fraction is the major carrier of heavy metals in the aquatic system except for Cu, Cr, and Zn, which occur mostly in the oxidizable or exchangeable fraction. A high value of Mn, Zn, and Cu in the exchangeable fraction indicates their ability to replace Ca in $CaCO_3$, whereas the presence of Cr, Ni, Cu, and Zn in the Fe—Mn hydroxide fraction indicates their absorption by Fe—Mn colloids. The sulfide fraction has a significant concentration of all metals except Mn. The study by Gupta et al. (2009) in Allahabad noted the abundance order of heavy metals as Zn > Pb > Cr > Cu > Cd. It has been noted that the metal accumulation in the sediment is associated with grain size and organic matter content, and found an increase in organic matter and decrease in particle size concentrates metal is sediments. The levels of Cu, Cr, Cd, Zn, and Ni were assessed in the Middle Gangetic Plain of Bhagalpur by Singh et al. (2012). The studied stretch marked high concentrations of Zn and Cr both in water and sediment, while the absence of Cd was recorded during 2006–2007. The maximum concentration of Zn (0.87 mg/g) in the sediment was recorded downstream of the burning site and Cr (0.14 mg/g) at the upstream side of Champanala. Automobile workshops, hospitals, sewage waste dying, printing, manufacturing, and storage of industrial waste have been suggested as some potential route sources of heavy metal contamination in the studied area.

1.4.2 Heavy Metals in the Sediment of River Ganga After 2010

The study conducted after 2010 not only assessed the heavy metal concentration in water and sediment of River Ganga but also used fractional and multivariate statistical techniques for the source characterization of heavy metals. The role of physicochemical characteristics, vertical distribution of heavy metals in soil, spatial–temporal and seasonal variations, microbes, and enzymatic activities are some major novelties included in heavy metals' assessments. An extensive review of these studies has been summarized in Table 1.3. Pandey et al. (2014) studied the effect of wastewater discharge on river water quality in Varanasi. The selected nine heavy metals (Cr, Mn, Fe, Co, Ni, Cu, Zn, Cd, and Pb) were analyzed for their concentration and abundance in geochemical fractionation. The predominance of Mn and Fe in the residual fraction suggests their lithogenic origin, and the presence of other metals in the reducible, exchangeable, and oxidizable fraction indicates some anthropogenic influence. The study also evaluated the relation between pH and heavy metals and found that slightly acidic ($pH = 6$) to highly basic ($pH = 14$) conditions favor the redox reaction of these metal ions, which enhances their interconversion from one form to another in the geochemical fraction. Heavy metal pollution in the sediment of River Ganga at Varanasi has been reported by Pandey et al. (2015b), Pandey and Singh (2017), and Jaiswal and Pandey (2018). Pandey et al. (2015b) studied the metal accumulation in the sediment of River Ganga using pollution indices such as CF, EF, and geo-accumulation index. The indices combinedly have revealed moderate-to-severe pollution in the sediment of the Ganga River, where EF, CF, and I_{geo} value have varied from 8.76 to 14.16, 1.86 to 3.43, and 0.31 to 1.19, respectively. The highest geo-accumulation index value was noted for Cd (7.27–8.97), followed by Pb (2.41–3.24). Mn, Fe, Ni, and Cu show negative geo-accumulation values, while Cr ranges from 0.31 to 1.19 in the studied area. A CF greater than one was observed for nearly all selected heavy metals except Mn, Fe, 27% of Ni, and 13% of Cu. The partial least squares regression test has revealed the proportional effect of pollutants on heavy metal enrichment arising from drains, affecting the entire studied stretch. Pandey and Singh (2017) assessed upstream and downstream urban influences on the sediment of River Ganga and found a rising trend in the downstream concentration of selected heavy metals under the influence of agricultural and urban untreated wastewater discharge to the riverine system. The abundance order of heavy metal was noted to be $Fe > Mn > Zn > Cr > Cu > Ni > Pb > Cd$. The study uses principal component analysis to group sites based on metal enrichment, which has clustered four different groups, where less-polluted sites are placed in one group and sites receiving high heavy metal pollution have been grouped separately. Correlation between metals has suggested common source origin and geochemical behavior of metals, while a high correlation of Mn and Cd with organic carbon ($r_{Mn/OC} = 0.93$, $r_{Cd/OC} = 0.87$) indicates the chelating behavior of these metals. Rajmohan et al. (2014) studied the vertical distribution of heavy metals in the soil of the Eastern Ganga Basin agricultural field. The study also marked a high correlation between Fe, Mn, Zn, and Ni, suggesting common geogenic source, while Pb, Cd, and Co barely showed a correlation with other heavy metals indicative of their anthropogenic inputs. The vertical distribution of pH revealed an increasing trend of pH with depth. A high pH value in soil enhances more negative charge resulting in high absorption and coprecipitation of heavy metals in sediment. Further, the study divides soil into clay and sand zones and has noted the presence of Fe, Mn, Zn, and Ni in the clay zone, Co in the middle, and high enrichment of Cd in the sand zone. The study categorizes sediments of the Eastern Ganga Basin into uncontaminated to moderately contaminated zone based on the values of CF, EF, and geo-accumulation index. Effective range low (ERL) and effective range medium (ERM) values were compared with the heavy metal concentration of the studied stretch to elucidate the degree of heavy metal contamination

Table 1.3 The concentration of heavy metals present in the sediment of River Ganga after 2010.

Study area	Sampling year	HMs concentration (in mg/kg)									References
		Fe	Mn	Zn	Pb	Cu	Cr	Ni	Co	Cd	
Varanasi	2012–2013	7175.5–9385.0	350.87–409.44	137.25–201.2	148.83–211.36	12.67–84.0	126.84–196.11	14.63–82.5	29.98–102.24	9.52–79	Pandey et al. (2014)
Varanasi	2012–2013	7493.91–10,343.0	322.43–439.75	185.15–278.61	151.85–269.38	15.3–70.7	133.75–247.05	18.38–97.1	—	30.01–128.13	Pandey et al. (2015a)
Varanasi	2012	21,924.07–41,170.13	316.81–529.08	41.1–92.5	10.94–44.89	12.71–36.68	39.05–93.28	11.77–43.02	—	0.94–2.86	Pandey and Singh (2017)
Vaishali	—	21,520–62,430	317–810	56.0–160.4	2.90–32.60	29.80–99.70	37.20–351	21.20–114.9	19.70–73.00	0.04–0.15	Rajmohan et al. (2014)
Allahabad	2015	516.48–687.41	28.68–85.96	—	5.02–8.49	—	2.41–6.40	—	—	0.74–1.85	Pandey et al. (2015b)
Allahabad	2016	—	—	87.0–111.0	18.0–29.0	48.0–62.0	121.0–142.0	—	—	0.47–0.91	Pandey et al. (2016)
Haridwar to Kanpur	2016–2017	—	—	—	1.5–58	3.3–17	3.6–83	—	—	BDL-9.7	Kumar et al. (2020)
Uttar Pradesh	2016–2017	17,389–49,568	139–2167	6.3–104.3	2.1–36.5	2.1–73.98	7.12–155.0	3.54–53.1	—	0.21–3.6	Siddiqui and Pandey (2019)
Haridwar to Varanasi	2016–2017										Jaiswal and Pandey (2018)
Kanpur	2018	5882.17	—	66.35	6.55	19.17	135.9	22.74	16.47	3.89	Kumar et al. (2019)
Uttar Pradesh	—	26,981–65,546	225.5–624.8	42.0–64.0	5.9–29.2	21.3–42.0	30.58–56.3	5.05–38.8	—	0.41–4.8	Verma and Pandey (2019)
Kannauj and Varanasi	2016–2019	—	—	44.59	31.06	30.60	52.12	35.63	—	0.17	Jaiswal and Pandey (2021)
Varanasi	—	—	355.622	—	158.455	—	154.328	—	—	46.717	Mishra et al. (2023)

to cause biological effects. The result has noted a lower value of Pb, Cd, and Zn than their respective ERL and ERM values, where only Ni has exceeded the ERM value, with 87% of samples having a chance to cause biological effects. Pandey et al. (2015b) reported the significance of CF and PLI in the heavy metal assessment of sediment. The study carried out in the region of Allahabad has revealed the dominance of Fe followed by Mn > Pb > Cr and Cd. A positive correlation of Fe was noted with Cr ($r = 0.970$), Cd ($r = 0.907$), and Pb ($r = 0.943$) and Pb with Cd ($r = 0.995$) and Mn ($r = 0.985$), while a negative correlation of Mn with Cd ($r = -0.075$) suggests their differences in geochemical behavior. CF > 1 was observed for Cd, which has varied from 2.972 to 6.16, suggesting moderate to considerable contamination of Cd in the sediment of River Ganga. Kumar et al. (2020) investigated the spatial variation of heavy metal contamination in the sediment of River Ganga from Haridwar to Kanpur and reported an alarming condition of Ganga River for heavy metal accumulation in sediment. The average concentrations of Pb, Cr, Cu, and Cd were found to be 216,500 mg/kg, 212,800 mg/kg, 169,100 mg/kg, and 23,650 mg/kg, respectively. A relatively high value of Cd was reported at the upstream site, while other heavy metals showed an increasing trend at their downstream sites. An assessment of nine heavy metals for their CF, ecological risk, PLI, and geo-accumulation index of the river sediment in Uttar Pradesh has been done by Siddiqui and Pandey (2019). The abundance order of heavy metals in the sediment was noted as Fe > Mn > Cr > Zn > Cu > Ni > Pb > Cd. None of the samples have exceeded the probable effective concentration (PEC) value for Cd, Pb, Cu, and Zn. However, 39.2 and 18.5% of samples have surpassed this value for Cr and Ni. The I_{geo} value has revealed the unpolluted status of River Ganga in its upper studied stretch while showed moderate-to-heavy pollution at downstream sites. It has also detailed the possible source of heavy metals in river sediment. It has suggested the geogenic origin of Fe and Mn and the role of anthropogenic factors in the elevated concentration of Cr, Cd, Cu, Ni, Pb, and Zn. A very similar trend of heavy metal abundance was also observed by Kumar et al. (2019). It revealed the trends of metal concentration in water, sediment, and consumers of different trophic levels and considered sediment as the "sink" for metal pollution in river water and "source" for bottom trophic consumers. Verma and Pandey (2019) have analyzed metal fractionation, risk assessment, and the relationship between metals and microbial enzymatic activities. Metal fractionation has revealed an inverse relationship of metals with depth and has found a significant concentration of Cr, Cu, Pb, Cd, Fe, and Mn in the residual fraction, while Zn and Ni in the reducible fraction. It has been noted that Cd, Cr, Cu, and Ni have exceeded the threshold effective concentration (TEC) value for 34.3%, 28.6%, 28.6%, and 22.9% of the samples, respectively. However, none of the samples exceeded any metal's PEC value. The study also considered sediment microbial biomass (C_{mic}) as a potential tool for investigating heavy metal pollution. It has found a clear relationship between organic carbon, C_{mic}, and heavy metal concentration. According to Anderson (2003), a higher level of organic carbon generally masks the impact of heavy metals on microbial biomass, and leads to higher concentration of heavy metals in the sediment. However, failure of the aforementioned relationship indicates increased levels of heavy metal pollution in sediment. It also has assessed metabolic quotient (qCO_2) as the ratio of basal respiration to substrate-induced respiration. An increase in the value of qCO_2 suggests heavy metal pollution. The study marked a positive correlation of qCO_2 with heavy metal ($r = 0.62$–0.86). A significantly negative correlation between

C_{mic} and enzyme activity with exchangeable fraction and a weak negative correlation with a reducible fraction of heavy metals suggest their role in reducing microbial enzymatic activities in sediment. Risk assessment (Er) evaluated by the work has reported low-to-medium risk for all metals except for Pb, which showed a high level of ecological risk. The geo-accumulation index has classified River Ganga as polluted for Pb, Cd, Ni, Cr, and Zn while unpolluted for Fe, Cu, and Mn despite being present at high concentrations. Moreover, areas with higher metal release showed a larger loss of microbial activity and showed a higher risk of metal contamination (Jaiswal and Pandey 2021). The study also reported that heavy metal release from Ganga River riverbed sediment increases with dissolved oxygen level below 1.0 mg/l. Dubey (2021) showed an increasing trend in selected five heavy metal contamination in the sediment collected from the Ganga River System at Kanpur. Spatial analysis has noted a differential degree of pollution in river sediment, which is consistently rising downstream of the River Ganges. Dhiman et al. (2022) evaluated the metal index (MI) for nine heavy metals in the river water, sediment, and suspended particulate matter in Ganga River Basin. The study revealed that 27% of the places were anthropogenically contaminated with heavy metals in river water, where Cr and Cd contamination contributes to the highest risk in suspended sediments around urban areas in the middle basin.

1.5 Comparative Assessment of Heavy Metal Pollution in Sediment

A comparative discussion on the heavy metal pollution in the sediment of River Ganga has been summarized in Table 1.4. The heavy metal concentration for this study was adopted from the findings of Singh et al. (2003) and Verma and Pandey (2019), which have worked on the sediment of River Ganga at Uttar Pradesh. The concentrations of Zn, Pb, and Cd were markedly below the TEC guideline value before 1995. Cadmium varied from moderate-to-high risk from 1993 to 2017 (Table 1.5). I_{geo} value revealed a rise in Cd pollution in River Ganga since 1995, as it has increased by 440% from 1993. The other metals showed negative geo-accumulation value denoted uncontaminated status after 2015, which was marked moderately contaminated for Cr and Mn before 1995. The sediment of Ganga River System has experienced an increasing level of Fe and Cd in the recent past, while river water was marked with moderate contamination of Pb in 2017 (Table 1.4). Mishra et al. (2023) have also noted a very high degree of contamination of Pb and Cd in the sediment of River Ganga, based on the geo-accumulation classification index value. A similar finding has been reported by Aggarwal et al. (2022) that projects a serious concern about the increasing concentrations of Cd and Pb in river sediment, which may lead to further degradation of sediment quality in the near future as a result of increased human activities in the Ganga River Basin. After 2015, potential ecological risk index was noted to be 209.35, suggesting a moderate risk of heavy metal pollution, which was 81.27 before 1995. Cd, Pb, Cu, Ni, and Fe are some major contributors to the rise of 157.59% potential ecological risk to Ganga River sediment. Siddiqui and Pandey (2019), working in the same region, conclude that the rise in the level of various heavy metals is under the influence of mixed sources. The region has hundreds of industrial units such as textiles, electroplating, power plants, perfumeries,

Table 1.4 Comparative analysis of different heavy metal concentrations in River Ganga between two decades with their average concentrations (Cn), contamination factor (CF), geo-accumulation index (I_{geo}), Ecological risk (Er), average shale concentration (ASV), Environmental Protection Agency sediment guideline value (EPA), threshold effective concentration (TEC), probable effective concentration (PEC), effective range low (ERL), effective range medium (ERM), and toxic response value (TRV) in Uttar Pradesh, India.

	Before 1995				After 2015										
	Cn	CF	I_{geo}	Er	Cn	CF	I_{geo}	Er	ASV[a]	EPA[b*]	TEC[c]	PEC[d]	ERL[e]	ERM[f]	TRV[g]
Cr	141.8	1.58	0.02	3.15	44.30	0.49	−0.48	0.98	90	<25	43.4	111	81	370	2
Mn	1885	2.22	0.17	2.22	445	0.52	−0.46	0.52	850	<300	—	—	—	—	1
Fe	39,705	0.84	−0.25	0.84	49,163	1.04	−0.16	1.04	47,200	—	—	—	—	—	1
Ni	48	0.71	−0.33	3.53	25.80	0.38	−0.60	1.90	68	<20	22.7	48.6	20.9	51.6	5
Cu	53.9	1.20	−0.10	5.99	28.40	0.63	−0.38	3.16	5	<25	31.6	149	34	270	5
Zn	102.15	1.08	−0.14	1.08	53.45	0.56	−0.43	0.56	95	<90	121	459	150	410	1
Cd	0.59	1.97	0.12	59.00	1.96	6.53	0.64	196.00	0.3	<6	0.99	4.98	1.2	9.6	30
Pb	21.9	1.10	−0.14	5.48	20.75	1.04	−0.16	5.19	20	<40	35.8	128	46.7	2018	5

Cn, ASV, EPA, TEC, PEC, ERL, ERM and TRV values are given in mg/kg.
[a] Turekian and Wedepohl (1961); [b] Ogbeibu et al. (2014); [c,d] Siddiqui and Pandey (2019); [e,f] Long et al. (1995); [g] Chen et al. (2020).

Table 1.5 Sediment quality indices, with their respective ranges of classification and references.

Indices	Value	Sediment quality	References
Contamination factor (CF)	<1	Low contamination	Hakanson (1980)
	1–3	Moderate contamination	
	3–6	Considerable contamination	
	>6	High contamination	
Enrichment factor (EF)	1.5–3	Minor enrichment	Hakanson (1980)
	3.0–5.0	Moderate enrichment	
	5.0–10.0	Severe enrichment	
	>10	Very severe enrichment	
Geo-accumulation index (I_{geo})	<0	Uncontaminated	Tomilson et al. (1980)
	0–1	Uncontaminated to moderately contaminated	
	1–2	Moderately contaminated	
	2–3	Moderately to highly contaminated	
	3–4	Highly contaminated	
	4–5	Highly to extremely contaminated	
	>5	Extremely contaminated	
Ecological risk (Er)	<40	Low risk	Hakanson (1980)
	40–80	Moderate risk	
	80–160	Considerable risk	
	160–320	High risk	
	>320	Very high risk	
Pollution load index (PLI)	<1	Unpolluted	Mohiuddin et al. (2010)
	1	Baseline level of pollution	
	>1	Polluted	
Potential ecological risk index (PERI)	<150	Low risk	Kowalska et al. (2016)
	150–300	Moderate risk	
	300–600	Considerable risk	
	>600	High risk	

sugar mills, distilleries, paper and pulp, and numerous others that directly contribute to pollution in the river system. Besides, agricultural runoffs, untreated wastewater disposal, tannery effluents, tributaries, and confluences are some reported sources of these trace metals in the main channel of River Ganga (Singh et al. 2013; Matta et al. 2018). Thus, the comparison from 1995 to 2015 clearly shows the impact of urbanization, industrialization, and other human activities on concentrating heavy metals in the sediment of Ganga River.

1.6 Mitigation Strategies

The Ganga River Basin harbors more than 600 million people making it the most important water resource worldwide. Rapid industrialization and population explosion postindependence contributed to extensive contamination of the river. It led to the introduction of the first Ganga Action Plan (GAP) by Prime Minister Rajiv Gandhi in 1985 to control the pollution of river water and to improve its quality. Numerous studies conducted by the Central Pollution Control Board (CPCB) along the stretch of River Ganga identified industrial wastes and domestic wastes as the prime factors deteriorating the quality of the river. At the end of this plan, the government of India claims that 64.77% of the sewage generated was treated prior to discharge in the Ganga River. However, the first GAP did not cover the majority of the polluted tributaries; therefore, the second phase of GAP was approved and launched to further improve the quality of water. This phase included states such as Delhi, Haryana, Uttarakhand, Uttar Pradesh, Bihar, Jharkhand, and West Bengal. The focus was to curb the industrial pollutants that served as major contributors of heavy metals and to repair and install sewage treatment plants along the stretch. The project received funding from the World Bank and also from the Japan International Cooperation Agency for specific projects in Varanasi. Despite all the efforts, heavy metal pollution continues to rise. The major shortcomings are the negligence of various industries in treating the effluents generated by them. CPCB (2016) reported that common effluent treatment plants of several tanning industries in Kanpur are ineffective in treating the effluents. NMCG (2022) states that only 25% of the total sewage generated in the basin is treated prior to discharge. Therefore, an updated version of the plan is required to tackle all the shortcomings. The license of the industries that are not complying with the norms should be suspended. General awareness programs related to the harmful effects of heavy metals should be initiated. Continuous monitoring of the river water should be conducted by various governmental agencies as well as researchers to determine changes in the status of river water and sediments.

1.7 Conclusion

The book chapter has precisely reviewed the status of heavy metal pollution in the sediment of River Ganga over the past three decades. Detailed analysis of several studies conducted from the 1990s to 2020 revealed that the majority of heavy metals exceeded the prescribed standards along the entire river stretch. The increasing level of iron, cadmium, and lead in the river bed of Ganga after 2015 at Uttar Pradesh marks the rise in anthropogenic protuberance in river water. The review then highlights the potential role of sediments as a major sink for various contaminants including the heavy metals in a riverine ecosystem. Besides, the sediment of River Ganga is one of the most important morphological attributes of the river, which shelters a diverse range of flora and fauna. This led to direct exposure to heavy metals from the sediments by benthic macroinvertebrates and several fish species, which consequently gets accumulated in the successive trophic levels. Further, the tendency of trace metal to get bioaccumulated in the sediment and biota of Ganga River System possess a greater risk to human health. Therefore, the chapter encourages the authorities concern to monitor and maintain a controlled disposal practice of these trace metals in River Ganga. Moreover, the study will be beneficial for the young researchers to have a glance at the status of metal contamination in the sediment of River Ganga, so that more can be added to conserve the heritage and steps can be taken to minimize the metallurgical load on River Ganga.

References

Abbas, M., Chand, N., Khan, R.U. et al. (2019). Public health risk of heavy metal residues in meat and edible organs of broiler in an intensive production system of a region in Pakistan. *Environmental Science and Pollution Research* 26 (22): 23002–23009.

Aggarwal, M., Anbukumar, S., and Vijaya Kumar, T. (2022). Heavy metals concentrations and risk assessment in the sediment of Ganga River between Kanpur and Prayagraj, UP, India. *Sādhanā* 47 (4): 1–11.

Ajmal, M. and Khan, A.U. (1987). Heavy metals in water, sediments, fish and plants of river Hindon, UP, India. *Hydrobiologia* 148 (2): 151–157.

Ajmal, M., Khan, M.A., and Nomani, A.A. (1987). Monitoring of heavy metals in the water and sediments of the Ganga River, India. *Water Science and Technology* 19: 107–117.

Akpambang, V.O.E., Ebuzeme, G.C., and Akinola, J.O. (2022). Heavy metal contamination of topsoil around a cement factory – a case study of Obajana cement plc. *Environmental Pollutants and Bioavailability* 34 (1): 12–20.

Alasfar, R.H. and Isaifan, R.J. (2021). Aluminum environmental pollution: the silent killer. *Environmental Science and Pollution Research* 28 (33): 44587–44597.

Ali, M.M., Hossain, D., Al-Imran, M. et al. (2021). Environmental pollution with heavy metals: a public health concern. In: *Heavy Metals-Their Environmental Impacts and Mitigation* (ed. M.K. Nazel and H. Zhao), 1–20. IntechOpen.

Alloway, B.J. and Jackson, A.P. (1991). The behavior of heavy metals in sewage sludge-amended soils. *Science of the Total Environment* 100: 151–176. https://doi.org/10.1016/0048-9697(91)90377-Q.

Anderson, T.H. (2003). Microbial eco-physiological indicators to asses soil quality. *Agriculture, Ecosystems and Environment* 98: 285–293. https://www.sciencedirect.com/science/article/abs/pii/S0167880903000884.

Aravinthasamy, P., Karunanidhi, D., Shankar, K. et al. (2021). COVID-19 lockdown impacts on heavy metals and microbes in shallow groundwater and expected health risks in an industrial city of South India. *Environmental Nanotechnology, Monitoring & Management* 16: 100472.

Balali-Mood, M., Naseri, K., Tahergorabi, Z. et al. (2021). Toxic mechanisms of five heavy metals: mercury, lead, chromium, cadmium, and arsenic. *Frontiers in Pharmacology* 12: 227. https://doi.org/10.3389/fphar.2021.643972.

Bordoloi, N., Tiwari, J., Kumar, S. et al. (2020). Efficiency of algae for heavy metal removal, bioenergy production, and carbon sequestration. In: *Emerging Eco-Friendly Green Technologies for Wastewater Treatment* (ed. R.N. Bharagava), 77–101. Singapore: Springer.

Cai, K. and Li, C. (2022). Ecological risk, input flux, and source of heavy metals in the agricultural plain of Hebei Province, China. *International Journal of Environmental Research and Public Health* 19 (4): 2288.

Central Pollution Control Board (CPCB). (2016). A report on Ganga matters. Retrieved from http://cpcb.nic.in/cpcbold/Report_on_Ganga_Matter_of_Uttarakhand_and_Uttar%20 Pradesh.pdf (accessed 26 September 2022).

Chen, H., Chen, Z., Chen, Z. et al. (2020). Calculation of toxicity coefficient of potential ecological risk assessment of rare earth elements. *Bulletin of Environmental Contamination and Toxicology* 104: 582–587. https://link.springer.com/article/10.1007/s00128-020-02840-x.

Chen, S.Y., Li, Z., Li, K., and Yu, X.Q. (2021a). Small molecular fluorescent probes for the detection of lead, cadmium and mercury ions. *Coordination Chemistry Reviews* 429: 213691.

Chen, Z., Huo, X., Chen, G. et al. (2021b). Lead (pb) exposure and heart failure risk. *Environmental Science and Pollution Research* 28 (23): 28833–28847.

Chheang, L., Thongkon, N., Sriwiriyarat, T., and Thanasupsin, S.P. (2021). Heavy metal contamination and human health implications in the Chan Thnal reservoir, Cambodia. *Sustainability* 13 (24): 13538.

Chow, Y.N., Lee, L.K., Zakaria, N.A., and Foo, K.Y. (2021). Integrated assessment of nickel electroplating industrial wastewater effluent as a renewable resource of irrigation water using a hydroponic cultivation system. *Frontiers in Plant Science* 12: 609396.

De Koning, H.P. (2020). The drugs of sleeping sickness: their mechanisms of action and resistance, and a brief history. *Tropical Medicine and Infectious Disease* 5 (1): 14. https://doi.org/10.3390/tropicalmed5010014.

Dhiman, A., Ramanthan, A.L., Macklin, M. et al. (2022). Heavy metal distribution in various environmental matrices and their risk assessment in Ganga River basin, India. *Human and Ecological Risk Assessment: An International Journal* 29: 1–30.

Dubey, I. (2021). Study on the toxic effects of heavy metals in sediments and water of river Ganga. *International Journal on Biological Sciences* 12 (2): 89–97.

Dustgeer, M.R., Zakria, H.S., Jilani, A. et al. (2023). Use of carbon based photocatalyst for metal removal. In: *Emerging Techniques for Treatment of Toxic Metals from Wastewater* (ed. A. Ahmad, R. Kumar, and M. Jawaid), 475–494.

Dwivedi, S., Mishra, S., and Tripathi, R.D. (2018). Ganga water pollution: a potential health threat to inhabitants of Ganga basin. *Environment International* 117: 327–338.

Dzvinamurungu, T., Rose, D.H., Chimwani, N., and Viljoen, F. (2022). Using process mineralogy as a tool to investigate blending potential of the pentlandite-bearing ores at the Nkomati Ni mine in South Africa. *Minerals* 12 (5): 649.

Ehis-Eriakha, C.B. and Akemu, S.E. (2022). Impact of heavy metal pollution on the biotic and abiotic components of the environment. *South Asian Journal of Research in Microbiology* 13: 38–54.

Feary, J. and Cullinan, P. (2022). *Encyclopedia of Respiratory Medicine*, 2e (ed. S.M. Janes). Elsevier Publications.

Fu, Z. and Xi, S. (2020). The effects of heavy metals on human metabolism. *Toxicology Mechanisms and Methods* 30 (3): 167–176. https://doi.org/10.1080/15376516.2019.1701594.

Genchi, G., Carocci, A., Lauria, G. et al. (2020). Nickel: human health and environmental toxicology. *International Journal of Environmental Research and Public Health* 17 (3): 679.

Gökalp, F.D., Doğanlar, O., Doğanlar, Z.B., and Güner, U. (2022). The genotoxic effects of a mixture of aluminum, arsenic, cadmium, cobalt, and chromium on the gill tissue of adult zebrafish (Danio rerio, Hamilton 1822). *Drug and Chemical Toxicology* 45 (3): 1158–1167.

Gondek, K., Mierzwa-Hersztek, M., Kopeć, M., and Spałek, I. (2021). Compost produced with addition of sewage sludge as a source of Fe and Mn for plants. *Ecological Chemistry and Engineering* 28 (2): 259–275.

Gupta, A., Rai, D.K., Pandey, R.S., and Sharma, B. (2009). Analysis of some heavy metals in the riverine water, sediments and fish from river Ganges at Allahabad. *Environmental Monitoring and Assessment* 157 (1): 449–458.

Haider, F.U., Liqun, C., Coulter, J.A. et al. (2021). Cadmium toxicity in plants: impacts and remediation strategies. *Ecotoxicology and Environmental Safety* 211: 111887.

Hakanson, L. (1980). An ecological risk index for aquatic pollution control. A sedimentological approach. *Water Research* 14 (8): 975–1001. https://doi.org/10.1016/0043-1354(80)90143-8.

Hameed, ERA, Shehata, MA, Waheed, H, Samie, OMA, Ahmed, HH, Sherif, LS, & Ahmed, A 2019, 'Heavy metals can either aid or oppose the protective function of the placental barrier', *Macedonian Journal of Medical Sciences*, 7(17), 2814. doi: https://doi.org/10.3889/oamjms.2019.709.

Hayder, R., Hafeez, M., Ahmad, P. et al. (2023). Heavy metal estimation and quality assurance parameters for water resources in the northern region of Pakistan. *Water* 15 (1): 77.

Hessel, E.V., Staal, Y.C., Piersma, A.H. et al. (2021). Occupational exposure to hexavalent chromium. Part I. Hazard assessment of non-cancer health effects. *Regulatory Toxicology and Pharmacology* 126: 105048.

Hu, Y., Wu, M., Liu, R., and Sun, W. (2020). A review on the electrochemistry of galena flotation. *Minerals Engineering* 150: 106272.

Hussain, S., Khan, M., Sheikh, T.M.M. et al. (2022). Zinc essentiality, toxicity, and its bacterial bioremediation: a comprehensive insight. *Frontiers in Microbiology* 13: 1912. https://doi.org/10.3389/fmicb.2022.900740.

IARC (2013). IARC monographs on the evaluation of carcinogenic risks to humans. Non-ionizing radiation. *Radiofrequency Electromagnetic Fields* 102: 287–406.

Iordache, A.M., Nechita, C., Zgavarogea, R. et al. (2022). Accumulation and ecotoxicological risk assessment of heavy metals in surface sediments of the Olt River, Romania. *Scientific Reports* 12 (1): 1–11.

Jaiswal, D. and Pandey, J. (2018). Impact of heavy metal on activity of some microbial enzymes in the riverbed sediments: ecotoxicological implications in the Ganga River (India). *Ecotoxicology and Environmental Safety* 150: 104–115.

Jaiswal, D. and Pandey, J. (2021). Human-driven changes in sediment-water interactions may increase the degradation of ecosystem functioning in the Ganga River. *Journal of Hydrology* 598: 126261. https://doi.org/10.1016/j.jhydrol.2021.126261.

Jang, Y.C., Somanna, Y., and Kim, H.J.I.J. (2016). Source, distribution, toxicity and remediation of arsenic in the environment–a review. *International Journal of Applied Environmental Sciences* 11 (2): 559–581.

Kang, C.U., Ji, S.E., Pabst, T. et al. (2021). Copper extraction from oxide ore of almalyk mine by H_2SO_4 in simulated heap leaching: effect of particle size and acid concentration. *Minerals* 11 (9): 1020.

Kayode, O.T., Ogunyemi, E.F., Odukoya, A.M., and Aizebeokhai, A.P. (2022). Assessment of chromium and nickel in agricultural soil: implications for sustainable agriculture. *IOP Conference Series: Earth and Environmental Science* 993 (1): 012014. IOP Publishing.

Keshavarz, S., Faraji, F., Rashchi, F., and Mokmeli, M. (2021). Bioleaching of manganese from a low-grade pyrolusite ore using aspergillus Niger: process optimization and kinetic studies. *Journal of Environmental Management* 285: 112153.

Kianoush, S., Balali-Mood, M., Mousavi, S.R. et al. (2012). Comparison of therapeutic effects of garlic and d-Penicillamine in patients with chronic occupational lead poisoning. *Basic and Clinical Pharmacology and Toxicology* 110 (5): 476–481. https://doi.org/10.1111/j.1742-7843.2011.00841.x.

Klotz, K., Weistenhöfer, W., Neff, F. et al. (2017). The health effects of aluminum exposure. *Deutsches Ärzteblatt International* 114 (39): 653.

Kowalska, J., Mazurek, R., Gąsiorek, M. et al. (2016). Soil pollution indices conditioned by medieval metallurgical activity – a case study from Krakow (Poland). *Environment Pollution* 218: 1023–1036. https://doi.org/10.1016/j.envpol.2016.08.053.

Kuan, S.H., Ghorbani, Y., and Chieng, S. (2020). Narrowing the gap between local standards and global best practices in bauxite mining: a case study in Malaysia. *Resources Policy* 66: 101636. https://doi.org/10.1016/j.resourpol.2020.101636.

Kumar, M., Gupta, N., Ratn, A. et al. (2019). Biomonitoring of heavy metals in river Ganga water, sediments, plant, and fishes of different trophic levels. *Biological Trace Element Research* 193 (2): 536–547.

Kumar, D., Malik, D.S., Kumar, N. et al. (2020). Spatial changes in water and heavy metal contamination in water and sediment of river Ganga in the river belt Haridwar to Kanpur. *Environmental Geochemistry and Health* 42 (7): 2059–2079.

Kumari, A., Sinha, S.K., Rani, N., and Sinha, R.K. (2021). Assessment of heavy metal pollution in water, sediment, and fish of the river Ganga at Varanasi, India. *Arabian Journal of Geosciences* 14 (22): 1–11.

Langcay, M.L., Clemente, E.D., and Arranz, C.G. (2021). Risk assessment of mercury in soil and surface water in Brgy. Santa Lourdes, Puerto Princesa City, Palawan. *IOP Conference Series: Earth and Environmental Science* 690 (1): 012043. IOP Publishing.

Laura, D., Anna, P., Nicola, F. et al. (2021). Stress granules in *Ciona robusta*: first evidences of tia-1-related nucleolysin and tristetraprolin gene expression under metal exposure. *Comparative Biochemistry and Physiology Part C: Toxicology and Pharmacology* 243: 108977. https://doi.org/10.1016/j.cbpc.2021.108977.

Levin, R., Vieira, C.L.Z., Rosenbaum, M.H. et al. (2021). The urban lead (pb) burden in humans, animals and the natural environment. *Environmental Research* 193: 110377.

Li, X., Wu, L., Zhou, J. et al. (2022). Potential environmental risk of natural particulate cadmium and zinc in sphalerite-and smithsonite-spiked soils. *Journal of Hazardous Materials* 429: 128313.

Liu, P., Wu, Q., Wang, X. et al. (2022). Spatiotemporal variation and sources of soil heavy metals along the lower reaches of Yangtze River, China. *Chemosphere* 291: 132768.

Long, E.R., MacDonald, D.D., Smith, S.L., and Calder, F.D. (1995). Incidence of adverse biological effects within ranges of chemical concentrations in marine and estuarine sediments. *Environmental Management* 19: 81–97. https://doi.org/10.1007/BF02472006.

Luo, Z., Shu, J., Chen, M. et al. (2021). Enhanced leaching of manganese from low-grade pyrolusite using ball milling and electric field. *Ecotoxicology and Environmental Safety* 211: 111893.

Mahurpawar, M. (2015). Effects of heavy metals on human health. *International Journal of Research -GRANTHAALAYAH* 530: 1–7.

Martinez, V.D., Vucic, E.A., Becker-Santos, D.D. et al. (2011). Arsenic exposure and the induction of human cancers. *Journal of Toxicology* 2011: 431287. https://doi.org/10.1155/2011/431287.

Matta, G., Kumar, A., Naik, P.K. et al. (2018). Assessment of heavy metals toxicity and ecological impact on surface water quality using HPI in Ganga river. *INAE Letters* 3 (3): 123–129.

Matta, G., Kumar, A., Tiwari, A.K. et al. (2020). HPI appraisal of concentrations of heavy metals in dynamic and static flow of Ganga River system. *Environment, Development and Sustainability* 22 (1): 33–46.

Mehdi, S.E.H., Amen, R., Ali, A. et al. (2021). Sources, chemistry, bioremediation and social aspects of arsenic-contaminated waters: a review. *Environmental Chemistry Letters* 19 (5): 3859–3886.

Mensah, M.B., Lewis, D.J., Boadi, N.O., and Awudza, J.A. (2021). Heavy metal pollution and the role of inorganic nanomaterials in environmental remediation. *Royal Society Open Science* 8 (10): 201485.

Mir, A.R., Pichtel, J., and Hayat, S. (2021). Copper: uptake, toxicity and tolerance in plants and management of cu-contaminated soil. *Biometals* 34 (4): 737–759.

Mishra, R., Singh, E., Kumar, A. et al. (2023). Metal pollution in marine environment: sources and impact assessment. In: *Metals in Water* (ed. S.K. Shukla, S. Kumar, and P.K. Mishra), 175–193.

Mohan, R., Deevakar, L., and Sivakumar, V. (2022). Towards holistic technology solution to chromite ore processing residue (COPR) challenge; global issue: review and analysis. *International Journal of Environmental Science and Technology* 19: 665–676. https://doi.org/10.1007/s13762-020-03097-x.

Mohiuddin, K.M., Zakir, H.M., Otomo, K. et al. (2010). Geochemical distribution of trace metal pollutants in water and sediments of downstream of an Urban River. *International Journal of Environmental Science and Technology* 7: 17–28. https://doi.org/10.1007/BF03326113.

Mold, M., Cottle, J., and Exley, C. (2019). Aluminium in brain tissue in epilepsy: a case report from Camelford. *International Journal of Environmental Research and Public Health* 16 (12): 21–29.

Movahednia, M., Maghfouri, S., Fazli, N. et al. (2022). Metallogeny of Manto-type cu-(ag) mineralization in Iran: relationship with neo-Tethyan evolution and implications for future exploration. *Ore Geology Reviews* 149: 105064.

Mukherjee, I., Singh, U.K., and Singh, R.P. (2021). An overview on heavy metal contamination of water system and sustainable approach for remediation. *Water Pollution and Management Practices* 255–277.

Nigam, A., Sharma, N., Tripathy, S., and Kumar, M. (2021). Development of semiconductor based heavy metal ion sensors for water analysis: a review. *Sensors and Actuators A: Physical* 330: 112879.

Nikkhou, F., Kartal, M., and Xia, F. (2021). Ferric methanesulfonate as an effective and environmentally sustainable lixiviant for Zn extraction from sphalerite (ZnS). *Journal of Industrial and Engineering Chemistry* 96: 226–235.

Nivetha, A., Sakthivel, C., and Prabha, I. (2021). Heavy metal contamination in groundwater and impact on plant and human. In: *Spatial Modelling and Assessment of Environmental Contaminants* (ed. P.K. Shit, P.P. Adhikary, and D. Sengupta), 233–246. Cham: Springer.

NMCG. (2022). National Mission of Clean Ganga, Pollution threat. https://nmcg.nic.in/pollution.aspx (accessed 26 September 2022).

Nriagu, J. (2007). *Zinc Toxicity in Humans*, 1–7. School of Public Health, University of Michigan.

Ogbeibu, A.E., Omoigberale, M.O., Ezenwa, I.M. et al. (2014). Using pollution load index and geoaccumulation index for the assessment of heavy metal pollution and sediment quality of the Benin River, Nigeria. *Natural Environment* 2 (1): 1–9.

Okrusch, M. and Frimmel, H.E. (2020). Sediments and sedimentary rocks. In: *Mineralogy*, 417–452. Berlin, Heidelberg: Springer.

Onukwuli, O.O., Udeh, B.C., Omotioma, M., and Nnanwube, I.A. (2021). Corrosion inhibition of aluminium in hydrochloric acid medium using cimetidine as inhibitor: empirical and optimization studies. *Anti-Corrosion Methods and Materials* 68: 385–395.

Oteyola, A.O. and Ola-Oladimeji, F.A. (2022). Heavy metal contamination from construction materials. In: *Ecological and Health Effects of Building Materials* (ed. J.A. Malik and S. Marathe), 113–131. Cham: Springer.

Paithankar, J.G., Saini, S., Dwivedi, S. et al. (2021). Heavy metal associated health hazards: an interplay of oxidative stress and signal transduction. *Chemosphere* 262: 128350. https://doi.org/10.1016/j.chemosphere.2020.128350.

Pandey, S. and Kumari, N. (2023). Impact assessment of heavy metal pollution in surface water bodies. In: *Metals in Water*, 129–154.

Pandey, J. and Singh, R. (2017). Heavy metals in sediments of Ganga River: up-and downstream urban influences. *Applied Water Science* 7 (4): 1669–1678.

Pandey, N. and Tiwari, A. (2021). Human health risk assessment of heavy metals in different soils and sediments. In: *Heavy Metals in the Environment* (ed. V. Kumar, A. Sharma, and A. Cedra), 143–163.

Pandey, M., Tripathi, S., Pandey, A.K., and Tripathi, B.D. (2014). Risk assessment of metal species in sediments of the river Ganga. *Catena* 122: 140–149.

Pandey, M., Pandey, A.K., Mishra, A., and Tripathi, B.D. (2015a). Assessment of metal species in river Ganga sediment at Varanasi, India using sequential extraction procedure and SEM-EDS. *Chemosphere* 134: 466–474.

Pandey, R., Raghuvanshi, D., and Dixhit, A. (2015b). Characterization of heavy metals in water and sediment of river Ganga through index analysis approach. *History* 40 (183): 158–166.

Pandey, R., Raghuvanshi, D., Sharma, P.K. et al. (2016). Evaluation of the status of heavy metal contamination in sediment of the river Ganga at Allahabad, India. *International Journal of Natural Sciences* 7: 852–859.

Parmar, M. and Thakur, L.S. (2013). Heavy metal Cu, Ni and Zn: toxicity, health hazards and their removal techniques by low-cost adsorbents: a short overview. *International Journal of Plant, Animal and Environmental Sciences* 3 (3): 143–157.

Patel, N., Chauhan, D., Shahane, S. et al. (2021). Contamination and health impact of heavy metals. In: *Water Pollution and Remediation: Heavy Metals* (ed. M.I.A. Inamuddin and E. Lichtfouse), 259–280. Cham: Springer.

Paul, D. (2017). Research on heavy metal pollution of river Ganga: a review. *Annals of Agrarian Science* 15 (2): 278–286.

Petranich, E., Predonzani, S., Acquavita, A. et al. (2022). Rapid thermoscanning technique for direct analysis of mercury species in contaminated sediments: from pure compounds to real sample application. *Applied Geochemistry* 143: 105393.

Plum, L.M., Rink, L., and Haase, H. (2010). The essential toxin: impact of zinc on human health. *International Journal of Environmental Research and Public Health* 7 (4): 1342–1365.

Prasad, S., Yadav, K.K., Kumar, S. et al. (2021). Chromium contamination and effect on environmental health and its remediation: a sustainable approaches. *Journal of Environmental Management* 285: 112174.

Purushothaman, P. and Chakrapani, G.J. (2007). Heavy metals fractionation in Ganga River sediments, India. *Environmental Monitoring and Assessment* 132 (1): 475–489.

Qader, A., Rehman, K., and Akash, M.S.H. (2021). Genetic susceptibility of δ-ALAD associated with lead (pb) intoxication: sources of exposure, preventive measures, and treatment interventions. *Environmental Science and Pollution Research* 28 (33): 44818–44832.

Rajmohan, N., Prathapar, S.A., Jayaprakash, M., and Nagarajan, R. (2014). Vertical distribution of heavy metals in soil profile in a seasonally waterlogging agriculture field in eastern Ganges Basin. *Environmental Monitoring and Assessment* 186 (9): 5411–5427.

Rakhimov, I.R., Saveliev, D.E., Rassomakhin, M.A., and Samigullin, A.A. (2022). Chromian Spinels from Kazanian-Stage Placers in the Southern Pre-Urals, Bashkiria, Russia: morphological and chemical features and evidence for provenance. *Minerals* 12 (7): 849. https://doi.org/10.3390/min12070849.

Ray, P.D., Yosim, A., and Fry, R.C. (2014). Incorporating epigenetic data into the risk assessment process for the toxic metals arsenic, cadmium, chromium, lead, and mercury: strategies and challenges. *Frontiers in Genetics* 5: 201. https://www.frontiersin.org/articles/10.3389/fgene.2014.00201/full.

Saidi, M. and Kadkhodayan, H. (2022). Experimental and simulation assessment to mitigate the emission of sulfide toxic gases and removing main impurities from Zn+ Pb+ Cu recovery plants. *Chemical Product and Process Modeling* 18 (2): 195–214. https://doi.org/10.1515/cppm-2021-0062.

Salama, S.A., Arab, H.H., Hassan, M.H., and Maghrabi, I.A. (2019). Cadmium-induced hepatocellular injury: modulatory effects of γ-glutamyl cysteine on the biomarkers of inflammation, DNA damage, and apoptotic cell death. *Journal of Trace Elements in Medicine and Biology* 52: 74–82.

Sanz, J., Tomasa, O., Jimenez-Franco, A., and Sidki-Rius, N. (2022). Cadmium (Cd) [Z= 48]. In: *Elements and Mineral Resources*, 43–45. Cham: Springer.

Sati, V.P. (2021). The Ganga Between Gaumukh and Uttarkashi. In the Ganges. *Cultural, Economic and Environmental Significance.* 71–92. https://doi.org/10.1007/978-3-030-79117-9_5.

Schaeffer, J.W., Adgate, J.L., Reynolds, S.J. et al. (2020). A pilot study to assess inhalation exposures among sugarcane workers in Guatemala: implications for chronic kidney disease of unknown origin. *International Journal of Environmental Research and Public Health* 17 (16): 5708.

Schindler, M., Santosh, M., Dotto, G. et al. (2022). A review on pb-bearing nanoparticles, particulate matter and colloids released from mining and smelting activities. *Gondwana Research* 110: 330–346.

Shah, K.N., Bhatt, R., Rotow, J. et al. (2019). Aurora kinase A drives the evolution of resistance to third-generation EGFR inhibitors in lung cancer. *Nature Medicine* 25 (1): 111–118. https://www.nature.com/articles/s41591-018-0264-7.

Sibanda, V., Khan, R., and Danha, G. (2019). The effect of chemical reagents on flotation performance of a pentlandite ore: an attainable region approach. *Powder Technology* 352: 462–469. https://doi.org/10.1016/j.powtec.2019.04.062.

Siddiqui, E. and Pandey, J. (2019). Assessment of heavy metal pollution in water and surface sediment and evaluation of ecological risks associated with sediment contamination in the Ganga River: a basin-scale study. *Environmental Science and Pollution Research* 26 (11): 10926–10940.

Singh, M., Müller, G., and Singh, I.B. (2002). Heavy metals in freshly deposited stream sediments of rivers associated with urbanisation of the Ganga plain, India. *Water, Air, and Soil Pollution* 141 (1): 35–54.

Singh, M., Müller, G., and Singh, I.B. (2003). Geogenic distribution and baseline concentration of heavy metals in sediments of the Ganges River, India. *Journal of Geochemical Exploration* 80 (1): 1–17.

Singh, L., Choudhary, S.K., and Singh, P.K. (2012). Pesticide concentration in water and sediment of River Ganga at selected sites in middle Ganga plain. *International Journal of Environmental Sciences* 3 (1): 260–274.

Singh, H., Yadav, S., Singh, B.K. et al. (2013). Assessment of geochemical environment from study of river sediments in the middle stretch of river Ganga at Ghazipur, Buxar and Ballia area. *Proceedings of the National Academy of Sciences, India Section B: Biological Sciences* 83 (3): 371–384.

Singh, H., Pandey, R., Singh, S.K., and Shukla, D.N. (2017). Assessment of heavy metal contamination in the sediment of the river Ghaghara, a major tributary of the river Ganga in northern India. *Applied Water Science* 7 (7): 4133–4149.

Singh, S., Paswan, S.K., Kumar, P. et al. (2023). Heavy metal water pollution: an overview about remediation, removal and recovery of metals from contaminated water. In: *Metals in Water*, 263–284.

Srivastava, D., Tiwari, M., Dutta, P. et al. (2021). Chromium stress in plants: toxicity, tolerance and phytoremediation. *Sustainability* 13 (9): 4629.

Tomilson, D.L., Wilson, J., Harris, C.R., and Jeffreyd, D.W. (1980). Problem in assessment of heavy metals in estuaries and the formation of pollution index. *Helgol Wiss Meeresunlter* 33: 566–575.

Tsai, T.L., Kuo, C.C., Pan, W.H. et al. (2017). The decline in kidney function with chromium exposure is exacerbated with co-exposure to lead and cadmium. *Kidney International* 92 (3): 710–720.

Turekian, K.K. and Wedepohl, K.H. (1961). Distribution of the elements in some major units of the Earth's crust. *Geological Society of America Bulletin* 72: 175–191.

Ulusoy, H.B., Sonmez, M.F., Kilic, E.S.R.A. et al. (2015). Intraperitoneal administration of low dose aluminium in the rat: how good is it to produce a model for Alzheimer disease. *Archives Italiennes de Biologie* 153 (4): 266–278.

Velusamy, S., Roy, A., Sundaram, S., and Kumar Mallick, T. (2021). A review on heavy metal ions and containing dyes removal through graphene oxide-based adsorption strategies for textile wastewater treatment. *The Chemical Record* 21 (7): 1570–1610.

Verma, K. and Pandey, J. (2019). Heavy metal accumulation in surface sediments of the Ganga River (India): speciation, fractionation, toxicity, and risk assessment. *Environmental Monitoring and Assessment* 191 (7): 1–21.

Viste, E. and Sorteberg, A. (2015). Snowfall in the Himalayas: an uncertain future from a little known past. *The Cryosphere* 9 (3): 1147–1167. https://doi.org/10.5194/tc-9-1147-2015.

Wang, Z., Wei, X., Yang, J. et al. (2016). Chronic exposure to aluminum and risk of Alzheimer's disease: a meta-analysis. *Neuroscience Letters* 610: 200–206.

Wani, A.L., Ara, A., and Usmani, J.A. (2015). Lead toxicity: a review. *Interdisciplinary Toxicology* 8: 55–64.

WHO. (2011). Aluminium in drinking-water. Background document for development of WHO Guidelines for Drinking-water Quality. https://apps.who.int/iris/bitstream/handle/10665/75362/WHO_SDE_WSH_03.04_53_eng.pdf?se (accessed 23 September 2022).

Wise, J.T., Wang, L., Xu, J. et al. (2019). Oxidative stress of Cr(III) and carcinogenesis. In: *The Nutritional Biochemistry of Chromium (III)* (ed. J.B. Vincent), 323–340. Elsevier https://doi.org/10.1016/B978-0-444-64121-2.00010-6.

Xie, H., Liu, Y., Rao, B. et al. (2021). Selective passivation behavior of galena surface by sulfuric acid and a novel flotation separation method for copper-lead sulfide ore without collector and inhibitor. *Separation and Purification Technology* 267: 118621.

Xie, F., Yu, M., Yuan, Q. et al. (2022). Spatial distribution, pollution assessment, and source identification of heavy metals in the Yellow River. *Journal of Hazardous Materials* 436: 129309.

Xu, Y., Fu, J., Wang, H. et al. (2019). Arsenic exposure and lifestyle-related diseases. In: *Arsenic Contamination in Asia, Biological Effects and Preventive Measures* (ed. H. Yamauchi and G. Sun), 83–118. https://doi.org/10.1007/978-981-13-2565-6_6.

Younas, F., Niazi, N.K., Bibi, I. et al. (2022). Constructed wetlands as a sustainable technology for wastewater treatment with emphasis on chromium-rich tannery wastewater. *Journal of Hazardous Materials* 422: 126926.

Zambelli, B. and Ciurli, S. (2013). Nickel and human health. In: *Interrelations between Essential Metal Ions and Human Diseases*, 321–357. https://doi.org/10.1007/978-94-007-7500-8_10.

Zhang, X., Yang, L., Li, Y. et al. (2012). Impacts of lead/zinc mining and smelting on the environment and human health in China. *Environmental Monitoring and Assessment* 184 (4): 2261–2273.

Zhang, Y., Hou, D., O'Connor, D. et al. (2019). Lead contamination in Chinese surface soils: source identification, spatial-temporal distribution and associated health risks. *Critical Reviews in Environmental Science and Technology* 49 (15): 1386–1423.

Zhang, L., Ni, Z., Li, J. et al. (2022). Characteristics of nutrients and heavy metals and potential influence of their benthic fluxes in the Pearl River Estuary, South China. *Marine Pollution Bulletin* 179: 113685.

Zhao, Q., Wang, S., and Wang, J. (2021). Chemical behaviours of Arsenium, Chromium, Mercury, Lead, and Strontium in aqueous system. In: *E3S Web of Conferences*, vol. 290 (ed. B. Van der Bruggen). EDP Sciences.

Zimik, H.V., Farooq, S.H., and Prusty, P. (2021). Source characterization of trace elements and assessment of heavy metal contamination in the soil around Tarabalo geothermal field, Odisha, India. *Arabian Journal of Geosciences* 14 (11): 1–14.

2

Synergistic Process of Weathering and Erosion: Techniques of Measurement and Their Significance

Ashutosh Kumar[1] *and Pooja Rani*[2]

[1] *Department of Geology, Earth Science, Indira Gandhi National Tribal University, Amarkantak, Madhya Pradesh, India*
[2] *Agriculture Science, Department of Vocational Education, Indira Gandhi National Tribal University, Amarkantak, Madhya Pradesh, India*

2.1 Introduction

As the world population has been increasing day by day, the demand for food has also been raised, especially in underdeveloped countries. Search of new and better productive regions needs scientific attention to evaluate the productivity potential of agricultural soil. Finding fertile and productive areas has simultaneously drawn attention toward the attributes of climatic factors that control the agricultural potential of soil (Wicke et al. 2010; Qi et al. 2011; Baudron et al. 2012; Tittonell et al. 2012; Kotz'e et al. 2016), especially in dry climatic regions where the intensity of chemical weathering is very low and requires the influence of parent rocks for further genesis of soil (Santos et al. 2017). The process of destruction and disintegration of rocks by several factors is known as weathering. This factor is due to physical, chemical, mechanical, or biological process. For example, physical or mechanical process may be due to lateral extrusion due to broad strike-slip fault, which results in huge collisional stress as in the north Almora fault of western Himalaya (Joshi et al. 2019). Also, when a dyke intrudes into metamorphic rocks, it weakens the surface and promotes the physical activities that leads to the initiation of weathering (Joshi et al. 2016). In general, it can be explained as a process in which the breaking of rocks occurs and broken-down materials are obtained. As the rocks undergo physical and chemical breakdown due to weathering, loose fragments are generated. These loosened fragments are then carried away by some natural agents like wind, water, and glacier from one place to another. This process is known as erosion. Soil is an indispensable as well as essential source for all living organisms. Without it, human life cannot be imagined (Amundson et al. 2015; García-Ruiz et al. 2015; Anache et al. 2017). But soil erosion processes have been designated as a major threat toward better productivity of land, carbon and nutrient cycling, and overall sustainable agriculture. In this context, erosion has proven to be a curse on agricultural productivity and potential. Some natural agents like wind, snow, ice,

Weathering and Erosion Processes in the Natural Environment, First Edition. Edited by Virendra Bahadur Singh, Sughosh Madhav, Naresh Chandra Pant, and Ravi Shekhar.

and rivers help in transportation of weathered material away from its origin place. As the products of weathering are transported to another site, the cycle of weathering with fresh rocks continues. Weathering is classified into three types: physical weathering, chemical weathering, and biological weathering. Physical weathering means breakdown of rock induced mechanically by means of fracturing by rocks, thawing and freezing, or during transportation by rivers. Chemical weathering is the weathering induced chemically between minerals present in rocks and external and natural agents (air and water). Oxygen helps in oxidization of minerals to reactant products. However, water helps in conversion of minerals to clays or dissolves them. However, biological weathering comprises breakdown of rocks due to organisms by means of mechanical disintegration, chemical reactions, or a combination of both, whereas erosional processes comprise three stages, namely, detachment, transportation, and deposition. However, the main types of erosion are splash erosion, sheet erosion, gully erosion, and rill erosion. In this chapter, we will go through the different weathering and erosional processes as well as the methods associated with their measurement and significance. The most important aspect of weathering is its role in soil formation. By understanding its measuring process, we can easily understand the ambience of soil and can take significant measures toward understanding better concept about soil productivity and potential. Whereas erosion control measures are proven concept, with their help we can subsidize reducing the erosion potential of soil by reducing surface runoff flow velocity by stabilizing the exposed soil.

2.1.1 Weathering and Erosion: Synergistic Approach

The process of weathering of rocks and erosion occurs synergistically. Erosion removes weathered materials and helps to continue the cycle of weathering. For example, in Almora Nappe, metamorphic rocks support physicochemical conditions for the occurrence of weathering (Joshi et al. 2013). The synergistic approach of weathering and erosion is explained by Robinson and Moses (2011) and Viles (2013) in detail. By understanding the measuring process of weathering and erosion, we can easily understand the significance of this process. Consequently, this chapter reflects on measurement methods mainly because of two reasons. One is to make an interconnection between existing and mostly used methods of measurements to invent new technology and explain their ways to simplify the understanding of surface weathering of rocks. And second, to provide information about advantages and disadvantages of measuring and monitoring methods of weathering of rock surfaces and removal of the material obtained by weathering.

2.2 Method of Measuring Rock Surface Change

2.2.1 Measurement with Quartz Resistant Mineral Vein as Natural Reference

This technique measures relatively with a surface of known age as a reference, such as quartz vein (Dahl 1967; André 2002; Nicholson 2008). The major advantage of using this technique lies in the determination of weathering rate of approximately 1000 years terms rate, i.e. of very long duration. This method can be used in remote areas also, where using such instruments is difficult. This technique limits the potential as the magnitude of the topographical features of reference has been weathered or damaged by weathering. Consequently, it gives least and unknown values (Trudgill 1983, 1986; Mottershead 1997, 2000; Dorn 2003; Goldie 2005).

2.2.2 Measurement with Anthropogenic Features as Reference

This technique used several anthropogenic features as references to measure the changes in rock surface. Gravestones are used as significant references to determine chronicling of historical weathering rate in urban areas. It has a dated surface and creates an environment that possesses similarities with microenvironment. Gravestones are of different ages and made up of similar types of rocks (Cooke et al. 1995).

Meierding (1993) studied tombstone weathering, the rate of which was determined by a simple method. The technique is based on the observation that the marble slab facing the ground was found to be least weathered whereas the marble slab above the ground at some higher place demonstrated loss of calcite grain. It may give false reading due to being warped over long time. It simply measures historic rate of 10–100 s of years and relative to rock art, lead lettering, and lead plugs (Trudgill et al. 1989; Inkpen and Jackson 2000).

2.3 Contact Methods

Contact measurements are used for the rock surface touching the base. It may be any reference stud or box fitted to the surface of rocks. Types of this technique comprise rock profiler, traversing micro-erosion meter (TMEM), micro-erosion meter (MEM) (Askin and Davidson-Arnott, 1981; Inkpen et al. 2012), and traversing erosion beam (TEB) (Charman et al. 2007). In contact method of measurement, the resolution capacity is very high but possesses a limitation on dissected surface. However, it offers short-term rates that permit investigations of sub-annual variation. Whereas techniques associated with laser scanning are highly sensitive and can measure change minutely even in granite, a highly resistant rock, techniques associated with TMEM and MEM involve replacement frequently as their potentiality limits on highly resistant and soft rocks. However, sites swamped with water are highly difficult to be measured. The designs of TEB are specific, and embedded vertically in the shore stage, and used specifically for glacial or London clay (cohesive and soft materials). Permanent reference datum is provided specifically by TEB. It is used to measure erosion which is on high water marks (Charman et al. 2007).

2.4 Noncontact Methods

Noncontact methods for measurement include various techniques. Terrestrial laser scanner (Armesto-González et al. 2010), 3D laser scanner (Barnett et al. 2005), laser camera scanner (Birginie and Rivas 2005), rapid prototype printing, and 3D triangulation scanning based on ground and close range (Bourke et al. 2008) are such laser-based scanning techniques used to measure rates of weathering and also produce maps showing changes in surface. These instruments are very exclusive and need expert operatives to handle them.

The topographical expression of several surfaces without assuming their form before can be cataloged and measured physically by using 3D laser scanner (Bourke et al. 2008). Its portable features make it friendly for use in the field as well as laboratory. Similarly, 3D scanning techniques like light detection and ranging have found their use gradually in studies of geomorphology (MacMillan et al. 2003; Nagihara et al. 2004; Barlow and Hopkinson 2005; Heritage and Hetherington 2007), and their potential in rock breakdown studies has been noted elsewhere (Bridges et al. 2004; Heslop et al. 2004).

2.4.1 Measurement of Surface Recession Rate Using Micro-catchement

Measurement of surface recession rate and output data of chemical weathering are provided by using micro-catchment techniques over a fixed area. It can be used in the field as well as in the laboratory. Saxicolous lichens were used to measure the rate of chemical weathering of rock surfaces (Aghamiri and Schwartzman 2002). It can be understood by introducing the term bioprotection (Carter and Viles 2005). It is the process that interconnects geomorphic system with biotic system. Other such processes are bioerosion, bioweathering, biotransformation, bioconstruction, and biostabilization. Bioweathering and bioerosion are the processes that are known for the erosion of rocks and minerals and their biological contributions to weathering. The three main processes, bioerosion, bioconstruction, and bioprotection, are found to be involved in biogeomorphic system (Naylor et al. 2002). Measurement of weathering and erosion directly or indirectly with the help of organic agencies comes under bioerosion. However, bioconstruction concludes the accumulation of sedimentary deposits by organic agents.

Bioprotection is defined in two ways: active bioprotection, in which a crust or a protective patina is produced on a rock and passive bioprotection, in which organisms' roles are determined in the earth surface process. *Verrucaria nigrescens* acts bioprotectively and *Verrucaria baldensis* biodeterioratively; these are two main species that are specific to micro-catchement study (Naylor et al. 2002). They both act in different ways from each other. There are various species that play significant roles in the biogeomorphological system. These organisms may be microorganisms (Viles 2000) or cyanobacteria (Schneider et al. 1999).

2.4.2 Measurement of Surface Recession Rate by Using Rock Exposure Trial

These processes measure the recession rate of surface. In this method, rock discs, tablets, and blocks are used. These all are portable, in which different microenvironments are compared and pretreatment of rocks is used to investigate the impact relating to the history of stress. Weathering rate can be easily determined if the density of the rock is known. Micromorphological studies regarding mineral species analyze the removal of discs from the fieldwork to expose the extensive development of voids on the mineral surface (Dixon et al. 2006). RRS (rock removal slab) has emerged as a new technology for the measurement of microtopographical variation in the rock surface. RRS is cost-effective and easy to handle. It permits the use of the TMEM for effective results. TMEM is used on the vertical rock surface. Collecting the result after measurement, it led to remove the slab vertically (Moses 2000; Furlani et al. 2010). However, Jaynes and Cooke (1987) explained the measurement of stone weathering, which occurs in a polluted area. Stone weathering refers to the deterioration of building stones. Such type of weathering is often the result of acid rain in polluted areas. The polluted atmosphere of urban areas leads to excessive decay of the building stone. Marbles, sandstone, and limestones are specifically weathered. One of the well-known examples of marble weathering is decay of the Taj Mahal, Agra (Gauri and Holdren Jr 1981). Fassina (1978) also reported about the weathering of various marble buildings. St. Paul's Cathedral in the United Kingdom and Westminster Abbey in London (Sharp 1982) are examples of such decay. Two important consequences are arrived at by observing these processes. The first one focuses on the lower weathering rate, which is relatively less in rural areas. And the other lies within pollution gradients, gradient of weathering between urban and rural areas. The decay rate is exclusively found greater in the urban areas due to vast arena of pollution. Carbonic acid was found to be one of the important weathering agents for decaying of stone buildings (e.g. monumental buildings).

Methods used for measuring the rate of weathering in these areas are summarized in three indices: weight change, surface roughness, and bulk weight loss. Weight change is determined by computing the initial as well as subsequent dry weights, whereas leached dry weights are effective for determining lower protected samples. The second one, surface roughness, was used only after the removal of samples from the programmed that were exposed. And the bulk weight loss was associated with the time as a percentage of original dry sample. These were the indices of weathering (Cooke 1979). However, these weathering measurement rate indices were considered not satisfactorily successful because their major consideration was the gross changes of surface rather than subsurface. They lack in assessing the weathering catastrophes on a large scale and reveal little about their specific method. But they are the most widely used methods of assessing surface weathering rates of stone (Jaynes and Cooke 1987). MEM is also used for measuring weathering rate of limestone (Smith et al. 1995).

2.4.3 Measurement Using Principle of Photogrammetry

Photogrammetric method of measuring weathering rate over temporal and spatial way in shore platform erosion has an advantage over measuring techniques like portable laser scanner and MEM. Although these two methods comprise the most accurate measurement techniques and can also be handled in specific locations for short time scale, they possess some difficulties in applying for temporal and spatial changes as well as meso-scale changes (Dornbusch et al. 2010). Automatic digital elevation model and photogrammetric method (Ackermann 1996) possess efficiency in representation of topographical platforms rising on shore. Polynomial textural mapping, close-range photogrammetry, and repeat photography are the other techniques based on the principle of photogrammetry. These techniques exclusively measure the changes in the form of weathering rate and also determine the alternation in weathered surface. These methods measure the rate at which surface loss occurs on various scales. Among these techniques, the close-range photography possesses a limitation with issue of limited focus (Pentecost 1991; Inkpen and Jackson 2000; Duffy 2013).

2.4.4 Light Source-Based Techniques for Measuring Rate of Weathering

The variation in surface change can be measured by two proficient techniques based on comprehensible source of light (Birginie and Rivas 2005). These two techniques are efficient in measuring the rate of weathering and erosion. The first technique comprises projection of line patterns or fringe on the rock surface, i.e. stone. In this manner, a 2D charged coupled device (CCD) sensor is able to capture a only digital image, which can be processed on a larger area of rock surface. Aftermath, the calculation is done with the help of an algorithm for computing the surface relief. These are called Moire or Young's interferometer techniques that have been effectively used for the detection of 3D distortion around cracks in a structure with deterioration (Martinez et al. 2003) and to assess the deterioration status of stone (Spagnolo et al. 2002) and micro-electronic card surface variation (Jin and Bao 2001). This method of assessing weathering rate is simple and precise (Pieraccini et al. 2001). Whereas white and black tints create some problems on the surface when initial calibration is used (Sajan et al. 1997).

The other technique relies on laser light, in which scanning of rock surface is done with it and the resulting laser points or lines are tracked on the consecutive surface images with the help of a CCD camera (Birginie et al. 1993, 2000; Lartigue et al. 2002). These techniques, based on the laser source, depict the production of stable and coherent light, enable it for topographical extraction from rock surface, and reflect it to be based on the property of triangulation (Realini et al. 1994). The tracking algorithm and the scanning method are contingent to determine the performance of this system over the resolution of image peak, detection algorithm, contrast, and brightness (Chen 2002) are the quality indicators and completeness, point density, and noise (Lartigue et al. 2002) are crucial factors in defining the camera quality in the technology associated with profilometry.

2.4.5 Measurement Using Cosmogenic Dating

This technique, based on the principle of cosmogenic dating, mainly measures the rate of weathering on a time scale of million (10^6) to kilo (10^3) years. It is expensive and requires specific facility for dating. Balco et al (2008) categorized the rate of erosion and exposure by using concentration of ^{10}Be (Bierman and Nichols 2004) and ^{26}Al on the surface of rocks and then presented the complete, direct, and single method, which reflects the calibration method of measurement. Erosion rate can be measured by this method. Bierman and Nichols reviewed cosmogenic nuclides in 2004 specifically ^{10}Be and deliberated their significance in measuring the rates of erosion carried away by rivers, sediment produced on landscapes, transported and stored occasionally, and finally deposited in oceans.

This cosmogenic dating using ^{10}Be is also used for measuring the rate of tafone weathering. Tafone weathering refers to cavernous weathering. Cavernous weathering results in the formation of animal or human face-shaped structure. Raster electron microscopy (REM) and coupled energy-dispersive X-ray (EDX) are two techniques that are used to measure weathering and erosion rates of rock. REM provides significant evidence for chemical and mechanical weathering by visibility of grains with surface structures. The geochemical approach of ^{10}Be is also used in this kind of weathering (Bierman and Nichols 2004).

2.5 Techniques of Measuring Subsurface Changes in Rock

Subsurface change in rocks is measured by using various techniques that are described in the following sections.

2.5.1 Perpendicular Cut Technique of Rock to Surface

To measure the rate of weathering, it is important to quantify the strength of rock material. It states the relationship between morphology of landforms and rock characteristics. The method is illustrated by cutting the rock perpendicular to the surface. This technique is based on the principle of inspection visually. The inspection can be done using a microscope. It can easily show changes with accurate depth inside the rock (Smith and McAlister 2000). One of the potential limitations of this technique is being destructive and does not show change consistently over long time.

2.5.2 Technique of Strength Testing

2.5.2.1 Measurement by Triaxial Hoek Cell and Point Load Tester

The strength of rock material may not be a single criterion to measure the rock resistance or competence to erosion. The strength of rock material is definitely a simple criteria to explain the external forces on intact rock (Attewell and Farmer 1976). Other than this, v rock elasticity property and the dynamic Young's Modulus are two important variables (Allison 1988).

Destructive technique of measuring rock strength comprises Point Load Tester and Triaxial Hoek Cell. They are used to measure bulk sample strength. According to the nature of this technique, it is difficult to approve over long time. Allison (1990) implemented the measurement of rock flexural vibration. It measures the duration of eight waves that pass through piezoelectric detector which are kept in contact with the sample. The rate at which decay of vibration occurs is found to be directly proportional to the rigidity and hardness of the rock sample. The natural frequency along the vibration is resulted by its density, weight, and dimensions, and by merging these data with concomitant vibration characteristics documents Young's Modulus of Elasticity.

2.5.2.2 Identification Test of Vickers Microhardness and Knoop Hardness

This test is nondestructive and based on the ultrasonic measurements. It works efficiently with monocrystalline rocks and is not portable for field conditions. In this approach, recording the signal is the first step, and then uniaxial compressive strength is acquired. This is known as the most accurate method (Benavente et al. 2007a; Benavente et al. 2007b).

2.5.2.3 Measurement of Strength by Schmidt Hammer Rebound Test

Schmidt hammer (SH) is used to evaluate the weathering rate by determining the strength of rock material. This is a nondestructive test. This instrument efficiently measures the distance between a rebounded controlled impact and a rock surface (Goudie 2006). There are various versions of SH: N type, P type, and L type. SH is sensitive toward discontinuities in rocks and also to moisture contents, specifically in weak rocks (Sumner and Nel 2002). This technique enables the collection of various data over a long range in a short time.

2.5.2.4 Measuring with Equotip Hardness Tester

Equotip hardness tester is an efficient technique to measure the hardness of rock. To evaluate the efficiency, a comparison test between SH and Equotip harness testers was done. Consequences of this test indicate about proficiency of Equotip hardness tester. It is more sensitive as compared to SH. The Equotip hardness tester was also found to be capable of measuring the strength of other internal and superficially weathered layers of rocks (Aoki and Matsukura 2007). Across very thin layer of surface, it can also measure the variations in rock hardness. It uses smaller area as compared to SH and consequently can easily gather many readings over large areas in a very short time span.

Equotip hardness tester is mainly of two types. One is standard type D, and the other is Piccolo. These are used to measure the hardness of rock in terms of the weathering of rock on several types of limestone, sandstone, dolerite, and basalt. They provide evidence regarding behavior of weathering on different rock surfaces. The measured value of hardness is expressed in Leeb hardness or Leeb number (L value). Leeb number is the ratio of the rebound velocity to the impact velocity multiplied by 1000. This instrument has automatic compensation for impact direction. It is light in weight and has low impact energy. Equotip Piccolo is a compact type of the Equotip hardness tester, recently produced. This is a lightweight instrument that weighs only 110 g. Value of impact energy is only 11 N/mm (Viles et al. 2010).

2.5.2.5 Measurement with Ultrasonic Testing GrindoSonic

Rock strength is measured by an equipment known as GrindoSonic. This equipment is capable of measuring compressive strength indirectly. This instrument works on the principle of theory of elasticity applied on rock masses (Attewell and Farmer 1976; Selby 1982). Along with this, Young's modulus and several other parameters are recognized. The fundamental vibrational frequency of known dimension of a rock sample is determined by this equipment, followed by shock excitation. And then, the calibration is done to determine the hardness of rock and strength by measured elastic material (Allison 1987). This instrument is based on the ultrasonic testing device (Neville 1981). It is nondestructive in nature and portable due to lightweight. One of the potential limitations of this instrument is that its reading value varies with fractures and moisture level with rocks.

2.5.3 Measurement for the Test of Porosity

2.5.3.1 Total Water Absorption

The measurement of total water absorption is an important parameter to measure the porosity of rock to determine its rate of weathering (Fitzner 1998; Robinson and Williams 1987, 1989). Anthropogenic sources show a strong and positive correlation between sodium (Na) and total dissolve solids present in the groundwater and thus contribute ions and enhance total water in the ground (Madhav et al. 2020).

2.5.3.2 Measurement Based on Microscopic Observation

Measurement based on the microscopic observation can easily table changes on rocks with accurate depth increments. These changes can be precisely quantified with the help of image analysis. This method of measurement is of destructive nature which makes it hard to table changes for a long time (Smith and McAlister 2000; Tuğrul 2004; Sanjurjo-Sánchez and Alves 2012; Sanjurjo-Sánchez and Vázquez 2013).

2.5.3.3 Measurement Based on Mercury Intrusion Porosimetry (MIP)

In the process of physical weathering caused by ice and salt crystallization, the properties are demonstrated by pore space of the rock surface. The pore spaces have a critical effect over the weathering properties of rock and its deteriorating nature (Ruedrich and Siegesmund 2006). MIP is a rapid method. It measures the rate of weathering over a wide range. This laboratory-based method is destructive in nature and can cause damaging of pores (Fitzner 1998; Nicholson 2001; Graue et al. 2011).

2.5.3.4 Measurement Based on Nitrogen Adsorption Isotherms

The study of pores present in the rock surface plays an important role in determining the rate of weathering. In this context, nitrogen adsorption isotherms play an important role. In this method, nitrogen adsorption at 196 C helps to determine the pores having diameter less than 1 lm (Iñigo et al. 2000). Pores having this range of diameter play a crucial role in weathering process (Winkler 1973). Pore spaces are not damaged by using this technique. It is of destructive nature and laboratory based technique (Warscheid and Braams 2000).

2.5.3.5 Measurement Based on Gas Permeameters

Measurement based on gas permeameters includes Autoclam Permeability System, Ergotech, and Tiny-Perm. This technique is portable to field. It is nondestructive in nature and collects data rapidly (Beggan et al. 1996; Rusell et al. 2002; McKinley et al. 2006; Warke et al. 2006; Buj et al. 2011; Alikarami et al. 2013; Filomena et al. 2014).

2.5.3.6 Measurement Based on Electrical Resistivity Tomography

Electrical resistivity tomography is an important measure of assessing the geo-electrical measurement. This works on the principle of resistivity. Contact electrodes are used to induce an electric current in the structure. Electric field is created due to this current. Then again, the electric potential is determined by using contact electrodes in the path of electric streamlines. Consequently, the electric potential as well as electric current data are stored. However, if any ununiformed transmission occurs in the electric current beyond the surface, a change can be seen in the distribution of electric streamline. This is responsible for a modification in electric potential distribution and resulted in delivering information in the uniform state of wall (Schueremans et al. 2003). This technique is portable in field. This method requires high level of expertise for clarification of measurements; otherwise, the quantitative assessment of electrical resistivity tomography can be challenging and difficult (Abu-Zeid et al. 2006).

2.5.3.7 Measurement Based on Acoustic Monitoring

The ice formation within rock is an important aspect of frost weathering (Hallet et al. 1991), and it causes weathering of rock surface at several meters (Murton et al. 2006). acoustic emission (AE) is found to be an efficient technique in high-altitude rock surface. This technique is field based having large datasets. AE is used to measure active fracturing. It also measures porosity in the rock indirectly (Krautblatter and Hauck 2007; Amitrano et al. 2012; Draebing and Krautblatter 2012; Menéndez and David 2012).

2.5.4 Measurement for the Test of Internal Stress/Strain

2.5.4.1 Measurement Based on Temperature Gradient Between Surface and Subsurface

Thermal expansion is one of the reasons of granular disaggregation of rocks because it creates thermal differences between minerals. Variation in surface temperature expands the thermal expansion that induces stress and leads to rock disintegration (Gomez-Heras et al. 2006). This method is an important measure to compute the rate of weathering. The method can be used in lab as well as in field. Internal stressing or fracturing is a clear indication of this method (Hall and Hall 1991; Hall 1999; Hall and Andre 2001, 2003; Hall et al. 2008).

2.5.4.2 Measurement Based on Optical Fiber Sensors

Optical fiber sensors are a novel technology that monitors the stress in the structure of reinforced concrete structures. Cracking developed in these structures can be easily detected by using optical fiber sensors (Lau 2003). These sensors have the potential to assess the cracks or damage in the service life of these structures. During the early stage of construction, these optical fiber sensors are placed on the structures. This method can be calculated in field as well as in lab. Optical fiber sensors also indicate the possibility of internal fracture (Martín-Pérez et al. 2010).

2.5.5 Measurement Based on Moisture Content or Its Distribution

2.5.5.1 Measurement Based on Water Absorption Tubes

Water absorption tubes are the techniques to characterize the rate of weathering based on moisture content. This method can be conducted in field as well as in lab. However, if the rock is porous, there has been a leakage while applying this technique (Robinson and Williams 1987, 1989; Török 2003; Seidel et al. 2011).

2.5.5.2 Measurement Based on Moisture Content in Laboratory

Moisture content in rocks or freezing and thawing play an important role in mechanical weathering and disintegration. From laboratory experiment, the content of water present in the rocks is determined. Saturation coefficient or *S*-value was found as an indicator for frost proneness (created due to moisture content). This technique is destructive in nature and provides efficient result (Hall 1986; Sass 2005).

2.5.5.3 Measurement Based on Time Domain Reflectometry, Microwave, Electrical Resistivity, and 2D Resistivity

The techniques based on time domain reflectometry, microwave, electrical resistivity, and 2D resistivity are used to estimate weathering in the cultural heritage site materials. Methods or standards that are used for commercial or industrial purposes are not used in cultural heritage sites due to its destructive nature and difficult calibration if nondestructive (Camuffo and Bertolin 2012). These methods efficiently show the spatial and temporal variations due to water or moisture content in the rocks or on the surface. This method can be used in lab as well as in field condition. However, this is an expensive technique and possesses practical difficulties like probe insertion (Murton et al. 2000; LaBreque et al. 2004; Sass 2005; Jouniaux et al. 2006; Murton et al. 2006; Sass and Viles 2010; Srinivasan et al. 2010; Smith et al. 2011; Eklund et al. 2013).

2.5.5.4 Measurement Based on Fiber Optic Sensors

Measurement and monitoring of building stone weathering due to moisture content can be determined by using a novel and innovative technique based on optical fiber sensor. This is based on relative humidity and temperature. This technique is based on fiber Bragg gratings. Temperature and moisture content are two parameters that help in weathering or deterioration of building stone. Differences of temperature in rocks minerals cause effect of gradient inside the stone and led to weathering. Moisture content is also a crucial factor for the deterioration of stone. Several techniques have been used so far, but this technique of optical fiber sensor is a direct assessment of moisture content in stone. Performance of this technique is found to be better than others. Deterioration and decay of building stone can be understood easily due to this. This technique is suitable in field as well as in lab condition. However, drilling is necessary in this method (Sun et al. 2012).

2.6 Techniques Based on Microscope for Measuring Rate of Weathering

2.6.1 Technique Based on Stereo/Dissecting Microscope

Microorganisms play an indispensable role in weathering of rock surfaces. They are involved in the deterioration and degradation of rocks (Eckhardt 1985). Two types of weathering (biogeophysical and biogeochemical) take place between the contact zones of substrate and lichen thallus. Microorganisms like cyanobacteria, bacteria, fungi, and algae play an important role in weathering (Ascaso and Wierzchos 1995). This technique is based on light microscopy (LM). It has low magnification potential. This method is capable of low magnificent view of grain shape or color and topographical structure of rock surface. This technique offers high-resolution microscopic method for selecting samples.

2.6.2 Measurement Based on Petrographic Microscopy of Thin Sections or Acetate Peels

The method is based on light microscopic technique having low magnification. The petrographic microscope is efficient in measurement and identification of minerals in terms of their optical properties. This technique is a fundamental technique based on research of mineralogy and petrology of approx. 120 years. This instrument has contributed much in this field in terms of determining the rate of weathering. This technique is elegant with its different models and ranges. Its function is based on scientific approaches (Kile 2003). Mineralogy and structure of rocks are evidently provided by this technique. It provides perpendicular sections and clearly specifies weathering.

2.6.3 Measurement Based on Confocal Scanning Microscopic Technique (CSML)

CSML is a novel and advanced emerging technology deal in the area of fluorescence imaging; hence, florescence microscopy is a promising approach. As compared to the electron microscopic technique, CSML is poor in resolution power but, apart from it, enriched with capacity of 3D imaging system, approach to vibrant cellular and molecular procedures (Canette and Briandet 2014). This technique provides all information related to topography with the help of digitized optic sensors. It possesses high magnification. The basic meaning of confocal is with the image which is obtained only on the focal plane (Canette and Briandet 2014).

2.6.4 Technique Based on Structured Light Illumination Microscopy (SLIM)

The technique of SLIM is a novel, advanced technique having superresolution power. The technique is capable of optical sectioning. It has visualization capacity of 3D (Ma et al. 2021). It provides highly magnified resolution.

2.7 Techniques Based on Infrared Microscopic Techniques

2.7.1 Technique Based on Fourier Transform Infrared Microscopy (FTIR)

FTIR plays an important role in weathering caused by biological agents like lichens. The biodeterioration zone of rock and lichens can be benefitted by using the direct observation technology of FTIR (Ascaso and Wierzchos 1995). Other techniques used in direct observation are transmission electron microscopy (TEM), LM, and scanning electron microscopy (SEM), along with other analytical techniques like EDX, infrared spectroscopy (IR), X-ray diffraction (XRD), atomic absorption spectrophotometry, and flame photometry (Adamo and Violante 2000). By using FTIR beam, samples are illuminated for the detection of elemental and molecular compounds. This technique has low resolution and easily detects low atomic numbers.

2.8 Techniques Based on Electron Microscopic Techniques

2.8.1 Technique of SEM

SEM is an innovative technique that provides topographical information. It has high magnification but requires dry samples. However, there are some difficulties in porous sampling under vacuum. During the procedure of SEM, careful handling is necessary with biological samples (Dorn 1998).

SEM is a vital technique, which is used to recognize and identify the composition of rocks and also the compositional difference between rock and soil.

2.8.2 Technique Based on Backscattered Electron Detection (BSE) or SEM with X-Ray Energy-Dispersive Spectroscopy (EDS)

BSE and EDS provide chemical and geographical information about rocks. This methodology of determining rate of weathering and erosion possesses high magnification. The requirement of dry samples is mandatory under vacuum. If samples are porous, it creates complications. But careful handling is needed in biological samples (de los Ríos and Ascaso 2005).

2.8.3 Technique Based on Cryo-SEM (Low-Temperature SEM)

The process of mineral weathering is an important phenomenon in the area of geology and comes under geochemical weathering. It occurs at the surface of the earth. This process is also important from the point of view of soil formation and productivity. Apart from it, the role of mineral weathering in the evolution of surface, ground, and seawater composition is indispensable. The role of mineral weathering in continental denudation, global climate and atmospheric composition cannot be ignored. A major percentage of rock weathering occurs due to biological agents (Huang and Schnitzer 1986; Banfield and Nealson 1997). From field as well as laboratory data, it is evident that microorganisms have the capacity of enhancing reactions of weathering by increasing alumino-silicate minerals. Microbes such as phosphate-solubilizing bacteria produce various kinds of organic and inorganic acids and form metal–complex ligands and several secondary minerals (Rani et al. 2019). Experiments of mineral dissolution by batch culture technique using soil culture and bacteria are used to monitor the effect of these bacteria on rate of weathering and erosion with the help of microscope. Organic and inorganic acids produced by these bacteria are effective in cation release from rock (e.g. biotite and feldspars). These colonies of microorganisms on mineral grains are examined by using the technique of cryo-SEM. Bacteria form colonies on mineral grain surfaces. These colonies are often seen along the cleavage and edge of mineral surface. This method of cryo-SEM has high resolution and low-voltage technology. Bacteria on these colonies form a biofilm over the surface of mineral. The surface of mineral surrounded by microbial colonies appears expansively scrapped than those of uncolonized areas. CSML is also used to detect the microbial colonies of rock surface and measure the effect of pH in these surroundings (Barker et al. 1998).

2.8.4 Technique Based on Environmental SEM/Variable Pressure SEM

The freezing–thawing cycles of water and the effect of acid rain and salt crystallization enhance the rate of natural weathering of stone. These techniques are used to measure the rate of weathering and erosion. Various historic buildings and monuments are weathered by a combination of these natural effects. This technique is more effective at normal pressure. Samples required for this technique should be uncoated. Organic and porous samples are excellent. This technique is good for in situ experiments (Rao et al. 1996).

2.8.5 Technique Based on TEM/High-Resolution TEM/FTIR Microscopy

This technique is good in electron passing and forms image of a sample. It has high-resolution power. Ultrathin sections are required by this technique (Ascaso et al. 2002).

2.8.6 Scanning Transmission Electron Microscopy (STEM)

STEM is very effective in measuring the rate of weathering. In this technique, electron beam is used. A narrow spot of sample is focused by an electron beam where it is scanned and detected effectively. This technique has high resolution.

2.9 Techniques Based on Force Microscopy

2.9.1 Method Based on Scanning Force Microscopy (SFM)/Atomic Force Microscopy (AFM)

This technique of SFM or AFM is very effective in determining the rate of weathering in water mineral interactions. Interaction of water–mineral edge is observed very precisely with the help of this technique. Very high resolution is perceived by this technique.

2.10 Technique Based on 3D X-Ray Microscopy Computed Tomography (CT)

2.10.1 Measurement Based on X-Ray CT/Micro-CT/Nano-CT

Technique based on X-ray CT/micro-CT/nano-CT is effective in measuring the rate of weathering. The nature of this technique is nondestructive. By using this technique, the rock sample shows 3D effect of visualization and clearly shows internal structure. Unlike other methodologies, in this technique, sample preparation is not necessary. Image obtaining process is very fast. This technique can be coupled with XRF to get chemical analyses of rock sample. It has high-resolution power (Cnudde and Jacobs 2004).

2.11 Conclusion

The study of rate of weathering and erosion is essential due to its major role in soil formation. By understanding the measurement of weathering and erosion through these techniques, we can easily determine the rate of weathering and can take essential steps to prevent the excessive loss of soil particles or erosion. Due to the increase in the rate of weathering and erosion, several natural factors are affected which are dangerous for human as well as animals. The worldwide overview of weathering and erosion is hazardous from ecological point of view. Mitigation of weathering and erosion can be done by using techniques depending on rock surface, subsurface, and microscopic techniques. These techniques are found effective in minimizing the rate of weathering and erosion. Techniques based on physical, chemical, and biological measures are explained in this chapter, which play an important role in minimizing the loss. Analysis of rate of weathering is a crucial parameter which is directly related to human growth and development. By reducing this loss, we can achieve our goal toward sustainable development.

References

Abu-Zeid, N., Botteon, D., Cocco, G., and Santarato, G. (2006). Non-invasive characterisation of ancient foundations in Venice using the electrical resistivity imaging technique. *NDT and E International* 39: 67–75.

Ackermann, F. (1996). Techniques and strategies for DEM generation. In: *Digital Photogrammetry: An Addendum to the Manual of Photogrammetry* (ed. C. Greve), 135–141. Washington: American Society of Photogrammetry and Remote Sensing.

Adamo, P. and Violante, P. (2000). Weathering of rocks and neogenesis of minerals associated with lichen activity. *Applied Clay Science* 16: 229–256.

Aghamiri, R. and Schwartzman, D.W. (2002). Weathering rates of bedrock by lichens: a mini watershed study. *Chemical Geology* 188 (3-4): 249–259.

Alikarami, R., Torabi, A., Kolyukhin, D., and Skurtveit, E. (2013). Geostatistical relationships between mechanical and petrophysical properties of deformed sandstone. *International Journal of Rock Mechanics and Mining Sciences* 63: 27–38.

Allison, R.J. (1987). Non-destructive determination of Young's modulus and its relationship with compressive strength, porosity and density. In: *Deformation of Sediments and Sedimentary Rocks*, vol. 29 (ed. M.E. Jones and R.M.F. Preston), 63–69. Geological Society Special Publication.

Allison, R.J. (1988). A non-destructive method of determining rock strength. *Earth Surface Processes and Landforms* 13: 729–736.

Allison, R.J. (1990). Developments in a non-destructive method of determining rock strength. *Earth Surface Processes and Landforms* 15: 571–577.

Amitrano, D., Gruber, S., and Girard, L. (2012). Evidence of frost-cracking inferred from acoustic emissions in a high-alpine rock-wall. *Earth and Planetary Science Letters* 341–344: 86–93;Amundson, R., Berhe, A.A., Hopmans, J.W. et al. (2015). Soil and human security in the 21st century. *Soil Science* 348: 1261071. https://doi.org/10.1126/science.1261071.

Anache, J.A.A., Wendland, E.C., Oliveira, P.T.S. et al. (2017). Runoff and soil erosion plot-scale studies under natural rainfall: a meta-analysis of the Brazilian experience. *Catena* 152: 29–39. https://doi.org/10.1016/j.catena.2017.01.003.

André, M.F. (2002). Rates of postglacial rock weathering on glacially scoured outcrops (Abisko-Riksgränsen area, 68°N). *Geografis ka Annaler* 84A: 139–150.

Aoki, H. and Matsukura, Y. (2007). A new technique for non-destructive field measurement of rock-surface strength: an application of an Equotip hardness tester to weathering studies. *Earth Surface Processes and Landforms* 32: 1759–1769.

Armesto-González, J., Riveiro-Rodríguez, B., González-Aguilera, D., and Rivas-Brea, M.T. (2010). Terrestrial laser scanning intensity data applied to damage detection for historical buildings. *Journal of Archaeological Science* 37: 3037–3047.

Ascaso, C. and Wierzchos, J. (1995). Study of the biodeterioration zone between the lichen thallus and the substrate. *Cryptogamic Botany* 5: 270–281.

Ascaso, C., Wierzchos, J., Souza-Egipsy, V. et al. (2002). In situ evaluation of the biodeteriorating action of microorganisms and the effects of biocides on carbonate rock of the Jeronimos monastery (Lisbon). *International Biodeterioration and Biodegradation* 49: 1–12.

Askin, R.W. and Davidson-Arnott, R.G.D. (1981). Micro-erosion meter modified for use under water. *Marine Geology* 40: 45–48.

Attewell, P.B. and Farmer, I.W. (1976). *Principles of Engineering Geology*. London: Chapman and Hall 1045 pp.

Balco, G., Stone, J., Lifton, N.A., and Dunai, T.J. (2008). A complete and easily accessible means of calculating surface exposure ages or erosion rates from 10Be and 26Al measurements. *Quaternary Geochronology* 3: 174–195.

Banfield, J.F. and Nealson, K.H. (1997). Geomicrobiology: interactions between microbes and minerals. *Mineralogical Society of America Reviews in Mineralogy* 35: 448.

Barker, W.W., Welch, S.A., Chu, S., and Banfield, J.S. (1998). Experimental observations of the effects of bacteria on aluminosilicate weathering. *American Mineralogist* 83: 1551–1563.

Barlow, J. and Hopkinson, C. (2005). Assessing rotation in detached blocks along the Niagara Escarpment using ground-based LiDAR. In *26th Canadian Symposium on Remote Sensing*. 14–16 June 2005, Wolfville, Nova Scotia.

Barnett, T., Chalmers, A., Díaz-Andreu, M. et al. (2005). 3D laser scanning for recording and monitoring rock art erosion. *International Newsletter on Rock Art* 41: 25–29.

Baudron, F., Tittonell, P., Corbeels, M. et al. (2012). Comparative performance of conservation agriculture and current smallholder farming practices in semi-arid Zimbabwe. *Field Crops Research* 132: 117–128. https://doi.org/10.1016/j.fcr.2011.09.008.

Beggan, J., Long, A.E., and Basheer, P.A.M. (1996). The permeability testing of masonry materials. In: *Proceedings of SWAPNET'95, Stone Weathering and Atmospheric Pollution Network Conference, the Queen's University of Belfast* (ed. B.J. Smith and P.A. Warke), 205–211. London: Dunhead Publishing.

Benavente, D., Cueto, N., Martínez-Martínez, J. et al. (2007a). Influence of petrophysical properties on the salt weathering of porous building rocks. *Environmental Geology* 52: 197–206.

Benavente, D., Martínez-Martínez, J., Cueto, N., and García-del-Cura, M.A. (2007b). Salt weathering in dual-porosity building dolostones. *Engineering Geology* 94: 215–226.

Bierman, P.R. and Nichols, K.K. (2004). Rock to sediment – slope to sea with 10Be – rates of landscape change. *Annual Reviews Earth Planet Science* 32: 215–255.

Birginie, J.M., Redarce, T., and Jutard, A. (1993). Using a camera-laser association to detect skin edges in a tannery. *RAPA, Revue d'Automatique et de Productique Applique'e* 6: 43–63.

Birginie, J.M. and Rivas, T. (2005). Use of a laser camera scanner to highlight the surface degradation of stone samples subjected to artificial weathering. *Building and Environment* 40: 755–764.

Birginie, J.-M., Rivas, T., and Auger, F. (2000). Valoracion de la resistencia a la alteracion por niebla salina de dos calizas utilizadas en la construccion mediante metodos ponderales, metodos acusticos y tratamiento de imagenes. *Materiales de Construcción* 50: 27–43.

Bourke, M., Viles, H.A., Nicoli, J. et al. (2008). Innovative applications of laser scanning and rapid prototype printing to rock breakdown experiments. *Earth Surface Processes and Landforms* 33: 1614–1621.

Bridges, N.T., Razdan, A., Greeley, R., and Laity, J.E. (2004). High resolution laser scanning techniques for rock abrasion and texture analyses on Mars and Earth. In *LPSC XXXV*, abs. 1897, Document ID 20040065953.

Buj, O., Gisbert, J., McKinley, J.M., and Smith, B. (2011). Spatial characterization of salt accumulation in early stage limestone weathering using probe permeametry. *Earth Surface Processes and Landforms* 36: 383–394.

Camuffo, D. and Bertolin, C. (2012). Towards standardisation of moisture content measurement in cultural heritage materials. *E-Preservation Science* 9: 23–35.

Canette, A. and Briandet, R. (2014). *Confocal Laser Scanning Microscopy*', Agro Paris Tech. Massy: Elsevier.

Carter, N.E.A. and Viles, H.A. (2005). Bioprotection explored: the story of a little known earth surface process. *Geomorphology* 67: 273–281.

Charman, R., Cane, T., Moses, C., and Williams, R. (2007). A device for measuring downwearing rates on cohesive shore platforms. *Earth Surface Processes and Landforms* 32: 2212–2221.

Chen, H.-H. (2002). Variation reduction in quality of an optical triangulation system employed for underwater range finding. *Ocean Engineering* 29: 1871–1893.

Cnudde, V. and Jacobs, P.J.S. (2004). Monitoring of weathering and conservation of building materials through non-destructive X-ray computed microtomography. *Environmental Geology* 46: 477–485.

Cooke, R.U. (1979). Laboratory simulation of salt weathering processes in arid environments. *Earth Surface Processes* 4: 347–359.

Cooke, R.U., Inkpen, R.J., and Wiggs, G.F.S. (1995). Using gravestones to assess changing rates of weathering in the United Kingdom. *Earth Surface Processes and Landforms* 20: 531–546.

Dahl, R. (1967). Post-glacial micro-weathering of bedrock surfaces in the Narvik district of Norway. *Geografisker. Annalar* 49A: 155–166.

de los Ríos, A. and Ascaso, C. (2005). Contributions of in situ microscopy to the current understanding of stone biodeterioration. *International Microbiology* 8: 181–188.

Dixon, J.C., Campbell, S.W., Thorn, C.E., and Darmody, R.G. (2006). Incipient weathering rind development on introduced machine-polished granite disks in an Arctic environment, northern Scandinavia. *Earth Surface Processes and Landforms* 31: 111–121.

Dorn, R.I. (1998). *Rock Coatings. Developments in Earth Surface Processes*, vol. 6, 429. Amsterdam: Elsevier.

Dorn, R.I. (2003). Boulder weathering and erosion associated with a wildfire Sierra Ancha Mountains, Arizona. *Geomorphology* 55: 155–171.

Dornbusch, U., Moses, C.A., Robinson, D.A., and Williams, R.B.G. (2010). Soft copy photogrammetry to measure shore platform erosion on decadal time scales. In: *Coastal and Marine Geospatial Technologies*, Coastal Systems and Continental Margins, vol. 13 (ed. D. Green), 129–137. Springer.

Draebing, D. and Krautblatter, M. (2012). The influence of ice-pressure on p-wave velocity in alpine low-porosity rocks: a modified time-average model. *Geophysical Research* Abstracts 14. https://doi.org/10.1029/2019GL081981.

Duffy, S. with contributions by Bryan, P, Graeme, E, Beale, G, Pagi, H and Kotoula, E(2013). *Multi-Light Imaging Techniques for Heritage Application: PTM Guidelines*. English Heritage.

Eckhardt, F.E.W. (1985). Solubilization, transport, and deposition of mineral cations by microorganism efficient rock weathering agents. In: *The Chemistry of Weathering* (ed. J.I. Drever), 161–173. New York: Reidel Publishing.

Eklund, J.A., Zhang, H., Viles, H.A., and Curteis, T. (2013). Using handheld moisture meters on limestone: factors affecting performance and guidelines for best practice. *International Journal of Architectural Heritage: Conservation, Analysis and Restoration* 7: 207–224.

Fassina, V. (1978). A survey of air pollution and deterioration of stonework in Venice. *Atmospheric Environment* 12: 2205–2211.

Filomena, C.M., Hornung, J., and Stollhofen, H. (2014). Assessing accuracy of gas-driven permeability measurements: a comparative study of diverse hassler-cell and probe permeameter devices. *Solid Earth* 5 (1): 1–11.

Fitzner, B. (1998). Porosity properties of naturally or artificially weathered stones. *VIth International Congress on Deterioration and Conservation of Stone*, Torun (September 12–14).

Furlani, S., Cucchi, F., and Odorico, R. (2010). A new method to study microtopographical changes in the intertidal zone: one year of TMEM measurements on a limestone removable rock slab (RRS). *Zeitschrift für Geomorphologie* 54 (2): 137–151.

García-Ruiz, J.M., Beguería, S., Nadal-Romero, E. et al. (2015). A meta-analysis of soil erosion rates across the world. *Geomorphology* 239: 160–173. https://doi.org/10.1016/j.geomorph.2015.03.008.

Gauri, K.L. and Holdren, G.C. Jr. (1981). Pollutant effects on stone monuments. *Environmental Science and Technology* 15: 386390.

Goldie, H.S. (2005). Erratic judgements: re-evaluating solutional erosion rates of limestones using erratic-pedestal sites, including Norber Yorkshire. *Area* 37 (4): 433–442.

Gomez-Heras, M., Smith, B.J., and Fort, R. (2006). Surface temperature differences between minerals in crystalline rocks: implications for granular disaggregation of granites through thermal fatigue. *Geomorphology* 78: 236–249.

Goudie, A.S. (2006). The Schmidt hammer in geomorphological research. *Progress in Physical Geography* 30: 703–718.

Graue, B., Seigesmund, S., and Middenorf, B. (2011). Quality assessment of replacement stones for the Cologne cathedral: mineralogical and petrophysical requirements. *Environmental Earth Sciences* 63: 1799–1822.

Hall, K. (1986). Rock moisture content in the field and the laboratory and its relationship to mechanical weathering studies. *Earth Surface Processes and Landforms* 11: 131–142.

Hall, K. (1999). The role of thermal stress fatigue in the breakdown of rock in cold regions. *Geomorphology* 31: 47–63.

Hall, K. and Andre, M.-F. (2001). New insights into rock weathering from high frequency rock temperature data: an Antarctic study of weathering by thermal stress. *Geomorphology* 41: 23–35.

Hall, K. and Andre, M.-F. (2003). Rock thermal data at the grain scale: applicability to granular disintegration in cold environments. *Earth Surface Processes and Landforms* 28: 823–836.

Hall, K., Guglielmin, M., and Stini, A. (2008). Weathering of granite in Antarctica: II. Thermal stress at the grain scale. *Earth Surface Processes and Landforms* 33: 475–493.

Hall, K. and Hall, A. (1991). Thermal gradients and rock weathering at low temperatures: some simulation data. *Permafrost and Periglacial Processes* 2: 103–112.

Hallet, B., Walder, J.S., and Stubbs, C.W. (1991). Weathering by segregation ice growth in microcracks at sustained sub-zero temperatures: verification from an experimental study using acoustic emissions. *Permafrost and Periglacial Processes* 2: 283–300.

Heritage, G.L. and Hetherington, D. (2007). Towards a protocol for laser scanning in fluvial geomorphology. *Earth Surface Processes and Landforms* 32: 66–74.

Heslop, E.A., Viles, H.A., and Bourke, M.C. (2004). Understanding rock breakdown on Earth and Mars: geomorphological concepts and facet mapping methods. *LPSC XXXV*, abst. 1445.

Huang, P.M. and Schnitzer, M. (1986). *Interactions of Soil Microbes with Natural Organics and Microbes*, vol. 17. Madison: Soil Science Society of America 606 pp.

Iñigo, A.C., Vicente, M.A., and Rives, V. (2000). Weathering and decay of granitic rocks: its relation to their pore network. *Mechanics of Materials* 32: 555–560.

Inkpen, R.J. and Jackson, J. (2000). Contrasting weathering rates in coastal, urban and rural areas in southern Britain: preliminary investigations using gravestones. *Earth Surface Processes and Landforms* 25: 229–238.

Inkpen, R.J., Viles, H., Moses, C., and Baily, B. (2012). Modelling the impact of changing atmospheric pollution levels on limestone erosion rates in Central London, 1980–2010. *Atmospheric Environment* 61: 476–481.

Jaynes, E.M. and Cooke, R.U. (1987). Stone weathering in Southeast England. *Atmosphere and Environment* 21: 1601–1622.

Jin, G.C. and Bao, N.-K. (2001). Surface detection and 3D profilometry for microstructure using optical metrology. *Optics and Lasers in Engineering* 36: 1–9.

Joshi, M., Devi, P.M., and Kumar, A. (2013). Metamorphism in Chaukhutia area and possible linkages of Almora nappe with the higher Himalayan Metamorphics. *Journal of Scientific Research* 57: 11–19.

Joshi, M., Kumar, A., Ghosh, P. et al. (2019). North Almora fault: a crucial missing link in the strike slip tectonics of western Himalaya. *Journal of Asian Earth Sciences* 172: 249–263. https://doi.org/10.1016/j.jseaes.2018.09.002.

Joshi, M., Kumar, A., and Rai, S. (2016). Petrotectonics of olivine dolerite intruding Almora nappe, Kumaun lesser Himalaya. *Journal of Scientific Research* 60: 1–10.

Jouniaux, L., Zamora, M., and Reuschl, T. (2006). Electrical conductivity evolution of nonsaturated carbonate rocks during deformation up to failure. *Geophysical Journal International* 167: 1017–1026.

Kile, D.E. (2003). The petrographic microscope: evolution of a mineralogical research instrument. *The Minerological Record* 34 (6): 5–39.

Kotz'e, E., Loke, P.F., Akhosi-Setaka, M.C., and Du Preez, C.C. (2016). Land use change affecting soil humic substances in three semi-arid agro-ecosystems in South Africa. *Agriculture, Ecosystems and Environment* 216: 194–202. https://doi.org/10.1016/j. agee. 2015.10.007.

Krautblatter, M. and Hauck, C. (2007). Electrical resistivity tomography monitoring of permafrost in solid rock walls', *journal of geophysical research-ALL series*, 112. *Journal of Geophysical Research* 112: F02S20.

LaBreque, D.J., Sharpe, R., Wood, T., and Heath, G. (2004). Small scale electrical resistivity tomography of wet fractured rocks. *Ground Water* 42: 111–118.

Lartigue, C., Contri, A., and Bourdet, P. (2002). Digitised point quality in relation with point exploitation. *Measurement* 32: 193–203.

Lau, K.T. (2003). Fibre-optic sensors and smart composites for concrete applications. *Magazine of Concrete Research* 55 (1): 19–34.

Ma, Y., Wen, K., Liu, M. et al. (2021). Recent advances in structured illumination microscopy. *Journal of Physics: Photonics* 3 (2): 024009.

MacMillan, R.A., Martin, T.C., Earle, T.J., and McNabb, D.H. (2003). Automated analysis and classification of landforms using high-resolution digital elevation data: applications and issues. *Canadian Journal of Remote Sensing* 29: 592–606.

Madhav, S., Kumar, A., Kushawaha, J. et al. (2020). Geochemical assessment of groundwater quality in Keonjhar City, Odisha, India. *Sustainable Water Resources Management* 6: 46. https://doi.org/10.1007/s40899-020-00395-7.

Martinez, A., Rodriguez-Vera, R., Rayas, J.A., and Puga, H.J. (2003). Fracture detection by grating moire′ and in-plane ESPI technique. *Optics and Lasers in Engineering* 39: 525–536.

Martín-Pérez, B., Deif, A., Cousin, B. et al. (2010). Strain monitoring in a reinforced concrete slab sustaining service loads by distributed Brillouin fibre optic sensors. *Canadian Journal of Civil Engineering* 37: 1341–1349.

McKinley, J.M., Warke, P., Lloyd, C.D. et al. (2006). Geostatistical analysis in weathering studies: case study for Stanton moor building sandstone. *Earth Surface Processes and Landforms* 31: 950–969.

Meierding, T.C. (1993). Marble tombstone weathering and air pollution in North America. *Annals of the Association of American Geographers* 83: 568–588.

Menéndez, B. and David, C. (2012). The influence of environmental conditions on weathering of porous rocks by gypsum: a non-destructive study using acoustic emissions. *Environmental Earth Sciences* 1–16.

Moses, C.A. (2000). Field rock block exposure trials. *Zeitschrift für Geomorphologie* 120: 33–50.

Mottershead, D.N. (1997). A morphological study of coastal rock weathering on dated structures, South Devon, UK. *Earth Surface Processes and Landforms* 22: 491–506.

Mottershead, D.N. (2000). Weathering of coastal defensive structures in Southwest England: a 500 year stone durability trial. *Earth Surface Processes and Landforms* 25: 1143–1159.

Murton, J.B., Coutard, J.-P., Lautridou, J.-P. et al. (2000). Experimental design for a pilot study on bedrock weathering near the permafrost table. *Earth Surface Processes and Landforms* 25: 1281–1294.

Murton, J.B., Peterson, R., and Ozouf, J.-C. (2006). Bedrock fracture by ice segregation in cold regions. *Science* 314: 1127–1129.

Nagihara, S., Mulligan, K.R., and Xiong, W. (2004). Use of a three-dimensional laser scanner to digitally capture the topography of sand dunes in high spatial resolution. *Earth Surface Processes and Landforms* 29: 391–398.

Naylor, L.A., Viles, H.A., and Carter, N.E.A. (2002). Biogeomorphology revisited: looking towards the future. *Geomorphology* 47: 3–14.

Neville, A.M. (1981). *Properties of Concrete*. London: Pitman 779 pp.

Nicholson, D.T. (2001). Pore properties as indicators of breakdown mechanisms in experimentally weathered limestones. *Earth Surface Processes and Landforms* 26: 819–838.

Nicholson, D.T. (2008). Rock control on microweathering of bedrock surfaces in a periglacial environment. *Geomorphology* 101: 655–665.

Pentecost, A. (1991). The weathering rates of some sandstone cliffs, central weald, England. *Earth Surface Processes and Landforms* 16: 83–91.

Pieraccini, M., Guidi, G., and Atzeni, C. (2001). 3D digitizing of cultural heritage. *Journal of Cultural Heritage* 2: 63–70.

Qi, S., Zheng, H., Lin, Q. et al. (2011). Effects of livestock grazing intensity on soil biota in a semi-arid steppe of Inner Mongolia. *Plant and Soil* 340: 117–126. https://doi.org/10.1007/s11104-010-0463-6.

Rani, P., Singh, M., and Dwivedi, M.K. (2019). Effect of combined inoculation of phosphate solubilizing bacteria and endomycorrhizae fungi on microbial population and NPK uptake by wheat (*Triticum aestivum* L.) crop. *Current Journal of Applied Science and Technology* 32 (5): 1–9.

Rao, S.M., Brinker, C.J., and Ross, T.J. (1996). Environmental microscopy in stone conservation. *Scanning* 18 (7): 508–514.

Realini, M., Toniolo, L., Zanetta, P., et al. (1994). A laser method for the study of salt crystallization on stone surfaces. *Proceeding of III International Symposium on the Conservation of Monuments in the Mediterranean Basin, Venice* (22–25 June), pp. 73–76.

Robinson, D.A. and Moses, C.A. (2011). Rock surface and weathering: process and form. In: *The SAGE Handbook of Geomorphology* (ed. K.J. Gregory and A.S. Goudie), 291–309. London: SAGE.

Robinson, D.A. and Williams, R.B.G. (1987). Surface crusting of sandstones in southern England and northern France. In: *International Geomorphology 1986*, vol. 2 (ed. V. Gardner), 623–635. Chichester: Wiley.

Robinson, D.A. and Williams, R.B.G. (1989). Polygonal cracking of sandstone at Fontainebleau, France. *Zeitschrift für Geomorphologie* 33: 59–72.

Ruedrich, J. and Siegesmund, S. (2006). Salt and ice crystallisation in porous sandstones. *Environmental Geology* 52: 225–249.

Rusell, M.I., Harmon, N.G., Curran, J.M. et al. (2002). Permeation properties of building stone: the autoclam permeability system. In: *Understanding and Managing Stone Decay* (ed. R. Prikryl and H.A. Viles), 75–84. Prague: Karolinum Press.

Sajan, M.R., Tay, C.J., Shang, H.M., and Asundi, A. (1997). TDI imaging and scanning moire′ for online defect detection. *Optics and Laser Technology* 29: 327–331.

Sanjurjo-Sánchez, J. and Alves, C. (2012). Decay effects of pollutants on stony materials in the built environment. *Environmental Chemistry Letters* 10 (2): 131–143.

Sanjurjo-Sánchez, J. and Vázquez, E.V. (2013). Characterizing weathering of granite buildings by multifractal analysis of mercury intrusion porosimetry. *Vadose Zone Journal* 12 (3): https://doi.org/10.2136/vzj2012.0204.

Santos, J.C.B., Le Pera, E., Souza Júnior, V.S. et al. (2017). Gneiss saprolite weathering and soil genesis along an east–west regolith sequence (NE Brazil). *Catena* 150: 279–290. https://doi.org/10.1016/j.catena.2016.11.031.

Sass, O. (2005). Rock moisture measurements: techniques, results, and implications for weathering. *Earth Surface Processes and Landforms* 30: 359–347.

Sass, O. and Viles, H.A. (2010). Wetting and drying of masonry walls: 2D-resistivity monitoring of driving rain experiments on historic stonework in Oxford, UK. *Journal of Applied Geophysics* 70: 72–83.

Schneider, J., Campion, L., and Alsumard, T. (1999). Construction and destruction of carbonates by marine and freshwater cyanobacteria. *European Journal of Phycology* 34: 417–426.

Schueremans, L., Van Rickstal, F., Venderickx, K., and Van Gemert, D. (2003). Evaluation of masonry consolidation by geo-electrical relative difference resistivity mapping. *Materials and Structures* 36: 46–50.

Seidel, H., Siegesmund, S., and Sterflinger, K. (2011). Characterisation of stone deterioration on buildings. In: *Stone in Architecture, 4the* (ed. S. Siegsmunde and R. Snethlage), 347–410. Berlin: Springer.

Selby, M.J. (1982). *Hillslope Form and Process*. New York: Oxford University Press 264 pp.

Sharp, A. (1982). Weathering of the balustrade of St. Paul's cathedral, London. *Earth Surface Processes and Lundjortns* 7: 387–389.

Smith, B.J. and McAlister, J.J. (2000). Sampling and pre-treatment strategies for the chemical and mineralogical analysis of weathered rocks. *Zeitschrift für Geomorphologie* 120: 159–173.

Smith, B.J., Srinivasan, S., McCabe, S. et al. (2011). Climate change and the investigation of complex moisture regimes in heritage stone: preliminary observations on possible strategies. *Materials Evaluation* 69: 48–58.

Smith, D.I., Greenaway, M.A., Moses, C.A., and Spate, A.P. (1995). Limestone weathering in eastern Australia. Part 1: erosion rates. *Earth Surface Processes and Landforms* 20: 451–463.

Spagnolo, G.S., Ambrosini, D., Paoletti, D., and Accardo, G. (2002). Fibre optic projected fringes for monitoring marble surface status. *Journal of Cultural Heritage* 1: 337–343.

Srinivasan, S., Basheer, P.A.M., Smith, B.J. et al. (2010). Use of fiber optic and electrical resistance sensors for monitoring moisture movement in building stones subjected to simulated climatic conditions. *Journal of ASTM International* 7: 1.

Sumner, P. and Nel, W. (2002). The effect of rock moisture on Schmidt hammer rebound: tests on rock samples from Marion Island and South Africa. *Earth Surface Processes and Landforms* 27: 1137–1142.

Sun, T., Grattan, K.T.V., Srinivasan, S. et al. (2012). Building stone condition monitoring using specially designed compensated optical fiber humidity sensors. *IEEE Sensors Journal* 12 (5): 1011–1017.

Tittonell, P., Scopel, E., Andrieu, N. et al. (2012). Agroecology-based aggradation-conservation agriculture (ABACO): targeting innovations to combat soil degradation and food insecurity in semi-arid Africa. *Field Crops Research* 132: 168–174. https://doi.org/10.1016/j.fcr.2011.12.011.

Török, Á. (2003). Surface strength and mineralogy of weathering crusts on limestone buildings in Budapest. *Building and Environment* 38 (9): 1185–1192.

Trudgill, S.T. (1983). Preliminary estimates of intertidal limestone erosion, one tree island, southern great barrier reef, Australia. *Earth Surface Processes and Landforms* 8: 189–193.

Trudgill, S.T. (1986). Limestone weathering under a soil cover and the evolution of limestone pavements, Malham District, North Yorkshire, UK. In: *New Directions in Karst. Proceedings of the AngloFrench Symposium 1983*, vol. 24 (ed. K. Paterson and M.M. Sweeting), 461–471. Norwich: Geo Books.

Trudgill, S.T., Viles, H., Inkpen, R.J., and Cooke, R.U. (1989). Remeasurement of weathering rates, St Paul's cathedral, London. *Earth Surface Processes and Landforms* 14: 175–196.

Tuğrul, A. (2004). The effect of weathering on pore geometry and compressive strength of selected rock types from Turkey. *Engineering Geology* 75: 215–227.

Viles, A. (2013). Synergistic weathering processes. In: *Treatise on Geomorphology*, Weathering and Soils Geomorphology, vol. 4 (ed. J.F. Shroder and G.A. Pope), 12–26. San Diego: Academic Press.

Viles, H.A. (2000). Mitteilungen des Verbandes der Deutschen Hohlen- und Karstforscher e.V. Munchen. *Micro-organisms and Geomorphology* 46: 116–121.

Viles, H.A., Goudie, A.S., Grab, S., and Lalley, J. (2010). The use of the Schmidt hammer and Equotip for rock hardness assessment in geomorphology and heritage science: a comparative analysis. *Earth Surface Processes and Landforms* 36: 320–333.

Warke, P.A., McKinley, J., and Smith, B.J. (2006). Variable weathering response in sandstone: factors controlling decay sequences. *Earth Surface Processes and Landforms* 31: 715–735.

Warscheid, T. and Braams, J. (2000). Biodeterioration of stone: a review. *International Biodeterioration and Biodegradation* 46: 343–368.

Wicke, B., Smeets, E., Watson, H., and Faaij, A. (2010). The current bioenergy production potential of semi-arid and arid regions in sub-Saharan Africa. *Biomass and Bioenergy* 35: 2773–2786. https://doi.org/10.1016/j.biombioe.2011.03.010.

Winkler, E.M. (1973). *Stone Properties, Durability Man's Environment, Applied Minerology*, vol. 4. Springer.

3

Comparison of Major Hydrogeochemical Processes in Coastal Sedimentary and Hard Rock Aquifers of South India

Amjad Al-Rashidi[1], Bedour Alsabti[1], Thilagavathi Rajendiran[2], Singaraja Chelladurai[3], and Chidambaram Sabarathinam[1]

[1] *Water Resources Development Management Program, Water Research Center, Kuwait Institute for Scientific Research, Kuwait City, Kuwait*
[2] *Department of Geology, Anna University, Chennai, India*
[3] *Department of Geology, Presidency College, Chennai, India*

3.1 Introduction

In most rural homes in India, groundwater remains the only drinking water source. India is heading toward a significant dilemma of excessive groundwater use and groundwater quality decline, with an estimation of 30 million groundwater structures (World Bank Report 2010). Groundwater is threatened by its mismanagement, along with other factors related to it.

The security of drinking water and Indian agriculture are both reliant on groundwater. Millions of farmers around the nation utilize it as a shared resource. The day-to-day dependency on groundwater resources is increasing rapidly due to the depletion and contamination of surface water sources and the increase in utility for agricultural and domestic purposes (World Bank Report 2010; Selvam and Sivasubramanian 2012).

India's land and water resources are made up of hard rocks, making up around two-thirds of the country, and unconsolidated and semi-consolidated sediments make up the remaining portion. India receives more rainfall on average each year than the rest of the globe, which is approximately 105 cm (Döll and Schmied 2012). However, 80% of the Indian territories have semiarid weather. It was noticed that groundwater levels were declining because fewer wells produced groundwater. In contrast, more wells are being drilled, bore wells are failing, dug wells are drying up, and electricity usage is rising (Imtiyaz and Rao 2008). The significant factors causing the decline in groundwater are the temporal and spatial coverage of rainfall, the monsoon's general variability, topographic differences, and the prevalence of rainfall across the country (Rai et al. 1995).

Tamil Nadu is represented by various hydrogeological formations, with hard rock aquifers, semi-consolidated rock, and consolidated rock, covering nearly 73% of the state's

Weathering and Erosion Processes in the Natural Environment, First Edition. Edited by Virendra Bahadur Singh, Sughosh Madhav, Naresh Chandra Pant, and Ravi Shekhar.

1,30,060 km^2 area. These formations are mostly restricted to the eastern part of the state, the coastal tract for about 1000 km. Along the eastern coast, the depth of groundwater level ranges between 2 and 5 m bgl (below groundwater level).

The hydrogeochemical properties of groundwater are primarily determined by natural and anthropogenic activities. The deterioration of groundwater quality in coastal areas is caused by several chemical systems, including evaporation, sea spray and oceanic airborne particles deposited on the topsoil, saltwater intrusion, groundwater interplay with brine, and the host matrix (Sanford et al. 2007). Manmade contaminants from industries and sewage wastewaters, landfills, wastewaters, and agricultural fertilizers contribute significantly to water contamination (Mondal et al. 2008; Mondal et al. 2011b; Selvam et al. 2013; Chidambaram et al. 2012d). Due to groundwater pollution, most of the industrial activities in these regions are located along the shore, aggravating the possibility of seawater intrusion and salt pan contamination.

The hydrogeochemistry of the recharging water, along with interactions between soil and gases, soil and water, the lithology of the unsaturated zone, subsurface residence period, and chemical interactions with the host rock govern the hydrochemistry of the underground water (Appelo and Postma 2004). The hydrogeochemical processes that affect the hydrogeochemistry of groundwater also include ion exchange processes, solubility, deposition, absorbance, and degradation (MacDonald et al. 2001).

Various experts have undertaken several studies on groundwater quality worldwide (Mishra et al. 2005; Çelik and Yıldırım 2006; Leung and Jiao 2006). Many academics have looked at the link between minerals and rocks and the hydrogeochemistry of subsurface water (Srinivasamoorthy et al. 2008). The hydrogeochemistry of groundwater relies on the composition of the rocks and soils in the saturated zone (Foster et al. 2000; Chidambaram et al. 2008a). A more excellent understanding of hydrogeochemistry helps us to determine the geochemical process along the groundwater pathways and information on the residence time (Janardhana 2007; Chidambaram et al. 2011).

The hydrogeochemistry of the weathered and hard rock terrains were investigated by previous researchers (Srinivasamoorthy et al. 2008; Prasanna et al. 2010; Manikandan et al. 2011; Wanda et al. 2011). In a hard rock aquifer, groundwater quality and availability are directly impacted by changes in the local topography and drainage system (Partey et al. 2010). Determining the existence and distribution of subsurface water in a hard rock aquifer is challenging as there is a significant variation in the hard rock aquifer parameters (e.g., porosity, joints, and fractures) that characterize the storage and flow of groundwater (Das and Sar 1988).

The coastal sedimentary aquifers are still more fragile as they are easily susceptible to natural and anthropogenic influences. Since the coastal areas are vulnerable, necessary precautions must be taken to manage them sustainably. The issue becomes more acute when the coastal alluvial aquifer caps the layered aquifer sequence because coastal aquifers are threatened by excessive drawdown, seawater ingression, salt pan contamination, industrial pollution, aquaculture, and other anthropogenic impacts.

The current focus of this chapter lies on comparing the geochemistry of groundwater from the Tuticorin hard rock aquifer and Cuddalore sedimentary coastal aquifer of south India by determining key factors governing the geochemistry along with their spatiotemporal variations.

3.2 Study Area

3.2.1 Sedimentary Aquifer, Cuddalore District

The coastal sedimentary aquifer of Cuddalore district is located in the northeastern part of Tamil Nadu, India (Figure 3.1a), bounded by latitudes 15°5′ to 12°35′N and longitudes 78°38′ to 80°00′E, covering a total extent of 3678 km^2. The Bay of Bengal bounds the east side and has a low, undulating topography with about 5 m elevation above mean sea level and a gradual sloping toward the coast.

Rainfall causes occasional flooding in the coastal region during the northeast monsoon season (October–December). From June through September, the southwest monsoon delivers rare rainfall. The long-term average rainfall in the Cuddalore district is 1160.36 mm. The northeast monsoon produces 53% of Cuddalore's total rainfall, which is more effective than the southeast monsoon. The remaining rains are in the winter (March to May) and summer (January to February). The study area's temperature is between 25 °C and 31 °C,

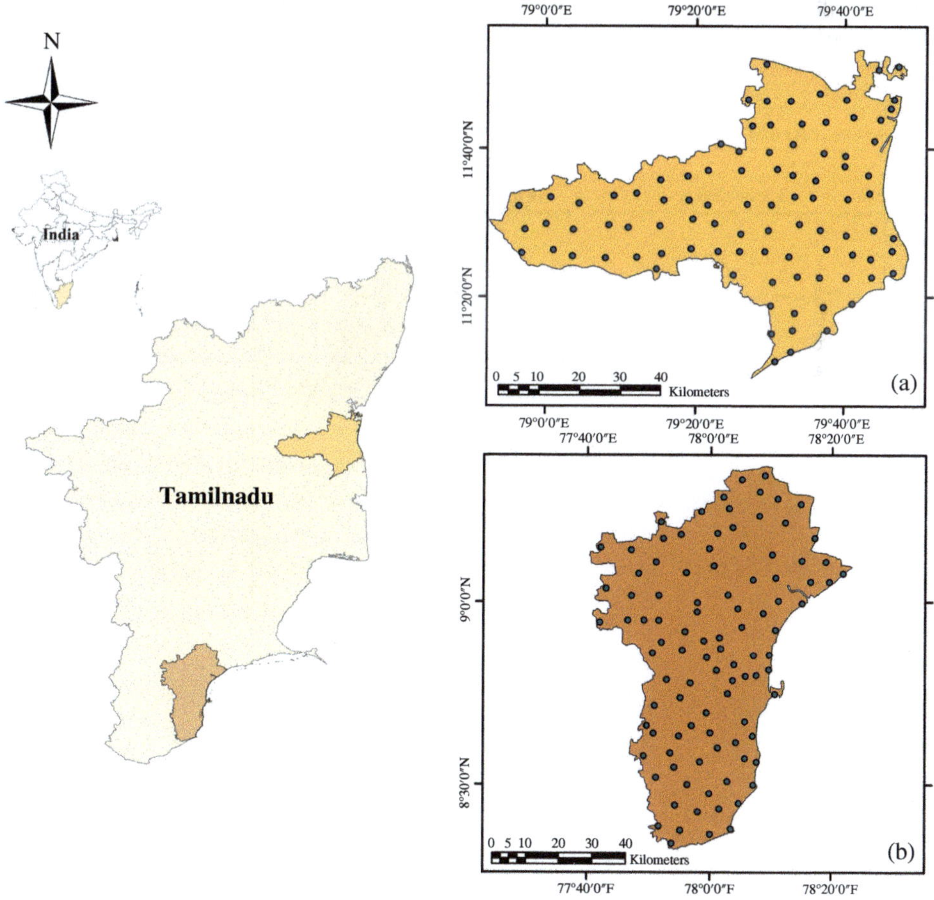

Figure 3.1 Study areas and sampling locations: (a) Cuddalore district and (b) Tuticorin district. Geologic maps of (c) Cuddalore district and (d) Tuticorin district.

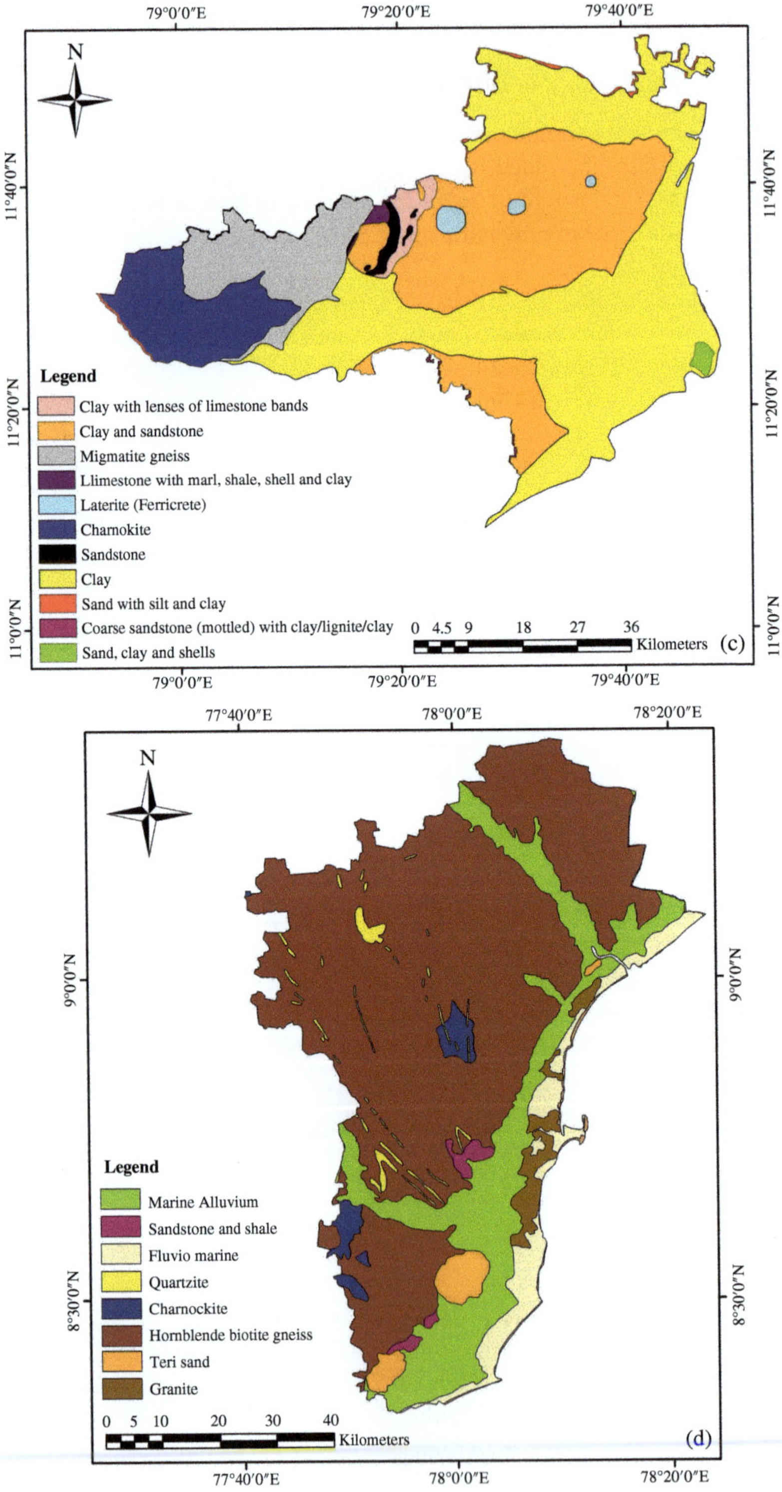

Figure 3.1 (Continued)

according to records of the Indian Meteorological Department. The highest temperatures were measured in April and May, while December and January recorded the lowest. The Ponnaiyar and Gadilam Rivers drain the northern portion of Cuddalore, while the Uppanar and Vellar Rivers drain the southern part. These rivers are all from the west to east sides and are ephemeral.

In the Cuddalore district, black, red, ferruginous, and arenaceous soils are the predominant soil types. Parts of Cuddalore, Virudhachalam, and Panruti taluks are covered by laterite and lateritic gravels, while red sandy soil covers the Cuddalore sandstone (Singaraja et al. 2012). Small patches of recent alluvial soil can be seen along the study area's stream and river courses.

The geological formations range from the Archaean age to the recent (CGWB 1993) and are divided into porous, fissured and fractured formations. The region's western portion is covered by fissured and fractured formations consisting of crystalline and hard-consolidated Archaean rocks (Figure 3.1c). There are economic deposits of lignite in the region and lignite is mined by open-cast mining.

The porous formation comprises semi-consolidated tertiary formations, which include the Cuddalore sandstone and Gopurapuram formation, and the unconsolidated Quaternary deposits like laterite, river, and coastal alluvium. The Cuddalore sandstone of Quaternary formation serves as a potential aquifer in the study area (CGWB 2009).

3.2.2 Hard Aquifer, Tuticorin District

The coastal hard aquifer is Tuticorin district, which is located in the south of Tamil Nadu, India (Figure 3.1b). The district falls between latitudes 8° 19′ and 9° 22′ N and longitudes 77° 40′ and 78° 23′E with an area of about 4621 km^2. Mannar Gulf borders the study area, and the districts of Tirunelveli, Virudhunagar, and Ramanathapuram, form its northern, eastern, and southern boundaries, respectively.

This district's overall terrain is undulating, with hill ranges. Along the banks of the rivers, flood plains are observed. These plains were created as a result of silt deposition during flood seasons. Both the northeast and southeast monsoons are influencial period of rainshowers in the area and most predominantly during north east monsoon. Most of the precipitation is from cyclonic storms in the Bay of Bengal depressions. Pre-monsoon (PRM) precipitation hardly registers in comparison to the very unpredictable southwest monsoon rains. During the research period of 2011–2012, average rainfall data reveals that the usual rainfall across the area ranges from roughly 2 mm to 217 mm. The region has a hot tropical climate; throughout the year, there is a high relative humidity ranging between 60 and 75%. The lowest and highest temperatures on average each year are 23 °C and 29 °C, respectively. Tamiraparani, Vaippar, and Kallar are the major rivers in the region.

Black cotton soil covers the western part of the district (Mondal et al. 2011b), scattered patches of red soil may be found on high land, and Teri soil can be found in the study area's southern region. Since the district has a lengthy coastal stretch of 163.5 km, most of the land is covered with sandy soils. Only the river floodplain and coastal areas have alluvial soil. Additionally, a few locations are represented by saline and alkaline soils.

Hard rock from the coastal region predominates throughout the study area. There are three main geological units: hornblende-biotite gneiss (HBG), fluvial, and alluvial marine. The HBG covers almost 70% of the Tuticorin district.

Most fluvial marine sediments dominate the coast and the eastern part of the Tuticorin district. Fluvial marine deposits are also found along the riverbanks. In the study area, there are four Charnockite patches, three located along the southern edge and one in the central region. The study area's western section may have two sand pockets and few small quartzite patches. Granite is observed along the east, while sandstone and shale are located in the central region (Figure 3.1d). The district's significant aquifer systems are made up of two types of formations: weathered and fractured crystalline rocks and unconsolidated to semi-consolidated formations.

3.3 Material and Methods

A total of 186 groundwater samples were collected from the sedimentary aquifer in the Cuddalore district (Figure 3.1a) from April 2015 to January 2016, groundwater samples were taken from the bore wells and hand pumps throughout the PRM and post-monsoon (POM) seasons. In addition, the groundwater sampling of the hard aquifer was carried out in the Tuticorin district with a total of 200 samples during July 2011 (PRM) and March 2012 (POM) (Figure 3.1b). Sampling was carried out with respect to lithology and spatial coverage.

The titrimetric method was used to analyze Ca^{2+}, Mg^{2+}, Cl^{2+}, and HCO_3^-, whereas flame photometer (Elico 378) was used to measure ions like Na^+ and K^+. The concentrations of SO_4^{2-}, PO_4^{2-}, NO_3^-, and H_4SiO_4 were measured using UV-Spectrophotometer (DR 6000, HACH). Fluoride and parameters like pH, electrical conductivity (EC), and total dissolved solids (TDS) were measured using a Thermo Orion electrode meter. Using an LED laser fluorimeter, uranium was analyzed using elements like Cu, Cd, Zn, and Pb by 797 VA Computrace. Standardized procedures were used to analyze all the measured parameters of groundwater samples along with major and minor ions (APHA 1995; Ramesh and Anbu 1996; Chidambaram 2000).

Aquachem and WATCLAST software were used for geochemical characterization (Chidambaram et al. 2003). SPSS was used for the statistical analysis, while ArcGIS was used to illustrate the geographical diagrams.

3.4 Results and Discussion

3.4.1 Water Chemistry

The maximum, minimum, and mean levels of groundwater chemical components for the two regions are shown in Table 3.1. Geochemical classifications of the obtained results for sedimentary and hard aquifers are presented in Table 3.2. Finally, the order of ion dominance during the POM and PRM seasons were provided for both research regions in Table 3.3.

Table 3.1 Maximum, Minimum, and Average of the different geochemical quality parameters of groundwater representing during PRM and POM for sedimentary and hard aquifers (all values in mg/l except EC in μS/cm and pH).

	pH	EC	TDS	Ca^{2+}	Mg^{2+}	Na^{+}	K^{+}	Cl^{-}	HCO_3^{-}	F^{-}	SO_4^{2-}	PO_4^{2-}	NO_3^{-}	H_4SiO_4	Temp°C
Sedimentary rock aquifer															
PRM															
Maximum	8.70	4890	2410	168	93.6	295	50.2	906	408	5.00	6.21	4.78	32.1	390	38.2
Minimum	5.59	95.7	50.1	8.00	2.40	1.20	0.20	17.7	24.0	0.02	0.06	0.01	0.44	5.00	15.1
Average	7.07	1284	610	49.2	26.5	67.2	6.19	166	172	0.61	1.20	0.28	27.9	63.9	30.7
POM															
Maximum	7.84	9660	3690	160	73.6	1045	41.5	1530	256	3.50	9.91	5.18	269	278	29.2
Minimum	5.10	169	72.4	12.0	1.30	2.80	0.10	48.7	3.66	0.00	0.08	0.00	6.24	0.30	27.0
Average	6.66	2365	786	50.6	21.1	141	6.43	314	54.9	0.74	2.32	0.30	58.9	12.9	27.5
Hard rock aquifer															
PRM															
Maximum	9.20	26240	16731	1600	1248	3980	521	10812	537	3.30	456	12.0	148	125	35.9
Minimum	6.80	309	205	4.00	4.80	14.8	0.50	35.5	12.2	BDL	1.00	BDL	0.51	1.25	27.6
Average	7.57	2878	1843	101	78.6	425	45.5	923	161	0.53	72.6	0.62	7.02	61.8	32.4
POM															
Maximum	8.80	30200	19366	608	559	6812	219	10990	2940	2.43	324	3.21	57.0	15.2	35.8
Minimum	6.95	372	254	4.00	2.40	14.7	0.83	35.5	36.6	BDL	1.00	BDL	0.23	2.00	29.3
Average	7.67	2358	1484	81.3	50	365	25.4	636	339	0.46	37.8	0.21	4.34	12.2	31.8

Table 3.2 Chemical characters of groundwater for PRM and POM obtained from the WATCLAST for sedimentary and hard aquifers.

		Sedimentary aquifer		Hard aquifer				Sedimentary aquifer		Hard aquifer			Sedimentary aquifer		Hard aquifer	
	Grade	PRM	POM	PRM	POM	Category	Grade	PRM	POM	PRM	POM	Category	PRM	POM	PRM	POM
Category	Na% Wilcox (1955)					USGS hardness						TDS Classification (USSL, 1954)				
Excellent	0–20	23	5	0	1	Soft	<75	1	3	3	5	<200	7	11	0	0
Good	20–40	38	32	24	24	Slightly hard	75–150	27	24	20	17	200–500	41	26	25	21
Permissible	40–60	21	23	32	32	Moderately hard	150–300	48	52	23	33	500–1500	43	47	50	45
Doubtful	60–80	11	27	34	33	Very hard	>300	15	14	54	45	1500–3000	2	9	15	16
Unsuitable	>80	0	6	10	10	*IBE Schoeller (1965)*						*Cation facies*				
Na% Eaton (1950)						(Na + K)rock- > Ca/Mg g.w.		17	4	33	48	Ca-Mg facies	21	4	0	2
Safe	<60	82	60	56	55	(Na + K)g.w.- > Ca/Mg rock		76	89	67	52	Ca-Na facies	72	89	100	97
Unsafe	>60	11	33	44	45	*Schoeller classification (1967)*						Na-Ca facies	0	0	0	1
S.A.R. Richards (1954)						Type I		93	92	75	98	Na facies	0	0	0	0
Excellent	0–10	93	87	84	85	Type II		0	0	3	0	*Anion facies*				
Good	10–18	0	5	7	8	Type III		0	1	22	2	HCO3 facies	0	0	0	0
Fair	18–26	0	1	4	2	Type IV		0	0	0	0	HCO3-Cl-SO4 facies	0	0	0	0
Poor	>26	0	0	5	5	*Corrosivity ratio (1990)*						Cl-SO4-HCO3 facies	81	13	42	81
R.S.C. Richards (1954)						Safe	<1	40	2	90	61	Cl-Facies	12	80	58	19
Good	<1.25	88	93	95	74	Unsafe	>1	53	91	10	39	*Hardness classification (Handa, 1964)*				

Medium	1.25–2.5	4	0	3	10	*Chloride classification (Stuyfzand, 1989)*					*Permanent hardness (NCH)*				
Bad	>2.5	1	0	2	16	Extremely fresh	0	0	0	0	A1	39	2	2	3
EC Wilcos (1995)						Very fresh	5	0	0	0	A2	5	44	36	19
Excellent	<250	6	5	0	0	Fresh	55	25	29	31	A3	12	46	50	32
Good	250–750	26	10	23	20	Fresh brackish	19	33	15	22	*Temporary hardness (CH)*				
Permissible	750–2250	49	43	49	54	Brackish	14	33	42	35	B1	3	0	0	5
Doubtful	2250–5000	12	27	18	18	Brackish-salt	0	2	13	11	B2	3	0	5	24
Unsuitable	>5000	0	8	10	8	Salt	0	0	1	1	B3	4	1	7	17
						Hyperhaline	0	0	0	0					

Table 3.3 The order of dominance of cations and anions in different seasons.

Aquifer	Season	Dominant anions	Dominant cations
Hard aquifer	PRM	$Cl^- > HCO_3^- > SO_4^{2-} > H_4SiO_4 > NO_3^- > PO_4^{3-} > F^-$	$Na^+ > Ca^{2+} > Mg^{2+} > K^+$
	POM	$Cl^- > HCO_3^- > SO_4^{2-} > H_4SiO_4 > NO_3^- > F^- > PO_4^{3-}$	$Na^+ > Ca^{2+} > Mg^{2+} > K^+$
Sedimentary aquifer	PRM	$Cl^- > HCO_3^- > H_4SiO_4 > NO_3^- > SO_4^{2-} > F^- > PO_4^{3-}$	$Na^+ > Ca^{2+} > Mg^{2+} > K^+$
	POM	$Cl^- > H_4SiO_4 > NO_3^- > HCO_3^- > SO_4^{2-} > PO_4^{3-} > F^-$	$Na^+ > Ca^{2+} > Mg^{2+} > K^+$

3.4.1.1 Sedimentary Aquifer

In the sedimentary aquifer of Cuddalore, the pH varies from 5.1 to 8.7, from acidic to alkaline. POM has the lowest pH, whereas PRM has the highest.

The change in EC during PRM is noted to range from 95.7 to 4890 μS/cm, whereas in the POM season, it varies between 169 and 9660 μS/cm. Higher EC values have been observed along the Gadilam and Coleroon river courses, which may be the result of sewage infiltration (Thivya et al. 2014), and along the coastal zone due to saline water ingression (Chidambaram et al. 2003).

TDS levels range from 68.9 to 3500 mg/l. TDS was observed to be less during PRM than POM, suggesting the dissolution and leaching processes in some seasons were followed by the dilution process.

Spatial Distribution of EC

Irrespective of seasons, the groundwater samples of the southeastern region reflected higher EC values (Figure 3.2). In the PRM season, EC ranged from 95.7 to 4890 μS/cm. Higher concentrations were recorded in the southern and southeastern parts of the Cuddalore district, whereas the lowest concentrations were recorded in the northeastern and central parts (Figure 3.2a).

In POM, it ranges from 169 μs/cm to 9660 μS/cm, with the highest concentrations being found in the southeastern and northwestern parts (Figure 3.2b). The majority of the POM samples representing the poor category of EC are noted in the eastern part, and few samples were distributed in the southern region. Higher EC values are observed during POM along the stretch, and few samples of moderate category fall in the southwestern and northern regions. The Coleroon River becomes polluted due to the infiltration of domestic sewage, as well as due to the leachates from landfills along the river and saline ingression. Therefore, it can be concluded that samples close to the river are impacted by sewage infiltration along the alluvial floodplain, resulting in elevated conductivity values. Higher values of EC are found near the river and decrease as the distance from the river increases; a similar case was reported in Sri Lankan urban areas (Ranjana and Naverathna 2011).

Furthermore, the occurrence of moderately to very salty groundwater may be attributed to inadequate replenishment due to precipitation and insufficient flushing of underground water (Tyagi et al. 2008), both of which are frequently observed in terrains with similar hard rock compositions (Chidambaram 2000; Srinivasamoorthy et al. 2005; Manivannan 2010).

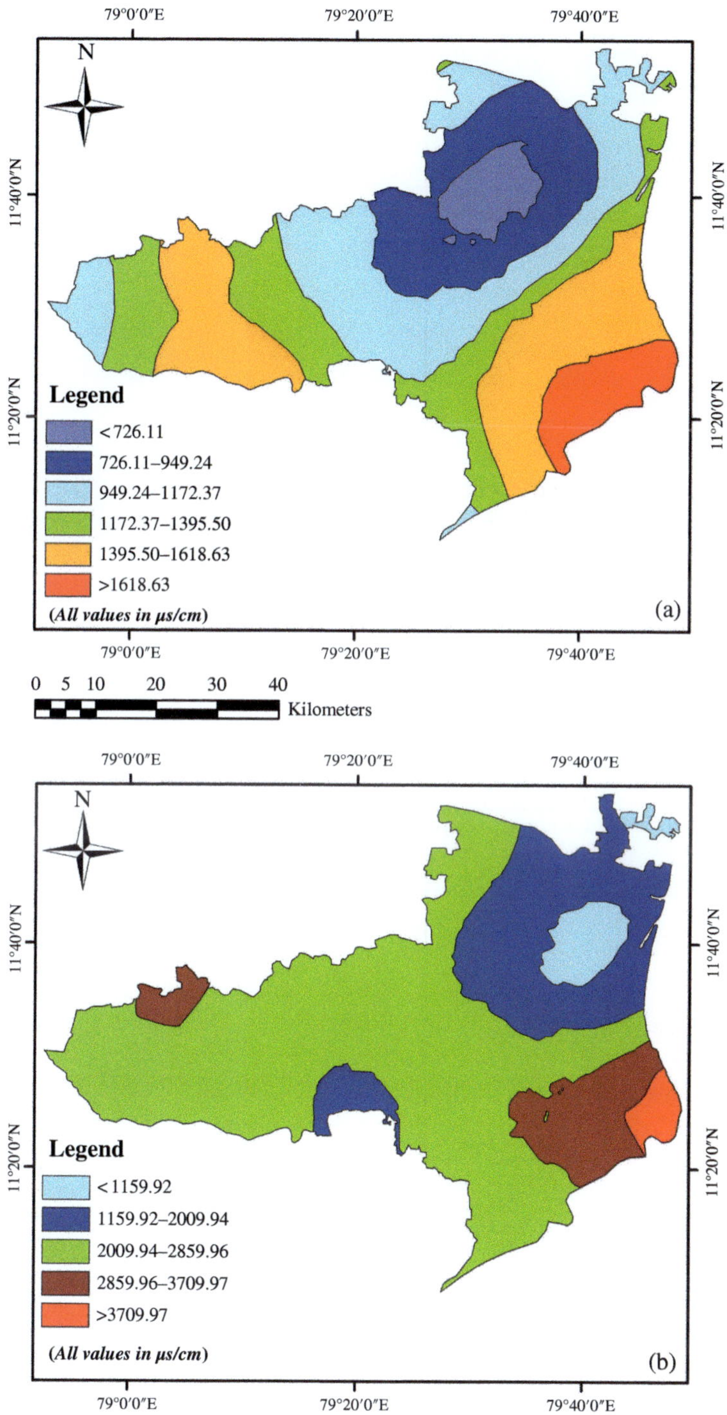

Figure 3.2 EC spatial variation of sedimentary aquifer in Cuddalore district, (a) PRM (b) POM, and Tuticorin district, (c) PRM and (d) POM.

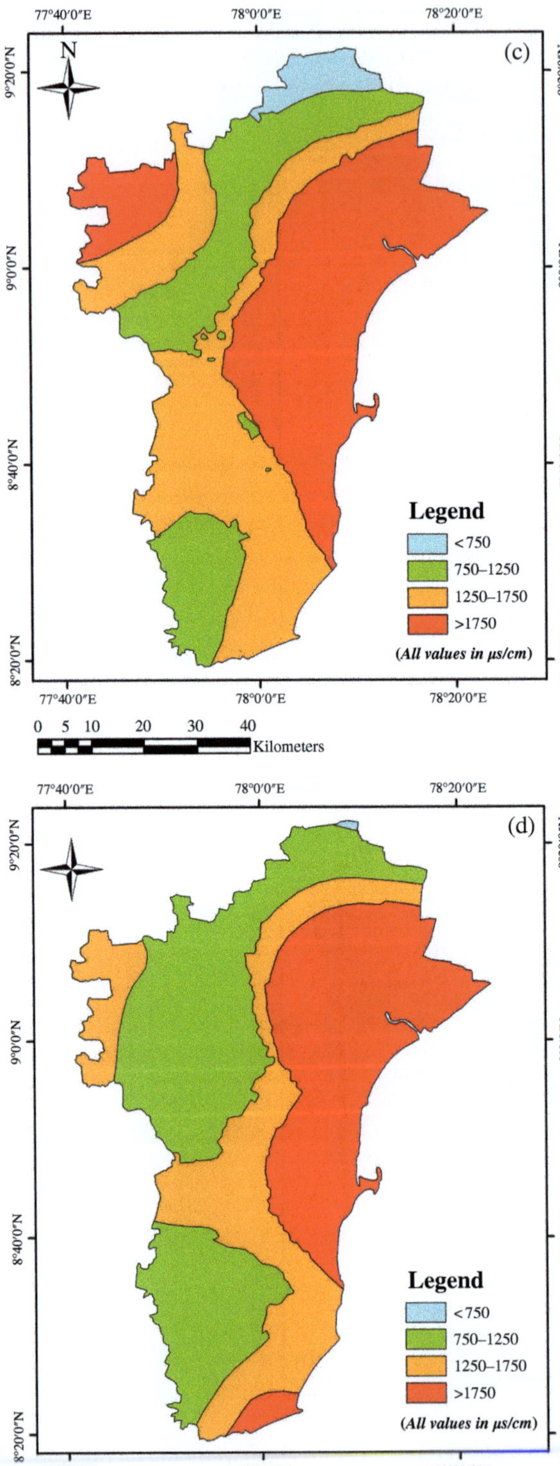

Figure 3.2 (Continued)

3.4.1.2 Hard Aquifer

In the hard aquifer of the Tuticorin district, the pH ranges between 6.8 and 9.2. The PRM recorded as both the lowest and highest pH values. pH is regulated by the balance of carbon dioxide, carbonate, and bicarbonate. Carbonic acid is created when CO_2 and water combine, and this acid influences the pH of the water. Therefore, higher pH levels of groundwater are noted in saltwater-intruded regions and along the salt pan boundaries.

The concentrations of EC in the Tuticorin district range from 309 to 30200 μS/cm. When compared to previous seasons, PRM had the highest EC value in the coastal area. This may be because of their proximity to the shore (Singaraja et al. 2013). However, the average values demonstrate that most samples had a value of 1250 μS/cm throughout the year.

The groundwater TDS ranges between 205 and 19366 mg/l. Although the highest TDS was observed during POM, groundwater extraction along the coastal region during the cropping season has resulted in the saltwater intrusion resulting in greater TDS levels in the study area's coastline region (Mondal et al. 2011a; Singaraja et al. 2012).

Spatial Distribution of EC

During PRM, the EC fluctuated from 309 μS/cm to 26240 μS/cm (Figure 3.2c). The eastern and northwestern parts of the Tuticorin district had the highest EC concentrations. The geographical distribution of the values demonstrates that EC in POM ranges from 372 μS/cm to 30200 μS/cm (Figure 3.2d). Northeastern and southeastern coastal regions showed higher concentrations of EC. This might be due to seawater intrusion along the shore (Prasanna et al. 2008; Singaraja et al. 2013).

3.4.2 Geochemical Classification

3.4.2.1 Sedimentary Aquifer

Most PRM samples cluster in the Na-Ca-HCO_3-Cl type (Figure 3.3a) and around the Na-HCO_3-Cl type, suggesting anthropogenic activities and ion exchange (Srinivasamoorthy et al. 2011) with some clustering in the Ca-HCO_3-Cl types. The Ca-HCO_3 type indicates groundwater replenishment through rainwater (Prasanna et al. 2010). A few occurrences of PRM samples were scattered in the mixed water type. Most POM samples represented the Na-Cl type, but only a few showed the Ca-Mg-Cl type, demonstrating the prevalence of seawater intrusion and the impact of anthropogenic activities. Samples of POM also fall in the Ca-Na-HCO_3 and Ca-Cl types. Most samples revealed cation and anion mixing, which might be attributed to additional leachates such as surface soil (Chidambaram et al. 2012a).

The association of Na-K-HCO_3-Cl also exists during PRM mostly because of weathering or ion exchange processes. Generally, ion exchange processes or feldspar weathering that occur during rock–water interactions are similar to this Na-K association. Three mechanisms were identified:

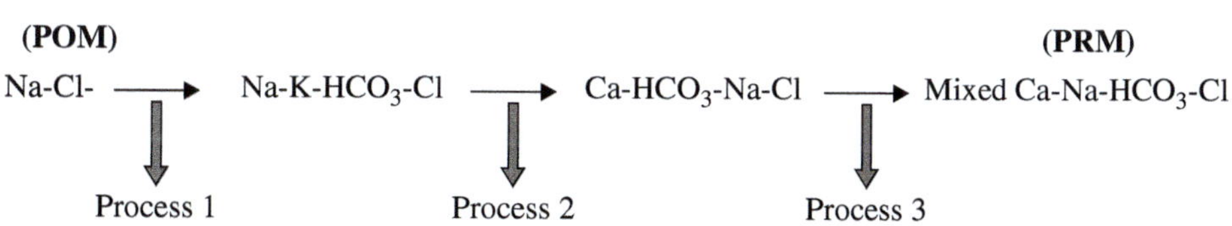

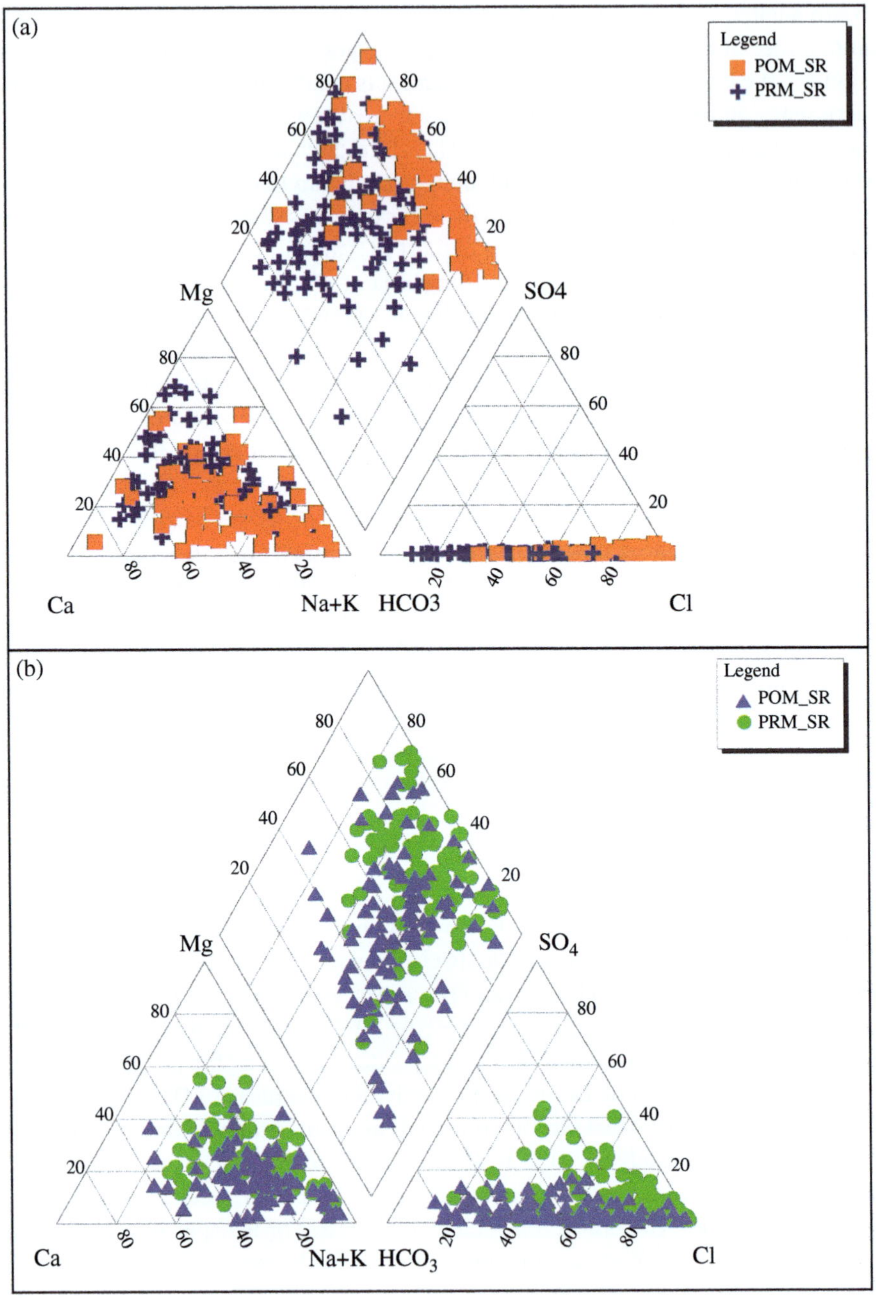

Figure 3.3 Piper plot of groundwater samples during PRM and POM seasons: (a) sedimentary aquifer and (b) hard rock aquifer.

However, with ion dissolution or leaching in groundwater, human activities prevail over weathering processes. Therefore, during POM season Na-Cl type water dominates.

In process 1, the release of K into groundwater or the Na absorption occurs due to mineral dissolution, and weathering resulted in the higher values of HCO_3.

Process 2 reflects the mechanism of weathering and ion exchange during PRM, thus resulting in a mixed water type of Ca-Na-HCO_3-Cl. It could be inferred that this type has resulted from mixing different surface water sources, notably in the region proximal to the mining sector with the pre-existing water type.

As a result of interactions with the aquifer matrix, groundwater mixing may cause the concentrations of Ca^{2+} to decrease relatively the concentrations of Na increase (Process 3). It could be noted that on a solid surface, Ca^{2+} ions are replaced by Na ions, as proven by the following equation:

$$Na - X + \frac{1}{2}Ca^{+} \rightarrow Na^{+} + \frac{1}{2}Ca^{+}$$

X is considered a topsoil exchanger, and water gets in touch with sodium originating from minerals or other anthropogenic sources, saltwater, or plagioclase feldspar in the aquifer. Through a process known as cation exchange, Na ions are released into the groundwater in substitution for the Ca^{2+} ions. As a result, Na ions build up in groundwater solutions (Kumar and Alappat 2005). The Cl^- is conservative and prevails in water during PRM due to the reverse ion exchange process, resulting in NaCl nature in POM season.

3.4.2.2 Hard Aquifer

The transition of the facies in the Tuticorin district is observed from a Na-HCO_3 type in POM to a Na-Cl type in PRM (Figure 3.3b). As a result, it makes the system less concentrated and less sensitive to the effects of Cl^-, which is the predominant anion during PRM. As a result, the system is diluted, and the influence of Cl^-, the predominant anion during PRM, is reduced. During POM, ion exchange reactions lead to the formation of water with the properties of Na^+, HCO_3^-, and Cl^- (Manivannan et al. 2011). The Ca-HCO_3 type, which indicates ion exchange and recharge processes in the near coastal area. Further, it is observed that during POM periods, cation exchange leads to the end member of the NaCl type. The predominant kind of ion exchange that occurs during weathering and natural recharging during the POM phase results in the formation of the NaCl type more than during the PRM phase. Though saline water or brackish water indicates the Na-Cl type, the change from one water type to another, with respect to season, indicates the ion exchange process.

3.4.3 Gibbs Method

3.4.3.1 Sedimentary Aquifer

The mechanisms that control the groundwater chemistry of the Cuddalore district were assessed using Gibbs ratio (Gibbs 1970). Most samples of PRM (Figure 3.4a) and POM (Figure 3.4b) represent the "weathering zone," with a very few samples representing the "evaporation zone." On the other hand, most POM data falls within the "weathering zone," with some samples falling along the evaporation-weathering boundary. Samples from both seasons occur outside the plot preview, which is an indication of anthropogenic activity.

While TDS and a few ion ratios form the basis of the Gibbs plot, specific samples that fall outside the boomerang plot and those with higher TDS values that fall in the evaporation zone may also be the result of anthropogenic causes in addition to natural ones like rain, weathering, and evaporation. This proves that, in addition to the natural source,

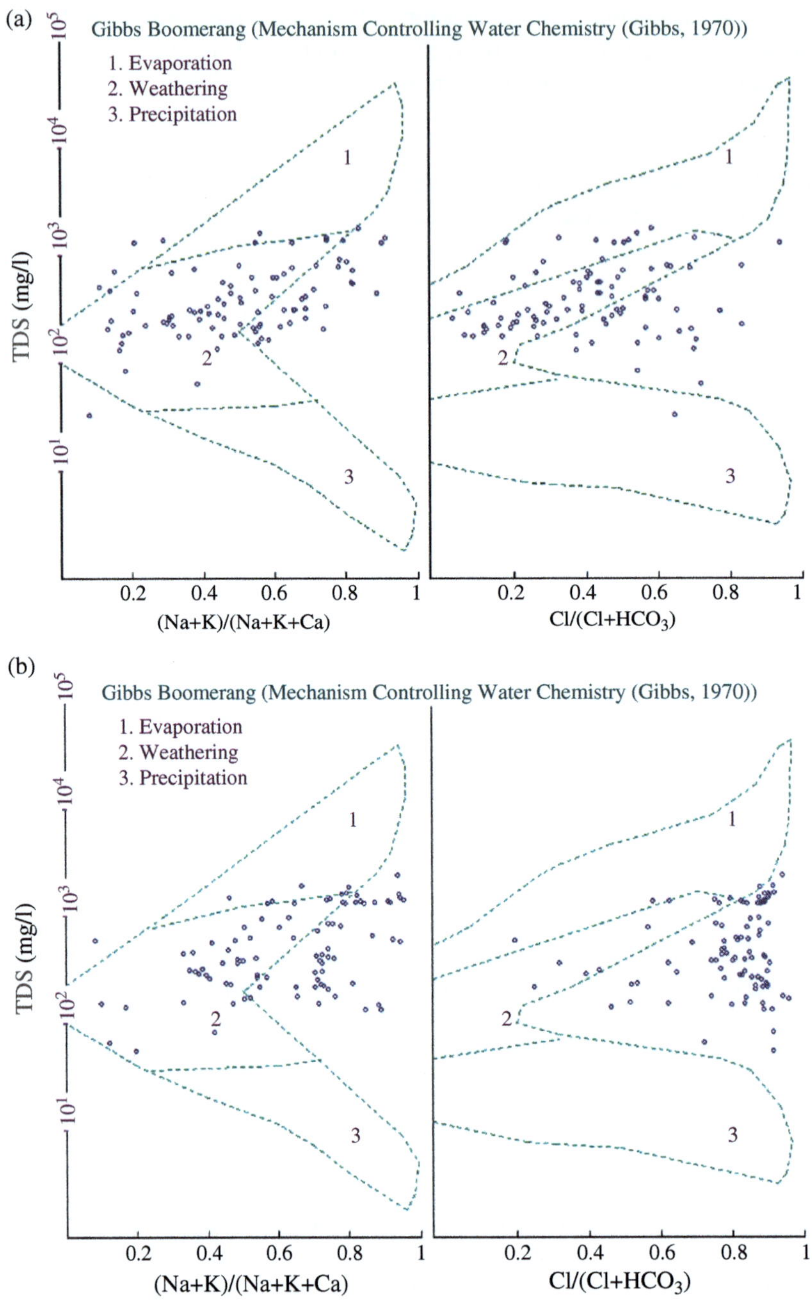

Figure 3.4 Gibbs plot for mechanism controlling the water chemistry for different seasons in sedimentary aquifer in Cuddalore district, (a) PRM and (b) POM, and hard aquifer in Tuticorin district, (c) PRM and (d) POM.

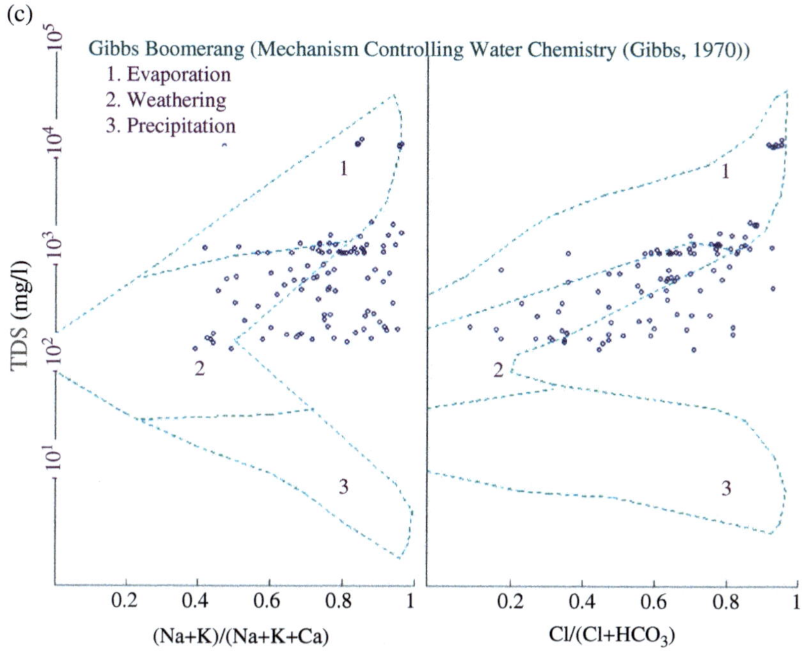

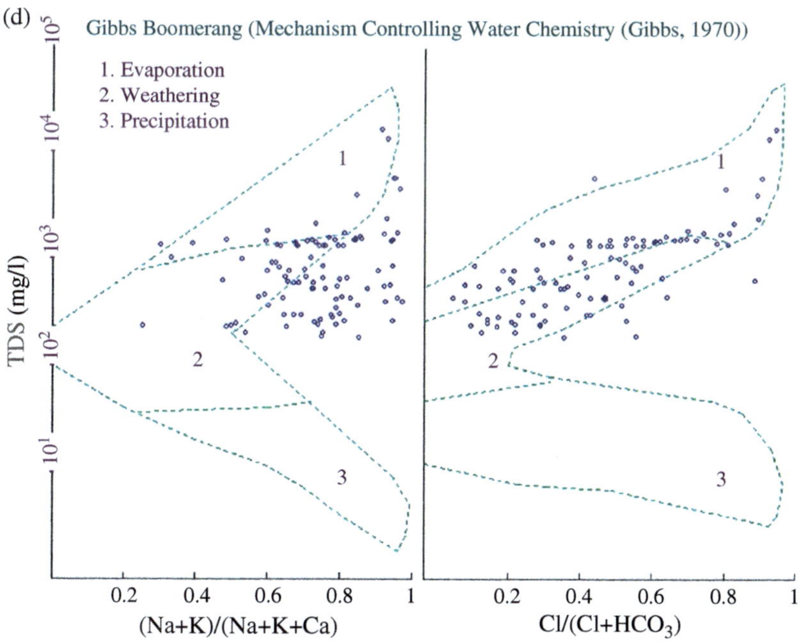

Figure 3.4 (Continued)

anthropogenic activity or artificial factors predominate the change in groundwater's chemical composition (Hem 1991; Karanth 1991). Most of the samples in this plot represent the zone predominant with the decomposition and disintegration of the mineral matrix (Chowdhury and Gupta 2011).

The samples represented in the plot show a scattered nature with no discrete pattern. Samples from the same season demonstrate a spatial variation in the geochemical processes. It's worth noting that the monsoon samples roughly correspond to the plot's periphery (i.e., during PRM, fewer samples have higher Ca^{2+} values expressed from the cation ratio, suggesting the predominance of Ca^{2+}, which is seen to decrease during POM). Given that these samples are outside the plot, the primary cause of this might be the influence of predominant weathering and leaching of ions during the monsoon.

A cluster of samples from PRM and POM seasons are noted along the margin of the three processes: evaporation, weathering, and precipitation. Hence, it's evident that most geochemical processes prevail in all other seasons with a variation in percentage. Most of the samples from the PRM season reflect the ion exchange process. It is noted that POM representations were observed in the boundary between evaporative and weathering zones, indicating the predominance of the mixing between these processes.

3.4.3.2 Hard Aquifer

During the POM and PRM seasons, hard rock aquifer samples fall outside the plot purview (Figure 3.4c,d). Few of the samples represent "the evaporation zone" during all seasons. Evaporation significantly impacts the amounts of ions produced by chemical weathering, which ultimately results in a rise in salinity. A rapid evaporation rate may lead to an increase in salinity and a higher value of Na:Ca in the water (Gibbs 1970). In the anion plot, most of the POM representations are observed in the weathering zone during POM (Figure 3.4d). However, during PRM (Figure 3.4c), the majority of the sample representation may be found outside Gibbs plot, which could be caused by anthropogenic sources rather than natural sources such as precipitation, weathering, or evaporation. A higher TDS content reflects that evaporation and precipitation processes are often responsible for controlling the surface water in these regions (Gibbs 1970). In addition, extensive agricultural activities (Zhang and Shan 2008) may influence this problem due to the fact that, excessive agricultural activities may increase soil weathering and erosion, leading to an increase in TDS value (Zhang et al. 1995). On the other hand, the initial Gibbs structure suggests that anthropogenic activities' effects on hydrogeochemistry are minor. The interaction between water and rock contributes in some way to the chemical characteristics of these samples (Chowdhury and Gupta 2011).

3.4.4 Statistics

3.4.4.1 Correlation Matrix

Several authors have indicated that a correlation matrix may be built using the Pearson correlation coefficient between different ions. It is the perfect standard for measuring correlation because it relies on the covariance technique (Swan and Sandilands 1995). The correlation coefficient provides details regarding the strength and direction of the correlation. When the correlation coefficient value is nearer to +1, it is claimed that the degree of correlation is perfect, though it ranges from −1 to +1. It is thought to have a strong and positive correlation for values ranging from +0.75 to +1, a fair degree of correlation between +0.25 and +0.75, and a minimal correlation between 0 and +0.25.

Sedimentary Aquifer

In PRM season, a strong association is established between Cl-SO_4; HCO_3-pH, EC; K-PO_4, Na-Cl, U; Cl^-, Na^+, Mg^{2+}, and Ca-Mg. Cl^-, SO_4 (Table 3.4), demonstrating the dominance of the decomposition and disintegration process along with dissolution. Secondary salt leaching is indicated by a strong correlation between chloride and Ca^{2+}, Na^+, and Mg^{2+}.

Chemical weathering is evident from the noteworthy correlation between EC, pH, and HCO_3 (Srinivasamoorthy et al. 2009). A positive weak correlation between NO_3, SO_4, and other ions is inferred due to the dilution process. A significant positive correlation between uranium, sodium, and magnesium implies that it could be present as a salt of these inorganic compounds. Decreased pH resulted in topsoil mineral dissolution, as shown by the poor positive correlation between pH and U. In addition, uranium enters the groundwater after dissolving from the mineral present in the host rock. The concentration of uranium in a solution increases as pH decreases.

In POM, Cl-Mg, Na^+, HCO_3^-, and H_4SiO_4 have a good correlation (Table 3.4). Mg^{2+} and Na^+ have poor correlations with other ions. Chloride exhibits a good correlation with magnesium and sodium, which mostly results from the impact of industrial effluents or secondary salt leaching that have formed along fractures or in formation permeable hotspots (Chidambaram et al. 2009). $Cl^-$and HCO_3^- ions have a strong association with magnesium and calcium indicating that this period is characterized by rock–water interaction and leaching (Prasanna et al. 2008). The principal ions have a moderate-to-high and good association with NO_3^- and other ions throughout the PRM and POM seasons, indicating that anthropogenic influence is high.

Hard Aquifer

In PRM, Cl-Ca, Cl-Mg, Cl-Na, Cl-K, Cl-SO_4^{2-}, Mg-Ca, Na-Ca, Mg^{2+}, SO_4^{2-}, and K-Ca have excellent correlations (Table 3.4). Chemical weathering is revealed by a significant association between HCO_3-F, NO_3-Na, Cl-SO_4^{2-}, SO_4^{2-}-Ca, and Mg^{2+} (Chidambaram et al. 2008b). Most ions are favorably related to Cl^-, especially K^+, Na^+, Mg^{2+}, and SO_4^{2-}(Table 3.4), which have a substantial correlation with Cl^- (Kim et al. 2003). This implies that these ions originated from the same substrate of salty fluids, dissolution, and disintegration followed by salt leaching (Chidambaram et al. 2009; Prasanna et al. 2010). The impact of agriculture, evaporation, insufficient runoff, and salt water on aquifers is also a notable factor (Subba Rao et al. 2012).

Cl-Ca, Cl-Mg, Cl-Na, Cl-HCO_3, Na-Ca, Ca-Mg, Mg^{2+}, SO_4^{2-}, and Cl-Ca, Mg^{2+}, Na^+, Cl^- have a significant correlation in POM (Table 3.4). Cl^- strongly correlates with Ca^{2+}, F^-, K^+, Na^+, Mg^{2+}, and HCO_3^- (Table 3.4), which suggests that salts are leaching. F-HCO_3 and K-Mg also exhibit significant correlations. Anthropogenic effect on the system is shown by the high connection between Ca^{2+}, Cl^-, Na^+, K^+, and SO_4^{2-}. Irrespective of season, Cl^- and SO_4^{2-} are the essential ions that show a moderate-to-good correlation with Na^+, Mg^{2+}, Ca^{2+}, and K^+ throughout both seasons. This is mainly owing to the effects of salinization (Singaraja et al. 2012), saltpan (Smith et al. 2004), especially along the coastline, and salt

Table 3.4 Correlation analysis and factor score of groundwater samples collected during PRM and POM for sedimentary and hard rock aquifers.

PRM	Ca^{2+}	Mg^{2+}	Na^{+}	K^{+}	F^{-}	Cl^{-}	HCO_3^{-}	SO_4^{2-}	PO_4^{3-}	NO_3	H_4SiO_4	U	pH	EC	PRM	1	2	3	4	
Sedimentary rock aquifer																				
Ca^{2+}	1.00														Ca^{2+}	0.70	−0.22	0.06	0.32	
Mg^{2+}	0.56	1.00													Mg^{2+}	0.82	0.13	0.00	0.03	
Na^{+}	0.37	0.59	1.00												Na^{+}	0.79	0.30	0.17	−0.07	
K^{+}	0.19	0.26	0.31	1.00											K^{+}	0.27	−0.04	0.84	−0.01	
F	−0.22	0.07	0.22	−0.08	1.00										F	−0.09	0.65	−0.14	−0.06	
Cl^{-}	0.71	0.74	0.80	0.38	−0.09	1.00									Cl^{-}	0.91	−0.18	0.17	0.09	
HCO_3^{-}	0.21	0.42	0.40	−0.01	0.41	0.09	1.00								HCO_3^{-}	0.31	0.81	0.17	0.09	
SO_4^{2-}	0.16	0.44	0.49	0.10	0.14	0.40	0.32	1.00							SO_4^{2}	0.55	0.33	−0.11	0.25	
PO_4^{3-}	−0.05	−0.14	0.06	0.54	−0.09	0.00	0.40	0.32	1.00						PO_4^{3-}	−0.11	0.02	0.87	−0.06	
NO_3^{-}	0.04	0.17	0.18	0.15	0.18	0.02	0.00	−0.03	−0.09	1.00					NO_3^{-}	0.02	0.45	0.28	0.23	
H_4SiO_4	0.16	−0.06	−0.05	−0.07	−0.01	0.03	0.02	0.30	0.02	0.07	1.00				H_4SiO_4	−0.03	0.03	−0.06	0.93	
U	0.30	0.52	0.56	0.10	0.04	0.48	0.27	0.32	−0.03	0.10	−0.07	1.00			U	0.68	0.13	−0.05	−0.20	
pH	0.06	0.25	0.33	0.11	0.24	0.07	0.62	0.26	0.02	0.15	−0.01	0.14	1.00		pH	0.02	0.71	0.07	−0.01	
EC	0.26	0.45	0.56	0.10	0.09	0.34	0.57	0.46	0.00	0.17	−0.04	0.48	0.41	1.00	EC	0.59	0.51	0.00	−0.11	
POM	**Ca^{2+}**	**Mg^{2+}**	**Na^{+}**	**K^{+}**	**F^{-}**	**Cl^{-}**	**HCO_3^{-}**	**SO_4^{2-}**	**PO_4^{3-}**	**NO_3**	**H_4SiO_4**	**U**	**pH**	**EC**	**POM**	**1**	**2**	**3**	**4**	**5**
Ca2+	1.00														Ca^{2+}	0.27	0.53	−0.54	0.03	−0.12
Mg^{2+}	0.35	1.00													Mg^{2+}	0.27	0.52	−0.25	0.26	0.03
Na^{+}	0.25	0.35	1.00												Na^{+}	0.56	0.43	0.15	0.50	0.23
K^{+}	0.18	0.14	0.24	1.00											K^{+}	0.58	−0.01	−0.22	0.02	0.02
F	−0.06	0.09	0.26	0.05	1.00										F	0.17	0.12	0.09	−0.18	0.82

Cl^-	0.42	0.52	0.94	0.25	0.22	1.00									Cl^-	0.53	0.52	−0.02	0.53	0.20
HCO_3^-	0.23	0.29	0.28	−0.01	0.09	0.31	1.00								HCO_3^-	−0.09	0.83	0.15	−0.02	0.03
SO_4^{2-}	−0.02	0.12	0.19	0.09	−0.09	0.20	0.04	1.00							SO_4^2	0.01	−0.07	0.02	0.74	−0.05
PO_4^{3-}	−0.20	0.00	0.18	−0.07	−0.09	0.14	0.12	0.00	1.00						PO_4^{3-}	−0.04	0.09	0.71	0.25	−0.11
NO_3^-	0.12	0.15	0.19	0.38	0.08	0.15	−0.10	−0.03	−0.05	1.00					NO_3^-	0.69	−0.15	−0.15	−0.20	−0.04
H_4SiO_4	0.08	0.04	0.00	−0.06	−0.02	−0.02	0.50	−0.03	0.15	−0.17	1.00				H_4SiO_4	−0.25	0.63	0.23	−0.20	−0.10
U	−0.07	−0.03	0.10	0.12	0.34	0.09	−0.06	0.05	0.05	0.06	−0.06	1.00			U	−0.04	−0.13	−0.05	0.15	0.78
pH	−0.14	0.00	0.13	−0.04	0.29	0.06	0.20	−0.95	0.23	0.01	0.12	0.01	1.00		pH	0.23	0.18	0.67	−0.33	0.20
EC	0.12	0.20	0.48	0.26	0.19	0.44	−0.04	0.14	−0.04	0.43	−0.13	0.04	0.26	1.00	EC	0.81	−0.01	0.16	0.09	0.07
PRM	**pH**	**Ca^{2+}**	**Mg^{2+}**	**Na^+**	**K^+**	**Cl^-**	**HCO_3^-**	**NO_3**	**PO_4^{3-}**		**F^-**	**SO_4^{2-}**	**H_4SiO_4**		**PRM**	**1**	**2**	**3**	**4**	
Hard rock aquifer																				
pH	1.00														pH	−0.26	−0.54	0.17	0.11	
Ca^{2+}	−0.30	1.00													Ca^{2+}	0.56	0.61	0.02	0.40	
Mg^{2+}	−0.28	0.75	1.00												Mg^{2+}	0.80	0.32	−0.01	0.27	
Na^+	−0.31	0.47	0.73	1.00											Na^+	0.91	0.15	0.02	−0.14	
K^+	−0.34	0.67	0.38	0.32	1.00										K^+	0.26	0.78	0.08	0.11	
Cl^-	−0.33	0.70	0.88	0.94	0.46	1.00									Cl^-	0.91	0.31	−0.02	0.06	
HCO_3^-	−0.07	0.12	0.11	0.24	0.16	0.18	1.00								HCO_3^-	0.15	0.17	0.80	−0.13	
NO_3^-	−0.18	0.05	0.17	0.39	0.20	0.32	0.13	1.00							NO_3^-	0.48	−0.01	−0.03	−0.50	
PO_4^{3-}	−0.10	0.00	−0.01	0.02	0.12	0.00	0.04	0.08	1.00						PO_4^{3-}	−0.21	0.54	0.02	−0.54	
F	0.21	−0.14	−0.08	−0.08	−0.10	−0.13	0.41	−0.06	−0.06		1.00				F	−0.05	−0.22	0.85	0.06	
SO_4^{2-}	−0.25	0.37	0.38	0.53	0.18	0.47	0.14	0.33	0.08		0.13	1.00			SO_4^{2-}	0.65	0.06	0.17	−0.21	
H_4SiO_4	−0.05	0.17	0.10	−0.13	0.01	−0.03	−0.12	−0.08	−0.15		0.01	−0.04	1.00		H_4SiO_4	−0.04	0.06	−0.06	0.69	

(Continued)

Table 3.4 (Continued)

POM	pH	Ca^{2+}	Mg^{2+}	Na^+	K^+	Cl^-	HCO_3^-	NO_3	PO_4^{3-}	F^-	SO_4^{2-}	H_4SiO_4	POM	1	2	3	4
pH	1.00												pH	−0.13	0.04	−0.87	0.01
Ca2+	−0.38	1.00											Ca^{2+}	0.81	−0.08	0.33	−0.02
Mg^{2+}	−0.16	0.72	1.00										Mg^{2+}	0.91	−0.05	0.13	0.03
Na^+	−0.12	0.69	0.77	1.00									Na^+	0.92	0.08	−0.02	−0.02
K^+	−0.34	0.27	0.43	0.21	1.00								K^+	0.29	−0.02	0.51	0.54
Cl^-	−0.16	0.75	0.88	0.97	0.28	1.00							Cl^-	0.96	0.00	0.03	0.00
HCO_3^-	−0.09	−0.02	0.01	0.19	0.04	0.11	1.00						HCO_3^-	0.06	0.81	0.11	0.08
NO_3^-	0.07	0.07	0.04	0.05	0.26	0.05	0.04	1.000					NO_3^-	0.09	0.07	−0.19	0.80
PO_4^{3-}	0.00	0.02	−0.04	−0.03	0.12	−0.04	0.00	0.13	1.00				PO_4^{3-}	−0.10	−0.07	0.11	0.56
F	0.13	−0.11	−0.07	−0.04	−0.11	−0.08	0.35	−0.05	−0.10	1.00			F	−0.04	0.81	−0.15	−0.13
SO_4^{2-}	−0.08	0.65	0.66	0.52	0.23	0.55	−0.04	0.18	−0.03	0.11	1.00		SO_4^{2-}	0.71	0.10	0.06	0.05
H_4SiO_4	−0.04	−0.23	−0.40	−0.47	−0.16	−0.48	0.05	−0.22	−0.01	0.15	−0.16	1.00	H_4SiO_4	−0.53	0.22	0.37	−0.30

leaching along fractures or permeability hotspots (Chidambaram et al. 2009). During PRM, the primary ion has a moderate-to-strong and good association with NO_3^- and few other ions, which reflect a higher proportion of manmade pollutants.

3.4.4.2 Factor Analysis (FA)

FA or principal component analysis is adopted to explain the origin of hydrogeochemistry of underground water (Gupta and Subramanian 1998) and a classification technique that aids in the origin, interpretation, and investigation of hydrochemical facies (Lawrence and Upchurch 1982). Additionally, it can be utilized in groundwater contamination studies (Jeong 2001).

A multivariate data set's covariance matrix can be understood through FA. In accordance with the criteria of Liu et al. (2003), we categorized the factor loadings as strong, moderate, and weak based on absolute loading values of >0.75, 0.75–0.50, and 0.5–0.30, respectively.

Sedimentary Aquifer

i) PRM

According to FA, four main contributors during PRM account for 64.8% of the total variance (TV) (Table 3.4). Na^+, Mg^{2+}, Ca^{2+}, SO_4^{2-}, Cl^-, and U loadings are observed to be significant in Factor I, indicating that high EC water may be due to salt leaching (Table 3.4). Based on this, we can conclude that the Na^+ and Cl^- in the area of investigation are originating from the salts (Ruiz et al. 1990; Voudouris et al. 1997). Factor II's coefficient of variation, at 16.5%, reflects the relevance of F^-, HCO_3^-, pH, and EC. It is the alkaline character of water with a high concentration of HCO_3^- that allows for greater mobility of F^- ions in the ground, leading to their accumulation in the water (Table 3.4).

The 11% of TV in factor III was indicated by enhanced potassium and phosphorus concentrations, suggesting that potash feldspar weathering was influenced by human activity, such as water softeners, septic systems, or fertilizers (Vengosh and Keren 1996). H_4SiO_4, which is equivalent to 7.7% TV, represents the silicate mineral weathering in Factor IV. The Na^+, K^+, F^-, PO_4, and pH, in factor IV, shows clearly the negative behavior.

ii) POM

Five significant factors were determined by FA explaining 64.76% of the POM TV (Table 3.4).

Factor I loadings are represented with 20.3% TV (NO_3, Cl^-, K^+, Na^+, and EC) suggesting leaching of secondary salt, and Na and Cl loading suggests a salty composition. Mg^{2+}. Ca^{2+}, Cl^-, HCO_3, and H_4SiO_4 are all represented in Factor II, accounting for 14.6% of the total variation; all these strong elements point to weathering, while HCO_3 and H_4SiO_4 specifically point to the disintegration and decomposition of silicate minerals.

With a TV of 10.2%, Factor III shows that the fertilizer affects water chemistry through PO_4 and pH enrichment (Table 3.4). With a TV of 10.0%, Factor IV is dominated by SO_4^{2-} and Cl^-, indicating a manmade impact by fertilizer application (Table 3.4). The combined uranium and fluoride TV in Factor V is 9.8%. There is a high link between uranium and fluoride, suggesting that these two elements are under comparable chemical constraints (Chidambaram et al. 2013).

Factor Score The calculation of factor scores (Dalton and Upchurch 1978) for each sample reveals a relationship between the factor scores for each sample and the intensity of the geochemical interaction related to the factor. Scores on the factors are evaluated further to determine where that factor is most strongly represented and how that zone vary spatially. Regression analysis is used to assess factor scores in the investigation. Indicators of a particular hydrogeochemical regime are highlighted in the positive regions, demonstrating where that factor is most prominent (Chidambaram et al. 2008a). The regions represented by factor I for the POM and PRM seasons indicated the hydrogeochemically active regions.

Factor score of the first factor during PRM and POM seasons superimposed on lithology, Figure 3.5a shows that during PRM, it is represented in the middle of the Cuddalore district, which is made up of sandstone and clay. The eastern and western portions of the Cuddalore district show typical representation of Factor 2, which consists of charnockite, sandstone, and clay. Sedimentary landforms in the eastern and central parts of the Cuddalore area, represented by Factors 3 and 4, consist of laterite, sandstone, clay, and shells (Figure 3.5a).

Factor I POM is located in the eastern part of the Cuddalore district, which is made up of sedimentary topography (sandstone and clay) that reflects the incursion of saltwater and the leaching of secondary salts, revealing the presence of agricultural and wetland along the coastline (Figure 3.5b). The northern and western regions of the Cuddalore district are represented by sandstone, clay, and migmatitic gneiss that represent Factor II. Cuddalore district's centeral region comprises of limestone bands intercalated with clay (Figure 3.5b), representing Factor III. The eastern part of the district, composed of clay and sandstone, is where Factor VI is most evident. Regardless of the season, they provide evidence of secondary salt leaching in the research area.

The major influencing factors in the Cuddalore district are seawater intrusion into coastal aquifers along with industrial and irrigational activities (Chidambram et al 2014). In the northwestern, central, and southern parts of the Cuddalore district, there are more areas with positive spatial representations of the factor.

The primary factor was the abundance of secondary salts – Ca^{2+}, Mg^{2+}, Cl^{-}, HCO_3^{-}, Na^{+}, and K^{+} – that seeped into the groundwater following the monsoon because of the surface water infiltration. Their comparatively large second-factor prevalence in POM and PRM seasons demonstrates the effects of human activities. During POM and PRM, the Cuddalore area's central and eastern sections have the highest proportion of positive portrayal (Figure 3.5a,b). For both seasons, Factors III and IV indicate anthropogenic dominance.

Hard Aquifer

i) PRM

Four significant variables, accounting for 67.78% of the PRM dataset's overall variation, were determined using FA. Calcium, magnesium, sodium, chlorine, nitric oxide, and sulfur dioxide are all associated (Table 3.4). This aquifer is dominated by salinization, as evidenced by the ionic associations. The increase in NO_3^- is an indicator of human influence, and it may be generated by the conversion of scrub, dunes, coastal sand (with or without vegetation), agricultural land, and mudflats into salt pans (Gangai and Ramachandran 2010). This trend significantly affected the groundwater table because these salt pans are fed by groundwater and marine (salt) water.

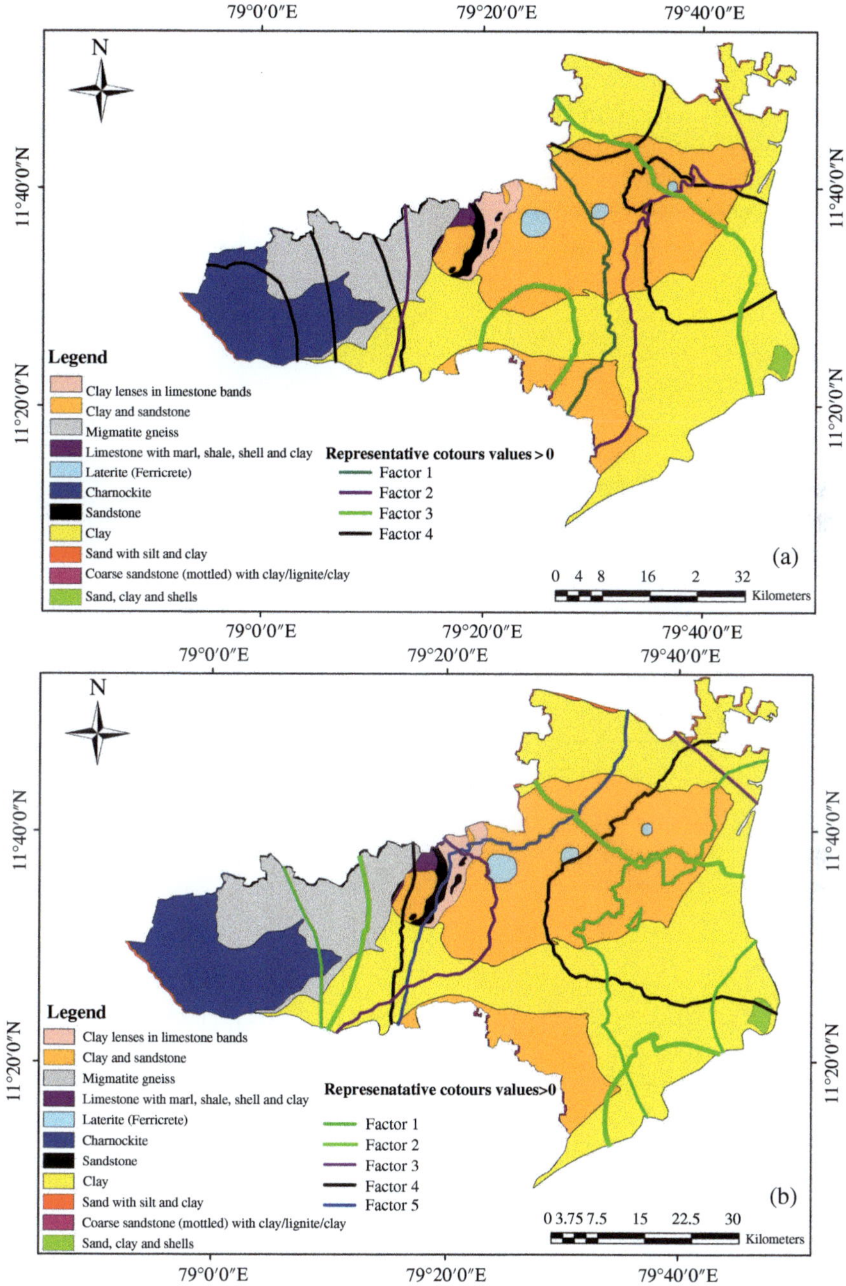

Figure 3.5 Spatial distribution of factor score correlated with lithology during (a) PRM, (b) POM for sedimentary aquifer, and (c) PRM, (d) POM for hard rock aquifer.

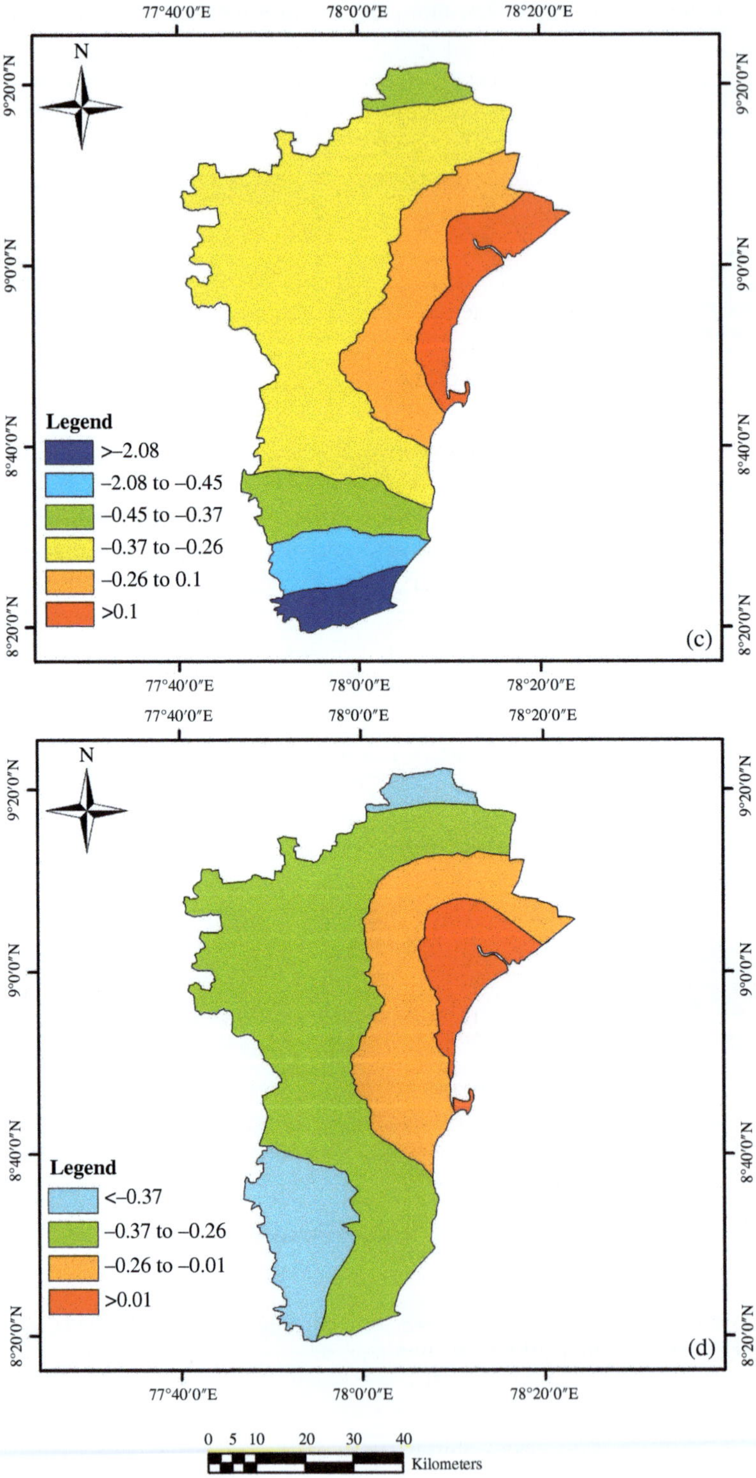

Figure 3.5 (Continued)

The leaching of salts, specifically Ca^{2+}, K^+, and PO_4^{2-}, represent Factor II (Prasanna et al. 2010). This factor may be caused by anthropogenic effects from fertilizer applications, particularly phosphatic fertilizers (Prasanna et al. 2010). The weathering process is characterized by Factor III (F^- and HCO_3^-). Similar findings were discovered in the groundwaters of Tamil Nadu's Dindigul district (Chidambaram et al. 2012b). Silicate mineral dissolution is indicated by Factor IV, which is represented by H_4SiO_4. Aluminosilicates, which make up most of the hard rock aquifer, when weathered, release silica into solutions (Chidambaram 2000).

ii) POM

From the POM data, four variables were extracted that accounted for 69.06% TV. Loadings of Ca^{2+}, Mg^{2+}, Na^+, K^+, Cl^-, and SO_4^{2-}indicate the influence of seawater in Factor I (Table 3.4) (Senthilkumar et al. 2008; Prasanna et al. 2010; Chidambaram et al. 2012c). The relatively low pH representation in Factor II samples prove that weathering has occurred and that fluoride has been released from the hydroxyl group of the mineral.

$$R(OH) + H_2O + CO_2 \rightarrow Clay + F + HCO_3$$

The R in the above equation signifies the cations in the clay, which are generally hydroxy alumino silicates. The studies by Nanyaro et al. (1984), Teotia et al. (1998) concluded that the Ca^{2+} and Mg^{2+} has been linked to high fluoride levels in the groundwater. They have also indicated that the solubility product of Ca and fluorite suggests an opposite link between $Ca^{2+}$and Mg^{2+} (Kundu et al. 2001) dissimilar to the concept proposed by Gaciri and Davies (1993). K^+, NO_3^-, and PO_4^{2-} all point to the anthropogenic effect of agricultural activities, especially phosphate and potash fertilizers, in Factors III and IV (Prasanna et al. 2010).

Factor Score The first factor in practically all four seasons, was influenced by Ca^{2+}, Mg^{2+}, Na^+, Cl^-, and SO_4^{2-}, and is well-represented in areas with more significant quantities of industrialization along the coast that are subjected to saltwater intrusion, including those under the impact of the State Industries Promotion Corporation of Tamil Nadu industrial area. Based on the factor score's spatial distributions, we can see that seawater and salt pan activity are concentrated along the coast in the hard rock aquifer region. There are places in this coastal region with SO_4^{2-} similar to typical seawater and those of salt pan acitivty (Figure 3.5c,d).

An important inland source of SO_4^{2-} was identified by Chivas et al. (1991) as the oxidation of organic Dimethyl Sulfide gas (DMS) released from extensive upwelling coastal waters. The elevated SO_4^{2-}/Cl^- ratios in the salt pan fluids may be due to the input of DMS that enhances groundwater's SO_4^{2-} concentration.

3.5 Conclusion

The hydrogeochemical study of the sedimentary and hard rock aquifer was carried out by collecting samples during PRM and POM seasons. The seasonal variation in hydrogeochemistry indicates that a higher conductivity was noted in the POM seasons along the aquifer in coastal region. This was impacted mainly by seawater, backwater infiltration, and weathering of rocks along the lithological contacts and in the hard rock regions (the western part in Cuddalore's case). In addition, higher EC values in the Cuddalore district have been observed along the boundaries of the Gadilam and Coleroon Rivers, indicating domestic sewage infiltration and landfill leachates.

In the Gibbs plot, Cuddalore's PRM and POM samples fall in the weathering zone in the cation plot. In contrast, in the anion plot, most of the sample is outside the plot, suggesting that anthropogenic activity predominates in the studied area. Similar representations are observed in the Tuticorin district. In both study areas, TDS was higher in POM than in PRM. This suggests a recharge accompanied by the dissolution and leaching of salts during POM, along with seawater intrusion and anthropogenic factors play a major role in the hydrogeochemistry. In all the seasons and the studied formations, Na was the principal cation, and Cl^- was the predominant anion.

The Piper plot of the Cuddalore district demonstrates that the majority of the samples fall in the Na-Cl facies during POM and mixed Na-Ca-HCO_3-Cl facies during PRM. Numerous factors, including weathering, ion exchange, dissolution, backwater contributions from evaporated surface water, anthropogenic inputs, and others, can affect the chemistry of groundwater, as shown by the mixing trend in the PRM. However, Tuticorin samples transitioned facies from a Na-HCO_3 type in POM to a Na-Cl type in PRM. The NaCl type is formed during the PRM phase as a result of seawater intrusion, ion exchange that takes place along with weathering and natural recharge during the POM season.

According to the FA for the data obtained, seawater intrusion, leaching of secondary salts, weathering, and anthropogenic influences are the key factors that govern the geochemistry of groundwater in both studied areas. However, the impact of the saltpan was observed to be higher in the Tuticorin region. Further, the sedimentary aquifers seem to have a more complex environment than the hard rock aquifers, owing to the aquifer properties and complex land use practices.

References

APHA (1995). *Standard Methods for the Examination of Water and Wastewater*. Washington: American Public Health Association, American Water Works Association and Water Environmental Federation.

Appelo, C.A.J. and Postma, D. (2004). *Geochemistry, Groundwater and Pollution*. CRC Press.

Çelik, M. and Yıldırım, T. (2006). Hydrochemical evaluation of groundwater quality in the Çavuşçayı basin, Sungurlu-Çorum, Turkey. *Environ. Geol.* 50: 323–330.

Central Groundwater Board (CGWB) (1993). *Ground Water Resources and Development Prospects in Pondicherry Region*. Union Territory of Pondicherry.

Central Groundwater Board (CGWB). (2009). South Eastern Coastal Region, District Groundwater Brochure, Thoothukudi. http://cgwb.gov.in/District_Profile/TamilNadu/thoothukudi.pdf (accessed 3 June 2023).

Chidambaram, S. (2000). Hydrogeochemical studies of groundwater in Periyar district, Tamilnadu, India. Unpublished Ph thesis. Dep. Geol. Annamalai Univ.

Chidambaram, S., Ramanathan, A., and Srinivasamoorthy, K. (2003). Lithological influence on the groundwater chemistry–Periyar district: a case study. Presented at the Conference on Coastal and Freshwater Issues, Anna University, Chennai, India. 8 December 2003.

Chidambaram, S., Ramanathan, A., Anandhan, P. et al. (2008a). A statistical approach to identify the hydrogeochemically active regimes in ground waters of Erode district, Tamilnadu. *Asian Journal of Water, Environment and Pollution* 5: 123–135.

Chidambaram, S., Ramanathan, A., Prasanna, M. et al. (2008b). Study on the impact of tsunami on shallow groundwater from Portnova to Pumpuhar, using geoelectrical technique-south east coast of India. *Indian Journal of Marine Sciences* 37 (1): 121–131.

Chidambaram, S., Senthil Kumar, G., Prasanna, M.V. et al. (2009). A study on the hydrogeology and hydrogeochemistry of groundwater from different depths in a coastal aquifer: Annamalai Nagar, Tamilnadu, India. *Environmental Geology* 57: 59–73.

Chidambaram, S., Prasanna, M., Karmegam, U. et al. (2011). Significance of pCO_2 values in determining carbonate chemistry in groundwater of Pondicherry region, India. *Frontiers in Earth Science* 5: 197–206.

Chidambaram, S., Anandhan, P., Prasanna, M.V. et al. (2012a). Major ion chemistry and identification of hydrogeochemical processes controlling groundwater in and around Neyveli lignite mines, Tamil Nadu, South India. *Arabian Journal of Geosciences* 6: 3451–3467.

Chidambaram, S., Bala Krishna Prasad, M., Manivannan, R. et al. (2012b). Environmental hydrogeochemistry and genesis of fluoride in groundwaters of Dindigul district, Tamilnadu (India). *Environment and Earth Science* 68: 333–342.

Chidambaram, S., Prasanna, M.V., Singaraja, C. et al. (2012c). Study on the saturation index of the carbonates in the groundwater using WATEQ4F, in layered coastal aquifers of Pondicherry. *Journal Geological Society of India* 80: 813–824.

Chidambaram, S., Karmegam, U., Prasanna, M., and Sasidhar, P. (2012d). A study on evaluation of probable sources of heavy metal pollution in groundwater of Kalpakkam region, South India. *The Environmentalist* 32: 371–382.

Chidambaram, S., Anandhan, P., Prasanna, M.V. et al. (2013). A study on the status of fluoride ion in groundwater of coastal hard rock aquifers of South India. *Arabian Journal of Geosciences* 6: 4167–4177.

Chivas, A., Andrews, A., Lyons, W. et al. (1991). Isotopic constraints on the origin of salts in Australian playas. 1. Sulphur. *Palaeogeography Palaeoclimatology Palaeoecology* 84: 309–332.

Chowdhury, A.K. and Gupta, S. (2011). Evaluation of water quality, hydro-geochemistry of confined and unconfined aquifers and irrigation water quality in Digha coast of West Bengal, India (a case study). *International Journal of Environmental Sciences* 2: 576–589.

Dalton, M.G. and Upchurch, S.B. (1978). Interpretation of hydrochemical facies by factor analysis. *Groundwater* 16: 228–233.

Das, S. and Sar, S. (1988). Groundwater exploration in Orissa. *Tech Annu. Inst Eng. Bhubaneswar* (1): 84–95.

Döll, P. and Schmied, H.M. (2012). How is the impact of climate change on river flow regimes related to the impact on mean annual runoff? A global-scale analysis. *Environmental Research Letters* 7: 014037.

Foster, S., Chilton, J., Moencg, M. et al. (2000). *Groundwater in Rural Development: Facing the Challenges of Supply and Resource Sustainability*. The World Bank.

Gaciri, S. and Davies, T. (1993). The occurrence and geochemistry of fluoride in some natural waters of Kenya. *Journal of Hydrology* 143: 395–412.

Gangai, I.P.D. and Ramachandran, S. (2010). The role of spatial planning in coastal management – a case study of Tuticorin coast (India). *Land Use Policy* 27: 518–534.

Gibbs, R.J. (1970). Mechanisms controlling world water chemistry. *Science* 170: 1088–1090.

Gupta, L. and Subramanian, V. (1998). Geochemical factors controlling the chemical nature of water and sediments in the Gomti River, India. *Environmental Geology* 36: 102–108.

Imtiyaz, M. and Rao, D. (2008). Influence of overexploitation on groundwater ecosystem in hard rock terrain. Presented at the Proceedings of International Groundwater Conference, March, 19–22, Jaipur, India. 88.

Janardhana, N.R. (2007). Hydrogeochemical parameters for assessment of groundwater quality in the upper Gunjanaeru River basin, Cuddapah District, Andhra Pradesh, South India. *Environmental Geology* 52: 1067–1074.

Jeong, C.H. (2001). Effect of land use and urbanization on hydrochemistry and contamination of groundwater from Taejon area, Korea. *J. Hydrol.* 253: 194–210.

Karanth, K. (1991). Impact of human activities on hydrogeological environment. *Journal of the Geological Society of India* 38: 195–206.

Kim, J.-H., Yum, B.-W., Kim, R.-H. et al. (2003). Application of cluster analysis for the hydrogeochemical factors of saline groundwater in Kimje, Korea. *Geosci. J.* 7: 313–322.

Kumar, D. and Alappat, B.J. (2005). Analysis of leachate pollution index and formulation of sub-leachate pollution indices. *Waste Management & Research* 23: 230–239.

Kundu, N., Panigrahi, M., Tripathy, S. et al. (2001). Geochemical appraisal of fluoride contamination of groundwater in the Nayagarh District of Orissa, India. *Environmental Geology* 41: 451–460.

Lawrence, F.W. and Upchurch, S.B. (1982). Identification of recharge areas using geochemical factor analysis. *Groundwater* 20: 680–687.

Leung, C.-M. and Jiao, J.J. (2006). Heavy metal and trace element distributions in groundwater in natural slopes and highly urbanized spaces in mid-levels area, Hong Kong. *Water Res.* 40: 753–767.

Liu, C.-W., Lin, K.-H., and Kuo, Y.-M. (2003). Application of factor analysis in the assessment of groundwater quality in a Blackfoot disease area in Taiwan. *Sci. Total Environ.* 313: 77–89.

MacDonald, A., Davies, J., and Ó Dochartaigh, B. (2001). Simple methods for assessing groundwater resources in low permeability areas of Africa. British Geological Survey Commissioned Report, CR/01/168N.

Manikandan, S., Chidambaram, S., Prasanna, M.V., et al. (2011). Hydrochemical characteristics and groundwater quality assessment in Krishnagiri District, Tamilnadu, India. *International Journal of Earth Sciences and Engineering*, 4(04), 623–632.

Manivannan, R. (2010). Hydrogeochemistry of groundwater in Dindigul district, Tamil Nadu, India. Unpublished Ph thesis. Dep. Earth Sci. Annamalai Univ., India.

Manivannan, R., Chidambaram, S., Anandhan, P. et al. (2011). Study on the significance of temporal ion chemistry in groundwater of Dindigul District, Tamilnadu, India. *E-Journal of Chemistry* 8: 938–944.

Mishra, P., Behera, P., and Patel, R. (2005). Contamination of water due to major industries and open refuse dumping in the steel city of Orissa – a case study. *Journal of Environmental Science & Engineering* 47: 141–154.

Mondal, N., Saxena, V., and Singh, V. (2008). Occurrence of elevated nitrate in groundwaters of Krishna delta, India. *African Journal of Environmental Science and Technology* 2: 265–271.

Mondal, N., Singh, V.S., Saxena, V., and Singh, V.P. (2011a). Assessment of seawater impact using major hydrochemical ions: a case study from Sadras, Tamilnadu, India. *Environmental Monitoring and Assessment* 177: 315–335.

Mondal, N.C., Singh, V.P., Singh, S., and Singh, V.S. (2011b). Hydrochemical characteristic of coastal aquifer from Tuticorin, Tamil Nadu, India. *Environmental Monitoring and Assessment* 175: 531–550.

Nanyaro, J., Aswathanarayana, U., Mungure, J., and Lahermo, P. (1984). A geochemical model for the abnormal fluoride concentrations in waters in parts of northern Tanzania. *Journal of African Earth Sciences* 1983 (2): 129–140.

Partey, F.K., Land, L.A., Frey, B., Premo, E. and Crossey, L. (2011). Final report on the geochemistry of Bitter Lake National Wildlife Refuge, Roswell, New Mexico. New Mexico Bureau of Geology and Mineral Resources, New Mexico Institute of Mining and Technology.

Prasanna, M., Chidambaram, S., Pethaperumal, S. et al. (2008). Integrated geophysical and chemical study in the lower subbasin of Gadilam River, Tamilnadu, India. *Environmental Geosciences* 15: 145–152.

Prasanna, M.V., Chidambaram, S., Shahul Hameed, A., and Srinivasamoorthy, K. (2010). Study of evaluation of groundwater in Gadilam basin using hydrogeochemical and isotope data. *Environmental Monitoring and Assessment* 168: 63–90.

Rai, S., Singh, U., and Singh, A. (1995). X-ray determination of magnetically treated liquid water structures. *Electro- and Magnetobiology* 14: 23–30.

Ramesh, R. and Anbu, M. (1996). *Chemical Methods for Environmental Analysis: Water and Sediment*. Macmillan India.

Ranjana, U., and Naverathna, M.P.C. (2011). River sand mining in southern Sri-Lanka and its effect on environment. Presented at the 11th International River symposium on "A Future of extremes" (Brisbane).

Rao, N.S., Rao, P.S., Reddy, G.V. et al. (2012). Chemical characteristics of groundwater and assessment of groundwater quality in Varaha River Basin, Visakhapatnam District, Andhra Pradesh, India. *Environmental Monitoring and Assessment* 184: 5189–5214.

Ruiz, F., Gomis, V., and Blasco, P. (1990). Application of factor analysis to the hydrogeochemical study of a coastal aquifer. *Journal of Hydrology* 119: 169–177.

Sanford, W., Langevin, C., Polemio, M., and Povinec, P. (2007). *A New Focus on Groundwater-Seawater Interactions*. IAHS Press.

Singaraja, C., Chidambaram, S., Anandhan, P. et al. (2013). Hydrochemistry of groundwater in a coastal region and its repercussion on quality, a case study – Thoothukudi district, Tamil Nadu, India. *Arabian Journal of Geosciences* 7: 939–950.

Singaraja, C., Chidambaram, S., Prasanna, M.V. et al. (2012). A study on the behavior of the dissolved oxygen in the shallow coastal wells of Cuddalore District, Tamilnadu, India. *Water Quality Exposure and Health* 4: 1–16.

Selvam, S. and Sivasubramanian, P. (2012). Groundwater potential zone identification using geoelectrical survey: a case study from Medak district, Andhra Pradesh, India. *Int. J. Geomat. Geosci.* 3: 55.

Selvam, S., Manimaran, G., and Sivasubramanian, P. (2013). Hydrochemical characteristics and GIS-based assessment of groundwater quality in the coastal aquifers of Tuticorin corporation, Tamilnadu, India. *Applied Water Science* 3: 145–159.

Senthilkumar, G., Ramanathan, A., Nainwal, H., and Chidambaram, S. (2008). Evaluation of the hydro geochemistry of groundwater using factor analysis in the Cuddalore coastal region, Tamil Nadu, India. *India Journal of Marine Sciences* 37 (2): 181–185.

Smith, S.J., Andres, R., Conception, E., Lurz, J., 2004. Historical sulfur dioxide emissions 1850-2000: methods and results. Pacific northwest National lab.(PNNL), Richland, WA (United States). https://doi.org/10.2172/15020102; https://www.osti.gov/servlets/purl/15020102.

Srinivasamoorthy, K., Chidambaram, S., Anandhan, P., and Vasudevan, S. (2005). Application of statistical analysis of the hydrogeochemical study of groundwater in hard rock terrain, Salem District, Tamilnadu. *J Geochem* 20: 181–190.

Srinivasamoorthy, K., Chidambaram, S., Prasanna, M. et al. (2008). Identification of major sources controlling groundwater chemistry from a hard rock terrain – a case study from Mettur taluk, Salem district, Tamil Nadu, India. *Journal of Earth System Science* 117: 49–58.

Srinivasamoorthy, K., Chidambaram, S., and Vasanthavihar, M. (2009). Application of multivariate statistical analysis in elucidation of hydrogeochemical data: a case study. *Int. J. Ecol. Econ. Stat.* 14: 98–109.

Srinivasamoorthy, K., Vasanthavigar, M., Chidambaram, S. et al. (2011). Characterisation of groundwater chemistry in an eastern coastal area of Cuddalore district, Tamil Nadu. *Journal of the Geological Society of India* 78: 549–558.

Swan, A.R. and Sandilands, M. (1995). Introduction to geological data analysis. Presented at the International Journal of Rock Mechanics and Mining Sciences and Geomechanics Abstracts, 32(8), p. 387A.

Teotia, M., Teotia, S., and Singh, K. (1998). Endemic chronic fluoride toxicity and dietary calcium deficiency interaction syndromes of metabolic bone disease and deformities in India: year 2000. *Indian Journal of Pediatrics* 65: 371–381.

Thivya, C., Chidambaram, S., Thilagavathi, R. et al. (2014). Identification of the geochemical processes in groundwater by factor analysis in hard rock aquifers of Madurai District, South India. *Arabian Journal of Geosciences* 7: 3767–3777.

Tyagi, S., Datta, P., Kulshreshtha, S., and Sharma, R. (2008). Isotopic and hydrochemical signatures in characterizing pollutants movement in overexploited groundwater aquifers of Delhi State. Presented at the 3rd WEPA International Forum on Water Environmental Governance in Asia, Putrajaya, Malaysia (October 23–24), pp. 1–8.

Vengosh, A. and Keren, R. (1996). Chemical modifications of groundwater contaminated by recharge of treated sewage effluent. *Journal of Contaminant Hydrology* 23: 347–360.

Voudouris, K., Lambrakis, N., Papatheothorou, G., and Daskalaki, P. (1997). An application of factor analysis for the study of the hydrogeological conditions in Plio-Pleistocene aquifers of NW Achaia (NW Peloponnesus, Greece). *Mathematical Geology* 29: 43–59.

Wanda, E., Monjerezi, M., Mwatseteza, J.F., and Kazembe, L.N. (2011). Hydro-geochemical appraisal of groundwater quality from weathered basement aquifers in northern Malawi. *Phys. Chem. Earth Parts ABC* 36: 1197–1207.

World Bank Report. (2010). Deep Wells and Prudence: Towards Pragmatic Action for Addressing Groundwater Overexploitation in India. The World Bank, *Washington Rep. No. 51676.*

Zhang, H. and Shan, B. (2008). Historical records of heavy metal accumulation in sediments and the relationship with agricultural intensification in the Yangtze–Huaihe region, China. *Sci. Total Environ.* 399: 113–120.

Zhang, J., Huang, W., Letolle, R., and Jusserand, C. (1995). Major element chemistry of the Huanghe (Yellow River), China-weathering processes and chemical fluxes. *Journal of Hydrology* 168: 173–203.

4

Textural and Mineralogical Signatures of Fluvial Sediments in Mountain Streams of Contrasting Climates in the Southern Western Ghats (India)

Jobin Thomas[1], Sabu Joseph[2], and Thrivikramji Kythavilakom Pillai[3]

[1] *Department of Geological and Mining Engineering and Sciences, Michigan Technological University, Houghton, MI, USA*
[2] *Department of Environmental Sciences, University of Kerala, Thiruvananthapuram, Kerala, India*
[3] *Centre for Environment and Development, Thiruvananthapuram, Kerala, India*

4.1 Introduction

Continental weathering and erosion rates are significant in sediment formation, redistribution, and elemental recycling and are vital to understand the geomorphic evolution of landscapes as the denudation processes signify the response of the rocks and minerals to a wide range of environmental drivers (McLennan 1993; Binnie and Summerfield 2013). Among the numerous factors, climate and topography (along with tectonics and lithology) are the primary determinants affecting denudation. While climate decides the water availability essential for mediating aqueous reactions in chemical weathering processes and sediment transport and controls the processes that are temperature-sensitive (e.g. chemical weathering of unstable minerals), topography modulates (orographic) rainfall patterns affecting the extent of denudation rates (White and Blum 1995; Roe et al. 2002; Dibiase and Whipple 2011; Kober et al. 2015; Tucker 2015; Raab et al. 2019). Further, climate affects the distribution of biotic factors (e.g. vegetation pattern) also, which, in turn, influences erosion and weathering processes (Corenblit et al. 2011; Schaller et al. 2018). Although the denudation rate strongly depends on topographic relief, hillslopes and stream channels in the areas characterized by relatively high relief exhibit a nonlinear relationship between relief and rate of denudation (Snyder et al. 2000; Montgomery and Brandon 2002; Binnie et al. 2007; Ouimet et al. 2009; Stock et al. 2009; DiBiase et al. 2010, 2012). A large body of evidence exists to demonstrate the definitive role of spatial variability in climate and physiography governing the physical characteristics, mineralogy, and geochemical signatures of fluvial sediments (Yang et al. 2004; Rajamani et al. 2009; Riebe et al. 2015; von Eynatten et al. 2016; Farooqui et al. 2021). Therefore, fluvial sediments originating from various biogeochemical and geophysical processes in the provenance and stream channel network constitute an archive of complex records that can be deciphered to model the environmental conditions in the source areas (e.g. Macklin et al. 2012; Demoulin et al. 2017).

Weathering and Erosion Processes in the Natural Environment, First Edition. Edited by Virendra Bahadur Singh, Sughosh Madhav, Naresh Chandra Pant, and Ravi Shekhar.

The fluvial system exhibits a variety of channel forms along with its profile reflecting changes in the energy conditions (Brussock et al. 1985). The mountain stream channels display quantitatively distinguishable channel geometry and morphological attributes than their lowland counterparts. The dominant alluvial stream channel morphologies in mountain river basins include cascade, step-pool, plane-bed, riffle-pool, and dune-ripple and exhibit characteristic topographic and channel attributes (Montgomery and Buffington 1997). The sediment supply and transport in mountain streams exert definitive controls on channel form, and consequently, the diversity in channel morphology reflects differences in sediment supply and transport rates (Mueller and Pitlick 2013). Although substrate size differs among the morphological types due to their characteristic conditions of formation and energy regimes, the pools are characteristically sites of accumulation of finer materials but mobilize during high flows (Wohl et al. 1993; Montgomery and Buffington 1997; Zimmermann and Church 2001; Buffington et al. 2002; Chin and Wohl 2005; Thompson 2013; Zimmermann 2013). Hence, characterization of sediments in mountain streams is critical to our understanding of climatic and topographic controls on channel sediments.

Since the seminal work of Garrels and Mackenzie (1971), the provenance and weathering of fluvial sediments have remained a focal area of river geochemistry. Although in the past decades much new knowledge and insights based on mineralogy and geochemistry have been gained on continental weathering and erosion by analyzing fluvial sediments (Martin and Meybeck 1979; Berner and Berner 1987; Savenko 2007; Viers et al. 2009), most of the studies had focused on large alluvial river systems of the world due to their significance in global sediment budget (e.g. Chakrapani and Subramanian 1990; Dupré et al. 1996; Gaillardet et al. 1999; Roy et al. 1999; Yang et al. 2004; Liu et al. 2007; Singh 2010; Maharana et al. 2018; Rousseau et al. 2019). But then, the classical findings of Milliman and Syvitski (1992) underscored the significance and role of small mountain rivers on the sediment fluxes to the oceans, and subsequently, numerous researchers (e.g. Lecomte et al. 2005; Goldsmith et al. 2015; Gardner et al. 2017; Lee et al. 2020; Wang et al. 2022) addressed the weathering processes and controls of these unique physical systems. Yet, most of these studies preferred the dissolved phase than the sediment phase for understanding the weathering processes and rates of denudation.

The orographic framework of the Western Ghats with a robust elevation gradient serves as a climate barrier of the Indian Peninsula, separating the humid western slopes from the rain shadow countryside. In addition, stream channels of the mountainous tracts of the Western Ghats feature significant morphologic diversity with unique physical and ecological processes, making them an ideal physical environment to demonstrate the role of climate and topography on the anatomy of fluvial sediments. Several researchers investigated the geochemical attributes and weathering processes of the rivers of the Western Ghats (e.g. Das and Krishnaswami 2007; Sensarma et al. 2008; Rajamani et al. 2009; Kessarkar et al. 2013; Mohan and Krishnakumar 2022) and demonstrated the significant roles of various factors influencing the composition of detritus. Deepthy and Balakrishnan (2005), while scrutinizing the clay mineralogy under contrasting climates of the Western Ghats, observed the dominance of 1:1 clays (kaolinite) and Fe—Al oxides (gibbsite/goethite) in the humid climate and 2:1 clays and kaolinite in the rain shadow region, implying the effect of climate on the kinetics of chemical weathering. Although the significance of climate variability and rain shadow effect on the hydrochemical facies and processes controlling stream water composition has been demonstrated (Thomas et al. 2015a, 2015b, 2015c; Sarath et al. 2022), texture, mineralogy, and geochemical characteristics of the fluvial sediments in the mountain rivers of the southern Western Ghats, as well as their climatic and topographic implications, are less explored and discussed. Hence, we analyzed the extent to which the detrital composition (in channels dominated by step-pool and riffle-pool morphologies) in two mountain river basins of the southern Western Ghats, India, which is greatly influenced by regional variations in climate and topography.

4.2 Study Area

The study area is two adjoining and comparably sized sixth-order river basins, viz., Muthirapuzha River Basin (MRB; 272 km^2) and Pambar River Basin (PRB; 289 km^2) in the highland physiographic zone of the southern Western Ghats, Kerala, India (Figure 4.1). Muthirapuzha is one of the major tributaries of the west-flowing Periyar River (Kerala), whereas Pambar joins the east-flowing Amaravati River – a tributary of Cauvery River (Tamil Nadu). The basin elevation of the MRB ranges between 740 and 2690 m above the mean sea level, while that of the PRB is between 440 and 2540 m above the mean sea level. The stream network of the MRB is mostly developed on the Munnar Plateau, while the

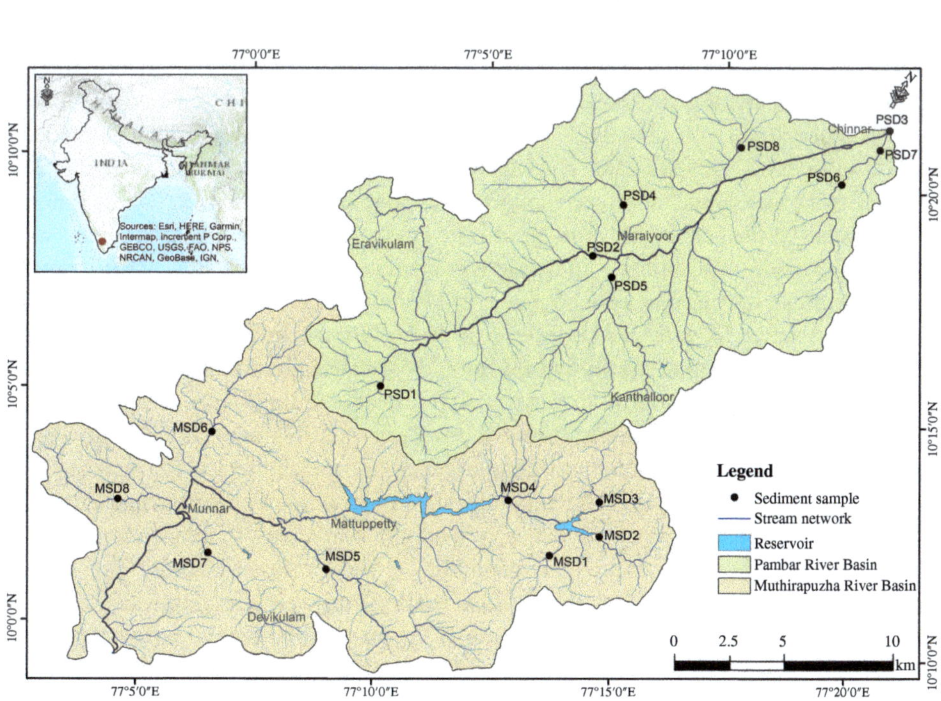

Figure 4.1 The mountain river catchments, viz., MRB and PRB in the southern Western Ghats, India. Sixteen sediment samples were collected from the study area.

PRB drains the northern scarps of the plateau. The geomorphic processes acting over different time scales resulted in a broad range of landforms in the region, where the dominant landforms of the basins include structural and denudational hills and terrain with concordant summits. Other geomorphological units, such as linear ridge, valley fill, valley flat, and local planation surface, add to the geomorphic diversity of the region. The slopes of the plateau and the terrain with concordant summits are characterized by the occurrence of escarpments. Stream channels of either of the basins demonstrate diverse channel morphological units dictated by the basin topography and channel characteristics. However, step-pool and riffle-pool morphological units dominate the mountain stream channels of the basins (Thomas 2012). Streamflow is regulated in both basins, where the MRB has an altered flow regime due to dams/weirs and stream channels of the PRB are regulated by weirs.

The river basins are set partly back to back in the southern Western Ghats, i.e. MRB on the windward slope and PRB on the leeward slope, to be exposed to contrasting climates. The mean annual rainfall of the MRB (1989–2009) is close to 3700 mm, whereas the PRB (1992–2008) receives only about 1100 mm. Although the annual rainfall distribution of the MRB does not exhibit a definite spatial pattern, the PRB shows a downstream decreasing trend. Some of the upstream locations in the PRB (e.g. Eravikulam National Park and Kanthalloor) receive comparable rainfall to that of the MRB (Jose et al. 1994; Chandrashekara and Sibichan 2006). The majority of the annual rainfall of the MRB and PRB is contributed by the Indian summer monsoon season (June–September), with the maximum incidence of rainfall in the month of July (949 mm and 178 mm, respectively). However, the contribution of the northeast monsoon (October–December) is significant in downstream of the PRB. The heterogeneity of the rainfall distribution in both basins indicates stronger climatic seasonality driven by the monsoon rainfall (Thomas et al. 2015a). The mean annual temperature of the MRB is about 18 °C but variable across the basin (16–20 °C) with cooler January and hotter May. The mean annual temperature of upstream areas of the PRB is about 23 °C, whereas downstream has a relatively higher temperature (>30 °C). In general, the temperature of the basins shows an altitudinal gradient, i.e. declining temperature with the rise in elevation. The rainfall and temperature data suggest that the MRB enjoys a monsoon-dominated tropical humid climate (Am), whereas the climate of the PRB shows a downstream variation from tropical humid (Am) to tropical wet and dry (or semiarid; Aw).

The lithological makeup of the region is dominated by high-grade metamorphic rocks of the Precambrian Southern Granulite Terrain of Peninsular India (Figure 4.2). Major rock types of the MRB are granite/granite gneiss, hornblende-biotite-gneiss, and calc-granulite, whereas hornblende-biotite-gneiss and granite gneiss are the chief rock types of the PRB (GSI 1992). The granite, a pink and medium-grained rock showing isotropic fabric, is mainly composed of K-feldspar, plagioclase, and quartz, while biotite, sphene, apatite, and zircon are the accessory minerals, and magnetite is the dominant opaque. The granite gneiss is also pink and medium-grained but with foliation due to parallel planar arrangement of flakes of biotite, prisms of hornblende and lenticular and flattened quartz grains, and occasionally aggregates of mica and feldspar (Thampi et al. 1993). Biotite and hornblende are the mafic minerals in both granite and granite gneiss, yet granite contains a lower mafic mineral content compared with granite gneiss. The hornblende-biotite-gneiss is a composite rock characterized by regular but alternating bands rich in quartzo-feldspathic and mafic

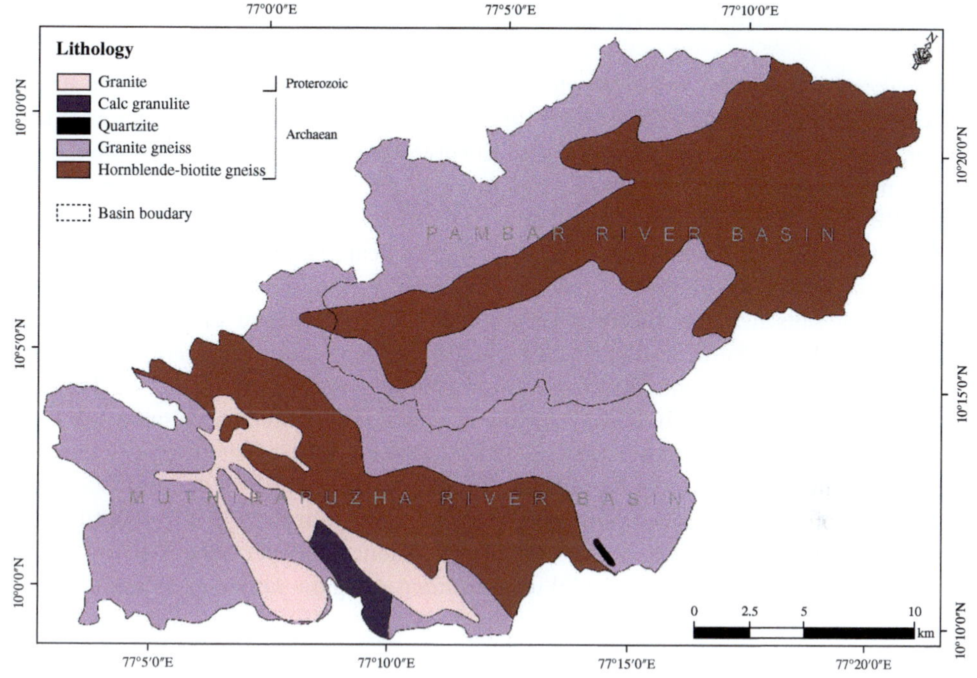

Figure 4.2 Major rock types of MRB and PRB (Adapted from GSI 1992).

minerals (e.g. quartz, plagioclase, K-feldspar, hornblende, clinopyroxene, and biotite). The calc-granulite is a medium-grained rock generally exhibiting faint layering due to the segregation of calc-silicate minerals and incipient development of metamorphic foliation defined by biotite and pyroxene (Thampi 1987). The dominant rock types of the region are accompanied by minor patches of quartzite, cordierite-bearing gneisses, charnockite, and crystalline limestone and intruded by pegmatite-, quartz-, syenite-carbonatite-, and aplite-veins as well as dolerite dykes. Laterite is exposed at many places in the MRB, particularly in the plateau, while such occurrences are uncommon in the PRB.

The major soil series in both basins is the Anamudy series (Typic Kandihumults; Ultisols). However, downstream of the MRB is characterized by soils of Pampadumpara (Ustic Kandihumults; Ultisols) and Venmany (Oxic Dystrustepts; Inceptisols) series, and the downstream of the PRB is covered by the Chinnar series (Typic Haplustolls; Mollisols) (SSO 2007; Soil Survey Staff 2014). The dominant land use/land cover types of the MRB are commercial (e.g. tea and cardamom) and eucalyptus plantations, covering about two-thirds of the basin area, whereas the PRB is dominated by the forested landscape. Monoculture plantations (e.g. tea and eucalyptus) and farmlands (paddy and sugarcane) cover roughly 25% of the area of the PRB. The southern montane wet temperate grasslands and southern montane wet temperate forests (i.e. shola forests) dominate among the diverse natural vegetation types of the MRB, while the natural vegetation of the PRB covers an entire spectrum ranging from the shola forests to dry scrubs. The marked differences in the temperature and rainfall regimes across the PRB along with topographic variations could be attributed to the diversity of the vegetation types (Thomas 2012).

4.3 Methodological Framework

We analyzed the textural and mineralogical attributes of fluvial sediments in the stream channels of the MRB and PRB to unravel the climate signatures and physiographic controls on the composition of the detritus. The majority of the samples were collected from the two dominant channel morphologies, viz., step-pools and riffle-pools. Eight samples each from the MRB (MSD1–MSD8) and PRB (PSD1–PSD8) were collected (including mainstream and chief tributaries) for granulometric and mineralogical analyses (Figure 4.3). A portion of the bulk sediment sample was repeatedly washed to remove the organic matter and other soluble salts and dried thereafter in a hot air oven. For textural analysis, the samples were split by Jone's Riffle Splitter for reducing the bulk to about 100 g. The sub-samples were accurately weighed and sieved at ½Φ intervals using ASTM sieves in a Ro-Tap sieve shaker for 30 minutes. Each sieved fraction was carefully stored in suitable vials and weights were recorded. From the sieve analysis data, statistical grain-size parameters (i.e. graphic mean, *Mz*; inclusive graphic standard deviation, σI; inclusive graphic skewness, SkI; and graphic kurtosis, KG) were estimated (Folk and Ward 1957). The identification and quantification of light and heavy minerals were done using permanent grain mounts. Approximately 300 grains (per sample) were counted using the James-Swift point counter following the line method (Carver 1971).

4.4 Results and Discussion

4.4.1 Spatial Variability in the Grain-Size Distribution of Fluvial Sediments

The textural data indicated that the sediment samples of the MRB and PRB are dominated (>60%) by medium to coarse sand fractions (i.e. 0–2Φ). The gravel content in the sediments varies between 1.8% and 5.9% (mean = 3.6%) in the MRB and 3.9 and 11.5% (mean = 6.8%) in the PRB. According to the textural classification of Folk (1980), the stream channel sediments of the MRB are dominated by slightly gravelly sand, while the sediments of the PRB are characteristically gravelly sand (Figure 4.4a). The sediments with gravel content varying between 0.01% and 5.0% are classified as slightly gravelly sand, whereas gravelly sand implies a gravel content of up to 30%. The dominance of the coarser fractions in the sediments is reflective of the mountainous topography as well as steeper channel gradients. In general, grain-size distributions show a discontinuity for fluvial sediments with partially overlapping coarse and fine components (Sun et al. 2002). The size frequency distribution of sand in the samples of the MRB and PRB (Figure 4.4b and c) suggests a bimodal to polymodal nature, with the primary mode summiting around 1Φ (coarse to medium sand) and the secondary mode at 3Φ (fine to very fine sand). These modes might be attributed to the different types of sediment transport or deposition as the sands with grain size ranging from 0 to 2Φ represent bedload transport, while 3–4Φ imply sediment moving in suspension (Ashley 1978; Valentine 2019). The polymodality of these sediments implies contributions from diverse sources, such as surface erosion on hillslopes, mass wasting, stream banks, tributaries, and anthropogenic disturbances (Hassan et al. 2005; Cienciala et al. 2021).

Figure 4.3 Photograph of the locations of MSD5 (top) and PSD7 (bottom) samples.

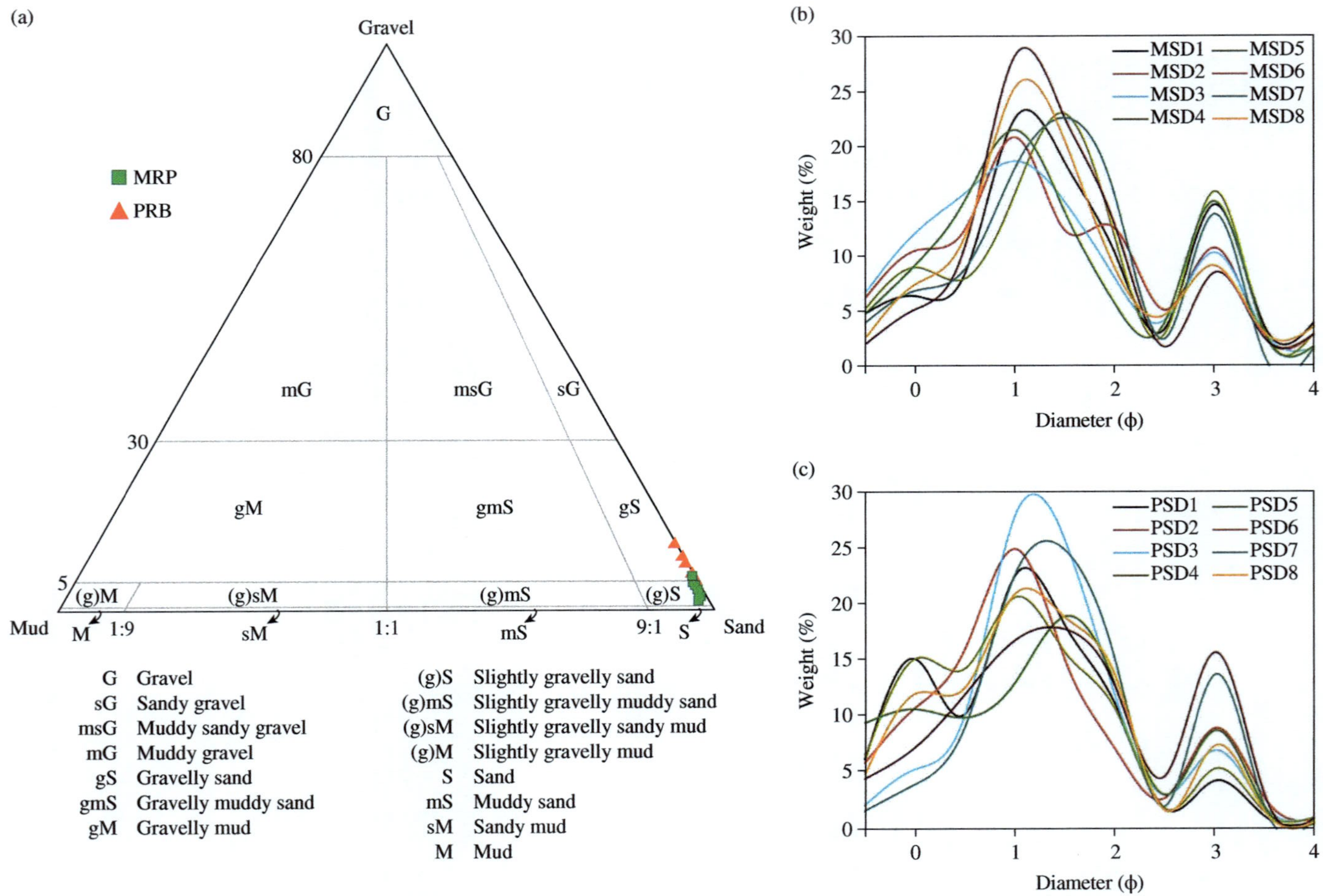

Figure 4.4 Textural characteristics of the sediments of the MRB and PRB. (a) Diagrammatic representation (after Folk 1980) showing the grain-size composition of the 16 samples based on the weight percentage of gravel, sand, and mud content. Sand: mud ratios are marked along the base of the diagram. Grain-size frequency distribution of sand in the sediments of the (b) MRB and (c) PRB.

Table 4.1 Textural parameters of sediments: MRB and PRB.

MRB						PRB					
Sample	M_z	σ_I	Sk_I	K_G	G/M	Sample	M_z	σ_I	Sk_I	K_G	G/M
MSD1	1.30	1.28	0.14	1.10	3.24	PSD1	0.61	1.09	−0.07	1.04	19.25
MSD2	1.06	1.27	0.14	0.98	4.29	PSD2	0.79	1.15	0.09	1.24	14.47
MSD3	0.94	1.26	0.15	1.03	4.61	PSD3	1.03	0.95	0.03	1.58	6.09
MSD4	1.29	1.25	0.05	1.07	1.16	PSD4	0.63	1.12	0.02	1.09	18.23
MSD5	1.08	1.32	0.15	0.97	10.98	PSD5	0.66	1.29	−0.10	0.91	26.72
MSD6	1.10	0.95	0.15	1.36	2.43	PSD6	1.22	1.26	0.02	1.00	14.31
MSD7	1.22	1.21	−0.01	1.23	4.63	PSD7	1.32	1.11	0.08	1.33	9.04
MSD8	1.18	1.15	0.21	1.27	2.85	PSD8	0.78	1.11	−0.04	1.13	15.51

M_Z-Graphic mean; σ_I-Inclusive graphic standard deviation; Sk_I-Inclusive graphic skewness; K_G-Graphic kurtosis and G/M-gravel-to-mud ratio. Units of M_z and σ_I are in Φ scale and others are dimensionless (Folk and Ward 1957).

The sediments of the MRB are characterized by a significantly larger mud fraction (compared with the PRB), whereas the sediments of the PRB have a relatively higher percentage of gravel fraction (than the MRB). Significantly higher gravel/mud ratios (6.09–26.72, Table 4.1) characterize sediments of the PRB, relative to MRB (1.16–10.98), and are chiefly attributed to the variabilities of climate and consequent weathering milieu. Hence, the gravel-to-mud ratio seems to be an effective discriminator of the sediments between the basins (Figure 4.5a), where the gravel: mud ratio of 6:1 demarcates the comparatively mud-poor (semiarid) PRB from the relatively mud-rich (humid) MRB.

Table 4.1 lists the statistical parameters of grain size (viz., *Mz*, *σI*, SkI, and KG) for the sediment samples of the MRB and PRB. The *Mz*, a measure of average grain size, of the samples in the MRB ranges from 0.94 (MSD3) to 1.30Φ (MSD1) and that in PRB varies from 0.61 (PSD1) to 1.32Φ (PSD7). The majority of the sediment samples of the MRB belong to medium sand ($n = 7$), whereas most of the sediments of the PRB are dominated by coarse sand ($n = 5$). The fluvial sediments fall in a narrow range of σI, ranging from 0.95 (MSD6) to 1.32Φ (MSD5) in the MRB and 0.95 (PSD3) to 1.29Φ (PSD5) in the PRB (Table 4.1), and most of the samples ($n = 7$) in either of the basins are poorly sorted. In general, most river sands have sorting values greater than 0.5Φ, and poor sorting of the sediments in the channels of the MRB and PRB is suggestive of the variability in the sediment sources as the addition of a new mode or mixing of sediments of diverse size classes in the main population can give rise to poor bimodal or polymodal sorting (Folk and Ward 1957; Friedman 1961). Here, in the MRB and PRB, the dominant sediment delivery processes are hillslope soil erosion, mass wasting, and stream bank erosion. Furthermore, a tributary with a weaker transport capacity mainly contributes to fine-grained sediments. Poor sorting of the sediments also reflects the significant physical weathering associated with steeper gradients of the stream channels on a high-relief terrain (Singh 2010; Maharana et al. 2018) and/or sediment transport for shorter distances. Tarras-Wahlberg

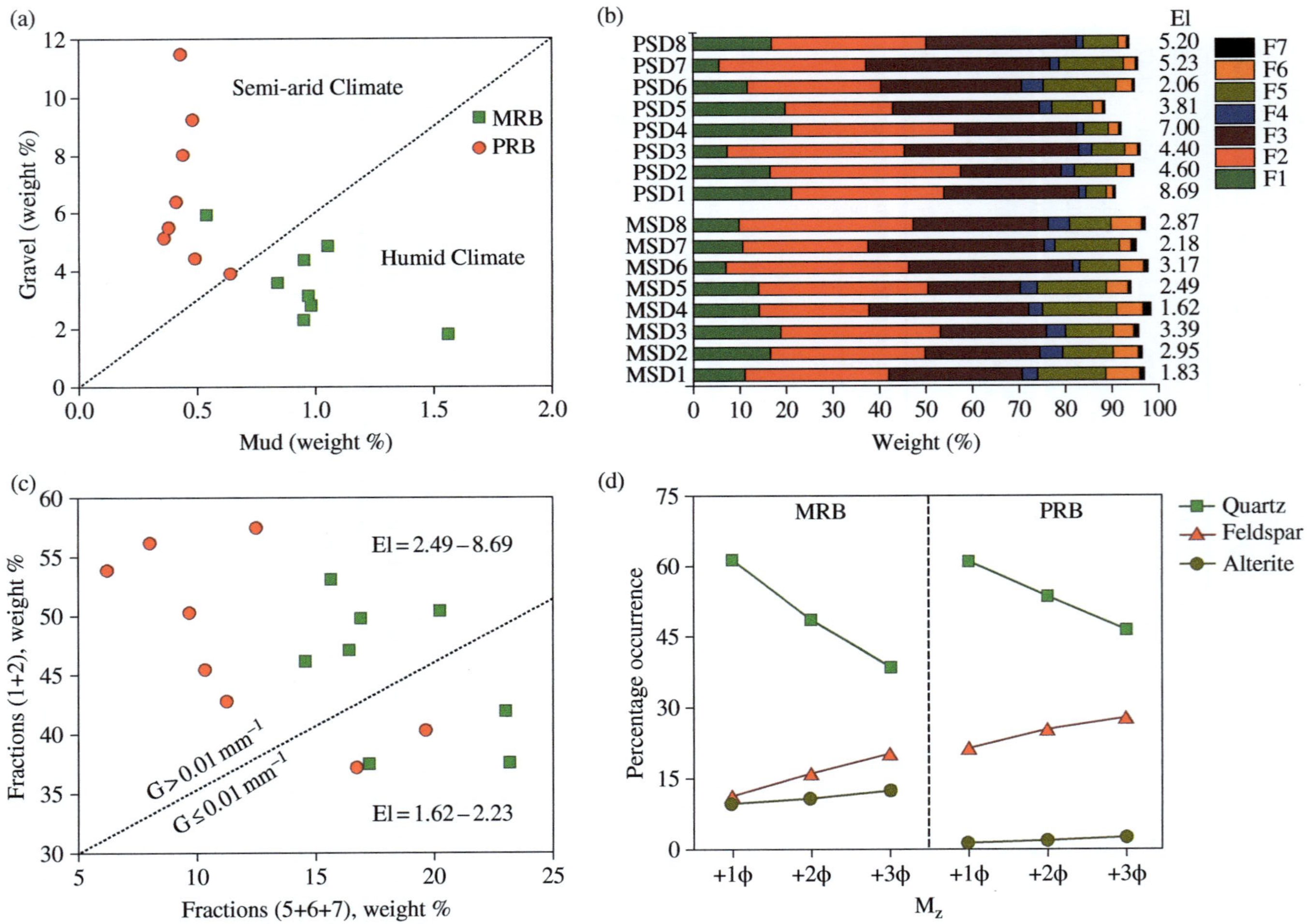

Figure 4.5 (a) Variability of gravel and mud fractions, (b) weight percentage of different fractions and corresponding energy index (EI), (c) bivariate relationship between the coarse and fine fractions, and (d) variation of quartz, feldspar, and alterite minerals in different fractions of the sediments of the MRB and PRB.

et al. (2002) also reported that immature riverine sediments occur in the channels characterized by steeper gradients that provide little time for sorting. Further, the episodic nature of river flow or a marked seasonal flow regime also influences the sorting process and polymodality, which is significant in the MRB and PRB due to the monsoon-driven seasonality and regulated hydrologic regime.

Skewness is a significant textural parameter, which is sensitive to the depositional environment and transport process (Folk and Ward 1957; Friedman 1961). The Sk_I varies from −0.01 (MSD7) to 0.21 (MSD8) in the MRB and −0.10 (PSD5) to 0.09 (PSD2) in the PRB (Table 4.1). The majority of the sediment samples of the MRB are finely skewed ($n = 6$), followed by symmetrical ($n = 2$), whereas most of the samples of the PRB are nearly symmetrical ($n = 7$), followed by coarsely skewed ($n = 1$). The fine skewed sediments of the MRB imply the relative dominance of the finer particles compared to the sediments of the PRB. The distribution of K_G values in natural sediments is strongly skewed (Folk 1980). The K_G ranges from 0.97 (MSD5) to 1.36 (MSD6) in the MRB and varies between 0.91 (PSD5) and 1.58 (PSD3) in the PRB (Table 4.1). Many of the samples in MRB and PRB are mesokurtic ($n = 5$ and 4, respectively) followed by leptokurtic ($n = 3$).

As the size distribution of the stream channel sediments exhibits significant differences between the basins, variation in the size distribution has been converted to an empirical EI (after Cohen et al. 2005) (Eq. 4.1):

$$\mathrm{EI} = \frac{\mathrm{Weight\%}\left(\mathrm{Fractions}\ 1+2\right)}{\mathrm{Weight\%}\left(\mathrm{Fractions}\ 5+6+7\right)} \tag{4.1}$$

where Fraction 1 = up to 0Φ, Fraction 2 = between 0 and 1Φ, Fraction 3 = between 1 and 2Φ, Fraction 4 = between 2 and 2.5Φ, Fraction 5 = between 2.5 and 3Φ, Fraction 6 = between 3 and 4Φ, and Fraction 7 = greater than 4Φ.

Figure 4.5b is the weight percentage of fractions 1–7 of the channel sediments of the MRB and PRB. The EI of the sediments of the MRB ranges from 1.62 (MSD4) to 3.39 (MSD3) with a mean of 2.56 ± 0.64, whereas the EI of the samples from the PRB varies between 2.06 (PSD6) and 8.69 (PSD1) with a mean of 4.75 ± 2.25. In addition, the EI as an energy-based discriminator of sediments is evident in Figure 4.5c, where the changes in the EI are coincident with the changes in stream channel gradient, morphological type, and bed character, reflecting the differences in the hydraulic energy conditions. Thus, the EI differentiates the channels with relatively higher channel gradients (i.e. ≥0.01 m/m) from the low-gradient channels as the fine (coarse) fraction shows a(n) depletion (enrichment) in higher (lower) stream channel gradients. The range of the EI differs between these two groups in that, channels with relatively higher channel gradients are characterized by an EI between 2.49 and 8.69, while the range of EI in the lower-gradient channels is only from 1.62 to 2.23.

4.4.2 Mineral Assemblage of the Fluvial Sediments

Grain mounts of coarse- (+1.0Φ), medium- (+2.0Φ), and fine-sand (+3.0Φ) fractions of the sediment samples of the MRB and PRB were subjected to mineralogical scrutiny and the results are summarized in Table 4.2. The abundances of the various detrital minerals in

Table 4.2 Mineral abundance (%) in coarse-, medium- and fine-sand fractions of the sediments of MRB and PRB.

	MRB							PRB						
	Coarse (+1.0Φ)		Medium (+2.0Φ)		Fine (+3.0Φ)			Coarse (+1.0Φ)		Medium (+2.0Φ)		Fine (+3.0Φ)		
Minerals	**Mean**	**σ**	**Mean**	**σ**	**Mean**	**σ**	**Grand mean**	**Mean**	**σ**	**Mean**	**σ**	**Mean**	**σ**	**Grand mean**
Quartz	61.23	3.36	48.57	3.87	38.64	4.78	49.48	61.08	2.21	53.58	2.60	46.41	3.03	53.69
Feldspar	11.48	1.86	16.07	0.89	20.34	2.13	15.96	21.50	2.56	25.33	3.06	27.85	2.96	24.89
Hypersthene	1.94	1.65	2.65	1.75	3.13	1.35	2.57	1.84	0.77	2.23	0.82	2.69	0.98	2.25
Hornblende	2.09	0.62	2.83	0.82	3.34	1.01	2.75	1.88	0.42	2.21	0.47	2.56	0.38	2.22
Biotite	4.62	1.34	5.63	1.12	6.35	1.38	5.53	5.33	1.32	5.71	1.37	6.37	1.50	5.80
Garnet	1.31	0.52	1.91	0.77	2.30	0.70	1.84	1.88	0.63	2.19	0.61	2.40	0.81	2.16
Zircon	0.25	0.17	1.14	0.67	1.92	0.68	1.10	0.39	0.38	0.85	0.23	1.35	0.32	0.86
Sillimanite	1.39	0.91	2.26	0.95	2.58	1.06	2.08	1.45	0.22	1.82	0.24	2.34	0.70	1.87
Monazite	0.12	0.06	0.59	0.20	0.97	0.31	0.56	0.00	0.00	0.36	0.43	0.71	0.47	0.36
Chlorite	0.15	0.19	0.23	0.25	0.38	0.35	0.25	0.04	0.06	0.10	0.08	0.25	0.12	0.13
Muscovite	0.04	0.05	0.12	0.06	0.19	0.13	0.12	0.04	0.09	0.05	0.03	0.13	0.05	0.07
Opaque	5.72	1.27	7.18	1.27	7.40	1.32	6.77	3.16	1.24	3.59	1.37	4.38	1.40	3.71
Alterite	9.67	2.70	10.82	2.84	12.48	3.03	10.99	1.42	1.31	1.99	1.29	2.58	1.68	2.00

σ: standard deviation.

the coarse sediment fraction (+1.0Φ) of the MRB suggest a dominance of quartz as a major phase (61.23%), followed by feldspar (11.48%), alterite (9.67%), opaque (5.72%), biotite (4.62%), and hornblende (2.09%) (Table 4.2). Similarly, the list of detrital species in the coarse sand fraction of the PRB is also capped by quartz (61.08%), followed by feldspar (21.50%), biotite (5.33%), and opaque (3.16%). Other minerals, viz., zircon, monazite, chlorite, and muscovite, occur only in minor amounts (<1.0%). Similar to the mineral abundance of the coarse fraction, quartz dominates in the medium sand fraction also, i.e. 48.57% in the MRB and 53.58% in the PRB (Table 4.2). Other dominant minerals in the medium sand fraction of the MRB are feldspar (16.07%), alterite (10.82%), opaque (7.18%), biotite (5.63%), hornblende (2.83%), hypersthene (2.65%), and sillimanite (2.26%), whereas, in the PRB, the presence of feldspar (25.33%), biotite (5.71%), opaque (3.59%), hypersthene (2.23%), hornblende (2.21%), garnet (2.19%), and alterite (1.99%) are also significant in the medium sand fraction. Other minerals, viz., monazite, chlorite, and muscovite, occur only in minor amounts (<1.0%). In the MRB, the fine fraction shows the dominance of quartz (38.64%; Table 4.2), closely followed by feldspar (20.34%), alterite (12.48%), opaque (7.40%), biotite (6.35%), hornblende (3.34%), hypersthene (3.13%), sillimanite (2.58%), and garnet (2.30%). The fine fraction of the PRB is also dominated by quartz (46.41%), followed by feldspar (27.85%), biotite (6.37%), opaque (4.38%), hypersthene (2.69%), alterite (2.58%), hornblende (2.56%), garnet (2.40%), and sillimanite (2.34%).

In general, quartz dominates the mineral suites with a grand mean of ~50% in the stream channel sediments of the MRB and ~54% in the PRB (Table 4.2), followed by feldspar (~16% and 25% each in the MRB and PRB, respectively). Although the presence of alterite is significant in the sediments of the MRB (~11%), it accounts only for 2% in the sediments of the PRB. Size-specific variations of mineral species are informative; in that, quartz content declines from coarse to fine fraction, whereas the content of feldspar and alterite increases with fining of grain size (Figure 4.5d). The percentage occurrence of heavy minerals also increases with the decrease in grain size (Table 4.2). Chemical weathering processes progressively transform the primary minerals, and especially feldspars, into clays, secondary oxides, and hydroxides and the degree of weathering affects the rate of transformation (Nesbitt et al. 1997). The susceptibility of different minerals to chemical weathering varies, wherein feldspars are less resistant than quartz. Hence, the abundance of feldspars in sediments reflects the weathering intensity. The proportion of feldspars is significantly lower in the sediments of MRB than in the PRB (Table 4.2), indicating the higher degree of chemical weathering conditions in the MRB. Compositional maturity is a measure of chemical alteration of the sediments during weathering. With increasing maturity, the abundance of quartz increases at the expense of feldspar, mafic minerals, and lithic fragments, and hence, the ratio between quartz and feldspar is an index for mineralogic maturity of detrital sands (Pettijohn 1957). The mean of the ratio between quartz and feldspar of the coarse-, medium-, and fine-sand fractions in the MRB is 5.47 (4.84–7.57), 3.03 (2.67–3.49), and 1.93 (1.45–2.57), respectively, whereas, in the PRB, the mean of the ratio is 2.88 (2.48–3.49), 2.15 (1.76–2.82), and 1.68 (1.41–2.01) (Table 4.3). Among different size ranges, the fine fraction is relatively immature (i.e. lower ratios of quartz-to-feldspar) in both the MRB and PRB. But the quartz-to-feldspar ratio of the sand fractions of the sediments of the MRB is significantly higher (implying relatively higher maturity) than that of the sediments of the PRB.

Table 4.3 Quartz/feldspar ratio and MIA of the sediments of MRB and PRB.

Sample	Quartz/Feldspar			MIA			Sample	Quartz/Feldspar			MIA		
	Coarse (+1.0Φ)	Medium (+2.0Φ)	Fine (+3.0Φ)	Coarse (+1.0Φ)	Medium (+2.0Φ)	Fine (+3.0Φ)		Coarse (+1.0Φ)	Medium (+2.0Φ)	Fine (+3.0Φ)	Coarse (+1.0Φ)	Medium (+2.0Φ)	Fine (+3.0Φ)
MSD1	4.84	3.06	2.10	82.86	75.35	67.71	PSD1	3.49	2.82	2.01	77.73	73.81	66.80
MSD2	4.85	3.49	2.57	82.91	77.71	72.01	PSD2	2.85	2.16	1.73	74.01	68.39	63.33
MSD3	6.70	3.37	2.37	87.01	77.14	70.29	PSD3	3.27	2.28	1.75	76.59	69.52	63.69
MSD4	4.91	2.67	1.72	83.09	72.77	63.20	PSD4	3.00	2.07	1.69	75.02	67.48	62.88
MSD5	7.57	2.90	1.89	88.33	74.34	65.45	PSD5	2.66	1.89	1.50	72.67	65.44	60.07
MSD6	4.95	3.11	1.84	83.18	75.67	64.76	PSD6	2.55	1.76	1.57	71.85	63.72	61.06
MSD7	5.01	2.83	1.45	83.35	73.90	59.20	PSD7	2.48	1.95	1.41	71.26	66.10	58.44
MSD8	4.96	2.81	1.51	83.21	73.73	60.23	PSD8	2.71	2.25	1.79	73.04	69.20	64.11
Mean	5.47	3.03	1.93	84.24	75.08	65.36	Mean	2.88	2.15	1.68	74.02	67.96	62.55

The mineralogical index of alteration (MIA), expressed as the ratio of quartz to the sum of quartz and feldspar (Eq. 4.2), is a handy tool in the assessment of mineralogical maturity and weathering intensity (Bahlburg and Dobrzinski 2011).

$$\mathrm{MIA} = \left[\frac{\mathrm{Quartz}}{\mathrm{Quartz} + \mathrm{Feldspar}}\right] \times 100 \tag{4.2}$$

The mean MIA of the sand fractions of the MRB sediments is 84.24, 75.08, and 65.36, respectively (Table 4.3), whereas the mean MIA for the PRB sediments is 74.02, 67.96, and 62.55, respectively, implying intense weathering conditions in the MRB.

4.4.3 Effect of Topography and Climate on Sediment Characteristics

The sediment size distribution and mineralogical composition in the channel network yield insight into weathering processes and hydrodynamics. Topography influences the erosion rates and thereby sediment residence times; in that, sediments on gentler slopes have longer exposure to weathering, leading to finer sediment supply to the stream channels (Norton and von Blanckenburg 2010; Riebe et al. 2015). Further, weathering also takes advantage of the available moisture and temperature in transforming the parent rock into sediments (White and Blum 1995). The sediment textural parameters show a dichotomy between the basins, wherein the sediments of the MRB are characterized by a relatively larger amount of fine fractions than the sediments of the PRB. The drainage network of the MRB is primarily developed on the Munnar Plateau, an uplifted erosional surface with a roughly southwesterly slope since the late Paleocene (Soman 2002), whereas the stream network of the PRB drains the northern scarps of the plateau. The topographic influence on the sediment texture is also evidenced by the EI (Figure 4.5c), where the variability in the channel gradient reflects the energy regime and thereby grain size. The M_z of the sediment samples, viz., PSD1, PSD4, PSD5, and PSD8, are significantly smaller (0.61–0.78; Table 4.1) than the rest of the samples. These samples represent the step-pool morphology, whereas the rest of the samples (except PSD3) were collected from riffle-pool reaches. The step-pool morphology in the MRB and PRB occurs in channels with steeper gradients (0.05–0.27 m/m), whereas the riffle-pool morphology dominates in channels with a gradient between 0.004 and 0.020 m/m (Thomas 2012). Although PSD2 has a relatively coarser grain size ($M_z = 0.79$), closer to the sediments from step-pool reaches, the sample represents a riffle-pool dominated channel reach characterized by relatively larger riffle bed material estimates (D_{50} and D_{84}). Indeed, the analysis demonstrated the significance of the channel gradient on the grain-size distribution of the mountain river sediments of the southern Western Ghats, which is also supported by the observations of Sebastian (1999). Further, the inferred variations in the grain-size parameters are correlated strongly with the channel characteristics and river basin topography.

The grain-size distribution of the sediments in the channels of the basins is not entirely controlled by the topography and hydrodynamics but is also influenced by the intensity of chemical weathering in the catchment area, which does not produce or yield coarser materials. Although the drainage systems of both the MRB and PRB are developed across the southern Western Ghats, the basins do experience an easterly climatic gradient, i.e. climate progressively transitions from tropical humid (in the MRB) to tropical semiarid (in the PRB). Different weathering regimes are controlled by the climate gradient, and consequently, differing weathering intensities are reflected by the grain-size fractions and mineral assemblages of the channel sediments. For example, the sediments of the PRB (MRB) have relatively larger (smaller) amounts of gravel fraction and a converse relationship for the mud fraction (Figure 4.5a). The dominance of physical weathering over chemical weathering may enhance the generation of coarser sediment fractions, while the dominance of chemical over physical weathering leads to the production of finer sediments (Riebe et al. 2015). This observation is consistent with the enhanced production of gravel fraction in the semiarid climate (i.e. in the PRB) and increased content of mud fraction in the humid climate (i.e. in the MRB). The variability in the detrital mineral composition also signifies the contrasting climates as the humid (semiarid) climate is characterized by relatively higher (lower) mineralogical maturity and mineral alterations described by the quartz-to-felspar ratio and MIA. In the tropical humid climate of the MRB, chemical weathering is deep and the topography is congenial providing longer residence times for the detritus than the PRB, leading to mineralogical maturity. Meanwhile, the grain size has a significant control on the mineral composition (Whitmore et al. 2004), which is reflected in the varying mineral abundance from coarse to fine-sand fractions (Table 4.3). However, the grain-size resolution in the mineralogical evaluation was limited only to three fractions (coarse, medium, and fine), and hence, interpretations are limited calling for the analysis of more samples. The analysis of grain-size distribution and mineralogical composition of sediments of the mountain stream channels hugely imply the significance of topography and climate but are insufficient to discriminate and generalize their independent roles. However, the MRB and PRB provide a typical example of mountain streams in which the climate and topography are undisputed determinants of fluvial sediment characteristics.

4.5 Summary and Conclusion

This study analyzed textural characteristics and mineralogical composition of channel sediments from two mountain river basins set in contrasting climate zones in the southern Western Ghats (India), viz., the MRB and PRB, to assess the role of physiography and climate on detrital composition. These basins are characterized by differing topography and contrasting climate (humid MRB vs. semiarid PRB). The grain-size analysis demonstrated the relative dominance of coarser sediments and lesser content of mud fraction in the PRB implying the significance of physical weathering compared to the finer sediments and larger mud fractions of the MRB. Consequently, the gravel: mud ratio efficiently discriminates the basins. The EI, derived from the grain-size distribution, is an effective index to differentiate steeper channels (i.e. ≥ 0.01 m/m) from the low-gradient channels. The mineralogical composition of the channel sediments suggests the dominance of quartz as a major phase in both the basins but variable fractions of feldspars and alterites. The significantly smaller content of feldspars in the sediments of the MRB than the PRB indicates intense chemical weathering conditions in the MRB, which is evinced by the higher quartz/feldspar ratios and MIA.

The analysis of the sediment characteristics of the MRB and PRB indicates that both the textural parameters and mineralogical composition vary markedly between the basins and appear to be controlled by topography and climate. The grain-size distribution of the sediments is primarily affected by the topographic variability and channel characteristics, whereas the contrasting climate setting is dominantly reflected in the detrital composition. However, these drivers (i.e. climate and topography) have mixed effects also on the sediment texture and mineralogy. Further, we assumed that the effect of lithology is a common denominator due to the similarity of the dominant lithologies between the basins. Although the variations in the sediment characteristics with respect to climate and topographic variability are robust, this analysis may be of limited value to discriminate and generalize their independent effects and warrants the study of a larger number of samples representing diverse topographic, lithologic, and climate settings. Such analysis can improve the understanding of how the climate and denudation jointly drive the evolution of drainage basins of the Western Ghats.

Acknowledgments

The first author (JT) is grateful to (late) Dr. R. Satheesh (SES, Mahatma Gandhi University, Kerala) for his moral support and mentoring during the early stages of his research career.

Acknowledgments are also due to the Kerala State Council for Science, Technology, and Environment, Thiruvananthapuram for the financial support (004/FSHP/05KSCSTE) and the Kerala Forest Department for permission and logistics for the field studies in the protected areas.

References

Ashley, G.M. (1978). Interpretation of polymodal sediments. *The Journal of Geology* 86 (4): 411–421. https://doi.org/10.1086/649710.

Bahlburg, H. and Dobrzinski, N. (2011). A review of the chemical index of alteration (CIA) and its application to the study of Neoproterozoic glacial deposits and climate transitions. In: *The Geological Record of Neoproterozoic Glaciations* (ed. E. Arnaud, G.P. Halverson, and G. Shields-Zhou), 81–92. London: Geological Society of London https://doi.org/10.1144/M36.6.

Berner, E.K. and Berner, R.A. (1987). *The Global Water Cycle: Geochemistry and Environment.* Prentice Hall.

Binnie, S.A. and Summerfield, M.A. (2013). Rates of denudation. In: *Treatise on Geomorphology* (ed. J.F. Shroder), 66–72. San Diego: Academic Press https://doi.org/10.1016/B978-0-12-374739-6.00149-4.

Binnie, S.A., Phillips, W.M., Summerfield, M.A., and Fifield, L.K. (2007). Tectonic uplift, threshold hillslopes, and denudation rates in a developing mountain range. *Geology* 35 (8): 743–746. https://doi.org/10.1130/G23641A.1.

Brussock, P.P., Brown, A.V., and Dixon, J.C. (1985). Channel form and stream ecosystem models. *Journal of the American Water Resources Association* 21 (5): 859–866. https://doi.org/10.1111/j.1752-1688.1985.tb00180.x.

Buffington, J.M., Lisle, T.E., Woodsmith, R.D., and Hilton, S. (2002). Controls on the size and occurrence of pools in coarse-grained forest rivers. *River Research and Applications* 18 (6): 507–531. https://doi.org/10.1002/rra.693.

Carver, R.E. (1971). *Procedures in Sedimentary Petrology*. New York: Wiley-Interscience.

Chakrapani, G.J. and Subramanian, V. (1990). Preliminary studies on the geochemistry of the Mahanadi river basin, India. *Chemical Geology* 81 (3): 241–253. https://doi.org/10.1016/0009-2541(90)90118-Q.

Chandrashekara, U.M. and Sibichan, V. (2006). Logs and snags in a shola forest of Kerala, India. *Journal of Mountain Science* 3 (2): 131–138. https://doi.org/10.1007/s11629-006-0131-8.

Chin, A. and Wohl, E. (2005). Toward a theory for step pools in stream channels. *Progress in Physical Geography: Earth and Environment* 29 (3): 275–296. https://doi.org/10.1191/0309133305pp449ra.

Cienciala, P., Melendez Bernardo, M., Nelson, A.D., and Haas, A.D. (2021). Sediment yield from a forested mountain basin in inland Pacific northwest: rates, partitioning, and sources. *Geomorphology* 374: 107478. https://doi.org/10.1016/j.geomorph.2020.107478.

Cohen, D.R., Dunlop, A.C., and Rose, T. (2005). Contrasting dispersion patterns for gold in stream sediments at Timbarra, NSW, Australia. *Journal of Geochemical Exploration* 85 (1): 1–16. https://doi.org/10.1016/j.gexplo.2004.08.002.

Corenblit, D., Baas, A.C.W., Bornette, G. et al. (2011). Feedbacks between geomorphology and biota controlling earth surface processes and landforms: a review of foundation concepts and current understandings. *Earth-Science Reviews* 106 (3–4): 307–331. https://doi.org/10.1016/j.earscirev.2011.03.002.

Das, A. and Krishnaswami, S. (2007). Elemental geochemistry of river sediments from the Deccan traps, India: implications to sources of elements and their mobility during basalt-water interaction. *Chemical Geology* 242 (1–2): 232–254. https://doi.org/10.1016/j.chemgeo.2007.03.023.

Deepthy, R. and Balakrishnan, S. (2005). Climatic control on clay mineral formation: evidence from weathering profiles developed on either side of the Western Ghats. *Journal of Earth System Science* 114 (5): 545–556. https://doi.org/10.1007/BF02702030.

Demoulin, A., Mather, A., and Whittaker, A. (2017). Fluvial archives, a valuable record of vertical crustal deformation. *Quaternary Science Reviews* 166: 10–37. https://doi.org/10.1016/j.quascirev.2016.11.011.

Dibiase, R.A. and Whipple, K.X. (2011). The influence of erosion thresholds and runoff variability on the relationships among topography, climate, and erosion rate. *Journal of Geophysical Research* 116 (F4): F04036. https://doi.org/10.1029/2011JF002095.

Dibiase, R.A., Whipple, K.X., Heimsath, A.M., and Ouimet, W.B. (2010). Landscape form and millennial erosion rates in the San Gabriel Mountains, CA. *Earth and Planetary Science Letters* 289 (1–2): 134–144. https://doi.org/10.1016/j.epsl.2009.10.036.

Dibiase, R.A., Heimsath, A.M., and Whipple, K.X. (2012). Hillslope response to tectonic forcing in threshold landscapes. *Earth Surface Processes and Landforms* 37 (8): 855–865. https://doi.org/10.1002/esp.3205.

Dupré, B., Gaillardet, J., Rousseau, D., and Allègre, C.J. (1996). Major and trace elements of river-borne material: the Congo Basin. *Geochimica et Cosmochimica Acta* 60 (8): 1301–1321. https://doi.org/10.1016/0016-7037(96)00043-9.

von Eynatten, H., Tolosana-Delgado, R., Karius, V. et al. (2016). Sediment generation in humid Mediterranean setting: grain-size and source-rock control on sediment geochemistry and mineralogy (Sila massif, Calabria). *Sedimentary Geology* 336: 68–80. https://doi.org/10.1016/j.sedgeo.2015.10.008.

Farooqui, S., Shah, A.P., Maurya, D.M. et al. (2021). Texture, mineralogy and geochemistry of late quaternary sediments of the Mahi River basin, western India: implications to climate and tectonics. *Applied Geochemistry* 134: 105088. https://doi.org/10.1016/j.apgeochem.2021.105088.

Folk, R.L. (1980). *Petrology of Sedimentary Rocks*. Austin: Hemphill Publishing Company.

Folk, R.L. and Ward, W.C. (1957). Brazos River bar [Texas]; a study in the significance of grain size parameters. *Journal of Sedimentary Research* 27 (1): 3–26. https://doi.org/10.1306/74D70646-2B21-11D7-8648000102C1865D.

Friedman, G.M. (1961). Distinction between dune, beach, and river sands from their textural characteristics. *Journal of Sedimentary Research* 31 (4): 514–529. https://doi.org/10.1306/74D70BCD-2B21-11D7-8648000102C1865D.

Gaillardet, J., Dupré, B., and Allègre, C.J. (1999). Geochemistry of large river suspended sediments: silicate weathering or recycling tracer? *Geochimica et Cosmochimica Acta* 63 (23–24): 4037–4051. https://doi.org/10.1016/S0016-7037(99)00307-5.

Gardner, C.B., Carey, A.E., Lyons, W.B. et al. (2017). Molybdenum, vanadium, and uranium weathering in small mountainous rivers and rivers draining high-standing islands. *Geochimica et Cosmochimica Acta* 219: 22–43. https://doi.org/10.1016/j.gca.2017.09.012.

Garrels, R.M. and Mackenzie, F.T. (1971). *Evolution of Sedimentary Rocks*. New York: Norton.

Goldsmith, S.T., Harmon, R.S., Lyons, W.B. et al. (2015). Evaluation of controls on silicate weathering in tropical mountainous rivers: insights from the isthmus of Panama. *Geology* 43 (7): 563–566. https://doi.org/10.1130/G36082.1.

GSI (1992). *District Resource Map, Idukki Distric, Kerala. Part-I. Geology and Mineralst*. Kolkata: Geological Survey of India.

Hassan, M.A., Church, M., Lisle, T.E. et al. (2005). Sediment transport and channel morphology of small, forested streams. *Journal of the American Water Resources Association* 41 (4): 853–876. https://doi.org/10.1111/j.1752-1688.2005.tb03774.x.

Jose, S., Sreepathy, A., Mohan Kumar, B., and Venugopal, V.K. (1994). Structural, floristic and edaphic attributes of the grassland-shola forests of Eravikulam in peninsular India. *Forest Ecology and Management* 65 (2–3): 279–291. https://doi.org/10.1016/0378-1127(94)90176-7.

Kessarkar, P.M., Shynu, R., Rao, V.P. et al. (2013). Geochemistry of the suspended sediment in the estuaries of the Mandovi and Zuari rivers, central west coast of India. *Environmental Monitoring and Assessment* 185 (5): 4461–4480. https://doi.org/10.1007/s10661-012-2883-7.

Kober, F., Zeilinger, G., Hippe, K. et al. (2015). Tectonic and lithological controls on denudation rates in the central Bolivian Andes. *Tectonophysics* 657: 230–244. https://doi.org/10.1016/j.tecto.2015.06.037.

Lecomte, K.L., Pasquini, A.I., and Depetris, P.J. (2005). Mineral weathering in a semiarid mountain river: its assessment through PHREEQC inverse modeling. *Aquatic Geochemistry* 11 (2): 173–194. https://doi.org/10.1007/s10498-004-3523-9.

Lee, Y.-J., Chen, P.-H., Lee, T.-Y. et al. (2020). Temporal variation of chemical weathering rate, source shifting and relationship with physical erosion in small mountainous rivers, Taiwan. *Catena* 190: 104516. https://doi.org/10.1016/j.catena.2020.104516.

Liu, Z., Colin, C., Huang, W. et al. (2007). Climatic and tectonic controls on weathering in South China and Indochina peninsula: clay mineralogical and geochemical investigations from the pearl, red, and Mekong drainage basins. *Geochemistry, Geophysics, Geosystems* 8 (5): Q05005. https://doi.org/10.1029/2006GC001490.

Macklin, M.G., Lewin, J., and Woodward, J.C. (2012). The fluvial record of climate change. *Philosophical Transactions of the Royal Society A - Mathematical Physical and Engineering Sciences* 370 (1966): 2143–2172. https://doi.org/10.1098/rsta.2011.0608.

Maharana, C., Srivastava, D., and Tripathi, J.K. (2018). Geochemistry of sediments of the peninsular rivers of the ganga basin and its implication to weathering, sedimentary processes and provenance. *Chemical Geology* 483: 1–20. https://doi.org/10.1016/j.chemgeo.2018.02.019.

Martin, J.-M. and Meybeck, M. (1979). Elemental mass-balance of material carried by major world rivers. *Marine Chemistry* 7 (3): 173–206. https://doi.org/10.1016/0304-4203(79)90039-2.

Mclennan, S.M. (1993). Weathering and global denudation. *The Journal of Geology* 101 (2): 295–303. https://doi.org/10.1086/648222.

Milliman, J.D. and Syvitski, J.P.M. (1992). Geomorphic/tectonic control of sediment discharge to the ocean: the importance of small mountainous rivers. *The Journal of Geology* 100 (5): 525–544. https://doi.org/10.1086/629606.

Mohan, U. and Krishnakumar, A. (2022). Geochemical aspects and contamination evaluation of major and trace elements in the sediments of Kallada river, southern Western Ghats, India. *Journal of Environmental Science and Health, Part A Toxic/Hazardous Substances and Environmental Engineering* 57 (4): 258–267. https://doi.org/10.1080/10934529.2022.2053450.

Montgomery, D.R. and Brandon, M.T. (2002). Topographic controls on erosion rates in tectonically active mountain ranges. *Earth and Planetary Science Letters* 201 (3–4): 481–489. https://doi.org/10.1016/S0012-821X(02)00725-2.

Montgomery, D.R. and Buffington, J.M. (1997). Channel-reach morphology in mountain drainage basins. *Geological Society of America Bulletin* 109 (5): 596–611. https://doi.org/10.1130/0016-7606(1997)109<0596:CRMIMD>2.3.CO;2.

Mueller, E.R. and Pitlick, J. (2013). Sediment supply and channel morphology in mountain river systems: 1. Relative importance of lithology, topography, and climate. *Journal of Geophysical Research - Earth Surface* 118 (4): 2325–2342. https://doi.org/10.1002/2013JF002843.

Nesbitt, H.W., Fedo, C.M., and Young, G.M. (1997). Quartz and feldspar stability, steady and non-steady-state weathering, and petrogenesis of siliciclastic sands and muds. *The Journal of Geology* 105 (2): 173–192. https://doi.org/10.1086/515908.

Norton, K.P. and von Blanckenburg, F. (2010). Silicate weathering of soil-mantled slopes in an active alpine landscape. *Geochimica et Cosmochimica Acta* 74 (18): 5243–5258. https://doi.org/10.1016/j.gca.2010.06.019.

Ouimet, W.B., Whipple, K.X., and Granger, D.E. (2009). Beyond threshold hillslopes: channel adjustment to base-level fall in tectonically active mountain ranges. *Geology* 37 (7): 579–582. https://doi.org/10.1130/G30013A.1.

Pettijohn, F.J. (1957). *Sedimentary Rocks*. New York: Harper and Row.

Raab, G., Egli, M., Norton, K. et al. (2019). Climate and relief-induced controls on the temporal variability of denudation rates in a granitic upland. *Earth Surface Processes and Landforms* 44 (13): 2570–2586. https://doi.org/10.1002/esp.4681.

Rajamani, V., Tripathi, J.K., and Malviya, V.P. (2009). Weathering of lower crustal rocks in the Kaveri river catchment, southern India: implications to sediment geochemistry. *Chemical Geology* 265 (3–4): 410–419. https://doi.org/10.1016/j.chemgeo.2009.05.007.

Riebe, C.S., Sklar, L.S., Lukens, C.E., and Shuster, D.L. (2015). Climate and topography control the size and flux of sediment produced on steep mountain slopes. *The Proceedings of the National Academy of Sciences* 112 (51): 15574–15579. https://doi.org/10.1073/pnas.1503567112.

Roe, G.H., Montgomery, D.R., and Hallet, B. (2002). Effects of orographic precipitation variations on the concavity of steady-state river profiles. *Geology* 30 (2): 143–146. https://doi.org/10.1130/0091-7613(2002)030<0143:EOOPVO>2.0.CO;2.

Rousseau, T.C.C., Roddaz, M., Moquet, J.-S. et al. (2019). Controls on the geochemistry of suspended sediments from large tropical south American rivers (Amazon, Orinoco and Maroni). *Chemical Geology* 522: 38–54. https://doi.org/10.1016/j.chemgeo.2019.05.027.

Roy, S., Gaillardet, J., and Allègre, C.J. (1999). Geochemistry of dissolved and suspended loads of the Seine River, France: anthropogenic impact, carbonate and silicate weathering. *Geochimica et Cosmochimica Acta* 63 (9): 1277–1292. https://doi.org/10.1016/S0016-7037(99)00099-X.

Sarath, P.K., Mangalaa, K.R., Cardinal, D. et al. (2022). Seasonal, weathering and water use controls of silicon cycling along the river flow in two contrasting basins of South India. *Chemical Geology* 604: 120883. https://doi.org/10.1016/j.chemgeo.2022.120883.

Savenko, V.S. (2007). Chemical composition of sediment load carried by rivers. *Geochemistry International* 45 (8): 816–824. https://doi.org/10.1134/S0016702907080071.

Schaller, M., Ehlers, T.A., Lang, K.A.H. et al. (2018). Addressing the contribution of climate and vegetation cover on hillslope denudation, Chilean coastal cordillera (26°-38°S). *Earth and Planetary Science Letters* 489: 111–122. https://doi.org/10.1016/j.epsl.2018.02.026.

Sebastian, B. (1999). Texture, mineralogy and chemistry of beach and river sands, southern parts of Kerala and Tamil Nadu. PhD thesis. University of Kerala, Thiruvananthapuram.

Sensarma, S., Rajamani, V., and Tripathi, J.K. (2008). Petrography and geochemical characteristics of the sediments of the small river Hemavati, southern India: implications for provenance and weathering processes. *Sedimentary Geology* 205 (3–4): 111–125. https://doi.org/10.1016/j.sedgeo.2008.02.001.

Singh, P. (2010). Geochemistry and provenance of stream sediments of the Ganga River and its major tributaries in the Himalayan region, India. *Chemical Geology* 269 (3–4): 220–236. https://doi.org/10.1016/j.chemgeo.2009.09.020.

Snyder, N.P., Whipple, K.X., Tucker, G.E., and Merritts, D.J. (2000). Landscape response to tectonic forcing: digital elevation model analysis of stream profiles in the Mendocino triple junction region, northern California. *Geological Society of America Bulletin* 112 (8): 1250–1263. https://doi.org/10.1130/0016-7606(2000)112<1250:LRTTFD>2.0.CO;2.

Soil Survey Staff (2014). *Keys to Soil Taxonomy*. Washington: United States Department of Agriculture, Natural Resources Conservation Service.

Soman, K. (2002). *Geology of Kerala*. Bangalore: Geological Society of India.

SSO (2007). *Benchmark Soils of Kerala*. Thiruvananthapuram: Soil Survey Organization, Department of Agriculture, Government of Kerala.

Stock, G.M., Frankel, K.L., Ehlers, T.A. et al. (2009). Spatial and temporal variations in denudation of the Wasatch Mountains, Utah, USA. *Lithosphere* 1 (1): 34–40. https://doi.org/10.1130/L15.1.

Sun, D., Bloemendal, J., Rea, D.K. et al. (2002). Grain-size distribution function of polymodal sediments in hydraulic and aeolian environments, and numerical partitioning of the sedimentary components. *Sedimentary Geology* 152 (3–4): 263–277. https://doi.org/10.1016/S0037-0738(02)00082-9.

Tarras-Wahlberg, H., Everard, M., and Harper, D.M. (2002). Geochemical and physical characteristics of river and lake [2pt] sediments at Naivasha, Kenya. *Hydrobiologia* 488 (1–3): 27–41. https://doi.org/10.1023/A:1023301808624.

Thampi, P.K (1987). Geology of Munnar Granite, Idukki district, Kerala, India. PhD thesis. University of Kerala, Thiruvananthapuram.

Thampi, P.K., Nair, P.K.R., and Balasubramonian, G. (1993). Petrochemistry and tectonic evolution of Munnar granite, Idukki district, Kerala. *Current Science* 64: 238–243.

Thomas, J. (2012). Channel characteristics of two upland river basins of contrasting climate: a study from Kerala. PhD thesis. University of Kerala, Thiruvananthapuram.

Thomas, J., Joseph, S., and Thrivikramji, K.P. (2015a). Discriminant analysis for characterization of hydrochemistry of two mountain river basins of contrasting climates in the southern Western Ghats, India. *Environmental Monitoring and Assessment* 187 (6): 365. https://doi.org/10.1007/s10661-015-4589-0.

Thomas, J., Joseph, S., and Thrivikramji, K.P. (2015b). Hydrochemical variations of a tropical mountain river system in a rain shadow region of the southern Western Ghats, Kerala, India. *Applied Geochemistry* 63: 456–471. https://doi.org/10.1016/j.apgeochem.2015.03.018.

Thomas, J., Joseph, S., and Thrivikramji, K.P. (2015c). Hydrogeochemical drivers and processes controlling solute chemistry of two mountain river basins of contrasting climates in the southern Western Ghats, India. In: *Environmental Management of River Basin Ecosystems* (ed. M. Ramkumar, K. Kumaraswamy, and R. Mohanraj), 355–396. Cham: Springer International Publishing https://doi.org/10.1007/978-3-319-13425-3_17.

Thompson, D.M. (2013). Pool-riffle. In: *Treatise on Geomorphology* (ed. J. Shroder and E. Wohl). San Diego: Academic Press https://doi.org/10.1016/B978-0-12-374739-6.00246-3.

Tucker, G.E. (2015). Landscape evolution. In: *Treatise Gn Geophysics* (ed. G. Schubert), 593–630. New York: Elsevier https://doi.org/10.1016/B978-0-444-53802-4.00124-X.

Valentine, P.C. (2019). Sediment classification and the characterization, identification, and mapping of geologic substrates for the glaciated Gulf of Maine seabed and other terrains, providing a physical framework for ecological research and seabed management, Scientific Investigations Report. US Geological Survey, Reston.

Viers, J., Dupre, B., and Gaillardet, J. (2009). Chemical composition of suspended sediments in world Rivers: new insights from a new database. *Science of the Total Environment* 407 (2): 853–868. https://doi.org/10.1016/j.scitotenv.2008.09.053.

Wang, R.-M., You, C.-F., Chung, C.-H. et al. (2022). Uranium isotopes in a subtropical mountainous river of Taiwan: insight into physical and chemical weathering processes. *Journal of Hydrology* 607: 127481. https://doi.org/10.1016/j.jhydrol.2022.127481.

White, A.F. and Blum, A.E. (1995). Effects of climate on chemical weathering in watersheds. *Geochimica et Cosmochimica Acta* 59 (9): 1729–1747. https://doi.org/10.1016/0016-7037(95)00078-E.

Whitmore, G.P., Crook, K.A.W., and Johnson, D.P. (2004). Grain size control of mineralogy and geochemistry in modern river sediment, New Guinea collision, Papua New Guinea. *Sedimentary Geology* 171 (1–4): 129–157. https://doi.org/10.1016/j.sedgeo.2004.03.011.

Wohl, E.E., Vincent, K.R., and Merritts, D.J. (1993). Pool and riffle characteristics in relation to channel gradient. *Geomorphology* 6 (2): 99–110. https://doi.org/10.1016/0169-555X(93)90041-Y.

Yang, S., Jung, H.-S., and Li, C. (2004). Two unique weathering regimes in the Changjiang and Huanghe drainage basins: geochemical evidence from river sediments. *Sedimentary Geology* 164 (1–2): 19–34. https://doi.org/10.1016/j.sedgeo.2003.08.001.

Zimmermann, A.E. (2013). Step-pool channel features. In: *Treatise on Geomorphology* (ed. J. Shroder and E. Wohl), 346–363. San Diego: Academic Press https://doi.org/10.1016/B978-0-12-374739-6.00245-1.

Zimmermann, A. and Church, M. (2001). Channel morphology, gradient profiles and bed stresses during flood in a step-pool channel. *Geomorphology* 40 (3–4): 311–327. https://doi.org/10.1016/S0169-555X(01)00057-5.

5

Crucial Interplay of Microbial Communities in Controlling the Geogenic Processes

Aseem Kerketta[1], *Joystu Dutta*[2,3], *Shristy S. Swarnkar*[1], *Amrita K. Panda*[1], *and Ashish Kumar*[1]

[1] *Department of Biotechnology, Sant Gahira Guru Vishwavidyalaya Sarguja, Ambikapur, Chhattisgarh, India*
[2] *Department of Environmental Science, Sant Gahira Guru Vishwavidyalaya Sarguja, Ambikapur, Chhattisgarh, India*
[3] *IUCN Commission on Ecosystem Management, Gland, Switzerland*

5.1 Introduction

Weathering describes the physical breaking or chemical dissolution or microbial action of rocks and minerals on the surface of the Earth. Different agents of weathering include different forms of water, acids and salts, flora, fauna, microbial activities, and temperature variations due to changing climate conditions. Erosion transports the weathered rocks and minerals from one place to another. The processes of weathering and erosion are unavoidable for a rock or a mineral on the surface of the Earth. Together, these processes craft various landmarks across the Earth, such as the Grand Canyon in the U.S. state of Arizona. The humongous canyon is 446 km long, 29 km wide, and more than 1600 m deep. Geogenic processes are dynamic that constantly alter and shape the landform features on Earth. The exposed surfaces of Earth, be it hilltops, mountain peaks, exposed rock-beds, valleys, or river channels, get worn away as a result of active processes of weathering. The time of exposure of a rock is directly proportional to its vulnerability to weathering. Extrusive Igneous rocks are often buried beneath rock layers and are lesser vulnerable to weathering and erosion agents as well as consequences of global climate change. The pedogenesis (soil formation mechanism) is initiated by weathering processes as it reduces larger rock and minerals to fine particles with due course of time. A principal factor for soil formation is the mixture of weathered minerals with plants, animal remains, fungi, bacteria, and other organisms to develop the organic matter. Monotypic weathered rock lacks essential minerals required for fertile soil formation, while end products of multiple rock clusters are much richer in mineral diversity, contributing to a fertile soil. Active weathering processes include formation of various soil types as diverse as loess, glacial till, and alluvial sediments. Rocks crushed under active geogenic processes are less susceptible to nutrient leaching. The process is more sustainable and cheaper than the production of conventional

Weathering and Erosion Processes in the Natural Environment, First Edition. Edited by Virendra Bahadur Singh, Sughosh Madhav, Naresh Chandra Pant, and Ravi Shekhar.

fertilizers (Fyfe et al. 2006; Hazra 2016; Manning and Theodoro 2020). During weathering processes, microbial fauna, such as root-associated bacteria, actively helps in extracting and releasing nutrients from different soil minerals thus actively contributing to soil fertility and sustainable agricultural practices. Scientists have identified that bacterial strain *Burkholderia glathei* PML1(12) not only promoted mineral weathering actively but also ensured consistent growth in plants. This is a sustainable and alternative method identified by agricultural microbiologists (Calvaruso et al. 2006). Available scientific literature points out that the physicochemical changes in the rhizospheric soil, induced by roots and their associated microflora, influence the weathering processes and nutrient uptake by plants (Marschner et al. 1986; Jones and Darrah 1994; Hinsinger 1998; Landeweert et al. 2001; Jones et al. 2003). This is an economically feasible way to improve the macronutrient availability from soil and facilitate the process of rock solubilization required for amendment procedures during elaborate soil-restoration mechanisms.

5.2 Mechanical/Physical Weathering

Physical rock breakdown popularly termed as mechanical weathering comprises all in situ rock breaking processes facilitated by several agents. The process is a crucial rate-limiting factor for a wide range of geomorphological processes (Heimsath et al. 1997; Kirchner et al. 2006; McAuliffe et al. 2006; McFadden 2013; Collins and Stock 2016; Lebedeva and Brantley 2017; Shobe et al. 2017). However, literatures with empirical examples and case studies regarding the drivers of mechanical weathering on Earth are limited and call for scientific attention. It is being widely considered as a natural Earth-building process by geologists across the globe. Water vapor is one of the principal agents of mechanical weathering in rocks. Water in its different forms, such as snowflakes, vapor, ice, and running flows, contributes to major shares of mechanical weathering on Earth surface. Scientists have established strong correlation between humidity variations and weathering as well as regolith production. Scientists have time and again tried to establish relationships between water and different types of weathering that shape landforms on Earth (Griggs 1936; Hall 1986; Matsukura and Takahashi 2000; Sass 2005; Burnett et al. 2008; Owen et al. 2011; Larsen et al. 2014; Haas et al. 2015). The direct mechanism of water associated weathering and erosion requires further scientific understanding (Hall and Hall 1996; Halsey et al. 1998; Sass 2005; Burke et al. 2007; Mol and Viles 2012). Freezing and thawing dominate mechanical weathering processes in temperate and tundra biomes (Gerber and Scheidegger 1969; Small et al. 1999; Wells et al. 2005; Humphreys and Wilkinson 2007; Elliott 2008; Moores et al. 2008; West et al. 2014). The liquid form of water enters into crevices of rocks. During lower temperature at night, water freezes. Freezing water due to increased volume expands the crevices and splits the rocks. Melting ice further facilitates transportation of weathered particles. The alternate freezing and thawing is a cyclical process termed as frost weathering or cryofracturing. The variations in diurnal and nocturnal temperature can also contribute to mechanical weathering in a process called thermal stress. Repeated heating and cooling for longer time period weakens the internal structure of a rock leading to its vulnerability to weathering and erosion. Rocky deserts are the most vulnerable landforms affected as a result of increasing temperatures during summers in tropics. Rocks get separated into thin sheets as a result of thermal stress leading to increasing fragility in a process called exfoliation. Curved plates of rocks are stripped from the mother rock during the exfoliation weathering process. Bornhardts considered to be weird landform features are formed as a result of exfoliation weathering. These are tall and isolated dome-shaped rock structures majorly found in rocky deserts. An iconic landmark in Rio de Janeiro, Brazil is Sugarloaf Mountain which is a bornhardt. Exfoliation weathering is also facilitated by variations in atmospheric pressure and tectonic movements. Absorption of water by clays near rock contributes to another form of mechanical weathering. Clay is more porous than rock and swell with water absorption. The unequal swelling and contraction result in rock breaking with time. Salt is also a major contributing factor in the process of weathering termed as haloclasty. Saltwater intrusion in the rock crevices and soil particles led to precipitation of the salt crystals and evaporation of the soil water with rising diurnal temperatures, resulting into further pressure in the internal rock structures leading to weathering of the rocks. Honeycomb weathering is closely associated with the process of haloclasty. The varied rock formations with innumerable number of pits formed by salt crystal growth result in honeycomb weathering which is very common in coastal areas and near to tidal creeks where save wave interaction with rock structures continuously results in weathering and fragility of the rocks. Salt upwelling expands underground salt domes, resulting in weathering of the overlying rock. The faults, joints, and bedding planes that form the foundation of geological structures in the ancient city of Petra, Jordan are made out of tectonic Earth movements. Salt upwelling from beneath the ground further adds to the destruction of ancient monuments in the area. Flora and fauna are known agents of mechanical weathering. Seeds falling in rock crevices grow into a mighty plant ultimately breaking large rocks. Bryophytes and Pteridophytes prominently enlarge rock crevices and faults as they grow. Various burrowing animals, like rats, moles, prairie dogs, and mongoose that tunnel underground, ensure mechanical weathering of rocks and minerals.

5.3 Chemical Weathering

Chemical weathering permanently alters the molecular arrangement of geological structures. The classic example being the atmospheric gas carbon dioxide present in the air or soil which chemically interacts with water in a process called carbonation leads to the formation of a weak acid known as carbonic acid which dissolves rock. Chemical weathering is the process of elemental fractionation ex-situ from parental bedrock on Earth surface. Interaction of chemicals leading to chemical reaction in rock changes the chemical composition of rocks and minerals. Water actively interacts with minerals leading to a series of chemical reactions and transformation of rocks and minerals. Metamorphic rocks are generally formed as a result of this. Chemical weathering is gradual and consistent process as rock mineralogy adjusts to the near-surface environment. The formation of secondary minerals from primary minerals is also a part of chemical weathering process. Oxidation and hydrolysis are the ubiquitous chemical processes that govern chemical weathering. The mechanism is further enhanced by various geological and biological agents as the acids produced by microbial communities and plant root metabolism.

Anthropogenic impacts leading to air pollution accelerates the frequency and intensity of weathering. Fossil fuels from vehicular exhaust, burning coal from industrial emissions, releases Nitrogen Oxides (NO_X) and Sulphur Oxides (SO_X) into the atmosphere. The chemical reaction of industrial chemicals with sunlight and moisture leads to acid formation and discharges into the atmosphere. The admixture of acids in rain water results in acid rain formation with pH lower than 5.6. Acid rain damages metamorphic rocks and sedimentary rocks. Ancient monuments and sculptures across the globe and majorly in industrial and polluted cities of the world are affected by acid rain. Acid rain corrodes marbles, sandstone, and limestone structures. A striking impact of acid rain and air pollution mediated damage of ancient monument is seen in one of the most prolific and iconic wonders of the world from India, the majestic Taj Mahal of Agra. The panoramic and exquisite marble structure developed yellow coloration and weathering as a result of joint interplay of air pollution and acid rain.

5.4 Biological Weathering

Weathering is a general term that describes the process of breakdown of the parent rock and its transformation into soil, clay, sand, and other substances (Zaharescu et al. 2020). Biological weathering, also known as biogenic weathering (Roger D. Finlay et al. 2020) or biotic weathering (Viles 1995) or sometimes also referred to as organic weathering (Frasca and Lama 2018), is a broad-range phenomenon that collectively involves the synergistic activity of plants, animals, and microbes to accomplish the process of weathering of rocks. Another process that is similar to biological weathering process is bioerosion; however, it involves breaking down and fractionation of any hard substance (rocks or non-rocks) into smaller constituent components (Davidson et al. 2018), and essentially, it's transportation to secondary location. Bioweathering, on the other hand, is mainly concerned with breaking down of hard substrata in situ. The overview of biological weathering processes is detailed in a schematic diagram depicted in Figure 5.1.

5.5 Weathering by Plants

Plants are considered as the most versatile agent in the weathering process as they inhabit a variety of rocks, thrive in diverse geo-climatic conditions, are part of many ecological processes and account largely for the process of weathering and mineralization (Arcand and Schneider 2006; Bai et al. 2012; Arkaduisz et al. 2014; Thakur et al. 2014; Dontsova et al. 2020; Harrison et al. 2020). Plants also significantly influence soil hydrology directly or indirectly and thus may alter the soil enzyme activity and may create a specific microclimatic condition different from the surrounding (Borken and Matzner 2009). The pressure exerted by slowly but steadily penetrating and expanding roots may result in weakening and rupturing of huge parent rock. Also, plant exudates result in concentration of organic matter in soil and increase the activity as well as diversity of soil microorganisms which further increases the rate of weathering (Zhao et al. 2013b).

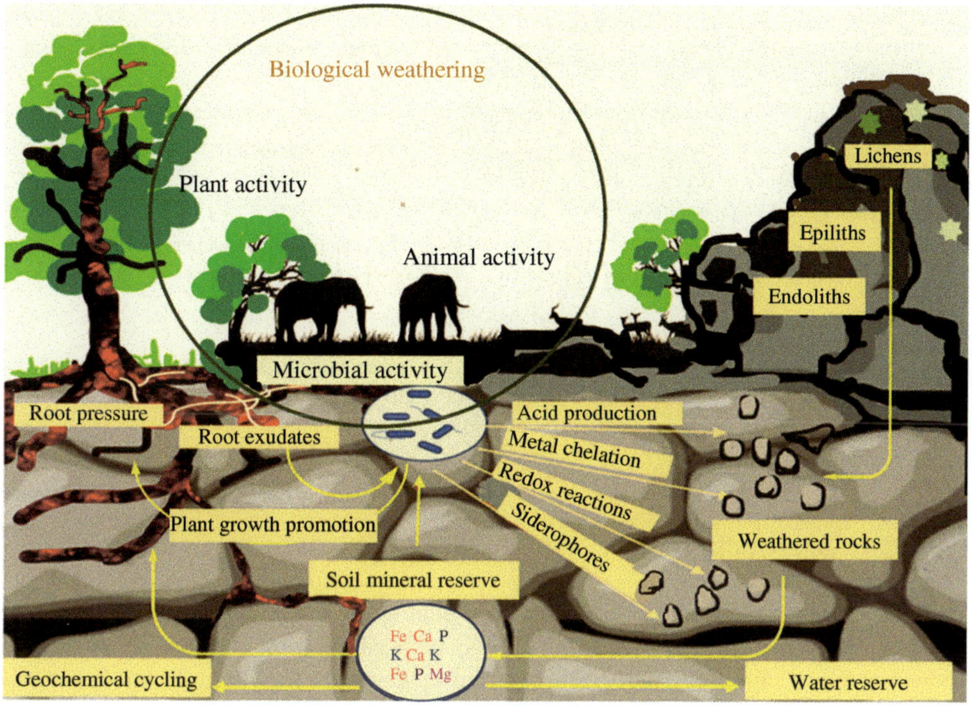

Figure 5.1 Overview of biological weathering process.

5.6 Weathering by Animals

The influence of animals on the process of weathering and erosion have been well accredited by researchers, both in terms of the wide range of animals involved and in terms of the different ways in which animals can mediate the process of weathering (Butler 1995). Excluding the impact of anthropogenic sources, the direct activities of animals responsible for weathering mainly involve changes in the landscape by burrowing and digging for food and shelter, grazing, stamping, urination and defecation, littering, etc. Indirectly, animals may affect the hydrodynamics of land by opening the ground to climatic and geomorphic influences e.g. rain water, wind erosion, and by increasing the microbial activity (Evans 1998). Major representative fauna includes digging and burrowing animals, such as rats, porcupines (*Hystrix* spp.), earthworms (*Lumbricus*), Grizzly bears (*Ursus arctos horribilis*), Marmots *(Marmota)*, Grooved-toothed rats (*Otomys* spp.), etc.

5.7 Microbial Weathering

Though microscopic in dimensions, microbes represent a multifarious group of organisms that play a crucial role in large-scale weathering of bedrocks. Microbiota mainly involved in weathering have been studied thoroughly and mainly involve lichen, fungi, bacteria,

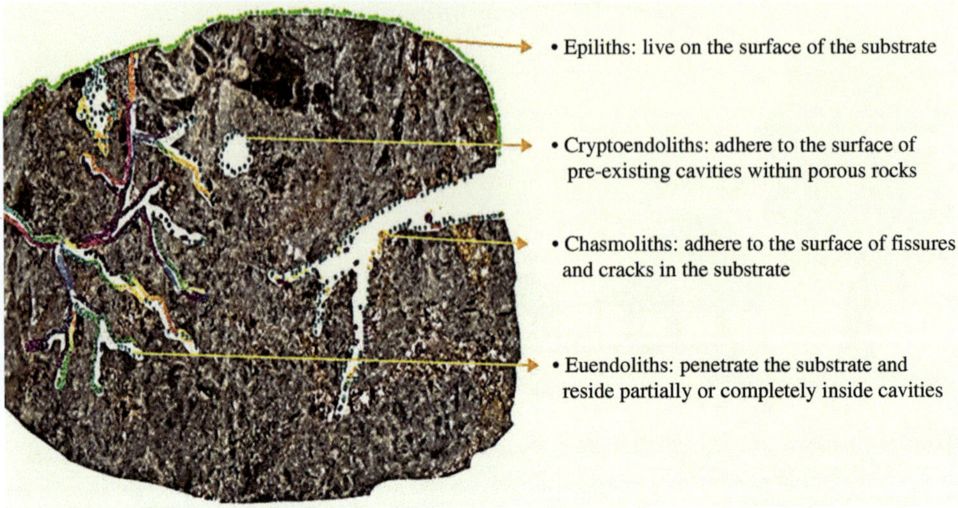

Figure 5.2 Types of lithobionts.

cyanobacteria, microalgae, and diatoms (Chen et al. 2000; Konhauser 2006; Mustoe 2018). The contribution of microbial weathering may be minimal from quantitative standpoint; however, microbial weathering and mineralization have very serious implications in terms of ecological service perspectives (Gadd 2007). Moreover, microbes play a very significant role in biogeochemical cycling (Gault et al. 2011).

Microbes surviving on rocks surfaces are technically defined as lithobionts (rock-dwellers) (Mojumdar et al. 2021) as depicted in Figure 5.2. There are many types of lithobionts further explained through this figure. Lithobionts may be restricted to the surface of rocks (epiliths) or may occupy niches under the rock surface (endoliths) (Golubic et al. 1981; Gaylarde et al. 2012). Some endoliths actively bore into rock (euendoliths), while others occupy preexisting pores and cavities (cryptoendoliths), and/or cracks and fissures connected to the surface (chasmoendoliths) (Dievart et al. 2022). Lithobionts are living on Earth since the Proterozoic era which dates back to some 2500 to 541 million years ago and have played an important role in the development and evolution of life, as well as in the production and destruction of carbonates. The study of microorganisms is an indispensable aspect to explain their role in weathering processes and is also clear to work out culturable and unculturable diversity of soil inhabitant microflora. Weathering of rock is one of the most significant geochemical processes which is taking place on the Earth and results into the pedogenesis, maintenance of soil productivity, regulating atmospheric composition, and global climate change.

5.8 Mechanisms of Microbial Weathering

For weathering of rocks and minerals, microbes employ various mechanisms with different specificity and efficacy on diverse geological substrates. Breaking down the mechanisms into categories will be useful for the better understanding of microbial interactions (Gadd 2010; Uroz et al. 2009). The overview of microbial weathering is depicted in Figure 5.3.

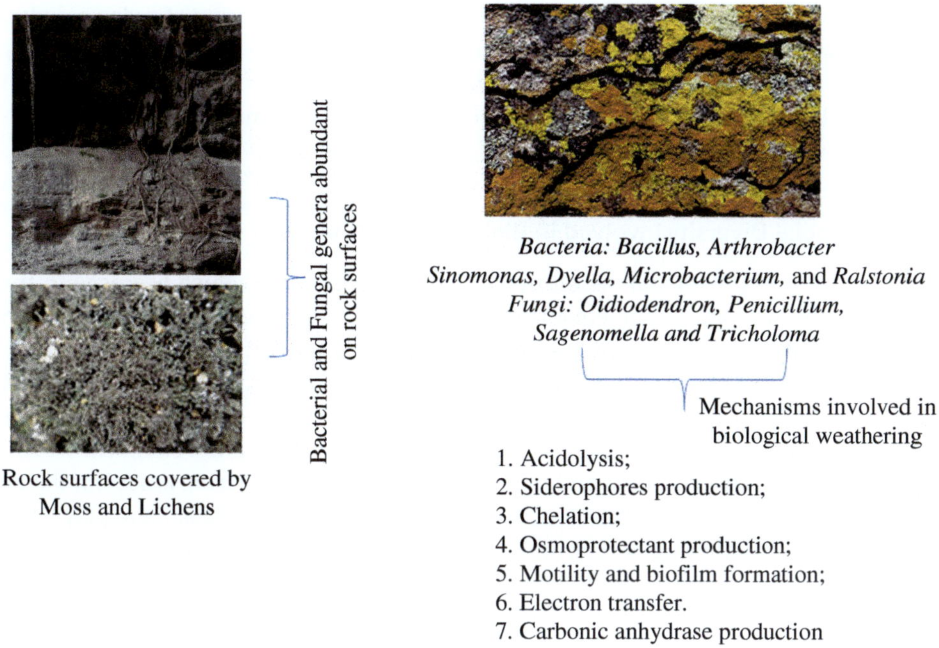

Figure 5.3 Microbial weathering mechanisms. *Source:* © American Geosciences Institute and used with their permission.

1) Change in pH neighboring the mineral, at either microscopic or mass scale, is a major endowment to mineral and rock weathering (Drever and Stillings 1997). pH of the surrounding environment can be substantially altered by the growth and activity of microorganisms. Both increase and decrease in pH can improve weathering processes. Organic acids produced as a by-product of cellular metabolism, such as citric acids, oxalic acids, amino acids, or phenolic acids, can make a direct contribution to mineral dissolution and rock weathering.
2) Chelation, elimination of heavy metals from core of a mineral or a bulk rock, is another prime weathering process (Liermann et al. 1999; Włodarczyk et al. 2015). Organic acids can also act as powerful chelators of heavy metals (Bennett et al. 2001). The potential of organic acids to attach to a heavy metal cation immobilizing it and potentially bring out as a mineral product is known as chelation. The secretion of organic acids, also known as organic ligands (in this context) into geological environments by microbes, can therefore facilitate the release of heavy metals from rocks and minerals (Ahmed and Holmström 2014; Bray et al. 2015; Saad et al. 2017).
3) Redox reactions muddled in weathering, oxidation, or reduction of a specific element within a mineral. For all biological metabolisms, energy is driven through redox reactions, the transfer of an electron between a redox couple, made up of an electron donor and an electron receiver. The electron is transferred from a higher energy state in the electron donor to a lower energy state in the electron receiver, and the loss of energy is utilized for the formation of adenosine triphosphate (ATP), the universal energy storage

molecule of life. Microbial metabolisms make use of a diverse array of redox couples, many of which are used by rock weathering organisms.

4) Breakdown of organic compounds, weathering of geological organic matter present abundantly in some sedimentary rocks, such as shales, is the final category of weathering mechanisms.

5.8.1 Microbial Diversity Involved in Biological Weathering

The process of breakdown and dissolution of rocks and minerals which begins with the determined physical agents and chemical processes, directing the formation of regolith (parent material), is known as weathering. A multiplex interplay of physical, chemical, and biological activities takes part in rock weathering processes. The rocks are broken down into smaller pieces in physical weathering, whereas the disintegration process continues through chemical reactions on the rock surfaces during the chemical weathering. The disintegration of rocks and minerals in biological weathering is accomplished mainly by living agents. Microbes associated with plant root and rhizosphere play a key role in biological weathering of rock and supply diverse inorganic compounds/elements which are prerequisite nutrients for plants (Chang and Li 1998). Microorganisms mediated weathering of crucial types of rocks results into the releasing various nutrients (Calvaruso et al. 2006). The expansion of plant root system creates big cracks in rock materials, where the microorganism forms biofilms and that eventually contributes to the weathering of the rocks. The mineral weathering by microorganisms is catalyzed by various mechanisms, such as acidolysis, siderophore production, chelation, osmoprotectant production, biofilm formation, etc. The potential role of carbonic anhydrase in the dissolution of wollastonite ($CaSiO_3$) was demonstrated by Xiao et al. 2014. A recent study reported the effect of a laboratory biofilm of the phototrophic cyanobacterium *Nostoc punctiforme* ATCC 29133 and the rock-inhabiting ascomycete *Knufia petricola* CBS 726.95 on mineral weathering (Seiffert et al. 2014). Mosses and lichens on rock surfaces also exhibit mutualistic symbiotic relation with microorganism present on rock surfaces, and on the basis of culture-independent analysis, it was found that the α-diversity of bacterial communities inhabiting rock surfaces increased considerably due to weathering of rock forming silicates (Wang et al. 2021a). *Actinobacteria*, *Acidobacteria*, and *Proteobacteria* have been reported as dominant phyla on rock surfaces, e.g. granite, apatite, marine basalts (Colin et al. 2017; Wang et al. 2021b), representing their adaptability to colonize the rock surfaces. Few most abundantly reported bacterial genera with their specific mineral weathering ability has now been known (Table 5.1). Biological weathering bacteria have been identified from various ecosystems: rhizosphere of many plants, rock surfaces, glaciers, etc. (Table 5.2). The role of fungi in determining soil fertility cannot be ignored and a large structural diversity of fungi is involved in weathering activities. Fungal biotransformation of vanadinite and mimetite by *Aspergillus niger* was reported by Cecia et al. (2015). Daghino et al. (2008) explained the involvement of *Paecilomyces lilacinus, Aspergillus fumigatus,* and *Verticillium leptobactrum* in the weathering of serpentine rocks. Another study also found the occurrence of *Trichoderma, Penicillium, Pestalotiopsis,* and *Fusarium* on limestone surfaces (Ortega-Morales et al. 2016).

Table 5.1 Few Bacterial Taxa reported mostly in mineral weathering.

Minerals	Bacteria	References
Sulfur	*Desulfosporosinus spp, A. ferrooxidans,* and *A. thiooxidans*	Mardanov et al. (2016)
Iron	*A. ferrooxidans, Leptospirillum* spp. *Acidimicrobiaceae (Acidimicrobium, idimicrobium, Ferrimicrobium,* and *Ferrithrix), Metallibacterium,* and *Ferroplasma. Acidiphilium, Acidisphaera, Acidimicrobiaceae,* and *Metallibacterium.*	Sánchez-Andrea et al. (2011); Johnson (1998); Hedrich and Schippers (2020); Kucera et al. (2020)
Lead and Zinc	*ferrooxidans* and *Leptospirillum* spp. (*Fp. acidiphilum,*)	Huang et al. (2011)
Cadmium and Nickel	*Clostridium* sp.	Francis and Dodge (1990)
Silicate	*Vermiculites* *Rhizobium* sp. Q32 *Paenibacillus* sp.	Ivanova and Chizhikova (2014); Zhao et al. (2013a); Liu et al. (2012)
Uranium	*Microbacterium* sp. *6–1*	Xia et al. (2020)
Copper	*Acidithiobacillus thiooxidans*	Hong and Valix 2014

Table 5.2 Habitat of few mineral weathering microorganisms.

Habitat	Microbial Taxa	Reference
Rhizosphere	*Boletus, Discosia, Exophiala, Mortierella, Penicillium, Pseudotomentella, Tricholoma,* and *Wilcoxina*	Anna Gałązka et al. (2022).
Mangrove rhizosphere	*Bacillus, Enterobacter* *Kluyvera, Paenibacillus* *Pseudomonas, Xanthobacter*	Vazquez et al. (2000)
Cactus rhizosphere	*Bacillus, Citrobacter*	Puente et al. (2004)
Mycorrhizosphere (Oak, beech, spruce)	*Caballeronia, Collimonas Paraburkholderia, Pseudomonas*	Nicolitch et al. 2016)
Rock crack	*Boletus, Exophiala, Oidiodendron, Penicillium, Sagenomella,* and *Tricholoma*	Anna Gałązka et al. (2022)
Rock surface	*Pantoea* and *Pseudomonas*	Wang et al. (2021b)
Glacier	*Polaromonas, Massilia* Glacier algae	Krauze et al. (2021); Tedstone et al. (2020)
Polar desert	*Trebouxia* sp., *Gloeocapsa* sp., *Hormathonema* sp., *Anabaena* sp., *Aphanocapsa* sp., *and Chroococcidiopsis* sp.	Omelon (2008)

5.8.2 Applications of Omics Technologies to Understand the Mechanism of Weathering

The evolutionary history, functional and ecological biodiversity, can be understood by genomic analysis of complex environmental samples. The main purpose of identification of unknown specimens is to generate vast DNA libraries which can be employed by DNA barcoding which further enrolls standardized species-specific genomic regions of DNA. 16S ribosomal RNA sequences are generally used for identification of bacteria (Welch and Huse 2011). However, for fungal studies, the internal transcribed spacer region of the nuclear ribosomal DNA is employed (Nilsson et al. 2008), while the regions of plastid DNA, including maturase K (matK) and ribulose-bisphosphate carboxylase (rbcl), required for the plant DNA barcoding (Burgess et al. 2011). The conventional sequencing method is most efficient for the development of huge DNA barcode reference libraries, but due to higher number of environmental samples, it is afar the scope of its capability. The use of Omics technologies simplifies many queries related to the genetic basis of mineral weathering process Figure 5.4. A recent study adopted quantitative proteomics technology to demonstrate that more oxidoreductases were secreted to enhance the transformation of serpentine and proved the role of various proteins to accelerate the weathering of serpentine (Liu et al. 2021). Another study used transcriptomics and proteomics to identify genes and proteins differentially expressed in the presence of minerals (Mg and Fe) and explained the mechanism of weathering of *Caballeronia mineralivorans* strain PML1(12) (Uroz et al. 2020).

Transcriptomics analysis also provides insights how various fungi interact with minerals and obtain required minerals from their respective environment through biological weathering. The transcriptomic sequencing of ectomycorrhizal fungus *Amanita pantherina* explained the upregulation of various genes, including K transporters during the

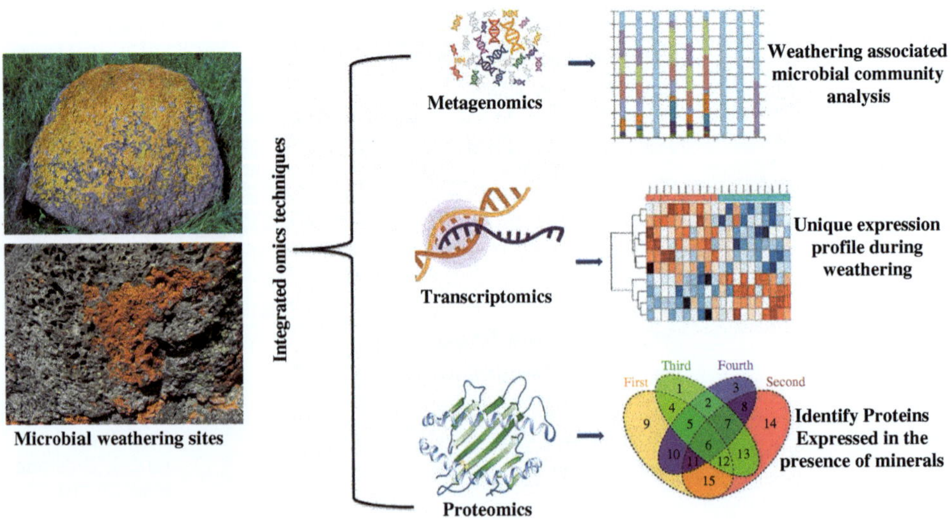

Figure 5.4 Omics approach to understand weathering microbial community. *Source:* Rosser1954 / Wikimedia Commons / Public domain; Andrewtappert / Wikimedia Commons / CC BY-SA 3.0.

weathering of K-containing feldspar and apatite (Sun et al. 2019). They also explained that due to this unique expression profile this fungus helps woody plants to uptake nutrients from forest facilitating augmented yield in sustainable forestry. An RNA sequencing study of *Aspergillus niger* unravels the mechanism of weathering of potassium feldspar and indicated the upregulation of genes Na^+, K^+-ATPase (447.6 multiples), cysteine synthase (9 multiples), cystathionine beta-synthase (5.6 multiples), and glutathione synthase (3.5 multiples) during the process (Wang et al. 2015).

5.9 Conclusion

The role of geomicrobiology is immense in ensuring the smooth functioning of geogenic processes. Microorganisms are omnipresent and play an important role in soil formation, nutrient recycling, and weathering by secreting organic acids and various enzymes. Microbial activities during element recycling maintain humus formation, stabilization, and particle aggregation of soils and endorse the soil fertility. Micobe added activities also help in release of nutrients from rocks and it pays a lot to our environment. In this chapter, we summarized mineral weathering process with emphasis on microbial weathering and functioning of microbial diversity in the mineral dissolution process. Mineral weathering bacteria have been studied and identified from various environments (rhizosphere, rock surface, rock crack, desert, glaciers, etc.), especially abundant on rhizosphere of trees and rock surfaces. The microorganisms live in symbiotic relationship in the rhizospheric region and interact with roots of many perennial and annual plants and ultimately help in nutrient mobilization. The acids produced by tree roots in nutrient limiting conditions causes dissolution of rocks and forms corrosion channels. Many studies reported the increased microbial activity in the rhizosphere region impacts directly on the mineral weathering process. The combinatorial approach of culture dependent and independent studies reported that the weathering potential of microbial communities increased by the presence of underneath lichens and mosses. The nutrients are released after rock weathering activities of microbes utilized by lichens and mosses and, in turn, they provide photosynthates to microbes on nutrient limiting rock surfaces. However, the role of microbes in such complex processes needs proper research and documentation. This chapter is a humble endeavor in that direction and reviewed the importance of Omics technology to disentangle the mineral and microbial interactions. By the advent of advanced transcriptomics and proteomics approaches, many differentially expressed genes have been identified which literally order the events activated by microbes during mineral weathering.

More experiment-based observations are required to understand the role of microbial communities, to understand weathering processes that is consistently occurring across the lithosphere and biosphere, thus dynamically influencing the interactions between geological and microbiological world. Mineral weathering microorganisms may be a better alternate to inorganic fertilizers and can improve crop production and plant growth sustainably by releasing nutrients sealed inside rocks in the inaccessible forms. The use of these efficient organisms reduces the risk of eutrophication related problems while using chemical fertilizers. Fungal mediated weathering has been emerged as a budding bioremediation strategy for asbestos contaminated soils. The exudates secreted from fungi reduce the toxicity of commonly used asbestos by the removal of iron content and also reduces the DNA damaging impacts of asbestos. A recent study also reported that the application of silicate to agriculture soil enhance carbon sequestration by silicate weathering. The application of easily weatherable silicates, such as olivine, volcanic rocks, and igneous rocks, etc. enhances the CO_2 consumption. Though few studies reported that enhanced weathering reduces the atmospheric carbon dioxide and increased agricultural sustainability through long-term carbon sequestration, the bioprospecting potential of weathering microbes in the field of carbon sequestration needed to be studied further.

References

Ahmed, E. and Holmström, S.J. (2014). Siderophores in environmental research: roles and applications. *Microbial Biotechnology* 7 (3): 196–208.

Arcand, M.M. and Schneider, K.D. (2006). Plant-and microbial-based mechanisms to improve the agronomic effectiveness of phosphate rock: a review. *Anais da Academia Brasileira de Ciências* 78: 791–807.

Arkaduisz, N., Nowak, S., Nobis, N., and Nobis, A. (2014). Vegetation of solid rock faces and fissures of the alpine and subnival zone in the Pamir Alai Mountains (Tajikistan, middle Asia). *Phytocoenologia* 44 (1–2): 81–101. https://doi.org/10.1127/0340-269X/2014/0044-0573.

Bai, J., Gao, H., Xiao, R. et al. (2012). A review of soil nitrogen mineralization as affected by water and salt in coastal wetlands: issues and methods. *CLEAN–Soil, Air, Water* 40 (10): 1099–1105.

Bennett, P.C., Rogers, J.R., Choi, W.J., and Hiebert, F.K. (2001). Silicates, silicate weathering, and microbial ecology. *Geomicrobiology Journal* 18 (1): 3–19.

Borken, W. and Matzner, E. (2009). Reappraisal of drying and wetting: effects on C and N mineralization and fluxes in soils. *Global Change Biology* 15: 808–824.

Bray, A.W., Oelkers, E.H., Bonneville, S. et al. (2015). The effect of pH, grain size, and organic ligands on biotite weathering rates. *Geochimica et Cosmochimica Acta* 164: 127–145.

Burgess, K.S., Fazekas, A.J., Kesanakurti, P.R. et al. (2011). Discriminating plant species in a local temperate flora using the rbcL+matK DNA barcode. *Methods in Ecology and Evolution* 2: 333–340.

Burke, B.C., Heimsath, A.M., and White, A.F. (2007). Coupling chemical weathering with soil production across soil-mantled landscapes. *Earth Surface Processes and Landforms* 32 (6): 853–873.

Burnett, B.N., Meyer, G.A., and McFadden, L.D. (2008). Aspect-related microclimatic influences on slope forms and processes, northeastern Arizona. *Journal of Geophysical Research* 113: F03002. https://doi.org/10.1029/2007JF000789.

Butler, D.R. (1995). *Zoogeomorphology: Animals as Geomorphic Agents*. Cambridge: Cambridge University Press 231 pp.

Calvaruso, C., Turpault, M.-P., and Frey-Klett, P. (2006). Root-associated bacteria contribute to mineral weathering and to mineral nutrition in trees: a budgeting analysis. *Applied and Environmental Microbiology* 72: 1258–1266. https://doi.org/10.1128/aem.72.2.1258-1266.2006.

Cecia, A., Kierans, M., Hillierd, S. et al. (2015). Fungal bioweathering of mimetite and a general geomycological model for lead apatite mineral biotransformations. *American Society for Microbiology Applied and Environmental Microbiology* 81 (15): 4955–4964. https://doi.org/10.1128/AEM.00726-15.

Chang, T.T. and Li, C.Y. (1998). Weathering of limestone, marble, and calcium phosphate by ectomycorrhizal fungi and associated microorganisms. *Taiwan Journal of Forest Science* 13: 85–90.

Chen, J., Blume, H.P., and Beyer, L. (2000). Weathering of rocks induced by lichen colonization – a review. *Catena* 39: 121–146.

Colin, Y., Nicolitch, O., Turpault, M.P., and Uroz, S. (2017). Mineral type and tree species determine the functional and taxonomic structure of forest soil bacterial communities. *Applied and Environmental Microbiology* 83: e02684-16.

Collins, B.D. and Stock, G.M. (2016). Rockfall triggering by cyclic thermal stressing of exfoliation fractures. *Nature Geoscience* 9: 395–400.

Daghino, S., Martino, E., Vurro, E. et al. (2008). Bioweathering of chrysotile by fungi isolated in ophiolitic sites. *FEMS Microbiology Letters* 285: 242–249. https://doi.org/10.1111/j.1574-6968.2008.01239.x.

Davidson, T.M., Altieri, A.H., Ruiz, G.M., and Torchin, M.E. (2018). Bioerosion in a changing world: a conceptual framework. *Ecology Letters* 21: 422–438. https://doi.org/10.1111/ele.12899.

Dievart, A.M., McQuaid, C.D., Zardi, G.I. et al. (2022). Photoautotrophic euendoliths and their complex ecological effects in marine bioengineered ecosystems. *Diversity* 14: 737. https://doi.org/10.3390/d14090737.

Dontsova, K., Balogh-Brunstad, Z., and Chorover, J. (2020). Plants as drivers of rock weathering. In: *Biogeochemical Cycles: Ecological Drivers and Environmental Impact* (ed. K. Dontsova, Z. Balogh-Brunstad, and G. Le Roux), 33–58. https://doi.org/10.1002/9781119413332.ch2.

Drever, J.I. and Stillings, L.L. (1997). The role of organic acids in mineral weathering. *Colloids and Surfaces A: Physicochemical and Engineering Aspects* 120 (1–3): 167–181.

Elliott, C. (2008). Influence of temperature and moisture availability on physical rock weathering along the Victoria land coast, Antarctica. *Antarctic Science* 20 (01): 61–67.

Evans, R. (1998). The erosional impacts of grazing animals. *Progress in Physical Geography* 22: 251–268.

Finlay, R.D., Mahmood, S., Rosenstock, N. et al. (2020). Biological weathering and its consequences at different spatial levels – from nanoscale to global scale. *Biogeosciences* 17: 1507–1533.

Francis, A.J. and Dodge, C.J. (1990). Anaerobic microbial remobilization of toxic metals coprecipitated with iron oxide. *Environmental Science & Technology* 24: 373–378.

Frasca, M.H.B.d.O. and Lama, E.A.D. (2018). Biological weathering. In: *Encyclopedia of Engineering Geology Encyclopedia of Earth Science Series* (ed. P.T. Bobrowsky and B. Marker). Cham: Springer https://http://doi.org/10.1007/978-3-319-73568-9-29.

Fyfe, W.S., Leonardos, O.H., and Theodoro, S.H. (2006). Sustainable farming with native rocks: the transition without revolution. *Anais da Academia Brasileira de Ciências* 78: 715–720. https://doi.org/10.1590/S0001-37652006000400007.

Gadd, G.M. (2007). Geomycology: biogeochemical transformations of rocks, minerals, metals and radionuclides by fungi, bioweathering and bioremediation. *Mycological Research* 111 (1): 3–49.

Gadd, G.M. (2010). Metals, minerals and microbes: geomicro- biology and bioremediation. *Microbiology* 156 (3): 609–643.

Gałązka, A., Marzec-Grządziel, A., Grządziel, J. et al. (2022). Fungal genetic biodiversity and metabolic activity as an indicator of potential biological weathering and soil formation – case study of towards a better understanding of earth system dynamics. *Ecological Indicators* 141: doi: 10.1016/j.ecolind.2022.109136.

Gault, A.G., Ibrahim, A., Langley, S. et al. (2011). Microbial and geochemical features suggest iron redox cycling within bacteriogenic iron oxide rich sediments. *Chemical Geology* 281 (1): 41–51.

Gaylarde, C., Gaylarde, P., and Neilan, B. (2012). Endolithic phototrophs in built and natural stone. *Current Microbiology.* 65: 183–188. https://doi.org/10.1007/s00284-012-0123-6.

Gerber, E. and Scheidegger, A. (1969). Stress-induced weathering of rock masses. *Eclogae Geologicae Helvetiae* 62 (2): 401–415.

Golubic, S., Friedmann, I., and Schneider, J. (1981). The lithobiontic ecological niche, with special reference to microorganisms. *Journal of Sedimentary Petrology* 51: 0475–0478.

Griggs, D.T. (1936). The factor of fatigue in rock exfoliation. *Journal of Geology* 44: 783–796.

Haas, T., Conway, S.J., and Krautblatter, M. (2015). Recent (late Amazonian) enhanced backweathering rates on Mars: paracratering evidence from gully alcoves. *Journal of Geophysical Research, Planets* 120: 2169–2189. https://doi.org/10.1002/2015JE004915.

Hall, K. (1986). Rock moisture content in the field and the laboratory and its relationship to mechanical weathering studies. *Earth Surface Processes and Landforms* 11 (2): 131–142.

Hall, K. and Hall, A. (1996). Weathering by wetting and drying: some experimental results. *Earth Surface Processes and Landforms* 21 (4): 365–376.

Halsey, D., Mitchell, D., and Dews, S. (1998). Influence of climatically induced cycles in physical weathering. *Quarterly Journal of Engineering Geology & Hydrogeology* 31 (4): 359–367.

Harrison, S., Spasojevic, M.J., and Li, D. (2020). Climate and plant community diversity in space and time. *PNAS* 117 (9): 4464–4470. https://doi.org/10.1073/pnas.1921724117.

Hazra, G. (2016). Different types of eco-friendly fertilizers: an overview. *Sustainability in Environment* 1 (1): ISSN 2470-637X (Print) ISSN 2470-6388 (Online). http://www.scholink.org/ojs/index.php/se.

Hedrich, S. and Schippers, A. (2020). Distribution of acidophilic microorganisms in natural and man-made acidic environments. *Current Issues in Molecular Biology* 40: 25–48.

Heimsath, A.M., Dietrich, W.E., Nishiizumi, K., and Finkel, R.C. (1997). The soil production function and landscape equilibrium. *Nature* 388 (6640): 358–361.

Hinsinger, P. (1998). How do a plant roots acquire mineral nutrients? Chemical processes involved in the rhizosphere. *Advances in Agronomy* 64: 225–265.

Hong, Y. and Valix, M. (2014). Bioleaching of electronic waste using acidophilic sulfur oxidising bacteria. *Journal of Cleaner Production* 65: 465–472.

Huang, L.-N., Zhou, W.-H., Hallberg, K.B. et al. (2011). Spatial and temporal analysis of the microbial community in the tailings of a pb-Zn mine generating acidic drainage. *Applied and Environmental Microbiology* 77: 5540–5544.

Humphreys, G.S. and Wilkinson, M.T. (2007). The soil production function: a brief history and its rediscovery. *Geoderma* 139 (1): 73–78.

Ivanova, E. and Chizhikova, N. (2014). Degradative crystal–chemical transformations of clay minerals under the influence of cyanobacterium-actinomycetal symbiotic associations. *Eurasian Journal of Soil Science (EJSS)* 3: 116.

Johnson, D. (1998). Biodiversity and ecology of acidophilic microorganisms. *FEMS Microbiology Ecology* 27: 307–317.

Jones, D.L. and Darrah, P.R. (1994). Role of root derived organic acids in the mobilization of nutrients from the rhizosphere. *Plant and Soil* 166: 247–257.

Jones, L.D., Dennis, P.G., Owen, A.G., and Van Hees, P.A.W. (2003). Organic acids behavior in soils—misconceptions and knowledge gaps. *Plant and Soil* 248: 31–41.

Kirchner, J.W., Riebe, C.S., Ferrier, K.L., and Finkel, R.C. (2006). Cosmogenic nuclide methods for measuring long-term rates of physical erosion and chemical weathering. *Journal of Geochemical Exploration* 88 (1): 296–299.

Konhauser, K.O. (2006). *Introduction to Geomicrobiology*. Wiley.

Krauze, P., Wagner, D., Yang, S. et al. (2021). Influence of prokaryotic microorganisms on initial soil formation along a glacier forefield on King George Island, maritime Antarctica. *Scientific Reports* 11 (1): 1–15.

Kucera, J., Lochman, J., Bouchal, P. et al. (2020). A model of aerobic and anaerobic metabolism of hydrogen in the extremophile *Acidithiobacillus ferrooxidans*. *Frontiers in Microbiology* 2020 (11): 610836.

Landeweert, R., Hoffland, E., Finlay, R.D. et al. (2001). Linking plants to rock: ectomycorrhizal fungi mobilize nutrients from minerals. *Trends in Ecology & Evolution* 16: 248–253.

Larsen, I.J., Almond, P.C., Eger, A. et al. (2014). Rapid soil production and weathering in the southern Alps, New Zealand. *Science* 343 (6171): 637–640.

Lebedeva, M.I. and Brantley, S.L. (2017). Weathering and erosion of fractured bedrock systems. *Earth Surface Processes and Landforms* 42: 2090–2108. https://doi.org/10.1002/esp.4177.

Liermann, L.J., Kalinowski, B.E., Brantley, S.L., and Ferry, J.G. (1999). Role of bacterial siderophores in dissolution of horn- blende. *Geochimica et Cosmochimica Acta* 64 (4): 587–602.

Liu, D., Lian, B., and Dong, H. (2012). Isolation of *Paenibacillus* sp. and assessment of its potential for enhancing mineral weathering. *Geomicrobiology Journal* 29 (5): 413–421.

Liu, H., Liu, X., Li, X. et al. (2021). The molecular regulatory mechanisms of the bacteria involved in serpentine weathering coupled with carbonation. *Chemical Geology* 565: 120069.

Manning, D.A.C. and Theodoro, S.H. (2020). Enabling food security through use of local rocks and minerals. *The Extractive Industries and Society* 7: 480–487. https://doi.org/10.1016/j.exis.2018.11.002.

Mardanov, A., Panova, I., Beletsky, A. et al. (2016). Genomic insights into a new acidophilic, copper-resistant *Desulfosporo sinus* isolate from the oxidized tailings area of an abandoned gold mine. *FEMS Microbiology Ecology* 92: 1–14.

Marschner, H., Romheld, V., Horst, W.J., and Martin, P. (1986). Root-induced changes in the rhizosphere: importance for the mineral nutrition of plants. *Zeitschrift für Pflanzenernährung und Bodenkunde* 149: 441–456.

Matsukura, Y. and Takahashi, K.I. (2000). A new technique for rapid and non-destructive measurement of rock-surface moisture content; preliminary application to weathering studies of sandstone blocks. *Developments in Geotechnical Engineering* 84: 47–54.

McAuliffe, J.R., Scuderi, L.A., and McFadden, L.D. (2006). Tree-ring record of hillslope erosion and valley floor dynamics: landscape responses to climate variation during the last 400 yr in the Colorado plateau, northeastern Arizona. *Global and Planetary Change* 50 (3): 184–201.

McFadden, L.D. (2013). Strongly dust-influenced soils and what they tell us about landscape dynamics in vegetated aridlands of the southwestern United States. *Geological Society of America Special Papers* 500: 501–532.

Mojumdar, A., Behera, T.H., and Ray, L. (2021). Lithobiontic ecology: stone encrusting microbes and their environment. In: *Environmental and Agricultural Microbiology: Applications for Sustainability* (ed. B.B. Mishra, S.K. Nayak, S. Mohapatra, and D. Samantaray), 341–360. Scrivener Publishing LLC. ISBN: 9781119525899 https://doi.org/10.1002/9781119525899.

Mol, L. and Viles, H.A. (2012). The role of rock surface hardness and internal moisture in tafoni development in sandstone. *Earth Surface Processes and Landforms* 37 (3): 301–314.

Moores, J.E., Pelletier, J.D., and Smith, P.H. (2008). Crack propagation by differential insolation on desert surface clasts. *Geomorphology* 102 (3): 472–481.

Mustoe, G.E. (2018). Biogenic weathering: solubilization of iron from minerals by epilithic freshwater algae and cyanobacteria. *Microorganisms* 6 (1): 8.

Nicolitch, O. et al. (2016). Soil type determines the distribution of nutrient mobilizing bacterial communities in the rhizosphere of beech trees. *Soil Biology and Biochemistry* 103: 429–445.

Nilsson, R.H., Kristiansson, E., Ryberg, M. et al. (2008). Intraspecific ITS variability in the kingdom fungi as expressed in the international sequence databases and ITS implications for molecular species identification. *Evolutionary Bioinformatics* 4: 193–201.

Omelon, C.R. (2008). Endolithic microbial communities in polar desert habitats. *Geomicrobiology Journal* 25 (7–8): 404–414.

Ortega-Morales, B.O., Narvaez-Zapata, J., Reyes-Estebanez, M. et al. (2016). Bioweathering potential of cultivable fungi associated with semi-arid surface microhabitats of Mayan buildings. *Frontiers in Microbiology* 7: 201. https://doi.org/10.3389/fmicb.2016.00201.

Owen, J.J., Amundson, R., Dietrich, W.E. et al. (2011). The sensitivity of hillslope bedrock erosion to precipitation. *Earth Surface Processes and Landforms* 36 (1): 117–135.

Puente, M.E. et al. (2004). Microbial populations and activities in the rhizoplane of rock-weathering desert plants. I. Root colonization and weathering of igneous rocks. *Plant Biology* 6: 629–642.

Saad, E.M., Sun, J., Chen, S. et al. (2017). Siderophore and organic acid promoted dissolution and transformation of Cr (III)–Fe (III)–(oxy) hydroxides. *Environmental Science and Technology* 51 (6): 3223–3232.

Sánchez-Andrea, I., Rodríguez, N., Amils, R., and Sanz, J.L. (2011). Microbial diversity in anaerobic sediments at Río Tinto, a naturally acidic environment with a high heavy metal content. *Applied and Environmental Microbiology* 77: 6085–6093.

Sass, O. (2005). Rock moisture measurements: techniques, results, and implications for weathering. *Earth Surface Processes and Landforms* 30 (3): 359–374.

Seiffert, F., Bandow, N., Bouchez, J. et al. (2014). Microbial colonization of bare rocks: laboratory biofilm enhances mineral weathering. *Procedia earth and planetary science* 10: 123–129.

Shobe, C.M. et al. (2017). *Field Evidence for the Influence of Weathering on Rock Erodibility and Channel Form in Bedrock Rivers*. Earth Surf: Processes Landforms https://doi.org/10.1002/esp.4163.

Small, E.E., Anderson, R.S., and Hancock, G.S. (1999). Estimates of the rate of regolith production using 10Be and 26Al from an alpine hillslope. *Geomorphology* 27 (1): 131–150.

Sun, Q., Fu, Z., Finlay, R., and Lian, B. (2019). Transcriptome analysis provides novel insights into the capacity of the ectomycorrhizal fungus *amanita pantherina* to weather K-containing feldspar and apatite. *Applied and Environmental Microbiology* 85 (15): e00719–e00719.

Tedstone, A.J., Cook, J.M., Williamson, C.J. et al. (2020). Algal growth and weathering crust state drive variability in western Greenland ice sheet ice albedo. *The Cryosphere* 14 (2): 521–538.

Thakur, D., Kaushal, R., and Shyam, V. (2014). Phosphate solubilising microorganisms: role in phosphorus nutrition of crop plants-a review. *Agricultural Reviews* 35 (3).

Uroz, S., Calvaruso, C., Turpault, M.P., and Frey-Klett, P. (2009). Mineral weathering by bacteria: ecology, actors and mechanisms. *Trends in Microbiology* 17 (8): 378–387.

Uroz, S., Picard, L., Turpault, M.P. et al. (2020). Dual transcriptomics and proteomics analyses of the early stage of interaction between *Caballeronia mineralivorans* PML1 (12) and mineral. *Environmental Microbiology* 22: 3838–3862.

Vazquez, P., Holguin, G., Puente, M.E. et al. (2000). Phosphate-solubilizing microorganisms associated with the rhizosphere of mangroves in a semiarid coastal lagoon. *Biology and Fertility of Soils* 30: 460–468.

Viles, H.A. (1995). Ecological perspectives on rock surface weathering: towards a conceptual model. *Geomorphology* 13: 21–35.

Wang, W., Lian, B., and Pan, L. (2015). An RNA-sequencing study of the genes and metabolic pathways involved in *aspergillus Niger* weathering of potassium feldspar. *Geomicrobiology Journal* 32: 689–700.

Wang, Q., Cheng, C., Agathokleous, E. et al. (2021a). Enhanced diversity and rock-weathering potential of bacterial communities inhabiting potash trachyte surface beneath mosses and lichens—a case study in Nanjing, China. *Science of The Total Environment* 785: 147357.

Wang, Q., Cheng, C., Agathokleous, E., Zang, S., & Sheng, X. (2021b). Bacterial communities in habitats of Dongchuan, China: their role in slate weathering in triggering and flowing areas of debris flow.

Welch, D.B.M. and Huse, S.M. (2011). Microbial diversity in the deep sea and the underexplored "Rare Biosphere". In: *Handbook of Molecular Microbial Ecology II: Metagenomics in Different Habitats* (ed. F.J. de Bruijn), 243–252. Wiley-Blackwell: Hoboken.

Wells, T., Binning, P., and Willgoose, G. (2005). The role of moisture cycling in the weathering of a quartz chlorite schist in a tropical environment: findings of a laboratory simulation. *Earth Surface Processes and Landforms* 30 (4): 413–428.

West, N., Kirby, E., Bierman, P., and Clarke, B.A. (2014). Aspect-dependent variations in regolith creep revealed by meteoric 10Be. *Geology* 42 (6): 507–510.

Włodarczyk, A., Stasiuk, R., Skłodowska, A., and Matlakowska, R. (2015). Extracellular compounds produced by bacterial consortium promoting elements mobilization from polyme- tallic Kupferschiefer black shale (fore-Sudetic monocline, Poland). *Chemosphere* 122: 273–279.

Xia, Q., Zhang, L., Dong, H. et al. (2020). Bio-weathering of a uranium-bearing rhyolitic rock from Xiangshan uranium deposit, Southeast China. *Geochimica et Cosmochimica Acta* 279: 88–106.

Xiao, L., Hao, J., Wang, W. et al. (2014). The up-regulation of carbonic anhydrase genes of *bacillus mucilaginosus* under soluble Ca^{2+} deficiency and the heterologously expressed enzyme promotes calcite dissolution. *Geomicrobiology Journal* 31: 632–641.

Zaharescu, D.G., Burghelea, C.I., Dontsova, K. et al. (2020). Biological weathering in the terrestrial system. In: *Encyclopedia of Earth Sciences Series* (ed. K. Dontsova, Z. Balogh-Brunstad, and G. Le Roux), 1–32. Washington: American Geophysical Union https://doi.org/10.1002/9781119413332.ch1.

Zhao, F., Qiu, G., Huang, Z. et al. (2013a). Characterization of rhizobium sp. Q32 isolated from weathered rocks and its role in silicate mineral weathering. *Geomicrobiology Journal* 30 (7): 616–622.

Zhao, Z.Q., Shahrour, I., Bai, Z.K. et al. (2013b). Soils development in opencast coal mine spoils reclaimed for 1-13 years in the West-northern loess plateau of China. *European Journal of Soil Biology* 55: 40–46.

6

Evolution of Soil Erosion and Sedimentation Vulnerability of Western Himalayan Lake Sukhna, India

Prabhat Semwal[1], Suhas Damodar Khobragade[1], and Neeraj Pant[1,2]

[1] *Hydrological Investigations Division, National Institute of Hydrology, Roorkee, India*
[2] *School of Engineering, Design and Built Environment, Western Sydney University, Sydney, Australia*

6.1 Introduction

Soil erosion and subsequent sedimentation are natural processes, active all throughout geological time, mainly influenced by water and wind impact on the surface of the earth. Over the last four decades, approximately 33% of the total worlds arable land has been lost due to deforestation and continues to be lost at a rate of ~10 million hectares per year (Pimentel et al. 1995; Gupta et al. 2010). Soil erosion not only reduces the water storage capacity in the upstream of the basin but also affects water quality and causes siltation in the downstream regions (Setyawan et al. 2019). Soil erosion and its eventual accumulation is one of the serious problems faced by many water bodies (lakes, ponds, and reservoirs), which is mainly responsible to reduce their potential and valuable lives (Semwal et al. 2017). Hence, assessment of the soil erosion and sedimentation can develop our understanding related to adaptation of appropriate conservation methods.

The Himalayan region is one of the most susceptible regions for soil erosion (Jain et al. 2001; Rawat et al. 2013; Kumar et al. 2014; Semwal et al. 2017; Pant et al. 2020), whereas, foothills or Siwalik region is also known for the highest rates of sedimentation (Central Water Commission 2015). The major factors that influence soil erosion are LULC (land use and land cover), soil properties, rainfall intensity, slope length, and steepness (Xu et al. 2017). It is widely agreed that both soil erosion and sediment yields are exacerbated by heavy rainfall and runoff events (Lecce et al. 2006; Vicente et al. 2007). The vegetation prevents soil erosion by shielding the soil from falling raindrops, increasing the water penetration capacity of the soil, and reducing the velocity of rainfall–runoff (Baver 1956). Apart from that, vegetation also enhances the binding property of the soil, maintains the roughness of the soil surface, and improves the physical, chemical, and biological properties of the soil (Alejandro and Omasa 2007). Soil properties like texture, moisture, roughness, and organic matter also affect and are responsible for the soil erosion. During monsoon, high-intensity rainfall–runoff process may accelerate the erosion of loose unconsolidated

Weathering and Erosion Processes in the Natural Environment, First Edition. Edited by Virendra Bahadur Singh, Sughosh Madhav, Naresh Chandra Pant, and Ravi Shekhar.

soil and cultivated land. In hilly areas, topographic factors such as slope and its length accelerate the runoff process, thereby causing increased erosion (Pant et al. 2020). Moreover, the variation in the LULC pattern has been studied by many authors and found to be most appropriate and responsible to explain the connectivity between soil erosion and sedimentation deposition in any basin (De Vente et al. 2007; Xu et al. 2017).

Sukhna lake is an important tourist attraction and center of recreation located at the foothills of the North Western Siwalik Himalayas. The lake was created by damming the Sukhna stream (choe), which is a seasonal or ephemeral in nature. The lake has been declared as a protected wetland of national importance by the Government of India. However, being located at the foothills of Siwalik, which is subjected to high erosion, the choe carries out high sediments, which gets deposited in the lake and directly affects the storage capacity of Sukhna lake. The man-made lake was built in 1958 and just after one decade about 38% of its storage space was depleted (Singh 2002). This alarming rate of siltation led to initiation of investigations into the problem. According to Shrimali et al. (2001), 32.9% of the catchment area of the lake is susceptible to high or very high erosion. During the 1970s, studies conducted by the Indian Agriculture Research Institute, New Delhi, and the Central Soil and Water Conservation Research and Training Institute (CSWCRTI), Chandigarh, recommended a range of steps to manage erosion in the catchment area. Subsequently, scientific actions were commenced for the conservation of the soil in the catchment (Singh 2002). The numerous initiatives undertaken included the planting around 1.6 million of locally suitable varieties of trees during 1989–1990, the installation of earthen silt retaining dams, masonry check dams, gully bed grade stabilizer, vegetative walls, mechanical spurs, and the development of contour trenches (CSWCRTI 1993; Singh 2002; Grewal 2009). Although it is known that the various soil conservation measures undertaken have significantly reduced siltation of the lake and increased its longevity (Khobragade et al. 2013). The earlier reported siltation studies on the lake (Bansal and Mishra 1982; Bansal and Grewal 1986; CSWCRTI 1993) mostly deal with the siltation problems in initial years of the lake. However, there is scarcely any recorded effort to systematically evaluate the effects or efficacy of the measures undertaken to reduce the erosion and sedimentation. Hence, in this study, investigations have been carried out to (i) assess sedimentation rate and expected life of the lake and (ii) understand the impact of various conservation measures on the storage capacity of the lake.

6.2 Study Area

The research was performed for Sukhna Lake, situated in the Union Territory of Chandigarh, India (Figure 6.1). It is a shallow kidney-shaped lake with an average depth of ~3.3 m and a maximum depth of ~6.5 m. The lake is about 2.32-km long and about 1.06-km wide. The man-made Sukhna lake was created as part of the Chandigarh city master plan prepared by the French architect Le-Corbusier. The lake catchment (fan-shaped) area lies between 30°49′7″N to 30°44′6.75″N and 76°48′16.79″E to 76°53′38.26″E in the foothills of the Siwalik Hill ranges, which form part of the fragile Himalayan ecosystem, called Kandi (Bhabhar) region of the Ghaggar river basin. The catchment area of the Sukhna lake is about 42.14 km^2. The catchment consists of primarily alluvial sandy soils embedded with

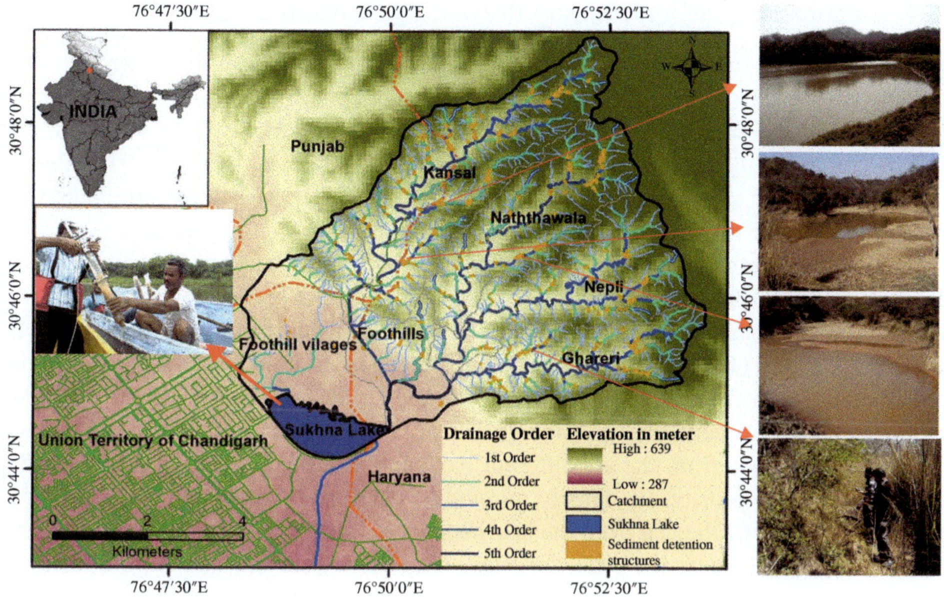

Figure 6.1 Location map of the study area representing Sukhna lake and its catchment. Sukhna choe stream and its drainages are also shown along with sediment detention structures. The inset represents collection of sediment core from the Sukhna lake.

clay deposits and is particularly vulnerable to soil erosion (Singh 2002). At the time of construction of the lake, the storage capacity was 10.74 MCM. According to the bathymetric survey carried out by IPRI (Irrigation and Power Research Institute) in 2013, the lake storage capacity was ~537.84 ha-m, (IPRI 2013). The lake water spread area was 188 ha after completion in 1958 and the average depth was 4.69 m with the lowest point at 343.2 m amsl (Singh 2002). The surface area at the end of the flood season of 2013 was reduced to 160.841 ha with the deepest point at 354.48 m amsl elevation (IPRI 2013).

6.3 Data Used and Methodology

6.3.1 Data Used

In this analysis, the topographic map of 1:25000 scale, prepared by the SOI (Survey of India) for Sukhna lake catchment was used to produce the different maps (i.e. elevation, slope, drainage density). Landsat 7 ETM+ image of 2001, 2006 and Landsat-8 images of 2013 have been used to prepare LULC for the study area. Rainfall data for the study period of 2010–2012 was collected from CSWCRTI, Chandigarh. The other data used in the study include bathymetric data, storage capacity data, and remote sensing data. Sediment cores were also collected from different location of lake to estimate dry density, bulk density, and soil moisture content and to determine the cesium activity in the lake sediments. The analysis of sediment cores has been carried out in the laboratory of NIH (National Institute of Hydrology), Roorkee.

6.3.2 Determination of Bulk Density, Dry Density, Moisture Contents

In order to understand the sedimentation aspects of the lake and soil erosion and conservation aspects of its catchment, various components have been studied namely (i) bulk density, dry density, soil moisture content, and ^{137}Cs activity of lake bed sediment; (ii) development of sediment deposition profile; (iii) sediment mass balance; (iv) modeling; (v) and analysis of various factors like LULC, silt detention structures, drainage, and drainage density of the study area. Sediment deposition profiles of the lake have been prepared to understand the sediment deposition pattern in the lake, while sediment mass balance approach has been applied to estimate rate of sedimentation for the lake during different time periods. Soil yield from the lake catchment has been estimated using the Garde and Kothyari (1987) model for the period of 2010–2012. Evaluation of the model has been carried out to determine its suitability for the study area, based on the validation carried out using sediment mass balance approach. The detailed methodology is as follows:

The sediment cores from the six different locations of the lake have been collected to determine the bulk density, dry density, and moisture contents in the lake bed sediments. Sediment cores were collected from the lake bed using a sediment corer having a diameter of 7.6 cm. Figure 6.1 shows the collection of sediment core from the lake. All the sediment cores were sliced into 2-cm sections, oven dried and powdered. The bulk density, dry density, and moisture content at various depths in each coring location have been determined.

The bulk density is calculated as:

$$\rho_b = \frac{W_w}{V} \tag{6.1}$$

where, ρ_b = bulk density of the sample (g/cm); W_w = wet weight of the sample (g); V = volume of the sample (cc)

The dry density is calculated as:

$$\rho_d = \frac{W_d}{V} \tag{6.2}$$

where, ρ_d = dry density of the sample (g/cm); W_d = dry weight of the sample (g); V = volume of the sample (cc)

The moisture content is determined by:

$$Q = \frac{\left(W_w - W_d\right)}{W_w} \times 100 \tag{6.3}$$

where, Q = moisture content in the sample (%); W_w = wet weight of the sample (g); W_d = dry weight of the sample (g).

6.3.3 Measurement of Cs-137 Activity

The use of ^{137}Cs is becoming very popular for the dating of sedimentation rate in last 60 years. The natural fallout of ^{137}Cs was found considerable during 1953–1954, 1957–1958, 1963–1964, 1978–1979, and 1986–1987 due to explosion of various atomic bomb tests and nuclear accidents. ^{137}Cs is strongly absorbed on tiny particles like clay materials, silt, and humic materials. Surface soils of the catchment with an adsorptive capacity will have a ^{137}Cs contents and therefore, be able to act as self-tracer. The rate of sedimentation can be calculated from the depth of two principle time horizon, i.e. 1954 and 1964, in the ^{137}Cs concentration profile in the lake sediment cores. Analysis of one sediment core has been carried out to measure Cs-137 activity. The procedures given in Procedure Manual (Ver. 1.0) for measurement of cesium-137 in sediment (NIH, Roorkee, India) have been followed. The sediment core has been analyzed for ^{137}Cs activity using Multi Channel Gamma-Ray Mass. Spectrometer at Nuclear Hydrology Laboratory of NIH, Roorkee, India. The detection limit for ^{137}Cs by this method is 0.25 m Bq/g (Kumar et al. 2007).

6.3.3.1 Sediment Deposition Profile

Bathymetry survey of the lake is being regularly undertaken by IPRI, Amritsar. These data have been collected for the years 1982, 2002, 2003, 2004, 2005, 2010, and 2012 and based on these data, the depositional pattern of the sediments in the lake has been determined. The data were available as elevation of bathymetry survey points at either 10 or 20 m interval for the lake. The data were processed using Arc GIS 9.3 software to get the lake bed profiles. The bathymetry data of two different periods were superimposed to get the differences in elevation profiles of two successive periods. For example, the bathymetry data of year 2002 have been superimposed on the data of 1982 to get the profile for the period 1982–2002.

6.3.3.2 Sedimentation Mass Balance

From the sediment mass balance equation, the sedimentation rates for the lake is determined using the following equation:

$$\Delta S = \Delta d - \Delta r \tag{6.4}$$

Therefore

$$\Delta d = \Delta S + \Delta r \tag{6.5}$$

where,

ΔS = change in storage capacity of the lake during a period; Δd = total sediment deposited in the lake during the period (sediment inflow); and Δr = total sediment removed from the lake during the period.

Two possible reasons behind the loss of sediments from the lake (sediment outflow) can be either in the form of sediment lost by outflow or sediment removal due to dredging from the lake. Here it can be noted that during the study period, no outflow has been recorded from the lake, whereas dredging was regularly conducted during this period by the Engineering Department, Chandigarh Administration. A shift in the lake storage capacity was extracted from the lakes bathymetric results and silt accumulated in the lake over the period of 2003–2012 has been measured using the sediment balance method.

6.3.3.3 Sediment Yield from the Lake Catchment

The sediment yield from the lake catchment has been estimated using the model developed by Garde and Kothyari (1987) for the period of 2010–2012. The Garde and Kothyari model is a very popular model in India and was developed based on the analysis of the data from 50 catchments spread across India. The model as described by Upadhyay et al. 2012) is:

$$S_a = 1.182 \times 10^{-6} \times P_3^{1.29} \times A^{1.03} D_d^{0.40} \times S^{0.08} \times F_c^{2.42} \tag{6.6}$$

$$F_c = \frac{0.8F_A + 0.6F_G + 0.3F_F + 0.1F_W}{A_a} \tag{6.7}$$

where S_a is the mean annual sediment yield (Mm^3), P_3 is the annual average rainfall (mm) (here monsoon average rainfall was taken), A_a is the total catchment area (km^2), D_d is the drainage density (km^{-1}), S is the slope in m/m, F_c is the vegetation cover factor, F_A is the area under agriculture land (km^2), F_G is the area under grassland (km^2), F_F is the area under forest (km^2), and F_W is the area under waste land (km^2).

The Garde and Kothyari model gives sediment yield from the catchment at the outlet which is the inlet point of the lake and hence it is comparable with the sediment volume inflowing to the lake as obtained from the mass balance model (Eq. 6.5). The sediment volumes were converted to mass volume using a mean bulk density of 1.56 t/m^3 as per Semwal (2017), to make the rates obtained from the two methods comparable. The mean bulk density was calculated by averaging the bulk densities of six sediment cores collected from the lake.

6.3.3.4 Analysis of Factors Affecting Sedimentation and Conservation Measures

The soil erosion from the lake catchment and corresponding lake sedimentation is major management issue in the lake. Therefore, in this study, the major factors influencing the soil erosion in the lake catchment have been analyzed. Conservation factors such as reforestation and construction of silt detention structures in the catchment and removal of sediment through dredging from the lake have been analyzed. For proper understanding of the conservation measures, LULC maps of the study area for the years 2001, 2006, and 2013 have been prepared. The various silt detention structures in the catchment have been marked for same years. The LULC mapping were conducted using Landsat satellite images while silt detention structures were marked using Google Earth images and field survey were done for validation at some selected places (Figure 6.1). The satellite images of study area were classified using semi-supervised classification technique using ERDAS Imagine 9.2 software. Digital elevation model of the study area has been prepared using SOI 1:25000 toposheet using Arc GIS 9.3 software. Slope, drainage, and drainage density maps of the study area were also prepared. The digital elevation model was used to prepare the slope map of the lake catchment. The drainage map was prepared using the SOI toposheet, digital elevation model, Google Earth images, and Landsat satellite images. In this study, according to Strahler (1964) stream ordering technique, drainage order was given to each drain to estimate drainage density in which the smallest fingertip tributaries are designated as first-order drain. The drainage density defines the ratio between the total length of streams with respect to total area of the watershed and demonstrates the proximity of channel spacing (Horton 1932). A relatively high stream density, weak or impervious sub-surface material, scanty vegetation, and mountain relief are indicated by higher drainage density values. Relatively low stream density, high resistant or permeable soil, densely covered forest, and low relief indicate lower drainage density values. The catchment drainage density map was developed using the line density function of Arc GIS 9.3 from the drainage map.

6.4 Results and Discussion

6.4.1 Bulk Density, Dry Density, and Moisture Contents

The bulk density and dry density of six cores, at different depths, collected from the lake bed are given in Figure 6.2a. Bulk density is a good indicator of soil compaction and associated with soil texture and organic matter contents. Generally, the bulk density of sediment increases with increasing depth due to compaction. Figure 6.2a shows that the bulk density is not uniform at

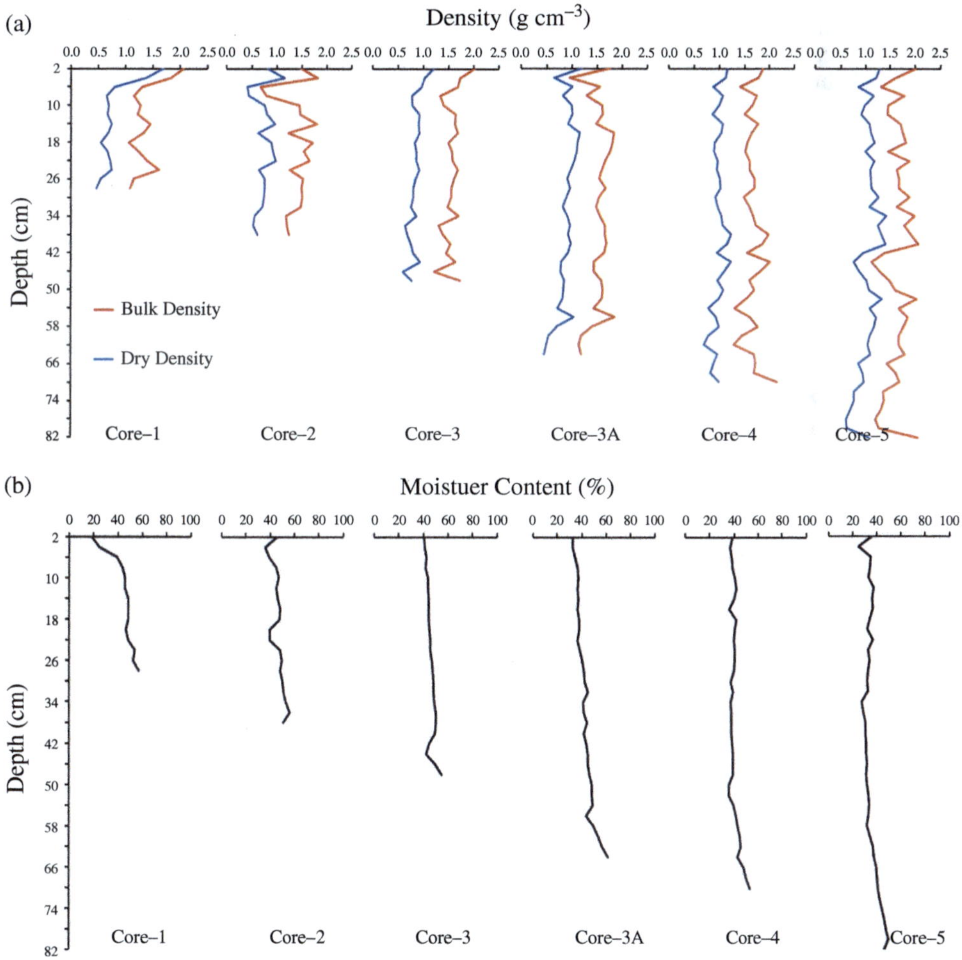

Figure 6.2 (a) Bulk density and dry density of sediment cores, (b) % moisture contents of sediment cores collected from lake bed at different depths.

various depths in the lake bed, indicating that the sediments are loose and unconsolidated at some depths while, at other depths, sediments are compact. It also indicates particle size are also varying at different depths and places in the lake. The bulk density of the lake bed sediment varies in the range of 0.68 g/cm^3–2.16 g/cm^3 with an average value of 1.56 g/cm^3. The dry density in the lake sediment varies in the range of 0.41 g/cm^3–1.70 g/cm^3 with an average value of 0.88 g/cm^3.

Determination of the percentage of moisture content can help in understanding the lithology and behavior of the lakebed at different locations. The variations in percent moisture content at different depths are given in Figure 6.2b. The moisture content varies in the range of 18.17–61.74% with an average value of 44.10%. Generally, as the depth increases, the soil becomes compact and moisture contents also get reduced but in the sediment cores of the lake moisture, contents are observed to be slightly increasing with depths for most of the cores. This could be due to the presence of sandy soil at lower depths aided by the hydraulic pressure that is high at these depths.

6.4.2 Caesium-137 Activity

Figure 6.3 shows the ^{137}Cs activity in the sediment core collected from the lake bed. The ^{137}Cs activity varies in the range of 13 m Bq/gm–51.86 m Bq/gm with the efficiency of 2.33%. The results obtained indicate that there is very little activity in the sediment sample obtained from the Sukhna lake. The highest peak obtained from the lake was 51.86 m Bq/gm at 6–8 cm depth from the lake bed. This is much less than the global peak of 130 Peta becquerels observed during 1964. Further, the activity pattern (Figure 6.3) also indicates that even the little activity that could be measured does not match the global fall-out pattern. This may be because desilting of the lake has been carried out in the past almost every year which has disturbed the whole sediment deposition pattern. For the analysis of the ^{137}Cs activity in the lake sediment, an undisturbed sediment core is needed, which cannot be collected for the lake. Moreover, the sediment rate of the lake is very high; hence, it is also possible that the ^{137}Cs deposits are contained in a deeper sediment layer. As the observations of ^{137}Cs in the present study were in less than 1-m core; therefore, the other techniques have been used for the study of lake sedimentation.

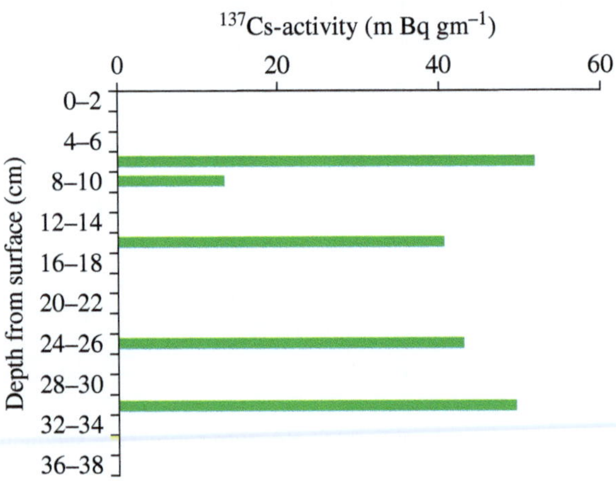

Figure 6.3 ^{137}Cs activity in sediment cores collected from Sukhna lake bed.

6.4.3 Sediment Deposition Profiles

The sediment deposition profiles of Sukhna lake have been presented in Figure 6.4. The negative value in the profile at a particular point indicates the removal of sediment from that point through dredging while positive value indicates deposition at that point through sedimentation in the lake. In general, the dredging has been carried out toward the eastern side of the lake, as well as along the northern periphery of the lake. As a result, the sediment deposition profile for different periods shows an increase in lake

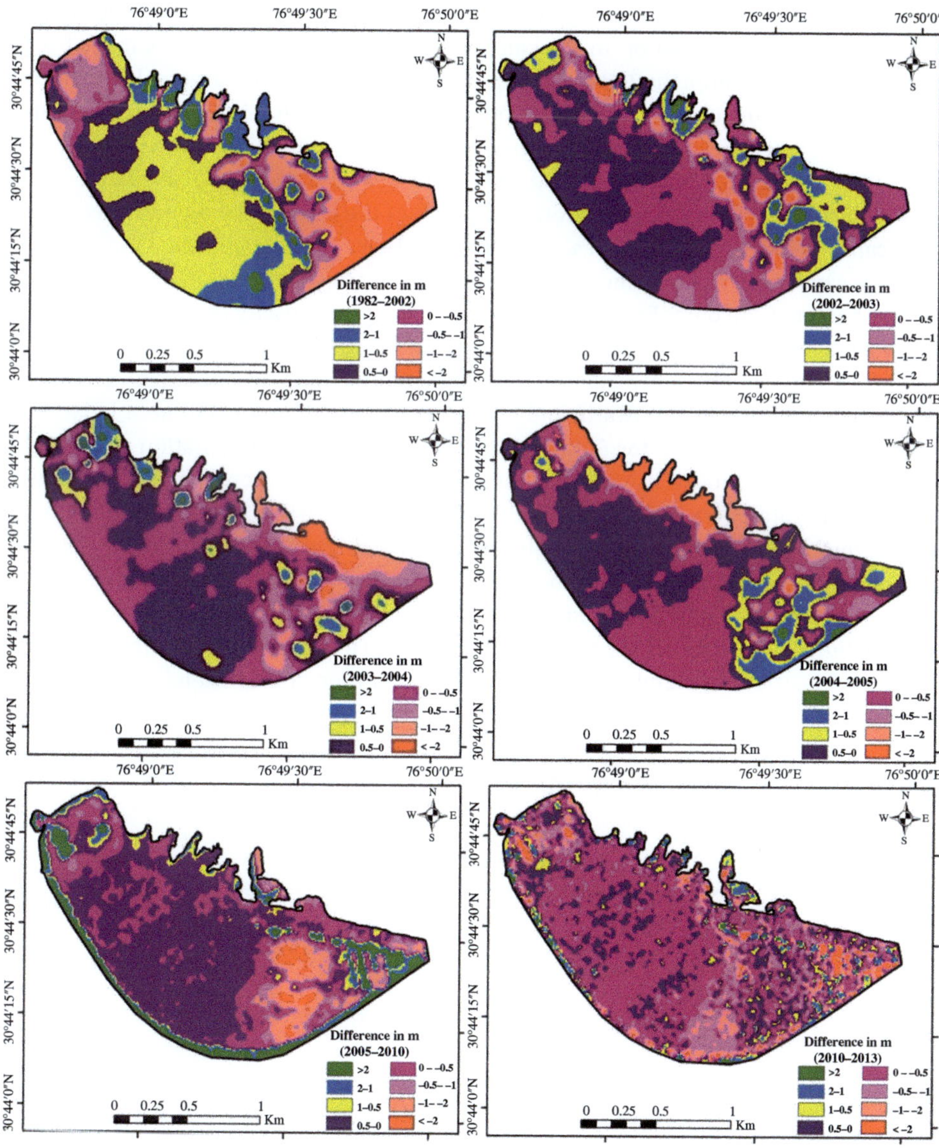

Figure 6.4 Sediment deposition profiles of Sukhna lake during different periods (1982–2013).

depths in these parts while reductions in lake depths have been observed in most other parts owing to sediment deposits (Figure 6.4). The variation in the depths clearly indicates variable sedimentation rates and sediment removal volumes during different periods.

6.4.4 Computation of Sedimentation Rate

During the 1960s and 1970s, the rate of sedimentation in the lake was very high. By 1971, approximately half of the lakes storage capacity was depleted, although the depletion of water storage capacity grew to about 68.5% by the year 1976 (Singh 2002). According to Singh (2002), mass deforestation in the catchment of the lake was the key cause of high sedimentation levels before the 1980s, leaving steep sloping bare hills vulnerable to extreme erosion. However, due to reforestation and other vegetative mechanical measures initiated after the 1970s, a significant reduction in the average sedimentation rate from 49.95 ha-m/y during 1959–1964 to 13.02 ha-m/y during 1984–1989 was observed (Singh 2002).

Figure 6.5 presents the changes in storage capacity of the lake during different years. The storage capacity of the lake reduced significantly till 1999. The available data indicate that the storage capacity of the lake increased significantly during 2002 by 170 ha-m due to the raising of height of the dam. From 2002 to 2013, the storage capacity of the lake remained more or less constant. Recent rate of sedimentation for the lake has been determined by sediment balance method considering the bathymetric data of the lake. Therefore, bathymetric data from 2003 onwards only have been considered for sediment balance to estimate sedimentation rates for Sukhna lake. The estimated sedimentation rates of the lake are presented in Table 6.1. Average sedimentation rate for the period of 2008–2010 was found to be negative (−8.13 ha-m/y). This indicates that during 2008–2010, the rate of sediment removed from the lake through dredging was higher than the rate of sediment deposited in the lake. However, for the period of 2010–2012, sedimentation was higher than the sediment removal. If we take the average of values for different time periods, the average sedimentation rate comes out to be 7.93 ha-m/y. However, if we consider the whole period of 2003–2012, the average value of lake sedimentation comes out to be 7.01 ha-m/y.

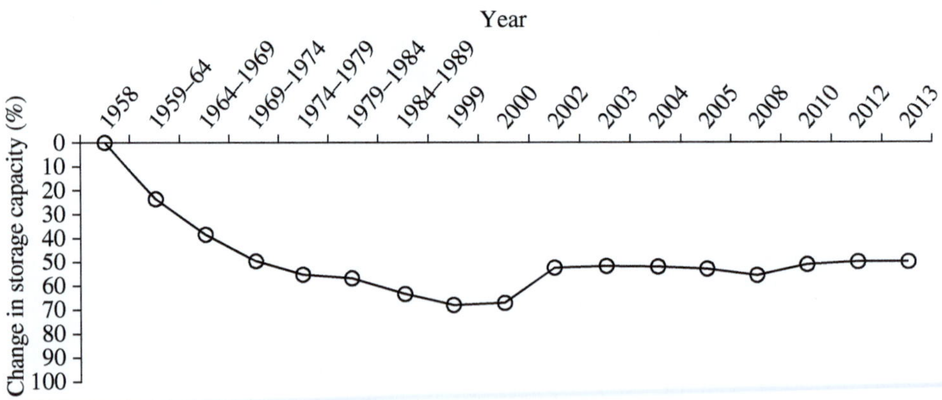

Figure 6.5 Change in storage capacity of the Sukhna lake during different years.

Table 6.1 Sediment deposition rates into Sukhna lake during different time periods.

Periods	Sediment removed from the lake (ha-m)	Sediment deposited into lake (ha-m)	Sediment deposited into lake (ha-m/y)
2003–2004	0	5.15	5.15
2004–2005	6.84	18.30	18.30
2005–2008	9.45	36.83	12.28
2008–2010	32.82	−16.26	−8.13
2010–2012	26.91	19.05	9.53
2003–2008	16.3	60.29	12.06
2004–2012	76.02	57.93	7.24
2003–2012	76.02	63.08	7.01

Thus, it is clear that compared to the initial decades of the 1950s–1980s, the recent sedimentation rates have reduced significantly, due to a number of conservation measures that were undertaken in the lake basin. Based on the estimated average sedimentation rate of 7.01 ha-m/y for recent time (2003–2012), the expected life of the lake comes out to be 76.7 years. Thus, the lake is likely to be completely filled with sediment by 2090, if the conditions remain unchanged.

6.4.5 Comparison with Garde and Kothyari (1987) Model

The Garde and Kothyari (1987) model requires various input parameters viz., drainage density, average slope, and vegetation cover factor, etc. as discussed below.

6.4.5.1 Elevation and Slope

The maximum and minimum elevations of catchment area are about 620 m and 350 m amsl respectively (Figure 6.6a), whereas the maximum relative relief is observed to be 269 m. The slope in the study area varies between 0° to 56° (average 11.74°) with standard deviation of 9.65° (Figure 6.6b). The gradient is found to be minimal in the direction of SW and rises in the direction of NE and SE. The Siwalik stretches from the NW-SE pattern to the north-eastern frontier of Chandigarh. The hilly catchment has a steep slope that becomes gentle toward the southeastern direction. The average slope for the catchment area is 0.21 m/m.

6.4.5.2 Drainage Characteristics

The drainage and drainage density of the study area is presented in Figure 6.6c and Table 6.2 indicating dendritic to sub-dendritic drainage pattern. However, at few locations, the drainages are showing parallel to sub-parallel pattern. The dendritic pattern indicates the homogeneity in texture and lack of natural (geological) structural control. The parallel pattern indicates that the study area has the gentle and uniform slope with less resistant bedrocks, producing reasonable environmental condition for the soil erosion in the study area. The average drainage density of the Sukhna lake catchment is estimated to be 5.56 km^{-1}.

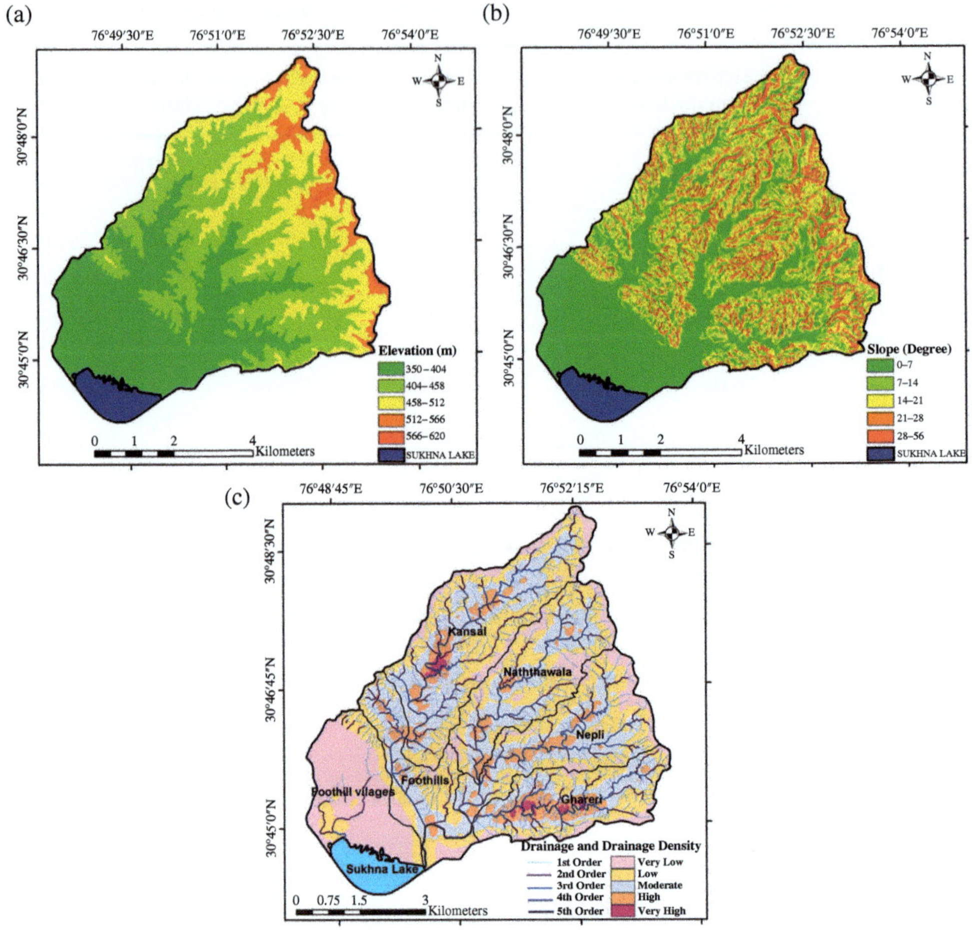

Figure 6.6 (a) Elevation, (b) Slope, and (c) drainage density map of the Sukhna lake catchment.

Table 6.2 Ordering of drainage in the sub-catchments of the Sukhna lake and drainage density.

	Order					Drainage density
Sub-catchment	**Ist**	**IInd**	**IIIrd**	**IVth**	**Vth**	**(km/km^2)**
Ghareri	114	32	7	1	—	7.57
Kansal	190	48	15	5	—	6.49
Naththawala	88	29	4	1	—	5.75
Nepli	93	31	5	3	—	6.09
Foothills	55	16	3	3	1	5.7
Foothill villages	26	9	1	—	—	1.78
Catchment	565	133	27	5	1	5.56

6.4.5.3 LULC

Figure 6.7 presents the LULC map of the study area for 2001, 2006, and 2013. In this study, the catchment area has been broadly classified into eight classes viz., ponds and check dams, lake, settlement, dense forest, moderately dense forest, open forest, barren/open land/scrub, and agriculture. The LULC types in the sub-catchments of the Sukhna Lake (Kansal, Naththawala, Nepli, and Ghareri) are mainly forest, river, check dams, and ponds. The foothills are frequently dominated by agriculture, and settlement while moderately dominated by dense forests. The shrubs are not classified separately in the LULC map but have been considered as part of the forested area in the catchment. Based on LULC, vegetation cover factor has been calculated. The average value of vegetation cover factor for catchment area comes out to be 0.34. The detailed study of change in LULC in the study area is given in subsequent section.

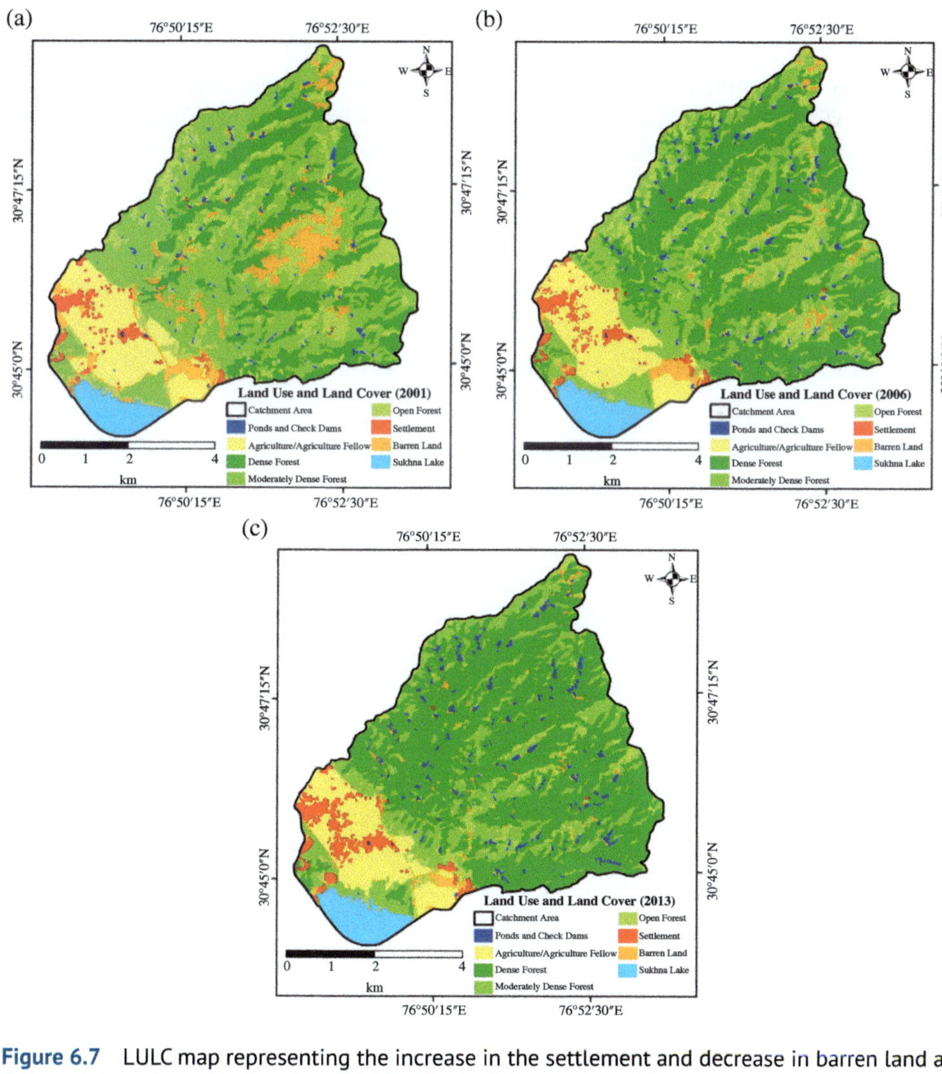

Figure 6.7 LULC map representing the increase in the settlement and decrease in barren land at Sukhna Lake catchment area during (a) 2001, (b) 2006, and (c) 2013.

Table 6.3 Sediment yield using Garde and Kothyari (1987) model for the period of 2010–2012.

Year	Rainfall (mm)	Sediment yield (m^3)	Sediment yield (tons)	Sediment yield (t/ha/y)
2010	1447.50	85679.00	133659.24	31.72
2011	756.10	38590.33	60200.91	14.29
2012	847.86	48553.77	75743.88	17.97
Average		38564.61	89868.01	21.33

6.4.5.4 Modeled Sediment Yield

Year-wise estimates of total sediment yield for 2010–2012 for the lake catchment as estimated using Garde and Kothyari (1987) model are presented in Table 6.3. The average value of vegetation cover factor, drainage density, and average slope for catchment area are 0.34, 5.56 km^{-1}, and 0.21 mm^{-1}, respectively, while the average sediment yield was found to be 21.33 t/ha/y. The observed total sediments deposited in the lake estimated from the sediment mass balance (Eq. 6.5) were 19.05 ha-m, which comes around 35.26 t/ha/y after converting the volume to the mass unit. This indicates that the sedimentation rate modeled by the Garde and Kothyari model grossly underestimates when compared with the observed estimates obtained from the sediment mass balance. The underestimation is as high as 39.52%. This clearly indicates that the Garde and Kothyari (1987) model is not suitable for the study area.

6.4.6 Analysis of Impact of Conservation Measures on Soil Erosion

6.4.6.1 Changes in LULC Pattern

LULC plays an important role in reducing direct runoff and thereby soil erosion leading to reduced sedimentation in the downstream lakes. Figure 6.7 represents the pattern of changes in LULC of the Sukhna lake and its catchment, while Table 6.4 shows the percentage changes in the major LULC in the catchment during 2001, 2006, and 2013, respectively. A total number of ~1.66 million seedlings were planted during 1989–2000 in the Sukhna lake catchment (Grewal 2009). It clearly indicates that a significant reforestation was carried out to reduce the soil erosion from the lake catchment. Yangchan et al. (2014) quantified forested area including shrubs during 1989 around 35.68 km^2 in the Sukhna lake catchment, while, in this study, the total forested area in the catchment is found to be about 33.33 km^2, 35.05 km^2, 34.83 km^2 during 2001, 2006, and 2013, respectively. It shows that the forested area of the lake catchment has not been changed significantly, but the change in classes of forested area from one class to another class is significant. For example, the moderately dense forest in the catchment is found to be 40.88%, 29.49%, and 21.32% during 2001, 2006, and 2013, respectively, while dense forest is found to be 23%, 38.74%, and 52.50% during 2001, 2006, and 2013, respectively. It clearly indicates that the moderately dense forest is converting into dense forest in the lake catchment. This increase in dense forest in the catchment is indicator of reducing soil erosion by improving the soil health, increasing the organic contents of the soil, and protecting the soil from the direct rainfall impact and armoring of the soil by the tree roots. Grewal (2009) reported the overall tree density improved from 162 trees/ha to 450 trees/ha, bushes density increased from 5977/ha to

Table 6.4 Change in major LULCs in the study area.

	2001		2006		2013	
Land use/land cover	**Area (km^2)**	**Total Area (%)**	**Area (km^2)**	**Total Area (%)**	**Area (km^2)**	**Total Area (%)**
Check dams and ponds	0.53	1.20	0.59	1.34	0.77	1.76
Settlement	0.78	1.77	0.81	1.84	1.26	2.89
Agriculture/agriculture fallow	4.60	10.48	4.51	10.28	3.81	8.73
Barren/open land	3.21	7.31	1.44	3.28	1.47	3.37
Open forest	5.30	12.07	5.11	11.65	2.60	5.96
Moderately dense forest	17.93	40.88	12.93	29.49	9.31	21.32
Dense forest	10.09	23.00	16.99	38.74	22.92	52.50
Sukhna lake	1.44	3.28	1.48	3.37	1.52	3.48
Total	43.66	100	43.86	100	43.66	100

8984/ha, the leaf litter accumulation improved from 2.05 t/ha to 4.49 t/ha, and overall organic carbon content increased from 0.44% to 0.70% in the Ghareri sub-catchment of the Sukhna lake. Semwal et al. (2017) reported that the 52% of study area is protected from soil erosion due to canopy cover in the catchment. The other classes like settlement, agriculture, and barren land have not been changed significantly in the study area.

6.4.6.2 Silt Detention Structures

Another important aspect that makes barrier to reach sediment into the lowland areas (Sukhna lake and foothills villages) is sediment detention structure (various check dams and ponds) (Figure 6.1). In the study area, a number of check dams and ponds have been constructed to conserve the soil and to reduce the sedimentation rates. About 113, 123, 161 check dams and ponds have been marked during 2001, 2006, and 2013, respectively. The area of these structures is found to be 0.53 km^2, 0.59 km^2, and 0.77 km^2 during 2001, 2006, and 2013, respectively. Table 6.5 shows the storage capacity of check dams and ponds in the sub-catchments of the Sukhna lake. Total storage capacity of check dams and ponds in the catchment is about 256.50 ha-m, while the storage capacity of Sukhna lake is about 537.84 ha-m during 2013. It means that the check dams and ponds retained about half the storage capacity of Sukhna lake. These check dams and ponds reduced the direct runoff and capture the sediments, thereby reducing the sedimentation rate.

Semwal et al. (2017) mentioned higher rate of soil erosion (more than 40 t/ha/y) in hilly catchment. Due to this, the storage capacity of check dams and ponds in the catchment is reducing year by year. Grewal (2009) observed that in the areas where series of check dams are constructed, the upper most check dam filled by sediments with higher rates compared to lower ones. It is clear from the Figure 6.1 that the some of the check dams and ponds of the catchment completely filled with sediment and cover with bushes or shrubs, due to which the storage capacity of the silt detention structures has been lost completely. It indicates that the check dams and ponds can reduce the rate of sedimentation and soil erosion

Table 6.5 Storage capacity of the check dams and ponds in sub-catchments of Sukhna lake.

Sub-catchment	Storage capacity (ha-m)
Ghareri	38.88
Kansal	112.55
Naththawala	52.23
Nepli	35.84
Lower foothills	11.83
Foothill villages	5.17

Table 6.6 Total number and area of check dams in different sub-catchments during the 2013.

Sub-catchment	Area of sub-catchments (Km^2)	No. of check dams	Total area of check dams (Km^2)	Area of check dams/ Sub-catchment area	No. of check dams/ Sub-catchment area
Ghareri	5.85	27	0.13	0.02	4.62
Kansal	10.84	52	0.29	0.03	4.80
Naththawala	7.56	40	0.18	0.02	5.29
Nepli	7.43	23	0.12	0.02	3.10
Lower Foothills	4.82	13	0.04	0.01	2.70
Foothill villages	5.87	6	0.01	0.00	1.02
Total	42.37	161	0.77	0.02	3.80

in the catchment for some years but it is not a long-term solution. Ultimately, these sediment detention structures will be filled by sediment in near future and again the rate of sedimentation will become high for the Sukhna lake. Therefore, dredging of these structures should be necessary to reduce the sedimentation rates into the Sukhna lake.

Table 6.6 shows total number and area of check dams in different sub-catchments, whereas Table 6.7 shows check dam's vis-a-vis stream length and drainage density of sub-catchments in the study area. There are about 161 check dams and ponds in the catchment covering an area of 0.77 km^2 in 2013. The maximum of 52 check dams are found in Kansal while minimum of six are found in foothill villages. This is obvious because while the Kansal topography is hilly with large number of streams, while the foothill villages are relatively much flat with few streams. However, Naththawala with 5.29 check dams per km^2 of the sub-catchment area has the densest network of check dams. Foothill villages with only 1.02 check dams per km^2 area has most sparse network of check dams. Analysis of the ratio of number of check dams to stream length (Table 6.7) indicates that the ratio is highest for Naththawala (0.92) and lowest for foothill villages but is relatively low for Nepli, although Nepli is at a much higher altitude and has larger stream length than Naththawala. Table 6.7 also indicates that the ratio of number of check dams to drainage density varies from sub-catchment to sub-catchment. It is high (8.01) for Kansal because of the high stream length. However, although Ghareri, Naththawala, and Nepli have more or less the same

Table 6.7 Check dam's vis-a-vis stream length and drainage density of sub-catchments during 2013.

Sub-catchment	Stream length (Km)	Drainage density (Km/Km^2)	No. of check dams/ Stream length	No. of check dams/ drainage density
Ghareri	44.07	7.57	0.61	3.57
Kansal	70.35	6.49	0.74	8.01
Naththawala	43.47	5.75	0.92	6.96
Nepli	45.31	6.09	0.51	3.78
Lower foothills	27.49	5.7	0.47	2.28
Foothill villages	10.49	1.78	0.57	3.37
Average		5.56	0.64	4.66

stream length but the ratio varies significantly. Thus, it appears that no scientific criteria have been followed while deciding the number of check dam for a sub-catchment.

As mentioned earlier, various conservation practices such as plantation of locally suitable varieties of trees, construction of earthen silt retention dams, gully bed grade stabilizer, vegetative barriers, mechanical spurs, and construction of contour trenches have also been applied in the lake catchment. Due to combined effect of the conservation practices, various geomorphological changes occurred in the Sukhna lake catchment. Changes in angle of repose, bed gradient, cross section of water courses, meandering patterns and changes in the composition of bed material have been noted by Grewal (2009). The main reason for these changes is construction of series of silt detention structures in the flowing streams. The silt detention structures reduced the runoff velocity, which retained the water and sediment behind the check dams and reduced the carrying capacity of the streams. As a result, the sediment load was shed along the banks and stream channels become straighter. This has also increased the deposition of the heavy sediment across the streams channel and behind the check dams, resulting in the growth of vegetation that allows only fine particles to flow over the check dams. Semwal et al. (2017) noted that even though conservation practices reduced the rate of erosion in the catchment but due to high slope and steepness of hilly catchment, the erosion rates are still high in the catchment, as a result reducing the storage capacity of the lake.

6.5 Summary and Conclusions

Soil degradation and its resulting accumulation in Sukhna lake is one of the serious problems, which is resulting to decline in its storage space and useful life. Therefore, it is necessary to understand various aspects related to sedimentation and erosion at regional scale. This study includes various aspect, studied at the catchment level including (i) understanding the density, soil moisture content, and ^{137}Cs activity of lake bed sediments at various depths; (ii) development of sediment deposition profile; (iii) sediment mass balance; (iv) modeling; and (v) analysis of various aspects like LULC, silt detention structures, elevation and slope, drainage, drainage density of the lake catchment.

The bulk density of the Sukhna lake sediment varies in the range of $0.68\,g/cm^3$–$2.16\,g/cm^3$ with an average value of $1.56\,g/cm^3$. Average bulk density of lake bed sediment indicates that there are no spaces for water to infiltrate from lake bed. The soil moisture content varies in the range of 18.17%–61.74% with an average value of 44.10%, indicating the presence of both loose and/or consolidated soil type at the lake bed. The highest ^{137}Cs activity obtained from the lake was 51.86 m Bq/gm which is much lower than the global peak of 130 Peta becquerels observed during 1964. The activity pattern does not match the global fallout pattern due to desilting of the lake bed sediments which has disturbed the whole sediment deposition pattern. Bathymetry and sediment removal data also reveal that huge amount of sediments has been removed from the Sukhna lake over the years through dredging to maintain the storage capacity of the lake, whereas the sediment deposition profiles show that the average rates of sediment removal were lower in comparison to average rates of deposition of the sediments in the lake. Analysis of the sediment deposit profiles also indicates that the sediments are mostly received from the eastern and northern periphery of the lake. However, the sediment yield predicted by the Garde and Kothyari (1987) model is found to be significantly lower; by 39.52%, indicating that the soil loss predicted using Garde and Kothyari (1987) model is not suitable for the study area.

The reported rate of sedimentation during the initial years of the construction of the lake (1958–1965) was about 50 ha-m/y, whereas the loss in storage capacity of the lake till 1980s was about 68.5%. However, the sediment mass balance carried out for different durations between 2003 to 2012 shows that the recent rate of sedimentation of the Sukhna lake varies in the range of 5.15 ha-m/y–18.30 ha-m/y with an average value of 7.01 ha-m/y for the period of 2003–2013. Thus, the study brings out that compared to the initial periods, the recent sedimentation rates have come down significantly. The sediment mass balance approach shows that the expected life of the lake is found to be ~76.7 years and lake will lose complete storage capacity by 2090, if the condition will remain same. However, analysis of conservation measures shows that the canopy cover area and organic matter contents in the soil are increasing significantly and reducing the rate of soil erosion in the study area. Silt detention structures are helpful to reduce the surface runoff and thereby reducing the rate of soil erosion. However, due to huge sedimentation, some of the silt detention structures are completely filled with sediment; hence, silt detention structures are not a long-term solution for the sedimentation problem in the Sukhna lake.

Acknowledgments

The authors are thankful to the IPRI, Amritsar, and Engineering Department, Chandigarh Administration, for providing the necessary data.

Author Contribution

The study conception and design were performed by SDK and PS. Material preparation, data collection, and analysis were performed by PS. The first draft of the manuscript was written by PS and NP redesign and drafted the final version of the manuscript.

References

Alejandro, M. and Omasa, K. (2007). Estimation of vegetation parameter for modeling soil erosion using linear Spectral Mixture Analysis of Landsat ETM data. *ISPRS Journal of Photogrammetry and Remote Sensing* 62 (4): 309–324.

Bansal, R.C. and Grewal, S.S. (1986). Studies on sedimentation of Sukhna lake; corrective measures and Hydrology in annual progress, report central soil and water conservation research and training institute, research center.

Bansal, R.C. and Mishra, P.R. (1982). Sedimentation of Sukhna Lake, Chandigarh. *Status Report*, 1–6.

Baver, L. D. (1956). Soil Physics (Vol. 81, No. 4). LWW.

Central Soil and Water Conservation Research and Training Institute (CSWCRTI). (1993). Working paper on Conservation of Sukhna Lake, Central Soil and Water Conservation Research and Training Institute Research Centre, Chandigarh.

Central Water Commission (2015). *Compendium on Silting of Reservoirs in India*. New Delhi: Central Water Commission, Government of India.

De Vente, J., Poesen, J., Arabkhedri, M., and Verstraeten, G. (2007). The sediment delivery problem revisited. *Progress in Physical Geography* 31: 155–178.

Garde, R.J. and Kothyari, U.C. (1987). Sediment yield estimation. *Journal Irrigation and Power (India)* 44 (3): 97–123.

Grewal, S.S. (2009). Study on impact of soil conservation measures in the catchment of Sukhna Lake on ground water, soil and geology. *Bulletin of Society for Promotion and Conservation of Environment (SPACE).*

Gupta, R.D., Arora, S., Gupta, G.D., and Sumberia, N.M. (2010). Soil physical variability in relation to soil erodibility under different land uses in foothills of Siwaliks in NW India. *Tropical Ecology* 51 (2): 183.

Horton, R.E. (1932). Drainage-basin characteristics. *Transactions, American Geophysical Union* 13 (1): 350–361.

IPRI (2013). *Sedimentation survey of Sukhna Lake Chandigarh after the Flood season of 2013.* Amritsar: Irrigation and Power Research Institute.

Jain, S.K., Kumar, S., and Varghese, J. (2001). Estimation of soil erosion for a Himalayan watershed using GIS technique. *Water Resources Management* 15 (1): 41–54.

Khobragade S., Semwal, P., Kumar, C. P., Kumar, S., Jain, C. K., and Singh, R. D. (2013). Integrated Hydrological Investigations on Sukhna Lake, Chandigarh for Its Conservation and Management. Draft final report of the Consultancy Project submitted to Chandigarh Administration.

Kumar, A., Devi, M., and Deshmukh, B. (2014). Integrated remote sensing and geographic information system based RUSLE modelling for estimation of soil loss in western Himalaya, India. *Water Resources Management* 28 (10): 3307–3317.

Kumar, B., Rai, S.P., Nachiappan, R.P. et al. (2007). Sedimentation rate in North Indian lakes estimated using ^{137}Cs and ^{210}Pb dating techniques. *Current Science* 00113891: 92(10).

Lecce, S.A., Pease, P.P., Gares, P.A., and Wang, J. (2006). Seasonal controls on sediment delivery in a small coastal plain watershed, North Carolina, USA. *Geomorphology* 73: 246–260.

Pant, N., Dubey, R.K., Bhatt, A. et al. (2020). Soil erosion and flood hazard zonation using morphometric and morphotectonic parameters in Upper Alaknanda river basin. *Natural Hazards* 103: 3263–3301. https://doi.org/10.1007/s11069-020-04129-y.

Pimentel, D., Harvey, C., Resosudarmo, P. et al. (1995). Environmental and economic costs of soil erosion and conservation benefits. *Science* 267: 1117–1123.

Rawat, J.S., Joshi, R.C., and Mesia, M. (2013). Estimation of erosivity index and soil loss under different land uses in the tropical foothills of Eastern Himalaya (India). *Tropical Ecology* 54 (1).

Semwal, P., Khobragade, S.D., and Nainwal, H.C. (2017). Modelling of recent Erosion rates in a Lake catchment in the North-Western Siwalik Himalayas. *Environmental Processes* 4 (2): 355–374.

Setyawan, C., Lee, C.Y., and Prawitasari, M. (2019). Investigating spatial contribution of land use types and land slope classes on soil erosion distribution under tropical environment. *Natural Hazards* 98 (2): 697–718.

Shrimali, S.S., Aggarwal, S.P., and Samra, J.S. (2001). Prioritizing erosion-prone areas in hills using remote sensing and GIS—a case study of the Sukhna Lake catchment, Northern India. *International Journal of Applied Earth Observation and Geo-information* 3 (1): 54–60.

Singh, Y. (2002). Siltation problems in Sukhna Lake in Chandigarh, NW India and comments on geo-hydrological changes in the Yamuna-Satluj region. *ENVIS Bulletin: Himalayan Ecology and Development* 10 (2): 18–31.

Strahler, A.N. (1964). Quantitative geomorphology of drainage basin and channel networks. In: *Handbook of Applied Hydrology*, vol. 4 (ed. V.T. Chow), 39–76. New York: McGraw-Hill Book Company *Section 4II*.

Upadhyay, R., Pandey, K., Upadhyay, S.K., and Avinash, B. (2012). Annual sedimentation yield and sediment characteristics of upper lake, Bhopal. *Journal of Chemical Sciences* 2 (2): 65–74.

Vicente, M.L., Navas, A., and Machín, J. (2007). Identifying erosive periods by using RUSLE factors in mountain fields of the Central Spanish Pyrenees. *Hydrology and Earth System Sciences Discussions* 12 (2): 523–535.

Xu, Y., Tang, H., Wang, B., and Chen, J. (2017). Effects of landscape patterns on soil erosion processes in a mountain–basin system in the North China. *Natural Hazards* 87 (3): 1567–1585.

Yangchan, J., Jain, A.K., and Tiwari, A.K. (2014). Spatial changing pattern in land use/land cover using GIS: a case study of Sukhna choe watershed, India. *International Journal of Scientific and Engineering Research* 5 (10): 1350–1400.

7

Geochemical Characterization and Baseline Determination of Trace Elements in Stream Waters from a Part of the Carajás Mineral Province, Brazil

Gessica da Silva[1], *Leandro S. Quaresma*[1], *Prafulla K. Sahoo*[2,3], *Gabriel N. Salomão*[2], *and Roberto Dall'Agnol*[1,2,4]

[1] *Programa de Pós-graduação em Uso Sustentável de Recursos Naturais em Regiões Tropicais, ITV, Belém, Brazil*
[2] *Instituto Tecnológico Vale (ITV), Belém, PA, Brazil*
[3] *Department of Environmental Science and Technology, Central University of Punjab, Bathinda, India*
[4] *Programa de Pós-graduação em Geologia e Geoquímica, Instituto de Geociências, Universidade Federal do Pará, Belém, Brazil*

7.1 Introduction

Water is the main medium of transport of chemical elements in surface environments, and its composition is fundamental for the cycle of chemical elements between soil–sediment–water systems (Kabata-Pendias and Mukherjee 2007; Manahan 2010). In surface waters, the concentration of chemical elements is influenced by geological setting, the nature of the soil/rock types, vegetation cover, topography, the actions of biological agents, and climate conditions (Keller 2012; Licht and Bittencourt 2013; Bundschuh et al. 2016; Gómez et al. 2017; Salomão et al. 2018; Sahoo et al. 2019). In hot and humid climates, chemical weathering predominates with a rise in temperature and precipitation, which brings more dissolved components to water bodies compared to the cold and dry regions. In addition, in hot and humid climates, the community of microorganisms and vegetation play an important role in the solubilization, mobilization, dispersion, and fixation of chemical elements in water (Licht and Bittencourt 2013). In natural conditions, the water bodies maintain a prestine conditions, although in some cases, this can be highly enriched due to the naturally enriched elements in rocks or ore deposits in the surrounding areas (Salomão et al. 2018; Sahoo et al. 2019). In other cases, the natural characteristics can be altered through the punctual release of various chemical contaminants from anthropogenic components such as effluents, sewage sludge, mine drainage, municipal waste, and fertilizers (Gaillardet et al. 2003; Sahoo et al. 2012; Sahoo et al. 2016; Ismaiel et al. 2018). Surface runoff also causes changes in stream/river water chemistry due to the entry of transported material into the water bodies in dissolved and/or suspended forms (Kabata-Pendias and Mukherjee 2007). To assess water quality and determine its source of contamination, it is important to define geochemical background or baseline values for chemical elements.

Weathering and Erosion Processes in the Natural Environment, First Edition. Edited by Virendra Bahadur Singh, Sughosh Madhav, Naresh Chandra Pant, and Ravi Shekhar.

In recent years, establishing geochemical background or baseline levels is becoming an important aspect in environmental geochemistry, especially in identifying anomalies and defining the level of soil and water contamination (Gałuszka 2007; Rodrigues and Nalini 2009; Galuszka and Migaszewski 2011; Osté et al. 2011; Salomão et al. 2018, Sahoo et al. 2019; Sahoo et al. 2020; Quaresma et al. 2022). Reimann and Garrett (2005) point out that the term geochemical background was originally used by exploration geochemists to distinguish "normal" (natural) concentrations of a given element from anomalous or discrepant concentrations. In environmental sciences, geochemical background is understood differently and is mainly used in the identification and differentiation of anthropogenic influences from natural concentrations (Galuszka and Migaszewski 2011). Matschullat et al. (2000) defined geochemical background as a relative measure that differentiates the concentrations of chemical elements in samples due to natural processes, from those of anthropogenic origin. Gałuszka (2007) defines it as the natural concentration of a substance or element in an environmental sample or in a specific medium, considering space and time factors. Especially in the case of surface waters, differentiating the levels of pollution or anthropogenic contamination from natural enrichment is a complex task, mainly due to the variability of conditions, climatic, hydrological, biological, and anthropic conditions to which this system is exposed (Gałuszka 2007; Arpine and Gayane 2016). In general, the chemical concentration supposedly coming only from natural processes, without anthropogenic interference, is characterized as the natural background level (Reimann and Garrett 2005; Gałuszka 2007). However, due to the difficulty in finding untouched areas, without traces of human interference, the term concentrations of environmental background was developed, specifically for areas that are under the influence of urban, agricultural, and industrial activities (Gałuszka 2007). For Reimann and Garrett (2005), it is appropriate to use the terms natural background, when the influence of natural processes on the data can be identified, and environmental background, when the interference of these processes is difficult to understand. The term "geochemical baseline" is commonly used as a synonym for "geochemical background," however there is some scientific differences between the two (Reimann and Garrett 2005), as the former best represents the chemical concentration of a substance in an environmental sample at a specific place and time, and allow the assessment of chemical changes of anthropic origin in the environment (Gałuszka 2007). However, the latter task is quite complex due to the fact that surface waters represent a thermodynamically open system and there are multiple potential sources of change in their environmental conditions, and this is more interesting for natural mineral provinces.

The Carajás Mineral Province is one of the main mineral provinces in the world and is located in the Eastern Amazon, in the southeast of Pará (DOCEGEO 1988; Vasquez et al. 2008). The largest Brazilian mineral exploration projects are located there, with active iron mines in Serra Norte (N4, N5, and N4WSul), Serra Leste, and Serra Sul (S11D); copper in Sossego and Salobo; manganese in Azul; and nickel in Onça-Puma. This region is also dynamic because it is located in an area of expansion of the agricultural frontier that marks the occupation of the Amazon (Souza-Filho et al. 2016). It encompasses the permanent challenges registered in different areas of the Amazon to harmonize population growth, expansion of activity, and economic production with environmental preservation of the rainforest and the environment in general. The Itacaiúnas River Watershed (IRW), located in the Tocantins-Araguaia hydrographic region, is the main drainage system in this region, and the consumptive uses of its water resources are predominantly made by the urban, livestock, and industrial sectors of the region, but non-consumptive uses such as navigation, sewage dilution, fishing, and recreation also stand out (Sedeme 2017). Considering the potential anthropogenic sources of modification of the natural conditions of the surface waters of the region, it is important to have an adequate study on the chemical quality of water resources and geochemical baseline levels of the main chemical elements in soils, stream sediments, and surface waters in the IRW region. In this region, there were no systematic studies of water quality, and geochemical baseline values for the main parameters of interest were made until the development of the Itacaiúnas Geochemical Mapping and Background (ItacGMB) project at Instituto Tecnológico Vale (ITV), Belém, Brazil. In this project, the water quality characteristics of the Vermelho and Sororó Rivers, Parauapebas River, and the entire IRW were studied by Salomão et al. (2018), Quaresma et al. (2022), and Sahoo et al. (2019), respectively. The present work focuses on the middle Itacaiúnas Subbasin (MISB), which will be a great supplement to the ItacGMB project, as it will help in understanding the factors that control the natural characteristics of surface waters, as well as it will be fundamental for the definition, in a more regional scale, of more consistent reference values for the effective evaluation of the water quality of this subbasin, since the limits provided for in the legislation (BRASIL 2005) are generalized and do not consider the particularities of each region.

Keeping in view the abovementioned points, the present study was carried out in the MISB with the following specific objectives: i) to assess the seasonality of the chemical quality of surface waters; ii) to compare the regional maps of the spatial distribution of variables measured in surface waters; iii) to understand the hydrogeochemical factors/processes controlling water quality; and iv) to define geochemical baseline values of the main elements present in surface waters. The results of this study would contribute to the future water-quality monitoring and sustainable management of water resources in the IRW and its subbasins.

7.2 Materials and Methods

7.2.1 Study Area

The MISB is located in the Carajás Mineral Province, more specifically in the IRW, which is located in the southeast of Pará (Figure 7.1a). The Itacaiúnas River begins in the municipality of Água Azul do Norte, in Serra da Seringa, and flows into the Tocantins River, in the municipality of Marabá (Possas et al. 2016). The IRW is composed of six subbasins (Figure 7.1b): River Itacaiúnas Subbasin (IRSB), River Parauapebas Sub-spiepr DHYPbasin (RPSB), River Sororó Sub-spiepr DHYPbasin (RSSB), River Cateté Sub-spiepr DHYPbasin (RCSB), River Sororó Sub-spiepr DHYPbasin (RSSB), and River Vermelho Sub-spiepr DHYPbasin (RVSB). Together, these subbasins drain an area of approximately 42 thousand km^2 (Souza-Filho et al. 2016). The MISB (Figure 7.1c) is located in the IRSB, at latitude 5°30′ and 6° S and at longitude 50°30′ and 50° W. In this region, the Itacaiúnas River has many tributaries: on its left bank, the Cinzento, Tapirapé, Preto, and the Salobo Igarapés,

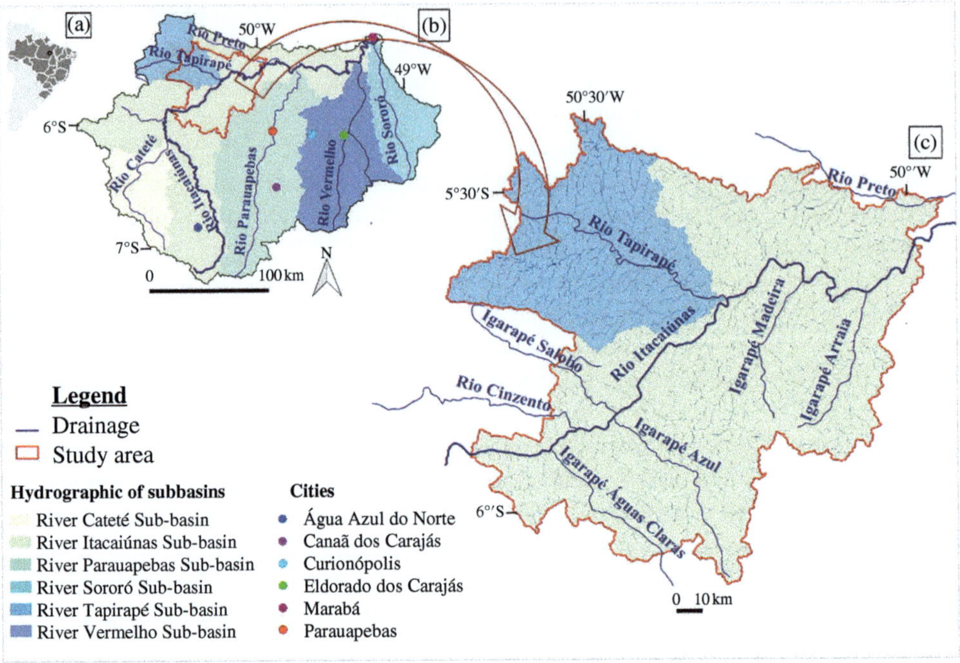

Figure 7.1 Location maps: (a) Itacaiúnas river basin in the state of Pará; (b) Highlighting the middle Itacaiúnas sub-basin (MISB) in the Itacaiúnas river basin (in red); (c) Detailed MISB (Modified from GAT 2007). Note: River Cateté Sub-basin (RCSB); River Itacaiúnas Sub-basin (IRSB); River Parauapebas Sub-basin (RPSB); River Sororó Sub-basin (RSSB); River Tapirapé Sub-basin (RTSB); River Vermelho Sub-basin (RVSB).

and, on its right bank, the Azul, Madeira, Águas Claras, and Arraia creeks (Figure 7.1c). Its drainage network is of the fifth order, non-densified, with a predominance of the mixed pattern, oriented mainly in the E–W direction and, secondarily, in the N–S, NW–SE, and NE–SW directions, with the shape of the basin moving away from the circular pattern, which favors surface runoff (Silva Júnior 2017).

7.2.2 Hydrographic Characteristics, Land Use, and Climate Conditions of the Region

The landscape of the study area is composed of mountain ranges with exposed and flattened tops (Figure 7.2a), with an altitude between 500 and 837 m, which differ from the milder lower regions that vary between 300 and 80 m in the other areas of this basin (Silva Júnior 2017; Zappi 2017). The dominant soils in the area (Figure 7.2b) are latosols and dystrophic red-yellow argisols (Schaefer et al. 2017). Dystrophic red-yellow ultisols have low infiltration capacity, high proportion of surface runoff, and are extremely susceptible to erosion, while red-yellow latosols are characterized as deep or very deep, well-drained, and acidic soils (ICMBIO 2007), being predominant in the plateau areas of the study region. In these higher regions, the soils are intensely drained, poor in nutrients, and have a low concentration of organic matter (Schaefer et al. 2017).

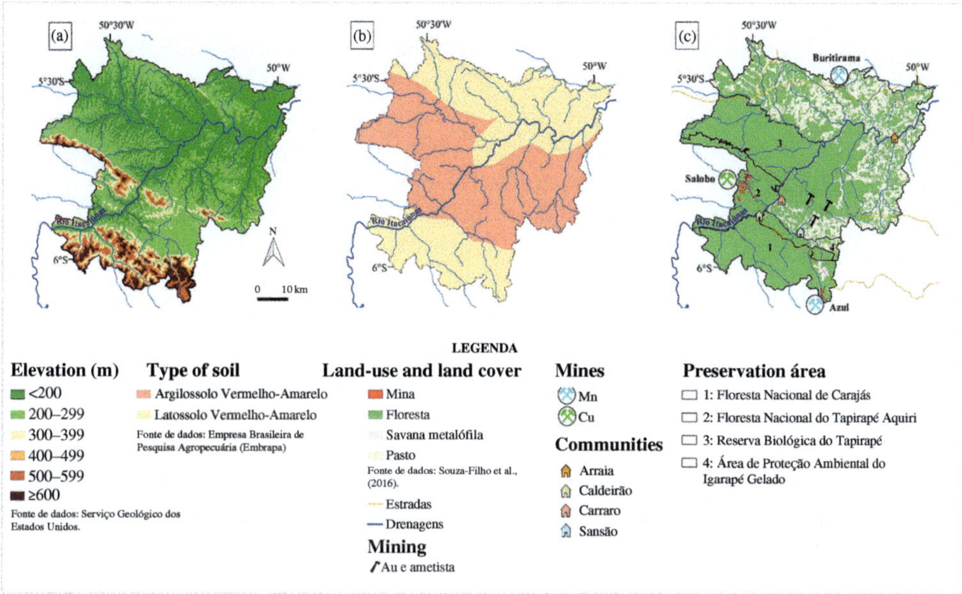

Figure 7.2 Simplified maps of the study area: (a) Elevation; (b) Types of soil (*Source:* Adapted from Schaefer et al. 2017); (c) Land use and land cover change (*Source:* Adapted from Souza-Filho et al. 2016). *Note:* 1: CNF; 2: Tapirapé-Aquiri (FNTA); 3: RBTA; 4: APAIG.

The tropical forest constitutes the original vegetation of the Carajás region (Figure 7.2c). It corresponds to a dense and open rainforest and is the predominant vegetation in the MISB. There are also local ferruginous rock fields or metallophilic savannas, known as "canga," which generally develop on iron-rich rocks of the Carajás Formation, which are directly related to iron ore deposits. The soils that develop on the canga are rich in Fe, poor in nutrients, and non-hydromorphic, presenting water deficiency in the dry season in the region (Schaefer et al. 2017; Zappi 2017). The soils of the MISB are predominantly preserved, as they are still covered by the region's original vegetation, which is protected by the Carajás Conservation Units mosaic. Among the conservation areas present in the research region (Figure 7.2c), there are parts of the Carajás National Forests (CNF) and Tapirapé-Aquiri National Forest (FNTA), Tapirapé Biological Reserve (RBTA), and the Igarapé Gelado Environmental Protection Area (APAIG). However, mainly due to deforestation, especially associated with the expansion of pastures (Souza-Filho et al. 2016), the soils of the north and east of the basin were intensely altered (Figure 7.2c). Among the main land uses (Figure 7.2c), are the activities developed by agriculture, especially pastures for cattle herds and cultivation of temporary monocultures (corn, rice, beans, cassava) or permanent (coconut from Bahia, cocoa, and coffee; ICMBIO 2009); mineral exploration of Cu and Mn; and roads and urban settlements (local communities), which generally lack a sewage collection and treatment system (Sedeme 2017). The anthropic terrains are essentially made up of dystrophic red-yellow latosols and, to a lesser extent, of argisols.

The climate is tropical monsoon, hot, and humid, with an average annual temperature of 27.2 °C and a relative humidity of 80% (Alvares et al. 2013). The annual rainfall in the region varies between 1420 and 2159 mm, and there is a clear rainfall seasonality, characterized by a rainy season (November to May), with an average annual rainfall of ~1452 mm and a less rainy or dry season (June to October), with an average annual rainfall of ~236 mm (Silva Júnior 2017). In the Serra Norte iron mines, in Carajás, the wettest period occurs between January and March, and the least wet between June and August (Teixeira 2016; historical averages between 1982 and 2015). The main water bodies in this region are perennial but show great variation in flows in the rainy and dry seasons (Silva Júnior et al. 2017).

7.2.3 Geological Setting of the Region

The geology of the IRW is relatively complex as it consists of three distinct provinces (Figure 7.3a,b): Carajás Province of Archean age and subdivided into the Carajás Basin (CB) in its central portion and into the Canaã dos Carajás, Sapucaia, and Rio Maria domains (CC), to the south; Trans-amazon Province of Archean to Paleoproterozoic, represented by the Bacajá Domain (BD), to the north; and the Araguaia Belt, of Neoproterozoic age, to the east (Sahoo et al. 2020; Salomão et al. 2021). The geology of the MISB comprises the BD to the north and the CB to the south (Vasquez et al. 2008, modified; Figure 7.3c). The exposed units in the BD are formed mainly by felsic granulites, with a predominance of high-grade charnockite rocks, and subordinate mafic granulites (Vasquez et al. 2008); by metamafic rocks in greenschist to amphibolite facies; and by metasedimentary sequences with

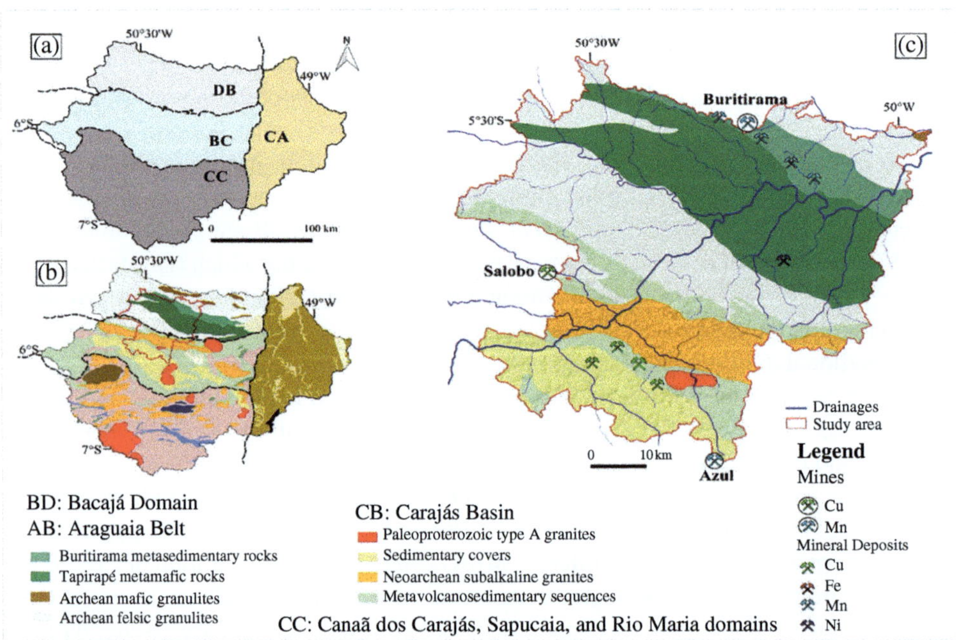

Figure 7.3 (a) Geological domains of IRW; (b) simplified geological map of the IRW highlighting the study area (MISB; in red; modified from Sahoo et al. 2020; Salomão et al. 2021); (c) simplified geological map of MISB (*Source:* Adapted from Vasquez et al. 2008).

associated manganese mineralizations. In general, the Tapirapé metamafic and Buritirama metasedimentary rocks are formed mainly by banded iron formations (BIF), metabasalts, dacites, amphibolites, metapelites, schists, and quartzites and present mineralizations of manganese, gold, and base metal sulfides (Zucchetti 2007). In turn, the Carajás Basin is composed essentially by Archean metavolcanosedimentary sequences, especially by mafic to intermediate and BIF metavolcanic rocks, with subordinate metagabbros and metapelites. These units host the Fe deposits and mineralization observed in the region (DOCEGEO 1988), in addition to IOCG (iron oxide copper gold) Cu mineralization (Moreto et al. 2015). Tapirapé metamafic and Buritirama metasedimentary rocks are formed by mafic and felsic rocks, mainly represented by iron formations, metabasalts, dacites, amphibolites, metapelites, schist, and quartz, and present mineralization of Mn, gold, and base metal sulfides (Zucchetti 2007). The metavolcanosedimentary sequences consist mainly of mafic to intermediate rocks, such as metapelites, amphibolites, and iron formations, which host the deposits and mineralization of Fe observed in the region, in addition to mineralizations of Cu oxides (Zn, Au, and Ag) (DOCEGEO 1988). Paleoproterozoic sedimentary covers partially cover Archean units and are composed mainly of clastic sedimentary rocks, such as metapelites, metasiltites, and meta-sandstones (Zucchetti 2007). Neo-Archaean and Paleoproterozoic granites still occur in the basin, geochemically related to type-A granites (Dall'agnol et al. 2017).

7.2.4 Sampling and Analytical Methods

The data used in this research were extracted entirely from the ItacGMB project (see Souza-Filho et al. (2022) for detailed information about the sampling strategy, sample collection, and analytical methods). Surface water samples were collected in two sampling campaigns during the year 2017, one in the rainy season (April to May) and another in the dry season (July to September), so that it was possible to evaluate the influence of seasonality. Samples were collected in previously defined microbasins (one per each watershed), preferably near the mouth of its main drainage, so that the sample represented the entire watershed. The results were entered and stored on tablets, which were also used to verify the location of sampling points, and served as a starting point for the organization of a geochemical database for the Background project. The sampling points were located with the support of georeferenced satellite images and based on a geographic information system. At MISB, 189 samples and 8 duplicates were collected, in which 95 samples and 5 duplicates were in the rainy season and 94 samples and 3 duplicates were in the dry season. Of all the 95 watersheds sampled in the rainy season, only one had a dry point in the dry season, which demonstrates that in this subbasin, there is practically no intermittence of drainage in the less rainy season, contrary to what was recorded in the subbasins of the Parauapebas (Quaresma et al. 2022) and Vermelho and Sororó (Salomão et al., 2018). Sampling and chemical analyses were performed by technicians from the Bioagri company (Merieux Group), which is a certified analytical laboratory. The samples were collected with a stainless steel bucket and transferred to 30 ml polyethylene bottles for the determination of anions, turbidity, and total dissolved solid (TDS) and 60 ml for the verification of chemical elements. The samples collected for the detection of chemical elements were acidified with nitric acid, under $pH < 2$, for their conservation. All samples were stored in ice boxes, at a temperature of

4 °C, for later analysis in the laboratory. In situ measurements and analytical tests were performed based on EPA (1993, 1999, 2013) and SMEWW (2017) standards. The physical–chemical parameters, temperature, hydrogenonic potential (pH), dissolved oxygen (DO), electrical conductivity (EC), and redox potential (Eh), were measured in the field using the Hanna multiparametric probe (model 98194). In the laboratory, the trace elements were determined by inductively coupled plasma mass spectrometry; anions, by ion chromatography; total phosphorus, by acid digestion; turbidity, by nephelometry; and TDS, by gravimetry. The chemical analyses were performed following the standard methods of SMWW (2012).

7.2.5 Data Preparation and Quality Control

The number of samples below the detection limit (DL) was verified for each analyzed parameter, in both sampling periods. The verification of this percentage is necessary to select the variables to be submitted to the statistical treatment and later to the calculation of the values of baseline. For this, the results of the chemical and physical–chemical analyses were organized in an Excel spreadsheet where the percentage of samples below the DL of the analytical method used was verified. The values of concentrations lower than the DL were replaced by half the DL of the method (0.5*DL) so that it was possible to consider all possible information, as adopted by Reimann and Filzmoser (2000). From this analysis, the duplicates of the samples were disregarded. To verify the variance between the results of the duplicate samples and the original samples, the relative percentage difference in the concentrations of the physico-chemical parameters, anions, and trace elements between the duplicate pairs was determined. The acceptable relative deviation for most of the parameters was ±15%.

7.2.6 Statistical Treatment of Data

To identify the variability, outliers, and distribution pattern of each variable, an exploratory and descriptive analysis of the data was carried out using the RStúdio Software (R Core Team 2018). The results were plotted on quantile–quantile (Q–Q) plots, boxplots, and exploratory data analysis (EDA) plots. Data normality was checked by applying the Shapiro–Wilk test (Shapiro and Wilk 1965), and as the results did not present a normal distribution (p-value > 0.05), nonparametric tests were used. To obtain the normality of the data distribution, the transformation to log scale was applied (Reimann and Filzmoser 2000; Reimann et al. 2005). To verify if there is a significant difference in the median concentrations between the two sampling seasons (rainy and dry), the Wilcoxon nonparametric test was applied (Wilcoxon 1945). This test compares the median values of a variable in relation to two groups. The null hypothesis (H0) is that there is no difference in concentration between sampling stations, while the alternative hypothesis (H1) is that there is a significant difference in concentration. Considering the significance limit of 0.05, that is, 5% probability of the H0 being true, if the calculated p-value is less than or equal to 0.05, the H0 is disregarded, which implies the difference. If it is greater than 0.05, it is concluded that there is no significant seasonal difference.

Furthermore, multivariate analyses such as Spearman's nonparametric correlation test (r), hierarchical cluster analysis (HCA), and principal component analysis (PCA) were applied. For this purpose, the packages "ggcorrplot," "dendextend," "FactoMiner," and "FactoExtra" from the RStúdio Software (R Core Team 2018) were applied in the same order. Correlation analysis allows the visualization of the most significant variables based on Spearman's correlation coefficient (*r*) for a significance level of 0.05 (5% error). According to Schober et al. (2018), the correlation coefficient (r) between the variables can be very strong (0.80–1), strong (0.70–0.79), moderate (0.40–0.69), weak (0.10–0.39), or negligible (0.00–0.10). HCA can be applied both to identify the similarity between samples, using the Euclidean distance (Q method), and between variables, through the correlation similarity coefficient (R method). However, only the similarity between the variables was identified, based on Spearman's correlation coefficient (*r*). Through this analysis, it was possible to identify the groups of variables that behave in a similar way. PCA allows the visualization of the components that best explain the variance of the dataset, as well as the weight (importance) that each variable has in each component, highlighting the correlation between the variables. That is, it shows the combination of the most interesting variables for the interpretation of results, allowing the reduction of the number of variables studied. Furthermore, a set of variables may be more suitable for identifying particular sources of anomalies than an assessment based on a single variable, since a geochemical landscape is better modeled from the interaction of several variables. As in the correlation analysis, the degree of correlation between the variables will be evaluated based on the Spearman correlation coefficient (*r*), using the classification scale described above, proposed by Schober et al. (2018).

7.2.7 Spatial Distribution Maps

The open-access software Quantum Gis (Quantum Gis Development Team 2009) with World Geodetic System 1984 (WGS84) datum was used to compose the maps. For the spatial representation of concentrations, six to seven class intervals were defined, depending on the individual results of each variable. The intervals were based on the minimum value, first quartile, median, third quartile, anomaly threshold, and maximum value.

7.2.8 Estimation of Geochemical Baseline Threshold Values

To calculate the geochemical baseline threshold (GB) values (upper limit), several statistical methodologies were widely applied (Ander et al. 2013; Reimann and Caritat 2017; Salomão et al. 2018; Sahoo et al. 2019), and based on the results of each one, the most appropriate techniques were selected for the present study. These techniques are cumulative frequency (CF) curve, 2σ iterative technique (2σ IT), distribution function (DF), Tukey's method (Tukey Inner Fence, TIF), median + 2 median absolute deviations (MMAD), and percentiles 5%, 75%, 95%, and 98%.

7.2.8.1 TIF

The TIF technique, based on Tukey (1977), was used to estimate the geochemical baseline values. To represent the results, a box plot was used, where 50% of the values were concentrated in the box, which was divided by a line that is equivalent to the median of the data. The stems extend to the highest or lowest value of the set of observations, being the data located beyond these stems characterized as outliers and represented by a special symbol

(Reimann et al. 2005). To calculate the TIF, it is necessary that the data is converted to log scale in base 10, to achieve symmetry, and then reconverted, so that the result can be used as a baseline value (Reimann et al. 2018). The geochemical baseline is represented by the upper limit of the interquartile range extended by 1.5 times the third quartile (Reimann et al. 2005; Rodrigues et al. 2013), according to the following equation:

$$\mathrm{TIF} = \mathrm{Q3} + 1.5 * \mathrm{IQR} \tag{7.1}$$

where Q3 represents the third quartile (75% percentile) and IQR the interquartile range, which corresponds to the third quartile (75th percentile) – first quartile (25% percentile). In the equation, 1.5 is a multiplication factor based on the supposed normal distribution of the data (Reimann et al. 2018). To determine this baseline value, the RStúdio Software (R Core Team 2018) was used.

7.2.8.2 CF Curves

As an approximation of Lepeltier's method (1969), the CF curves allow the calculation of the geochemical baseline through the accumulated frequency of each element on a linear scale, thus allowing the identification of different populations of data through inflection points (Matschullat et al. 2000). The baseline value will correspond to the upper limit of the most homogeneous dataset within each defined population, that is, by the most representative set of the total data (Rodrigues et al. 2013). This technique was applied using Minitab® software.

7.2.8.3 2σ IT and DF

These techniques were applied through the Visual Basic (VB) macro created by Nakić et al. (2007), i.e. called VB Background® (MS Excel). The macro provides, in addition to the baseline values, graphs with histogram and curves of normal distribution and accumulated frequency, for the original data and for the baseline values. The values with higher frequency and lower concentration are assumed to be representative of baseline values, while those with lower frequency and higher concentration would possibly be due to anthropogenic influence (Urresti-Estala et al. 2013). The 2σ IT calculates the standard deviation and the mean of the dataset and then establishes a range of values from the mean $\pm 2\sigma$, discarding values outside this range, until reaching the normality of the data (Matschullat et al. 2000; Rodrigues et al., 2013; Urresti-Estala et al. 2013). This procedure is repeated until all values fall within the average range $\pm 2\sigma$, which will represent the baseline range (Matschullat et al. 2000). The geochemical baseline will then be represented by the upper limit of this range (Matschullat et al. 2000; Rodrigues et al. 2013). The calculated DF uses the median to filter the data. Values above the calculated median are eliminated. From the final data, the mean and standard deviation are calculated to obtain the interval [mean $\pm 2\sigma$]. As in the IT, values beyond this range are disregarded. With the new mean and standard deviation, the accumulated normal distribution is calculated. The values within the range by both techniques were submitted to the Lilliefors statistical test of normality, which provides the T and the critical T (Tcrit) that certify the validity of the baseline values. For these methods to be considered adequate for the baseline estimation and not provide overestimated values, it is necessary that the T is smaller than the Tcrit (Urresti-Estala et al. 2013).

7.2.8.4 Percentiles (5%, 75%, 95%, and 98%) and MMAD

These methods, like the TIF, require the data to be transformed to log scale at base 10 and then calculate the MMAD and percentiles (Reimann et al. 2018). Then, the results are transformed again, being used as a baseline value. The calculations were performed using the R Studio Software (R Core Team 2018).

7.3 Results

7.3.1 Basic Statistics of the Water Quality Data

In most cases, the concentrations of As, Ag, Be, Ce, Cs, Cd, Hf, La, Sb, and Tl were below the DL, while in the cases of Ga, Hg, La, Mo, Pb, and W, more than 90% of the watersheds had levels below the DL. As a result, these elements will neither be considered in the statistical analyses nor treated in detail in the discussions of this study. In relation to anions, in the rainy season, more than 85% of the samples present levels of F^- and NO_3^- below the DL, and for more than 60% of them, the same is true for SO_4^{2-}. In the dry season, the picture is similar, as only 25% of the samples showed detectable levels of SO_4^{2-}, 33% of F^-, and 11% of NO_3^-. Cl^-, in turn, was the only anion detected in all samples collected in the wettest season and in 99% of those representative of the dry season. It was also observed that during the rainy season, a greater number of watersheds showed detectable concentrations of SO_4^{2-}, unlike the F^- that provided more samples with content above the DL in the dry season. Based on this evaluation, twenty metallic elements (Al, B, Ba, Ca, Co, Cr, Cu, Fe, K, Mg, Mn, Na, Ni, P, Rb, Sn, Sr, Ti, V, and Zn) and the 3 anions (Cl^-, SO_4^{2-}, and F^-) were selected for this study as they present a minimum of 30% of samples with concentrations above the DL. Tables 7.1 and 7.2 present the results of the basic statistical treatment obtained for the physical–chemical parameters plus anions and for the chemical elements, respectively. These tables also contain the number and percentage of samples below the DL per sampling period, as well as the results of the Shapiro–Wilk and Wilcoxon tests, and the logarithmic transformation (the first and last only for metals). The maximum values allowed (Maximum permitted value [VMP]) by CONAMA resolution 357/2005 for Class II waters are still available. Most of the variables showed a non-normal distribution (p-value* < 0.05) and high standard deviation (Tables 7.1 and 7.2), demonstrating a large dispersion in relation to the mean of the data. Also, the logarithmic transformation was not able to make all distributions normal (p-value log10 > 0.05), since several elements continued to show p-value* < 0.05. Figure 7.4 shows, as an example, the behavior of Fe, before and after the logarithmic transformation. The variables that showed a significant difference (p-value* < 0.05) between the sampling stations were the chemical elements Fe, Mn, Al, Ca, Zn, V, B, Sn, P, Cr, and Ni, the parameters physicochemical properties OD, CE, Eh, and turbidity, and the anion F^-. Most of the analyzed elements presented concentrations lower than the DL of the method or low average contents, when compared to the average values of chemical elements normally found in surface waters (cf. Table 7.1; based on Kabata-Pendias and Mukherjee 2007). To obtain an integrated and synthetic visualization of the data, EDA and Q–Q graphs were constructed for the different parameters. These graphs allow observing the distribution and possible presence of extreme values present in the dataset.

Table 7.1 Basic statistics of physical–chemical parameters and anions in both periods of MISB.

Parameter	Seasons	Unit	DL	N°(%) <DL	Min.	Q1	Med.	Q3	Max.	Mean	SD	*p*-value*	VMP
DO	Rainy	mg/l	0.1	0	1.00	5.10	5.70	6.40	8.00	5.57	1.42	< 0.05	>5
	Dry				1.40	5.40	6.15	7.00	8.10	6.05	2.11		
pH	Rainy		—	—	5.09	6.55	6.95	7.24	8.97	6.93	0.57	> 0.05	6–9
	Dry				5.60	6.77	7.07	7.28	8.96	7.02	0.44		
Temperature	Rainy	°C	—	—	20.10	23.80	25.40	26.30	37.00	25.20	1.97	> 0.05	—
	Dry				20.30	23.88	25.20	28.33	30.80	25.94	2.61		
EC	Rainy	μS/cm	1	0	12.00	33.00	46.00	55.00	236.00	49.38	33.34	<0.05	—
	Dry				9.00	21.75	34.00	54.25	752.00	46.77	76.52		
Eh	Rainy	mV	—	—	109.60	257.30	361.20	480.20	588.20	356.90	145.90	<0.05	—
	Dry				187.20	247.78	298.65	378.00	469.40	312.66	75.85		
TDS	Rainy	mg/l	5	0	12.00	40.00	51.00	64.00	346.00	56.91	42.45	>0.05	500
	Dry				9.00	38.00	49.00	60.00	440.00	54.12	45.32		
Turbidity	Rainy	mg/l	0.1	0	0.46	9.11	16.10	19.70	119.00	17.83	15.32	<0.05	100
	Dry				0.94	5.04	10.60	21.02	55.10	14.95	13.44		
Cl^-	Rainy	mg/l	0.5	0	1.00	1.92	2.68	3.87	40.00	3.56	4.44	>0.05	250
	Dry			1 (1%)	<0.5	2.04	3.22	3.91	79.10	4.13	8.03		
F^-	Rainy	mg/l	0.05	83 (87%)	<0.5	<0.5	<0.5	0.03	9.00	0.20	1.04	<0.05	1.4
	Dry			64 (68%)	<0.5	<0.5	<0.5	0.06	12.70	0.18	1.31		
SO_4^{2-}	Rainy	mg/l	0.5	62 (65%)	<0.5	<0.5	<0.5	0.74	128.00	3.07	14.53	>0.05	250
	Dry			70 (74%)	<0.5	<0.5	<0.5	0.54	121.00	2.65	14.06		

Min: Minimum; Max: Maximum; Med: Median; Q1: First Quartile; Q3: Third Quartile; SD: Standard Deviation; TDS: Total Dissolved Solid; DL: Detection limit; VMP: maximum values allowed by CONAMA resolution 357/2005; *p-value: Wilcoxon Test; N°(%) < DL: number (percentage) of samples below the detection limit.

Table 7.2 Basic statistics of chemical parameters in both periods of MISB.

Elements	Seasons	Unit	DL	N°(%) S<DL	Min.	Q1	Med.	Q3	Max.	Avg.	SD	*p*-value*	VMP
Al	Rainy	mg/l	0.001	0	0.01	0.07	0.13	0.22	3.16	0.23	0.427	<0.05	—
	Dry				0.02	0.04	0.07	0.12	3.68	0.14	0.381		
B	Rainy	μg/l	1	34 (36%)	<1	<1	2.80	4.98	209.0	6.78	23.77	<0.05	500
	Dry			76 (81%)	<1	<1	<1	<1	499.0	6.61	51.37		
Ba	Rainy	μg/l	1	0	4.09	37.6	50.7	84.7	199.0	61.4	0.040	>0.05	700
	Dry				7.08	28.8	44.0	80.4	207.0	57.1	0.043		
Ca	Rainy		0.001	0	0.56	2.46	3.61	4.60	20.60	4.00	2.587	<0.05	—
	Dry	mg/l		1	<0.001	1.62	2.56	3.20	24.40	3.08	2.884		
Co	Rainy	μg/l	1	66 (69%)	<1	<1	<1	1.22	6.86	0.89	0.89	>0.05	50
	Dry			73 (78%)	<1	<1	<1	<1	8.89	0.88	1.07		
Cr	Rainy	μg/l	1	53 (56%)	<1	<1	<1	1.43	62.00	1.75	6.32	<0.05	50
	Dry			77 (82%)	<1	<1	<1	<1	10.40	0.76	1.07		
Cu	Rainy	μg/l	1	42 (44%)	<1	<1	1.30	3.34	108.0	4.70	13.79	>0.05	—
	Dry			43 (46%)	<1	<1	1.18	1.77	17.50	1.51	1.98		
Fe	Rainy	mg/l	0.001	0	0.13	1.67	2.27	3.03	5.00	2.31	1.159	<0.05	—
	Dry		0.005		0.13	0.64	1.21	1.82	5.81	1.36	1.003		
K	Rainy	mg/l	0.001	0	0.16	1.14	1.94	2.50	35.60	2.54	4.535	>0.05	
	Dry				0.14	0.83	1.48	2.65	86.90	2.65	8.854		—
Mg	Rainy	mg/l	0.001	0	0.11	1.44	2.09	2.52	5.07	2.16	1.004	>0.05	—
	Dry				0.72	1.46	2.00	2.60	7.05	2.30	1.265		
Mn	Rainy	mg/l	0.001	0	0.004	0.05	0.09	0.13	0.85	0.12	0.143	<0.05	0.1
	Dry				0.005	0.03	0.07	0.10	0.72	0.09	0.110		
Na	Rainy	mg/l	0.001	0	0.71	2.56	4.13	5.37	29.00	4.48	3.771	>0.05	—
	Dry				0.99	2.64	4.05	5.09	86.60	4.98	8.738		
Ni	Rainy	μg/l	1	57 (60%)	<1	<1	<1	1.60	35.70	1.58	4.00	<0.05	25
	Dry			76 (81%)	<1	<1	<1	<1	5.03	0.72	0.61		

Min: Minimum; Max: Maximum; Med: Median; Avg: Average; Q1: First Quartile; Q3: Third Quartile; SD: Standard Deviation; TDS: Total Dissolved Solid; DL: Detection limit; VMP: maximum values allowed by CONAMA resolution 357/2005; **p*-value: Wilcoxon Test; N°(%) < DL: number (percentage) of samples below the detection limit.

7.3.2 Seasonal Variation of Water Quality Variables

7.3.2.1 Physico-chemical Parameters

The median, minimum, and maximum values, as well as possible outliers, of the physical–chemical parameters and of the anions in both seasons are presented in the boxplots (Figure 7.5). The limits established in CONAMA Resolution 357/05, for Class II waters, are presented by a red line (maximum value) and by a black line (minimum value). The rectangle is delimited by the 25% and 75% percentiles and is cut by a line that corresponds to the

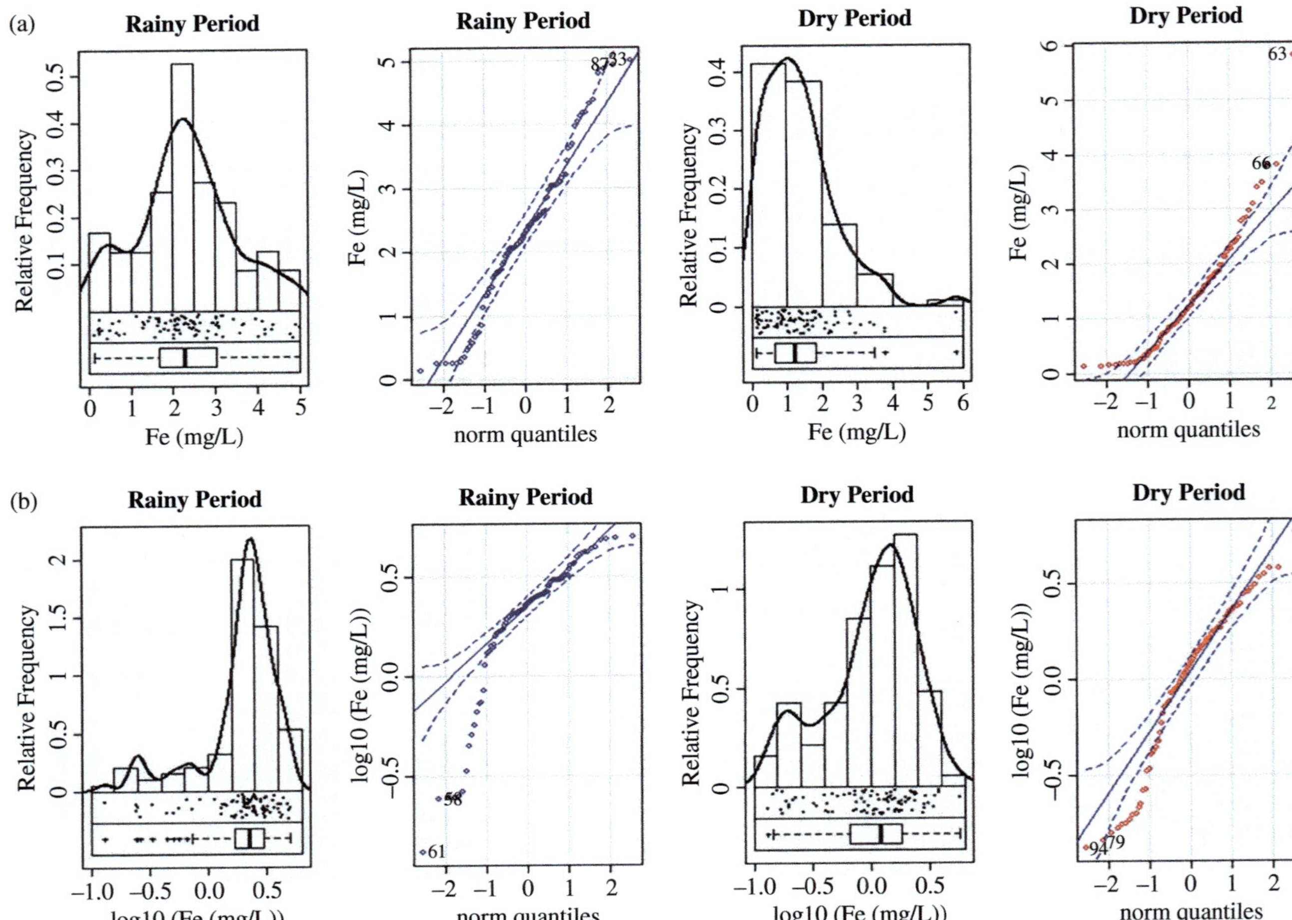

Figure 7.4 EDA and Q–Q plots of (a) not normalized and (b) log normalized of Fe concentration in both periods of MISB.

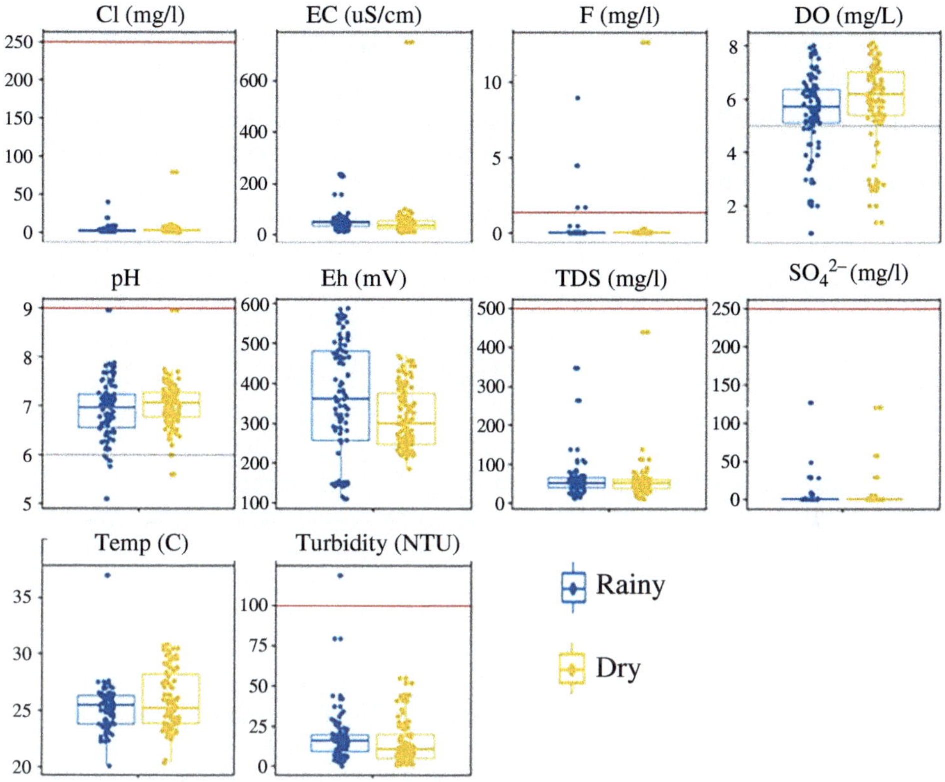

Figure 7.5 Boxplots of physical-chemical parameters and anions of surface waters of the middle Itacaiúnas Subbasin in both seasons. Note: the red line and the black line indicate, respectively, the maximum value and the minimum value allowed, according to CONAMA Resolution 357/05, for Class II Waters.

median of the data. The pH of the waters showed no significant difference between the two seasons and indicated, in general, almost neutral to slightly alkaline surface waters. The measured values remained, in most cases, within the pH range of 6 to 8 (Figure 7.5); however, some isolated values approached 9 and others were between 5 and 6, therefore weakly acidic. In both seasons, the median and mean pH are very close to 7 (Table 7.1). No watersheds with pH values higher than 9 were identified and values lower than 6 were measured in six watersheds in the rainy season and in only one in the dry season. Unlike the pH, the Eh of the river waters varied significantly with the change of season, being more oxidizing in the period of greater precipitation (Figure 7.5). During this period, the Eh ranged from 109.6 to 588.21 mV, with a median of 361.20 mV, while in the dry season, it ranged between 187.2 and 469.4 mV, with a median of 298.65 mV (Table 7.1). Surface waters were significantly more turbid in the rainy season, with turbidity values ranging from 0.46 NTU to 119 NTU and a median of 16.10 NTU (Table 7.1, Figure 7.5). In the dry season, in addition to lower values, ranging between 0.94 NTU and 55.10 NTU with a median of 10.60 NTU, the range of variation was much more restricted (Table 7.1). It was observed that only one watershed in the rainy season presented turbidity above 100 NTU (Figure 7.5). There was no significant change in the median levels of TDS with the change of season, unlike the EC which indicated slightly more conductive waters in the rainy season, with the exception of an extreme value of 752 μS/cm measured in the dry season (Table 7.1, Figure 7.5). In the rainy season, TDS showed concentrations between 12 mg/l and 346 mg/l with a median of 51 mg/l, and the conductivity varied between 12 μS/cm and 236 μS/cm with a median of 46 μS/cm, while that in the dry period the TDS levels ranged between 9 mg/l and 440 mg/l with a median of 49 mg/l, and the conductivity between 9 μS/cm and 752 μS/cm with a median of 34 μS/cm (Table 7.1). It can be seen that the TDS values are almost always lower than 100 NTU and do not exceed the VMP of 500 NTU indicated by the CONAMA Resolution (Figure 7.5). The temperature, as well as the TDS levels, also did not show significantly different median values between the seasons (Figure 7.5), being equal to 25.4 °C in the rainy season and 25.2 °C in the dry season (Table 7.1). The lowest temperature was equal to 20.1 °C and was measured in the rainy season, as well as the maximum of 37 °C (entirely isolated value), while in the dry period, the minimum was 20.3 °C and the maximum of 30.8 °C (Table 7.1). In terms of DO, in the rainy season the concentrations varied between 1 mg/l and 8 mg/l, with a median value of 5.7 mg/l, while in the dry season, it fluctuated between 1.4 mg/l and 8.1 mg/l, with a median content of 6.15 mg/l, presenting, therefore more oxygenated waters in the dry season. In general, the waters of the microbasins were well oxygenated in both periods, since the 25% quartile levels in both seasons were above 5 mg/l (Figure 7.5), the minimum level considered adequate for the development and maintenance of aquatic life in Class II waters. On the other hand, there is an expressive number of results with DO <5 mg/l and great dispersion of the values obtained, so their causes should be investigated in more detailed studies. Evaluating the distribution of the selected anions, it was found that Cl^- presented levels above the DL in practically all samples, while more than two-thirds of the SO_4^{2-} and F^- analyses were below the DL in both stations (Table 7.1, Figure 7.5). The Cl^- concentration ranged from 1 to 40 mg/l with a median of 2.68 mg/l in the rainy season and from 0.25 to 79.1 mg/l with a median of 3.22 mg/l in the dry season, being, therefore its values higher in the period of less precipitation. It is difficult to evaluate the behavior of F- and SO_4^{2-} due to the dominance of values below the DL. However, the results obtained do not point to a significant contrast between the two stations. The Cl^- and SO_4^{2-} contents are always below the VMP, while two samples from the rainy season and four from the dry season showed F^- levels higher than the VMP (Figure 7.5).

7.3.2.2 Chemical Elements

The boxplots of the trace elements in the samples analyzed in both dry and rainy seasons are represented in Figure 7.6. The possible outliers can also be visualized, as well as the legal maximum value, when available, which is represented by a red line. The evaluation of the dataset reveals that, in general, the contents of chemical elements in surface waters are higher in the rainy season than in the dry season. Another point worth mentioning is the fact that they are very restricted, with the exception of Mn, which will be discussed separately, the analyses that showed values above the VMP provided for in CONAMA resolution 357/2005. Except for Mn, only five samples were observed with a content higher than the VMP, two with Cr, two with Ni, and one with Zn (Figure 7.6), all measured in the rainy season. High values of Fe were also verified, but there is no definition of VMP for the total concentration of this element. The behavior of alkaline elements (Na, K, and Rb) and

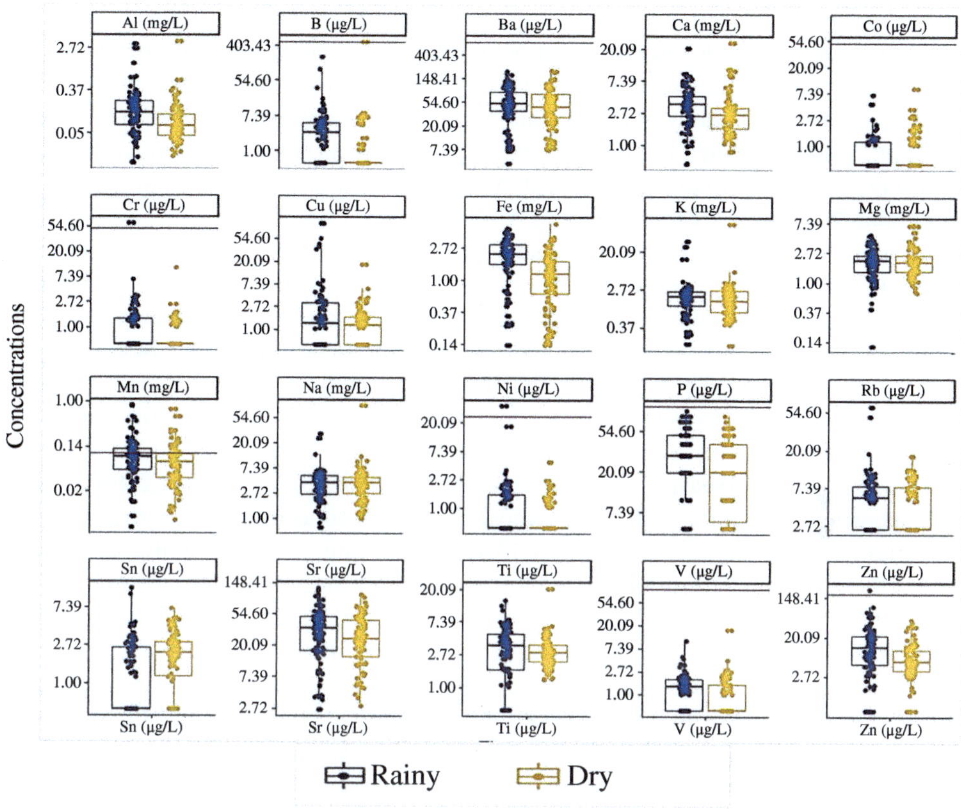

Figure 7.6 Boxplots of chemical elements from surface waters of the middle Itacaiúnas Subbasin. Note: the red line indicates the maximum value allowed for Class II waters, according to CONAMA Resolution 357/05.

alkaline earth elements (Mg, Ca, Sr, and Ba), with the exception of Ca, was uniform in the subbasin in both seasons. Unlike the Ca contents, whose medians (3.61 mg/l in the rainy season; 2.56 mg/l in the dry season) varied significantly between seasons (Table 7.2), with its content being higher in surface waters during the dry season. There were no significant differences with the change of season in the concentrations of Sr (34 μg/l; 25 μg/l), Mg (2.09 mg/l; 2.00 mg/l), K (1.94 mg/l; 1.48 mg/l), Na (4.13 mg/l; 4.05 mg/l), Rb (62.9 μg/l; 17.1 μg/l), and Ba (50.07 μg/l; 44 μg/l). It is also verified that the detected concentrations of Ba were lower than the VMP of 700 μg/l.

The Fe contents showed significant differences in the transition from the rainy season to the dry season (Figure 7.6, Table 7.2), being higher in the period of highest rainfall. In the rainy season, the median levels of Fe were equal to 2.27 mg/l, decreasing to 1.21 mg/l in the dry season (Table 7.2). There is no VMP for total Fe in the CONAMA resolution. Mn also tends to present higher levels during the rainy season, but the median values of 0.09 mg/l in the rainy season and 0.07 mg/l in the dry season (Table 7.2) are much closer than in the case of Fe. Another relevant aspect is that there are numerous analyses that showed values above the VMP of 0.1 mg/l and they occur in both seasons (Figure 7.6), where 29 samples correspond to the rainy season and 22 to the dry season. The median Al contents were significantly different between seasons, being higher in the rainy season (Figure 7.6). In this season, the median concentrations were equal to 0.13 mg/l, decreasing in the less rainy season to 0.07 mg/l (Table 7.2). Contrary to Al, the median levels of Ti did not change significantly between seasons, although they remain higher in the rainy season, being equal to 3.54 μg/l in the rainy season and 2.85 μg/l in the dry season (Table 7.2, Figure 7.6). Like Fe and Al, P also varied significantly between seasons, being more abundant in the rainy season. In that season, the median was 30 μg/l and 20 μg/l in the dry season, never exceeding the MPV of 100 μg/l predicted in the resolution (Table 7.2, Figure 7.6). More than 50% of the levels of Ni, Cr, and Co in the waters of the rainy season and more than 75% of the levels of the dry season were below the DL (Table 7.2, Figure 7.6). These proportions decrease but remain high for V (31% and 57%) and Cu (44% and 46%) in the rainy and dry seasons, respectively. Co does not vary significantly between the two seasons, its maximum content in the rainy season being 6.86 μg/l and 8.89 μg/l in the dry season, not exceeding the VMP of 50 μg/l (Table 7.2, Figure 7.6). The opposite is true for Cr, Ni, and V whose maximum concentrations in the rainy and dry seasons were, respectively, 62 and 10.4 μg/l, 35.7 and 5.03 μg/l, and 10.6 μg/l and 16.6 μg/l. All V concentrations are notably lower than the VMP (100 μg/l; Figure 7.6), while Cr and Ni showed two results each above the VMP (Figure 7.6). The Cu values between seasons are strongly discrepant, since in the rainy season its median was equal to 1.30 μg/l, while in the dry season, it was equal to 1.18 μg/l (Table 7.2, Figure 7.6). There is no VMP available for this element. Zn differs from Cr, Ni, V, Cu, and Co because it presents almost all the analyses with values above the LD (Table 7.2) but also exhibits higher median contents in the rainy season, when its median concentration was 11.9 μg/l, while in the dry season, it was 6.06 μg/l. The results obtained are lower than the VMP of 180 μg/l, with the exception of only one sample from the rainy season (Table 7.2, Figure 7.6). Sn is a notable exception in the set of elements analyzed, because, unlike the others, its median (<1 μg/l in the rainy season and 2.20 μg/l in the dry season) and average concentrations (1.68 in the rainy and 2.31 in the dry season) were higher in the dry season (Figure 7.6). There is no MPV for Sn, as well as for the elements Al, Ca, Cu, Fe, K, Mg, Na, Rb, Sn, Sr, and Ti (Table 7.2, Figure 7.6). In contrast to Sn, the total levels of B in the waters of the less rainy period decreased significantly, being detected in only 20% of the watersheds (Table 7.2, Figure 7.6). The maximum content in the dry season (499 μg/l) approached, but did not exceed, the MPV of 500 μg/l, while that of the rainy season was much lower (209 μg/l; Table 7.2, Figure 7.6).

7.3.3 Spatial Distribution of Water Quality Variables

The spatial distribution of physico-chemical parameters and trace elements was evaluated using geochemical maps. This can be seen for selective parameters in Figures 7.7 and 7.8. Geochemical maps were not prepared to present the distribution of sulfate (SO_4^{2-}) and fluoride (F^-) anions, because they had 70% or more concentrations below the DL (Table 7.1). Similarly, the chemical elements of which more than 50% of the analyses were below the DL were not considered (Cr, Ni, Co, cf. Table 7.2; as well as the elements discarded in the initial stages: Ce, Pb, Ga, Mo, Se, W, Hg, Cs, La, Tl, Cd, Sb, As, Be, Ag, and Hf).

The waters of the vast majority of watersheds showed values of the DO parameter above 5 mg/l, the minimum value admitted in CONAMA Resolution 357/2005. However, the waters of twenty watersheds in the rainy season and fifteen in the dry season showed DO concentrations lower than 5 mg/l. It is observed that values below 5 mg/l of DO are

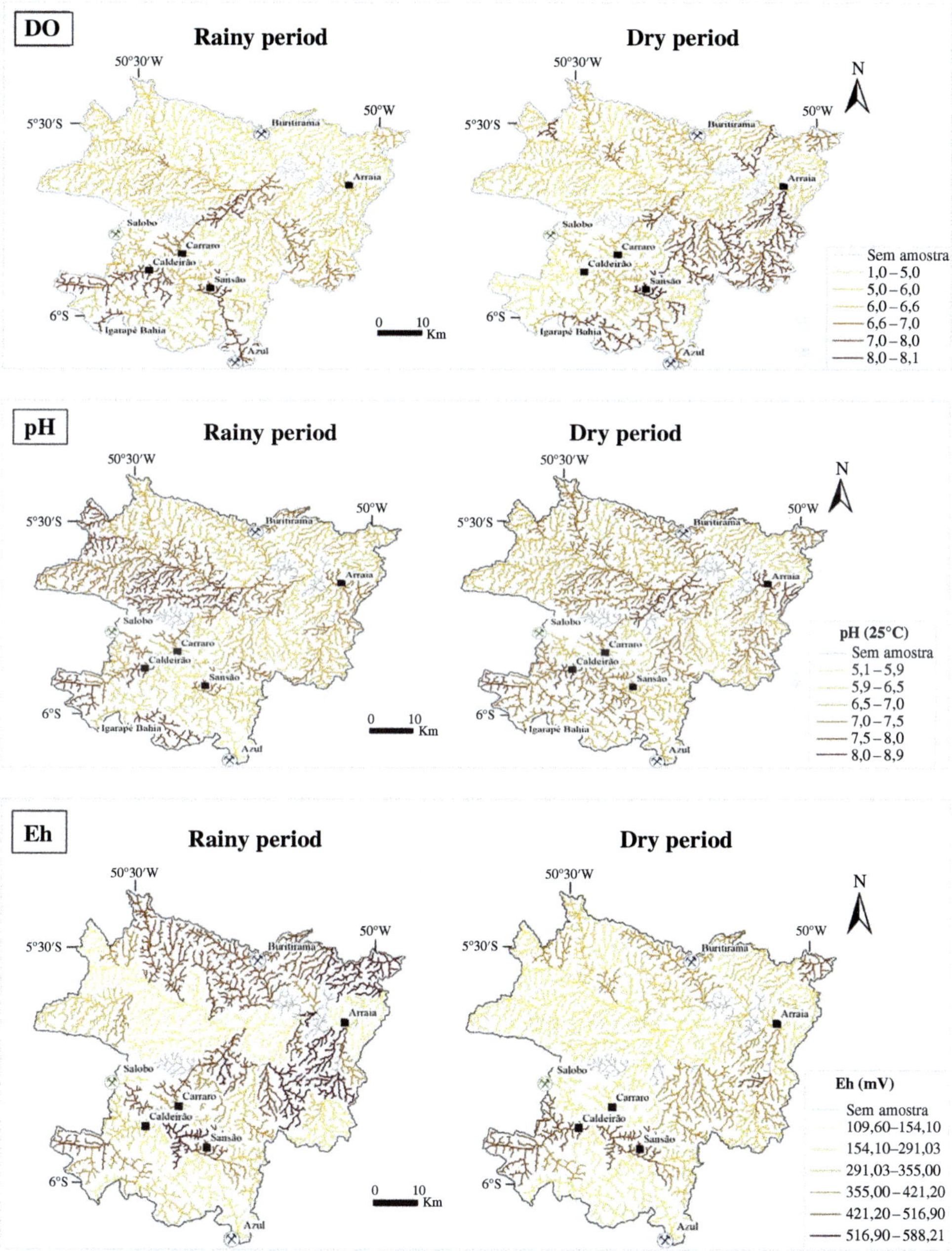

Figure 7.7 Spatial distribution map of physico-chemical parameters in two seasons of the middle Itacaiúnas Subbasin.

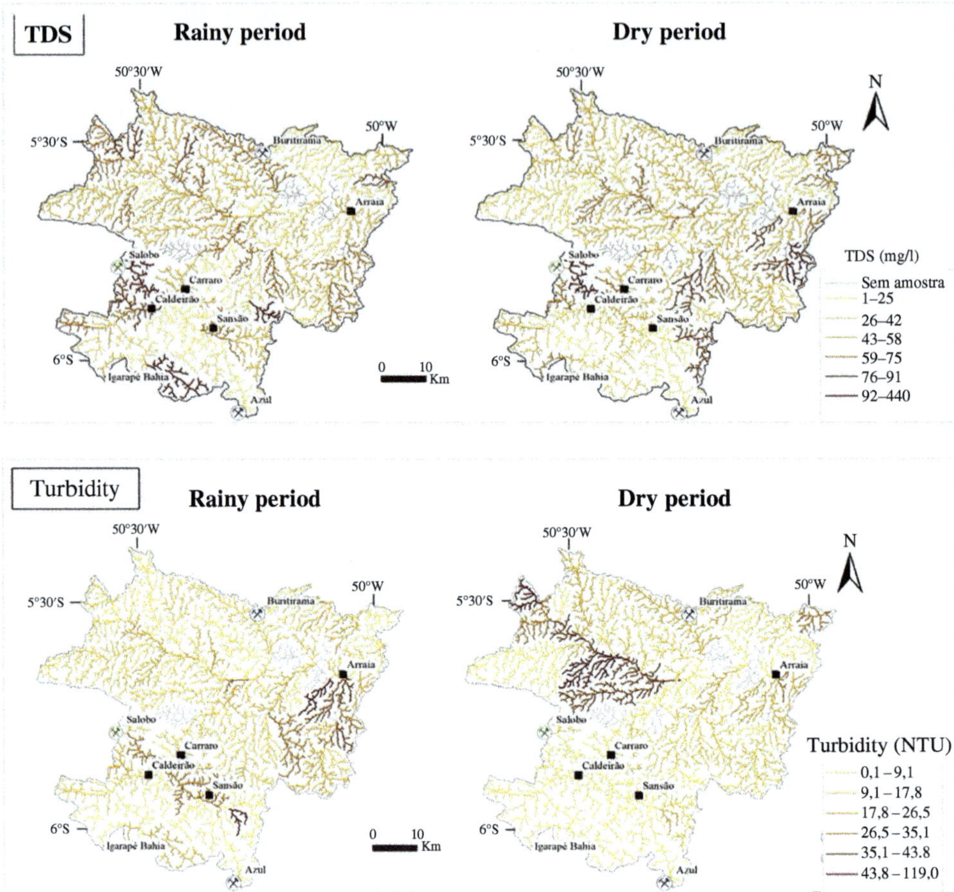

Figure 7.7 (Continued)

predominantly in the microbasins that drain the north and southeast of the subbasin, regions where there is less forest cover (Figure 7.2), although there are some values in disagreement with the legislation. (BRASIL 2005) in areas considered preserved. There is no great contrast in terms of TDS in the two stations (Figure 7.7). The highest values of TDS are located along the Salobo stream in both seasons, in the areas of the Águas Claras stream and in the APA of the Igarapé Gelado in the rainy season, and in the headwaters of the Arraias stream in the dry season. The lowest TDS values were recorded along the Azul stream in the south of the subbasin, in tributaries of the Preto River in the north and west of the area in streams that drain into the Tapirapé River. The spatial variation of the turbidity values in the surface waters of the subbasin can be seen in Figure 7.7. When comparing the two periods, it can be seen that the turbidity in the waters showed significant differences, and in the rainy season, the highest values were recorded in the Arraia, Azul, and Salobo streams, and in the upper portion of the Madeira stream, while in the season of lower rainfall they were concentrated in the Tapirapé River and its tributaries on the right bank, in the Arraia stream and in streams that flow into the Preto River, at the extreme northeast of the area.

It is noteworthy that adjacent to these drainages there are mining activities, farms, and rural communities and part of them are located in areas affected by intense deforestation (Figure 7.2), which probably may have contributed to these waters being more turbid than the waters of the other microbasins. The possible influence of the Salobo mine on the waters of the homonymous stream cannot be ruled out either. In the case of the Azul

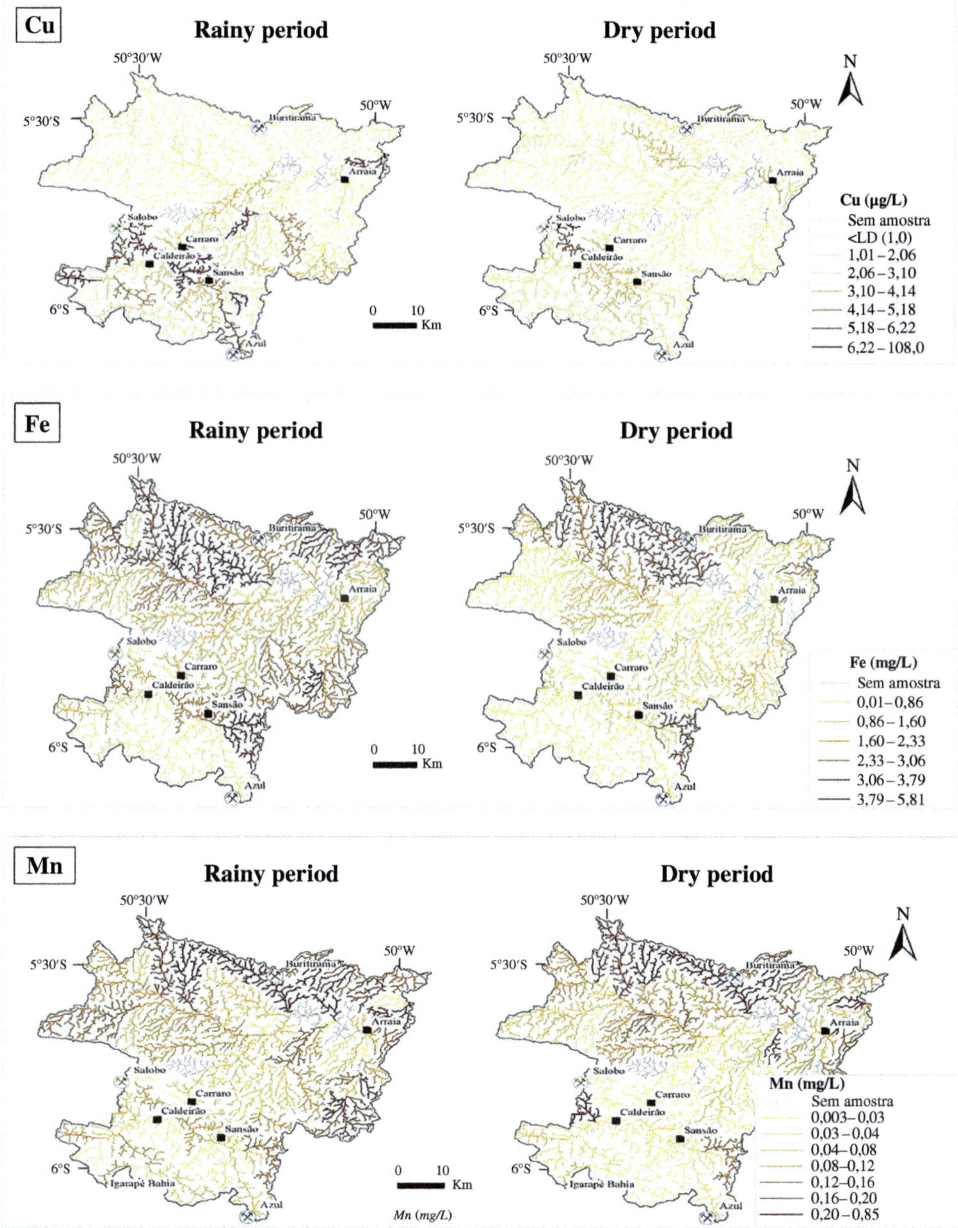

Figure 7.8 Spatial distribution map of chemical elements in two seasons of the middle Itacaiúnas Subbasin.

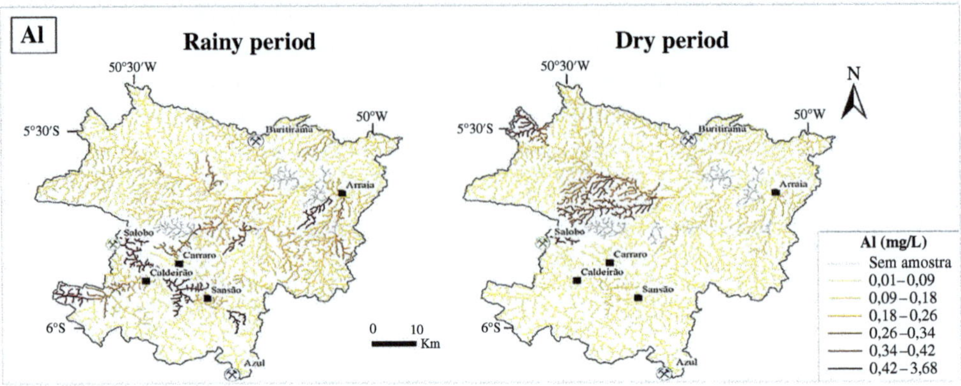

Figure 7.8 (Continued)

stream, the influence of the manganese mine located at its sources seems less likely, since the drainages with greater turbidity are located far downstream of the mine. In any case, it is worth noting that a turbidity value above the VMP of CONAMA Resolution 357/2005 was observed in only one microbasin during the rainy season (Figure 7.7, Table 7.1). When evaluating the spatial distribution of EC, it is noted, with the exception of an anomalous value in the dry season, a significant reduction in EC values with the change of season from the rainy season to the dry season. In the rainy season, the highest values tended to be concentrated in the Salobo stream and in the eastern portion of the subbasin, while the lowest values were located along the Águas Claras stream and in the north of the area. In the dry season, low EC values predominate in the south, center, and north of the subbasin. As in the spatialization of the EC, there was also a significant difference in the behavior of Eh of the waters of the sampled watersheds. Waters with more oxidizing characteristics were found in the east of the basin, possibly due to greater aeration of the rivers, since in this region, the landscape is altered, with a predominance of undergrowth, where the water courses are fully exposed to the atmosphere. In Figure 7.7, it is possible to observe the spatial distribution of the pH parameter. It is verified that the waters of the subbasin in the two sampling stations present a uniformity, being on average close to 7. The values below the minimum value (pH 6), corresponding to slightly acidic waters, were recorded along the stream. Blue and in the north of the area. As well as the pH parameter, the spatial distribution of the temperature parameter and of the Cl^- anion did not show marked variations between the two seasons. In the case of the Cl^- anion, there is a concentration, especially in microbasins of the Salobo and Arraias streams and in those that drain into the Preto River in the north of the area, but in general, its contents in the surface waters of the subbasin are very low.

The Fe content in the surface waters of the subbasin is markedly higher in its northern and eastern portions (Figure 7.8), which coincide with areas affected by intense deforestation (Figure 7.2), than in areas covered by tropical forest, which generally correspond to protected areas. It is also observed that the Fe contents are also higher in the rainy season than in the dry season. This contrast can be explained by the increase in leaching and Fe transport to drainage, especially in deforested areas, where surface flow during the rainy season is favored (Souza-Filho et al. 2016; Levy et al. 2018; Salomão et al. 2018). The spatial distribution of Mn (Figure 7.8) is similar to that of Fe in the rainy season but differs from it because it does not show a marked contrast between the two seasons. Furthermore, the sharp increase in its concentration in the microbasins of the northern portion of the studied area, within the BD, is notable. There is no evidence of Mn enrichment in the waters of the microbasins that drain into the Azul stream, suggesting that there is no Mn contamination that could be related to this mine. On the other hand, there are indications of Mn enrichment in the microbasins located close to the Buritirama mine. Al (Figure 7.8) during the rainy season shows a higher content in microbasins related to the Salobo, Azul streams, and streams in the central portion of the subbasin that flow into the Itacaiúnas River, while in the dry season, it is concentrated, especially in the northwest of the subbasin, on the Tapirapé River and its right bank tributaries. However, in general, the Al content in surface waters is very low in both seasons.

Regarding the distribution of Ca, Mg, K, and Na, it is found that these elements do not show very significant differences in the two stations, with the highest levels concentrated in microbasins downstream of the Salobo mine and, with the exception of K, in the area of the Arraia stream. The waters of the microbasins of the southern portion of the studied area tend to show lower contents of these elements. This behavior can be partially explained by the influence of reactive limestone barriers used to neutralize the acid drainages of the Salobo copper mine. On the other hand, as these elements showed a significant correlation with each other (in general, $r > 0.5$) in both periods, this indicates the influence of weathering of feldspar-rich silicatic rocks. Ba (Figure 7.8), unlike Sr, did not show marked differences in the two seasons and, in terms of spatial distribution, both show higher concentrations in the surface waters of the northwest portion of the subbasin, within the BD. The levels of these elements in the waters of the central and southern microbasins are, in general, very low. The distribution of Rb differs from Ba and Sr because it has many watersheds with contents lower than the DL. It tends to show higher concentrations in the center of the subbasin during the wet season and in more varied areas during the dry season. There is a contrast in the distribution of elements Ti and Zn. Both present a predominance of values above the DL. While the highest concentrations of Ti in the rainy season occurred in the central and northern portions of the subbasin, and in the dry season in its northwest portion, Zn shows higher levels in surface waters in the northwest portion of the subbasin in the rainy season and relatively regular distribution in the dry season. V has many values below DL; however, its spatial distribution is very similar to Ti. Cu (Figure 7.8) differs from the other elements because it shows notable enrichment, especially in the rainy season in the southern portion of the subbasin, coinciding with the northern copper belt, while in the northern portion, its contents are low and often smaller than the DL. Sn also exhibits a particular behavior, as it tends to present higher levels during the dry season and, above all, in the northwestern and southern portions of the studied area. Boron provided many results below the DL but showed higher concentrations during the rainy season in the northern and southern portions of the subbasin, the latter possibly related to the intense hydrothermal processes that mark the copper belt. In the dry season, almost all of the water analyses revealed B contents lower than the LD. Ni, Cr, and Co show analogies in their spatial distribution and in the dominance of levels below LD. Ni and Cr concentrations tend to be higher in the rainy season, and the microbasins with the highest levels are located in the central and northern portions of the area for Ni and central and southwest for Cr. Co, on the other hand, does not show much difference between the two seasons and is concentrated in both of them in an accentuated way in the northern part of the subbasin.

7.3.4 Multivariate Statistics

Correlation matrices were created to represent the correlation between the variables selected in the rainy season and in the dry season (Figure 7.9a,b). The shades of red color indicate a positive correlation between the variables and those of blue color

(a)

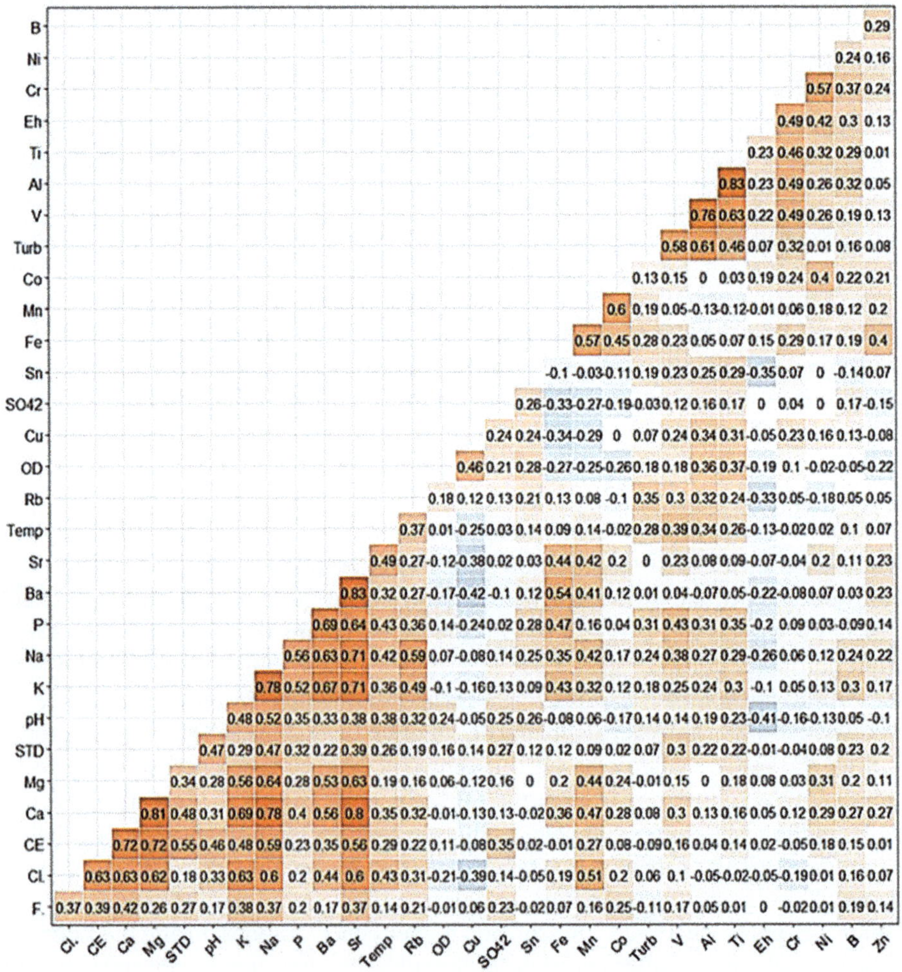

Figure 7.9 Spearman correlation matrix (r) of water quality parameters in two seasons (a: rainy season; b: dry season) of the middle Itacaiúnas Subbasin. (Abbreviations: Temp: Temperature; Turb: Turbidity; TDS: total dissolved solid; EC: electrical conductivity; DO: dissolved oxygen).

(b)

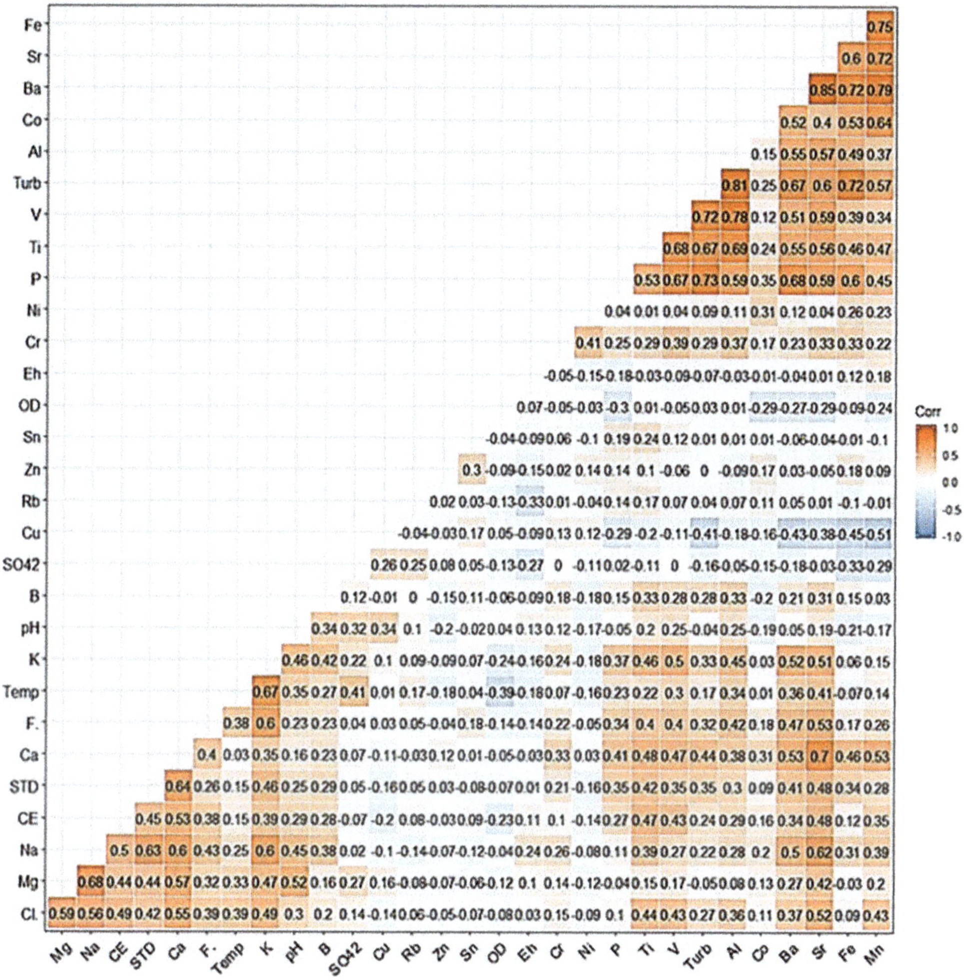

Figure 7.9 (Continued)

represent a negative correlation. The more intense the tonality, the more expressive the correlation. In both sampling periods, all strong correlations were positive. In the rainy season, the strong correlations between alkaline and alkaline earth elements, as well as those with EC, stand out. Also noteworthy are the strong correlations of Al with V and Ti. In the dry season, among the main correlations are those of Al, P, V, and Fe with turbidity, with less emphasis on alkaline and alkaline earth. In the rainy season, strong correlations occurred between the variables Al–Ti, Al–V, EC–Ca, EC–Mg, Ca–Mg, Ca–Na, Ca–Sr, K–Na, K–Sr, Na–Sr, and Ba–Sr (Figure 7.9a), while in the dry season, it occurred between Al–V, Ba–Sr, Ba–Fe, Ba–Mn, Sr–Ca, Sr–Mn, Mn–Fe, Turb–Fe, Turb–P, and Turb–V (Figure 7.9b). The physico-chemical parameters did not present a strong correlation, neither among them nor with the other variables, except in the already mentioned cases of EC with Ca and Mg in the rainy season and of turbidity with Al, Fe, P, and V in the dry season (Figure 7.9). A large number of variables showed a moderate positive correlation in both periods, while moderate negative correlations are scarcer, being limited to Ba–Cu and pH–Eh in the rainy season and Cu with Ba, Fe, and Mn and turbidity in the rainy season, dry season (Figure 7.9). The groups formed by the cluster analysis highlighted and confirmed the correlation between the variables verified in the Spearman correlation analysis (r), where the closest variables (smaller distance) present a greater correlation with each other. The following groups were formed in the rainy season (Figure 7.10a): Co–Fe–Mn, TDS–pH–Rb–Temp–P–K–Na–Ba–Sr–F–Cl–EC–Ca–Mg, DO–Cu–SO_4–Sn, Turb–V–Al–Ti–Eh–Cr–Ni–B–Zn. While in the dry period (Figure 7.10b), the following groups were formed: Cr–Ni, DO–Eh–Cu–Zn–Sn–SO_4^{2-}–Rb, pH–Mg–B–Temp–K–F, EC–Cl^-, and Na–TDS–Ca–Co–Fe–Mn–Ba–Sr–Ti–V–Al–Turb–P.

From the PCA analysis (Figure 7.11), it was possible to reduce the dataset and verify the variables that present the greatest contribution to explain, considering the parameters studied, the characteristics of the waters. Seven principal components with eigenvalues ≥ 1 were generated, which explained more than 70% of the data variability at each sampling station, with 44% of the total variance explained by the first two principal components. Considering a weight greater than 0.45, the first two main components selected the variables with the greatest contribution to the characteristics of the basin waters during the rainy season (Al, B, Ba, Ca, EC, Cl^-, Cu, K, Fe, K, Mg, Mn, Na, DO, P, pH, Rb, Sr, TDS, Temperature, Ti, and V) and in the dry season (Al, B, Ba, Ca, EC, Cl-, Cu, K, Fe, K, Mg, Mn, Na, DO, P, pH, Rb, SO_4^{2-}, Sr, TDS, Temperature, Ti, Turbidity, and V). In these two main components, due to their high weights, the chemical elements Ca, K, Na, and Sr, in the rainy season, and the chemical elements Al, Fe, K, Na, Sr, TDS, Ti, and V, in the rainy season, stand out of drought. The variables analyzed are represented by directional arrows in the correlation circle (Figure 7.11), and their degree of importance in the two main components is indicated by the size of the vectors. The closer the variable is to the circle, the more important the variable is to the PC. Based on their degree of affinity, these variables were grouped and represented by four clusters. In the rainy season, clusters 1, 2, 3, and 4 grouped, respectively, the variables: DO–Sn–Ti–Al–Rb–V–Turbidity–Temperature; SO_4^{2-}–pH–TDS–EC–Cl–K–Ca–Na; Ba–Sr–Zn–P–Mg–Fe–Mn–Mg; Cr–Cu–Ni–Eh. In the dry period, the variables were grouped in clusters 1, 2, 3, and 4, respectively: SO_4^{2-}–pH–Mg–Ca–Rb–B–Cl–EC; Ti–Zn–Sr–Ba–P–Fe–Mn; Cr–Cu–Ni–Eh; and DO–Sn–Na–K–V–Al–Turbidity–Temperature. In general, the groups presented the same variables in both periods.

7.3.5 Estimated Geochemical Baseline Values

The estimated geochemical baseline threshold (GB) values obtained by each technique and the number of concentrations that exceed these values are presented in Table 7.3, as well as the VMP of the elements provided for in CONAMA Resolution 357/05, for Class II waters, and the DL of the method. Due to undetected concentrations (<LD), it was not possible to establish GB values through the 2σ IT and DF techniques for elements B, V, Zn, Rb, Ti, Ni, Cr, Cu, and Sn, as well as for P in the dry season. Baseline levels varied

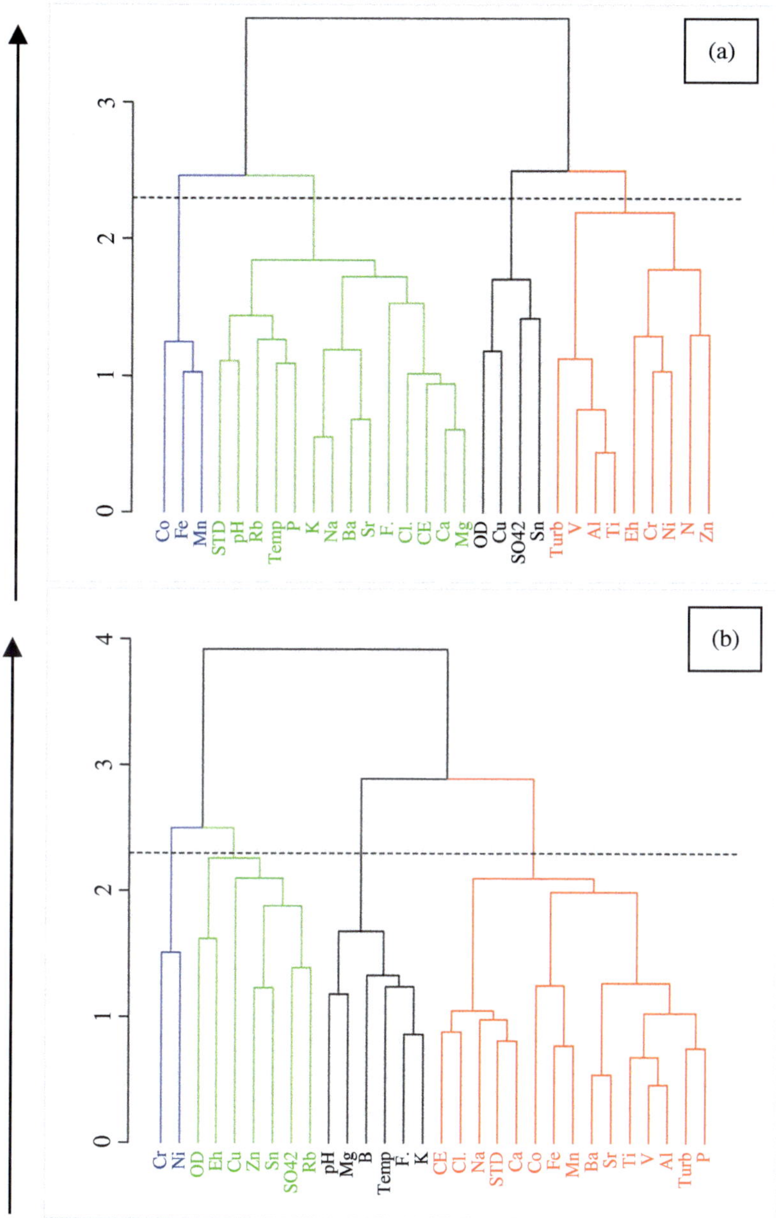

Figure 7.10 Cluster of water quality variables in two seasons (a): rainy season; (b): dry season) of the Middle IRSB. (Abbreviations – Temp: Temperature; Turb: Turbidity; TDS: total dissolved solid; EC: electrical conductivity; DO: dissolved oxygen).

considerably between techniques as well as between sampling stations. Some elements, such as B and Zn, presented very different values in the two stations, even using the same methodology. Comparing the results obtained by the different methods, it is observed that, leaving aside the values indicated by the fifth percentile (P5) and CF, the lowest baseline

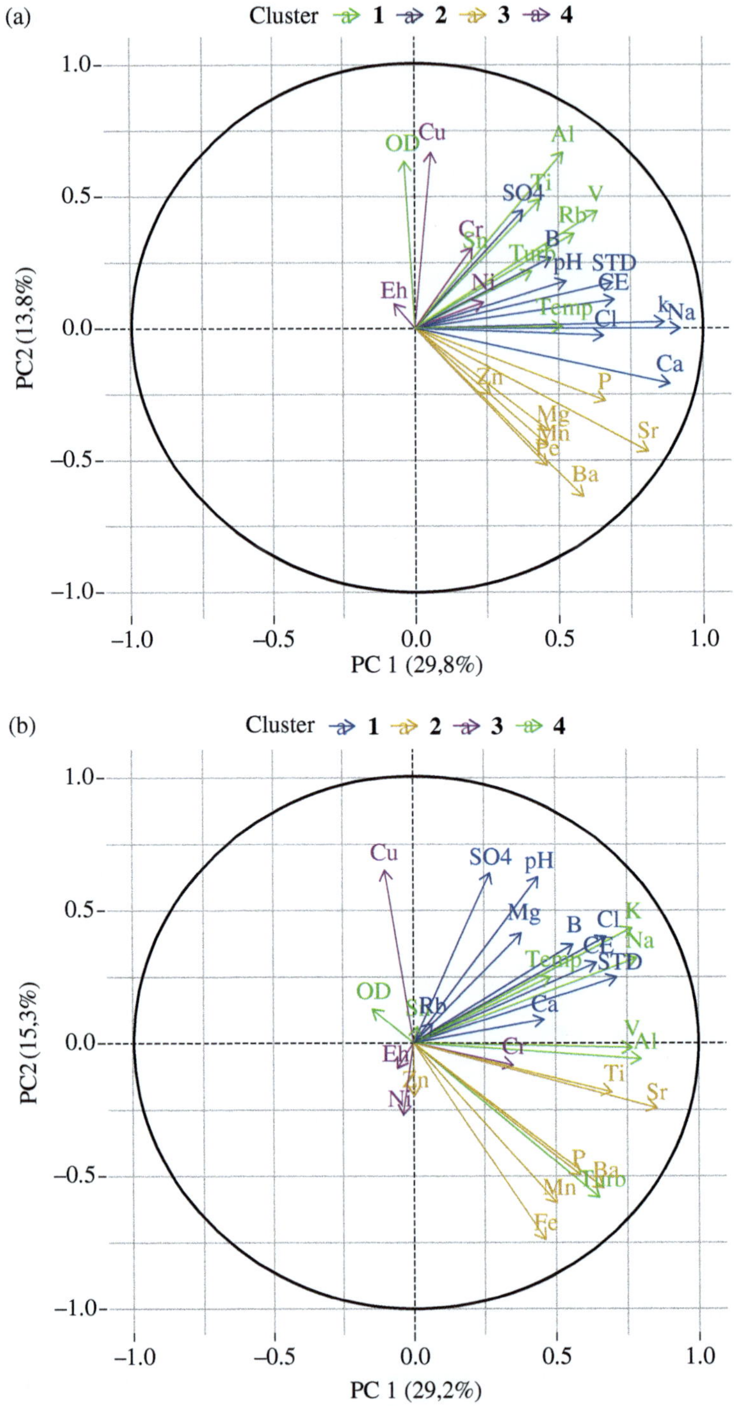

Figure 7.11 PCA water quality parameters in two seasons (a) rainy season; (b) dry season of the Middle IRSB. (Abbreviations – Temp: Temperature; Turb: Turbidity; TDS: total dissolved solid; EC: electrical conductivity; DO: dissolved oxygen).

Table 7.3 Geochemical baseline threshold (GB) values and number of samples above the GB for chemical elements, determined by: 2σ IT, DF, CF curve, percentiles (P5, P75, P95, and P98), MMAD, and TIF.

					GB values									No. of values > GB								
Element	Period	Unit	VMP	DL	P5	CF	P75	2σTI	DF	MMAD	P95	P98	TIF	P5	FC	P75	TI2σ	DF	MMAD	P95	P98	TIF
P	R	μg/l	100	10	<10	<10	**50.12**	100	100	100	79.43	***79.43***	199.53	78	95	20	0	0	0	9	9	0
	D				<10	<10	**39.81**	–	–	158.49	60.26	***61.66***	699.84	78	94	25	–	–	0	2	1	0
Fe	R	mg/l	—	0.001	0.26	2.64	**3.02**	3.30	4.50	5.50	4.37	***4.90***	7.41	72	29	25	15	4	0	5	2	0
	D			0.005	0.17	3.20	**1.82**	2.40	2.60	5.01	3.16	***3.80***	8.04	70	5	23	11	10	1	5	1	0
Mn	R	mg/l	0.1	0.001	0.02	1.84	0.13	0.10	0.20	0.38	0.43	0.55	0.52	70	0	22	29	10	6	4	2	3
	D				0.01	0.10	**0.10**	0.10	0.10	0.43	0.28	0.51	0.51	75	22	22	22	22	3	4	3	3
Al	R	mg/l	—	0.001	0.03	0.26	**0.22**	0.20	0.30	0.65	0.65	***1.41***	1.23	71	16	23	28	13	4	4	2	3
	D				0.03	0.12	**0.12**	0.10	0.10	0.30	0.31	***0.43***	0.52	75	23	23	28	28	6	5	2	2
Zn	R	μg/l	180	1	<1	19.6	**22.39**	–	–	91.20	57.54	***75.86***	204.17	71	25	23	–	–	2	0	2	1
	D				<1	6.76	**10.72**	–	–	26.30	33.88	***36.31***	50.70	69	35	23	–	–	9	6	2	0
Cu	R	μg/l	—	1	<1	<1	**3.24**	–	–	22.39	12.88	***45.71***	53.09	53	53	24	–	–	5	5	2	2
	D				<1	<1	**1.74**	–	–	14.79	3.72	***4.37***	11.22	51	51	26	–	–	1	5	2	1
Ni	R	μg/l	25	1	<1	<1	**1.58**	–	–	<1	2.57	***4.57***	8.91	38	38	24	–	–	38	5	2	2
	D				<1	<1	**0.50**	–	–	<1	1.66	***2.24***	0.50	18	18	18	–	–	18	5	2	18
Cr	R	μg/l	50	1	<1	<1	**1.41**	–	–	<1	3.09	***3.80***	6.68	42	42	24	–	–	42	5	5	2
	D				<1	<1	**<1**	–	–	<1	1.55	***1.91***	0.50	17	17	17	–	–	17	5	2	17
V	R	μg/l	100	1	<1	<1	**1.95**	–	–	4.37	3.47	***4.79***	14.96	66	66	23	–	–	3	5	2	0
	D				<1	<1	**1.58**	–	–	<1	2.51	***3.16***	8.91	40	40	23	–	–	40	5	2	1
Ti	R	μg/l	—	1	<1	<1	**5.01**	–	–	12.88	7.59	***10.96***	24.55	71	82	25	–	–	1	5	2	0
	D				1.51	3.54	**3.63**	–	–	6.17	5.62	***6.31***	7.76	70	24	24	–	–	3	5	2	1

(Continued)

Table 7.3 (Continued)

					GB values									No. of values > GB								
Element	Period	Unit	VMP	DL	P5	CF	P75	2σTI	DF	MMAD	P95	P98	TIF	P5	FC	P75	TI2σ	DF	MMAD	P95	P98	TIF
Ca	R	mg/l	—	0.001	1.07	4.78	**4.57**	5.40	6.70	7.94	7.94	***9.12***	11.61	72	19	24	16	12	5	5	3	1
	D				1.07	3.55	**3.16**	3.90	4.70	6.46	7.08	***10.00***	8.32	70	16	25	14	12	5	5	2	5
Mg	R	mg/l	—	0.001	0.81	2.52	**2.51**	3.10	3.80	4.57	4.07	***4.68***	5.75	71	23	24	15	8	3	6	2	0
	D				1.05	2.51	**2.51**	2.60	3.20	4.79	4.27	***6.92***	5.56	70	23	23	23	19	4	5	2	3
Na	R	mg/l	—	0.001	1.17	7.05	**5.37**	7.20	7.80	10.47	6.92	***13.80***	16.22	71	3	23	3	3	3	5	2	2
	D				1.26	6.01	**5.01**	6.70	7.50	8.91	7.59	***10.23***	12.74	70	9	24	8	5	4	5	2	1

					GB values									No. of values > GB								
Element	Period		VMP	DL	P5	CF	P75	2σTI	DF	MMAD	P95	P98	TIF	P5	CF	P75	2σTI	DF	MMAD	P95	P98	TIF
K	R	mg/l	—	0.001	0.50	0.82	**2.51**	3.40	4.00	4.68	3.47	***14.79***	8.13	71	75	22	5	4	3	5	2	3
	D				0.51	0.67	**2.63**	3.40	2.90	8.13	3.47	***4.47***	14.29	70	75	23	5	16	1	5	2	1
Ba	R	µg/l	700	1	10.00	<1	**85.11**	100	100	186.21	131.83	***169.82***	285.10	71	95	23	17	17	1	5	2	0
	D				8.71	<1	**75.86**	100	100	181.97	138.04	***194.98***	323.59	71	94	24	14	14	3	6	2	0
Sr	R	µg/l	—	1	3.98	<1	**50.12**	100	100	141.25	89.13	***102.33***	254.10	70	95	22	2	2	0	5	2	0
	D				5.25	<1	**44.67**	100	100	138.04	67.61	***81.28***	251.19	69	94	24	1	1	0	6	2	0
Rb	R	µg/l	—	5	<5	<5	**7.76**	-	-	28.84	11.48	***20.89***	42.17	59	59	23	-	-	2	5	2	2
	D				<5	<5	**7.59**	-	-	<5	11.75	***12.59***	39.81	41	41	25	-	-	41	7	2	0
B	R	µg/l	500	1	<1	<1	**4.90**	-	-	29.51	13.80	***39.81***	149.62	61	61	24	-	-	3	5	2	1
	D				<1	<1	**0.50**	-	-	<1	7.08	***7.76***	<1	18	18	18	-	-	10	3	2	18
Sn	R	µg/l	—	1	<1	<1	**2.57**	-	-	<1	4.37	***5.37***	29.85	43	43	23	-	-	43	5	2	0
	D				<1	<1	**2.95**	-	-	6.31	5.13	***5.50***	11.35	70	73	24	-	-	1	5	3	0

Note: E.Q: chemical element. R: Rainy period. D: Dry period. VMP: maximum value allowed in CONAMA resolution 357/2005 for Class II waters. DL: detection limit. "—": no VMP. '-': baseline value not calculated. Italics (T > Tcrit). **Bold:** conservative geochemical baseline threshold (CGB). ***Bold italics***: environmental geochemical baseline threshold (EGB). Underline: baseline greater than the VMP.

values were estimated through the seventy-fifth percentile (P75). The higher GB values were estimated by the TIF, 98% percentile (P98), and MMAD. The values, provided by the 5% percentile (P5), are very low values, which characterize them as extremely conservative values, being, therefore, viable, in the case of this study, only to characterize the lower limit of the geochemical baseline.

Although the definition of a natural geochemical baseline (NGB) is of great interest, in the case of surface waters, it is hampered by the fact that it is an open thermodynamic system, subject to the influence of many variables and, therefore, difficult to control. The surface water sample reflects a unique moment, as any change in precipitation and flow, or contamination by external agents, may modify existing conditions and introduce changes in the geochemical composition of the water. This becomes even more acute in areas subject to anthropic actions, as is the case of the studied region, where there are permanent mining and livestock activities. In this context, according to Sahoo et al. (2019), it is advisable to define a conservative geochemical baseline (CGB), which would correspond to the values calculated by the P75 technique, and that would be less influenced by anthropogenic impacts and closer to the natural baseline, but not exactly corresponding to it, because it may have some unnatural contribution. On the other hand, the values corresponding to the values estimated by P98 can be admitted as an environmental geochemical baseline (EGB) level, thus representing the chemical concentrations that were influenced by natural sources and diffuse anthropogenic sources (Sahoo et al. 2019). According to Reimann and Caritat (2017) and Reimann et al. (2018), the 98th percentile is one of the safest methods to indicate areas that need attention, as it provides a fixed number of variables that need care. On the other hand, more conservative values can produce a high number of cases that would need attention (Reimann et al. 2018).

Other techniques, such as Med + 2 MAD, P95, and TIF, can also be used to establish estimated EGB values (Sahoo et al. 2019). However, as seen in Table 7.3, this option implies, in the case of the Med + 2 MAD and P95 methods, in general, a greater number of samples with concentrations higher than the estimated baseline, resulting in a higher number of areas seen as problems (Reimann and Caritat 2017). The TIF method, in turn, tends to provide higher values than those indicated by the other methods and, therefore, tends to minimize the number of possibly impacted areas. Therefore, the 98th percentile (P98) becomes, comparatively, more suitable to define the EGB. From the results obtained by the different techniques, it is observed that, with the exception of the geochemical baseline values estimated by the Med + 2 MAD techniques for the elements P, Mn, Mg, and K; TIF for P, Fe, K, and Ba; FC to Fe; P75 for Sn; P95 for Mg, Na, Ba, and Sn; P98 for Ca, Mg, Ba, and Sn; all other techniques estimated higher geochemical baseline values for the rainy season.

It is also noted that, among the elements with VMP, only the DF, MMAD, P95, P98, and TIF techniques estimated GB values higher than the maximum limit allowed for Class II waters provided for in CONAMA resolution 357/05. This happened with Mn, in the case of DF, MMAD, P95, P98, and TIF techniques, with P, in the case of MMAD and TIF techniques, and with Zn, only with the TIF technique.

7.4 Discussion

7.4.1 Physicochemical Characteristics and Seasonal Variation of the Chemical Composition

The results obtained in the analysis of surface waters of the subbasin of the middle Itacaiúnas River showed that the seasonal variation between the rainy season (sampling in April and May) and the dry season (sampling in July, August, and September) resulted in a decrease in the levels of Fe, Mn, Al, Ca, Zn, V, B, P, Cr, and total Ni and the increase in the levels of total Sn in the surface waters of the subbasin, making them less turbid, less conductive and less oxidizing. Such facts can be attributed to the decrease in the volume of rainfall and surface flow and, consequently, the smaller amount of leached and transported load, in particulate or dissolved form, to the water bodies. It is worth remembering that in the Itacaiúnas River Basin, surface runoff and River flow, especially in water courses located in pasture areas, are higher during the rainy season (Souza Filho et al. 2016; Silva Júnior 2017), and this is also true in other areas of the Amazon region (Nóbrega et al. 2018; Levy et al. 2018). As a result, more particulate and dissolved material, such as sand, silt, clay, colloids, organic substances, algae, and minerals, can be transported to rivers due to soil leaching, rock weathering, erosion of river banks, or even by remobilization of the bottom sediments of water courses (Queiroz et al. 2009; Manahan 2010; Keller 2012; Siqueira et al. 2012; BRASIL 2014; da Conceição et al. 2015; Gómez et al. 2017; Ismaiel et al. 2018). In the rainy season, there was a moderate-to-strong correlation between water turbidity and total Al, Ti, and V contents, and relevant weights were assigned to the variables turbidity, Fe, and Cr in PC3. However, in the dry season, in addition to Al, Ti, and V, the elements Ba, Sr, Fe, and Mn also showed strong-to-moderate correlations with turbidity and are associated with it in PC1, together with TDS and EC, all with relevant weights. This allows us to infer that the turbidity was mainly due to the presence of these trace elements in the water. However, considering the higher levels of Fe and Mn in the waters, it is likely that both have exerted a greater influence on turbidity than the other elements mentioned. In addition, other factors not considered in this study, such as dissolved organic matter, may also have influenced this parameter.

In general, natural waters have a pH between 4 and 9 (Foley et al. 2009; Vasconcelos et al. 2009), and rainwater has an average pH of 5 (Szikszay 1993). The subbasin of the middle Itacaiúnas River is characterized by the marked predominance of the approximately neutral to slightly alkaline character of the waters, with rare analyses indicating pH values between 5 and 6. Similar results were obtained from the analysis of surface waters from areas considered natural of N3 and N4WSUL, located in the FNC, where the pH largely varied between 6 and 7.5 (Teixeira 2016), and in the subbasins of the Vermelho, Sororó, and Parauapebas Rivers (Salomão et al. 2018; Quaresma et al. 2022). Waters with a pH between 5 and 6.5 generally reflect the presence of weak acids, such as carbonic acids, humic and fulvic acids, while pH values between 3 and 5 are usually caused by the oxidation of sulfides (Vasconcelos et al. 2009). Although Cu and SO_4^{2-} are together in the cluster analysis, indicating oxidation of sulfide minerals, the weak correlation between Cu and SO_4^{2-} suggests that the lowest pH values were probably not produced by the oxidation of sulfide minerals, but possibly due to the decomposition of organic matter, especially in the rainy season, when the waters had significantly less DO.

The approximately neutral to slightly alkaline pH of river water, as well as the high concentrations of Ca, especially in surface waters during the rainy season, may reflect the influence of the dissolution of silicate minerals. This hypothesis can be corroborated by the moderate positive correlation between pH and Na, K, and Mg, as well as by the high weights given to the parameters pH, Cl–, Ca, K, Na, and Sr in PC1 of the station and to the variables Na, K, Al, and Sr in PC1 of the dry period, and may indicate the possibility of common sources for these cations, which would most likely be the feldspars essential constituents of the felsic and mafic silicate rocks present in the basin. The presented hypothesis could be tested using the Gibbs graph (Figure 7.12), where the ionic compositions of the water samples from the subbasin were plotted. This graph is divided into three fields: rock weathering dominance (carbonates, silicates, and sulfides); dominance of evaporation and crystallization; and dominance of rainfall (Chakrapani 2005). It is concluded that the weathering of rocks, especially feldspathic rocks, followed to a lesser extent by atmospheric precipitation, was the dominant control process of water chemistry. These processes justify the absence of sharp variations in pH between the two seasons and allowed to maintain, regardless of the amount of rainfall, the almost neutral to slightly alkaline pH of the waters. Also according to Vasconcelos et al. (2009), waters with a pH between 8 and 10 are related to the influence of carbonates, silicates, and aluminosilicates. In this sense, it is worth noting that the only drainages that presented a pH close to 9 are close to the limestone barriers used to mitigate acid drainage from the Cu do Salobo mine. This was possibly due to the leaching of Ca and Mg elements from these barriers to adjacent water bodies. Evidence of this is that the conductivity in the rainy season, compared to the dry season, especially in the Salobo

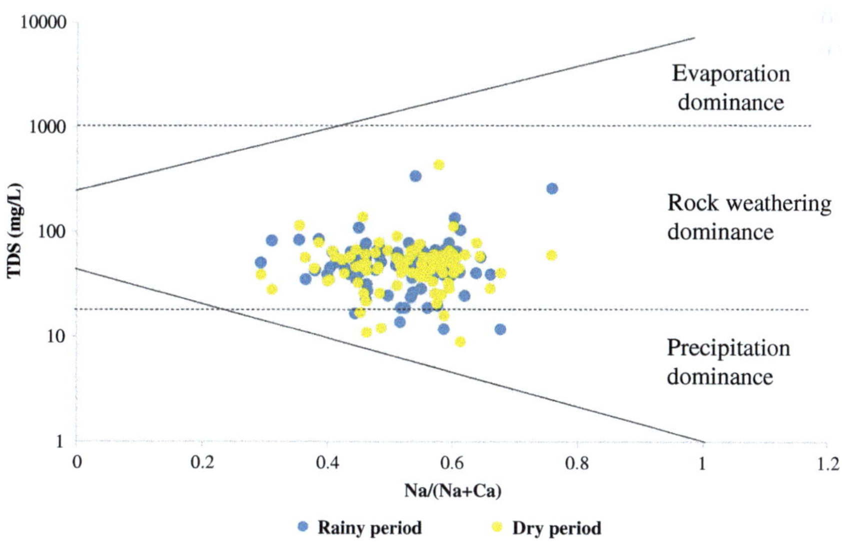

Figure 7.12 Gibbs plot showing the hydrogeochemical process controlling major water chemistry of MISB.

stream microbasin, reached significantly higher values. This process is also indicated by the strong correlation between EC and the elements Ca and Mg and the significant weights in PC 1, of the rainy season, granted to EC, and also to Ca, Na, K, Sr, P, V, TDS, and to the anion Cl^-, suggesting that the conductivity and TDSs in water were highly influenced by these trace elements and the presence of the Cl^- anion. There are no limits in the legislation for the EC parameter, but in general, natural waters have an EC between 10 and 100 μS/cm, while in waters under the influence of industrial and domestic effluents, the EC can reach up to 1000 μS/cm (BRASIL 2014; da Conceição et al. 2015). Although DO values above 5 mg/l largely predominate in the samples of the microbasins studied, which are adequate according to current legislation, there is a significant number of DO results below the minimum limit in both stations and this needs to be evaluated. Like several rivers in the Amazon (Silva et al. 2013; Queiroz et al. 2009; Silva et al. 2008), several rivers in the subbasin, even those exempt from the influence of anthropic activity and with preserved native forest, held DO levels less than 5 mg/l. A situation similar to that of the MISB was observed by Teixeira (2016) in water bodies located in areas without anthropic interference in the FNC, where most concentrations measured were above the minimum limit, but some DO values below 5 mg/l were also found The DO content in water depends especially on temperature, atmospheric pressure, and salinity, due to their influence on oxygen solubility. At sea level, the solubility of O_2 in water at 25 °C is 8.6 mg/l, decreasing with increasing atmospheric pressure, temperature, and salinity (Fiorucci and Benedetti 2005). This is because, with increasing altitude, there is a decrease in atmospheric pressure and O_2 pressure, while the increase in temperature causes an increase in the mobility of molecules, which can escape into the atmosphere when they reach the water surface. The reduction in DO levels may be due, among other reasons, to evaporation, aerobic oxidation of organic matter, abiotic chemical oxidation, biological respiration, and nitrification (Fiorucci and Benedetti 2005; Manahan 2010). There is no clear explanation for the spatial variation of DO in the MISB. However, the high positive weights attributed, in the CP of the rainy season, to the variables DO and Cu, the moderate positive correlation of DO and Cu, as well as the grouping of DO and Cu with SO_4^{2-}, suggest that the DO concentrations may have been smaller in this season due to chemical oxidation of sulfide minerals. However, the weak correlation between Cu and SO_4^{2-} indicates that this natural process is not the main control of DO levels in MISB waters. According to De Laune and Reddy (2005) and Fiorucci and Benedetti (2005), in unpolluted aquatic systems, organic substrates of biological origin, such as those from dead plants and animal remains, are usually the main materials oxidized by DO. In view of this and the fact that no concentrations of chemical elements indicative of human contamination of the surface waters of the MISB were detected, as well as due to the significant number of samples with concentrations of NO_3^- below the LD and the detected concentrations of total P being lower than the VMP, the lowest level of DO in the waters of the rainy season, together with the lowest pH values, is attributed, above all, to the oxidation of organic matter carried to the surface waters of the subbasin. Eh is widely used in water quality, limnological and geochemical studies and is considered a parameter that allows characterizing, through electron activity (e^-), the degree of reduction, or oxidation of the aquatic environment, and can be reductive under low pressure, activity, or oxidant in the face of high electron activity (Manahan 2010). Unlike lake water, the Eh of river water is always positive, due to the continuous aeration of surface water by the dissolution of atmospheric oxygen. It is also usually more stable throughout the water column, but it is subject to marked variations depending on the type of river and topographic conditions (Sondergaard 2009). According to the identification of the agents responsible for the redox reactions is fundamental for the processes of sorption (retention), mobility, and toxicity of trace elements in the hydrosphere, since they are the basis for the processes of chemical transformation of inorganic substances and are important for chemical reagents that participate in electron transfer or absorption. The main oxidizing agents are Fe, O_2, NO_3^-, and SO_4^{2-}, and the reducing agents are organic substrates. Mn and Fe were also highly concentrated in the waters of the northern portion of the subbasin during the rainy season, but there is no significant correlation between these elements and Eh, which weakens the hypothesis of a direct relationship between these parameters. On the other hand, Eh showed a moderate negative correlation with pH and a positive correlation with Ni, suggesting that the release of H^+ in the waters, as well as the oxidation of sulfide minerals, could be contributing to the more oxidizing character of the waters during the rainy season.

7.4.2 Geogenic and/or Anthropic Influence on the Chemistry of Subbasin Surface Water

The spatialization of the results of the physical–chemical analyses showed that in the general framework of the MISB, during the period of the sampling, there were no significant changes in the natural characteristics of the waters in most of the subbasin, but rather punctual changes located in a few microbasins, which apparently may be reflecting particular geological scenarios or effects of anthropogenic actions to which these microbasins were vulnerable. Water temperature is an important condition for the sustainability of water bodies, as it can influence aquatic life, physical–chemical and biological processes, as well as the final use of water resources. Its variability is part of natural climate control, but it can be affected by human activities. In the MISB, this parameter is generally close to 26 °C, suggesting little human influence, such as the release of wastewater. It is important to note that tropical rivers can have temperatures of up to 38 °C (Fiorucci and Benedetti 2005). It can be seen that a significant number of samples with concentrations below the DL of anions F^-, NO_3^-, and SO_4^{2-} and the occurrence of rare values above the VMP of F^- and NO_3^- demonstrate that there was no strong influence of anthropic activities in the subbasin, such as sewage disposal and application of agricultural inputs, nor from natural sources such as the alteration of fluorine-enriched minerals, such as biotite, siderophyllite and fluorite present in Paleoproterozoic type-A granites (Dall'agnol et al. 2005). Likewise, the weak correlation between pH and SO_4^{2-} and Cu, and the predominance of SO_4^{2-} concentrations below the DL and the legal maximum limit (250 mg/l), suggest that the subterranean water basin was not significantly affected by acid mine drainage. The concentrations well below the VMP of Cl^- in the water indicate the limited influence in the subbasin of anthropogenic sources, such as domestic and industrial effluents. The Cl^- present in the water may come mainly from atmospheric deposition (Gaillardet et al. 2003). Another possible source would be the weathering of relatively Cl-enriched amphiboles from Neoarchean granites and copper mineralizations present in the southern portion of the region (Dall'agnol et al. 2017). The low concentrations of F- in the waters may be due to the low content of F in the Archean rocks that are predominant in the study area. In general, there is evidence that the geochemical pattern of the distribution of elements in surface waters was strongly controlled by the geological environment and lithological composition of the basin. This is most evident in the distribution of the elements Fe, Mn, Ba, Cu, and Sr. The concentrations Fe, Mn, Ba, and Sr show higher in the north of the subbasin, in the BD, where metamafic and metasedimentary rocks predominate, while Cu presents, in general, contents below the DL in the microbasins of the north and is notably concentrated in those in the south, in the Carajás Basin, where metavolcanosedimentary rocks stand out and there are important mineralizations associated with the northern copper belt. According to Kabata-Pendias and Mukherjee (2007), iron can be present in the hydrosphere in oxidized (insoluble Fe III, ferric iron ion) or reduced (soluble Fe II, ferrous iron ion) forms, depending on environmental conditions, such as Eh, temperature, pH, water–rock interaction, water sediment, microbiological reactions, and binding with humic or non-humic complexes. Also according to these authors, Mn is chemically similar to Fe and is a relatively abundant element in the lithosphere, occurring mainly in mafic and metasedimentary rocks. In surface waters predominantly in the form of colloids, it can also be found in the form of hydrated Fe III and Fe II ions (Kabata-Pendias and Mukherjee 2007; Manahan 2010; Szikszay 1993). Among the parameters that govern the chemical behavior of Fe, Eh is the most important (Kabata-Pendias and Mukherjee 2007). It is easily solubilized in acidic and reducing waters and precipitated in different forms of oxides in oxidizing and alkaline or approximately neutral waters (Kabata-Pendias and Mukherjee 2007). The Fe content in the surface waters of a basin can be highly variable, reflecting the activities developed along the water bodies as well as the geological environment through which the waters circulate. Studies carried out in the Rio Vermelho and Sororó Basins (Salomão et al. 2018), and in natural regions around the mineral bodies of N3 and N4WSUL (Teixeira 2016) and integrated into the basin of the Itacaiúnas River (Sahoo et al., in impression), found that the waters of the Itacaiúnas River Basin are neutral to slightly acidic, generally with high total Fe content, possibly of geogenic origin. In the MISB, accentuated enrichment of Fe and Mn was observed, especially in the microbasins of the lower course of the subbasin, drained mainly by the Tapirapé River and its tributaries, and coincident with occurrences of metasedimentary rocks (quartzites, micaschists, iron, marble, and calc-silicate rocks) and with the main known Mn deposits within the BD. Everything indicates that the enrichment in Mn and Fe in this portion of the subbasin was intensified by the accentuated deforestation and the predominance of pastures that favored the superficial erosion of rocks and ferruginous crusts. The higher Fe and Mn contents in the surface waters of the rainy season indicate that this natural process was more marked in the subbasin during this period. The weak correlation of Fe and Mn concentration with DO and Eh indicates that the redox condition and DO variations are not factors that significantly influenced the concentration of these elements in the subbasin. On the other hand, the strong-to-moderate positive correlation of Fe and Mn, as well as the attribution to these variables of high positive weights in PC 1, in the rainy season, and negative weights in PC 2, in the dry season, together with similar behavior of Fe and Mn, suggest a possible common source for both elements, which can be mafic and metasedimentary rocks, as well as iron formations, therefore being considered of geogenic origin. The lowest concentrations of Mn and Fe occurred in the microbasins linked to the Águas Claras and Azul streams and the upper reaches of the Itacaiúnas

River and were certainly influenced by the preservation of native vegetation, as forests preserve the quality of water resources, by increasing the infiltration and decrease surface runoff, erosion, and sediment production, in addition to filtering pollutants (Keller 2012; Ramos et al. 2018). Although most concentrations of total Mn were lower than the maximum acceptable limit (0.1 mg/l) by legislation, in many watersheds the levels were above this value. However, it is considered that this enrichment is associated with the geomorphological and geological conditions of the subbasin, as well as with climatic conditions and land use and occupation in the region, since similar behavior has already been verified in other areas of the IRW (Teixeira 2016; Salomão et al. 2018; Sahoo et al. 2019; Quaresma et al. 2022). It is worth mentioning that there is no record in the microbasins of the southeastern portion of the subbasin of an increase in Mn content in surface waters due to the influence of the Mn do Azul mine, located immediately south of the subbasin boundary. In many watersheds, not only the total concentration of Mn, but also of other trace elements, such as Cr and Ni, or even physical–chemical parameters, such as pH, is in disagreement with the established limits, for waters Class II, by CONAMA. This is explained by the fact that the standards adopted were not prepared considering the regional characteristics of Brazilian watersheds, which has already been questioned in several studies (Vasconcelos et al. 2009; Silva et al. 2015; Gloria et al. 2017). According to Reimann and Garrett (2005), any determination of regulatory value must be accompanied by an explanation of the factors that were taken into account, as well as the methods used for its formulation. These authors emphasized that values considered natural never cause ecosystem problems. The rivers of the middle Itacaiúnas River Basin are examples of how local factors such as soil type, lithology, climate, vegetation, and changes in land use and occupation can influence the geochemical composition of water. The complementary evaluation, carried out by the chemical water quality index, showed that more than 97% of the watersheds had water with chemical quality between good and excellent, being only very locally considered bad or very bad. This demonstrates that the contribution of anthropogenic activities in the chemical quality of the surface waters of the basin, although it exists, was not intense and generalized enough to modify the dominant scenario that points to good-quality waters. This implies that there are probably only point and localized sources of alteration in the environmental quality of surface water resources in this subbasin. It should be noted that this picture is valid for the period in which the sampling was carried out, and it is possible that the situation in the microbasins in which low chemical water quality was observed has evolved positively. However, the possibility of a decline in chemical quality in some watersheds cannot be ruled out, but it is unlikely that there have been very sharp changes in the picture obtained, as there have been no facts, from 2017 to date, to justify this. In any case, it is evident that continuing to monitor the water resources of the subbasin is imperative to ensure a clear and consistent view of the state of its surface waters.

7.4.3 Comparison Between the Geochemical Signature in the Waters of MISB and Other Subbasins of IRW

The geochemical distribution of trace elements in the surface waters of the rivers of the middle IRSB reflects the heterogeneity of the local geological composition, as well as the different environmental conditions of the basin, especially the climatic season and the types of land use and cover. In general, in the surface waters of the middle IRSB, the highest concentrations of the studied trace elements occurred in the period of greatest precipitation. A similar behavior was also observed in the subbasins of the Parauapebas, Vermelho, and Sororó Rivers, as well as in the surface waters of drainages located in the CNF (Teixeira 2016; Salomão et al. 2018; Quaresma et al. 2022). Such facts reinforce the premise that seasonal climate variations in the region play an important role in the chemical composition of river waters. The geochemical patterns seem to be strongly linked to the geology of each region and were actively influenced by the conservation status of the soils. The surface waters that drain the MISB and River Parauapebas Subbasin (RPSB) have low levels of Co, Cr, Ni, Cu, and V. In these regions, in both periods, the median contents of the first three elements are lower than the DL of the method (<1 μg/l), as well as the median contents of V, in the dry season. On the other hand, the waters of the rivers located in the River Vermelho Subbasin (RVSB) are comparatively richer in Co, Ni, and V, and the waters of the River Sororó Subbasin (RSSB) River in Cr, Ni, and V, especially in the region's rainy season. Unlike the other elements mentioned, Cu is comparatively more abundant in surface waters, both in the MISB and in the RPSB, RVSB, and RSSB, although in the latter subbasin, the median content detected in the waters during the dry season is less than 1 μg/l. However, the median concentrations of Cu in the MISB waters are comparatively lower than the median levels recorded in the RPSB, RVSB, and RSSB Rivers (Salomão et al. 2018; Quaresma et al. 2022). From this information, it is possible to notice the interference of the geological context on the geochemical signature of these elements in each subbasin. The chemical elements Co, Cu, Ni, and V tend to be more abundant in surface waters under the influence of the Carajás Province, such as the waters of the western portion of the SBRV, due to the affinity of Cu and Co for metavolcanosedimentary units, as well as Ni and V with mafic-ultramafic units (Salomão et al. 2018). The contents of alkaline elements (K, Na, and Rb) and alkaline earth elements (Ca, Mg, Sr, and Ba) are equally high in the surface waters of the MISB and RPSB. However, it was found that the median and average contents found in the MISB are lower when compared to the median concentrations of the PRSB. It was also observed that, as in MISB, all SBRP watersheds also showed detectable concentrations of elements Ba, K, Mg, and Na. Boron is more abundant in surface waters during the rainy season, both in MISB and RPSB. However, when comparing their respective median concentrations, it was observed that in the surface waters of the MISB, during the rainy season, the median content detected is significantly higher than the median content detected in the RPSB, while in the dry season the median concentrations are lower than the DL (1 μg/l), in both subbasins.

Differently from the previously mentioned elements, the Zn and Sn elements presented higher occurrence in the dry season, both in MISB and in RVSB, RSSB, and RPSB. During the rainy season, the median levels of Zn in the surface waters of the MISB were lower than the median levels found in RSSB and higher than those found in RVSB and RPSB, while the median levels of Sn were higher than DL (1 μg/l) only in RPSB. On the other hand, during the dry season, the median levels of Zn in the MISB are notably lower than the median levels found in the other subbasins, while the median levels of Sn in the MISB exceed the median levels found in the RPSB and RVSB, but not the median levels of RSSB. In MISB, as well as in RPSB, Al levels higher than DL were detected in all watersheds in both periods. However, it was observed that the median levels found in the MISB, during the rainy season, are considerably lower than the median concentrations detected in the RPSB and approximately equal during the dry season. Unlike Al, Ti was 100% detected only in the dry season and particularly in the MISB. On the other hand, as well as the Al contents, the median Ti contents in the surface waters of MISB and RPSB are higher than LD, but higher, in both stations in RPSB. Comparing the distribution of P in the MISB and in the RPSB, it is observed that the concentrations detected in the MISB are comparatively lower and, in many samples, the contents are lower than the DL (1 μg/l). Differently from the results obtained in the RPSB, all the values detected in the MISB do not exceed the VMP for this parameter. This fact may be a reflection of the importance of preserving its water bodies, since the MISB, as described in the previous sections, differs from the other subbasins of the IRW, especially for its preservation areas, which are fundamental for the renewal and preservation of surface waters. High levels of Fe and Mn are the predominant geochemical signature in the MISB, as well as in the subbasins of the Parauapebas, Vermelho, and Sororó Rivers. When comparing the median concentrations of Fe and Mn measured in the MISB with the medians obtained in the RPSB, RVSB, and RSSB, it is observed that, in general, the highest median concentrations occur in the rainy season and, normally, the maximum concentration of Mn in the surface water exceeds the VMP of 0.1 mg/l, defined in CONAMA Resolution 357/05, for Class II waters. A similar behavior was observed in the surface waters of regions without human interference, adjacent to the mineral bodies of N3 and N4WSUL in the FNC (Teixeira 2016). The median contents of these elements in the surface waters of the MISB are comparatively lower than the median contents found in the RPSB, RVSB and RSSB. The conversion of areas originally covered by forests into pastures seems to be actively contributing to the increase in Fe and Mn levels in the surface waters of these subbasins (Salomão et al. 2018, Sahoo et al. 2019; Quaresma et al. 2022). This hypothesis is strongly reinforced when observing that in the MISB there is a marked concentration of Fe and Mn in water bodies that are located in regions of the subbasin that are intensely deforested, while lower levels are normally found in the water courses, water from regions considered pristine, such as the southern and eastern preservation areas of MISB.

7.4.4 Geochemical Baseline Levels in the Surface Waters of the Subbasin

The results of the geochemical baseline threshold (GB) value calculation revealed that most of the methods estimated higher values for the rainy season, which are almost lower than or equal to the VMP (CONAMA resolution 357/2005 for freshwater of Class II). Similar results were also observed in other regions of the basin (Salomão et al. 2018; Sahoo et al. 2019). However, some estimated GB values were higher than the legal thresholds established in that resolution. Although the establishment of lower limits for the geochemical baseline may be important in certain situations, in the present study, as a large number of samples with values below the DL were recorded, it makes no sense to try to discuss such values, even because in most situations they do not can be fixed. As demonstrated by the spatial distribution maps, there is a significant variation in the concentrations of trace elements and physico-chemical parameters between the two periods, as well as between the regions of the subbasin. The rainy season and the north of the subbasin presented, in general, higher concentrations for most of them, probably due to natural and anthropic influences discussed in the previous sections. On the other hand, in the dry season and in the south of the basin, the concentrations are comparatively lower.

According to Reimann and Garrett (2005), Reimann et al. (2018), and Reimann et al. (2018), background levels can vary in the same area or in different areas, due to peculiarities present in each area. This demonstrates that it may not be suitable to determine constant values of geochemical baseline for the entire subbasin studied nor for the two climatic seasons. It is therefore recommended to establish geochemical baseline values separately by period and, if possible, by region. In the case of the MISB, a separate geochemical baseline estimate was made for each of the seasons. As for EGB, for an appreciable number of chemical elements, including P, Fe, Mn, V, Ti, Ca, Mg, Na, Ba, Sr, Rb, and Sn, the variations were not very significant, with Ca, Mg, Ba, and Sn had slightly higher EGB values in the dry season. On the other hand, Al, Zn, Cu, Ni, Cr, K, and B showed great variation in values, with EGB being much higher in the rainy season, with the growth of the order of 2 to 10 times. When evaluating the CGB variations between the two stations (P75), it appears that the differences are much less pronounced, with marked variations only for Zn, Ni, and B. It is also verified that only K and Sn showed higher EGB values in the dry season.

Deciding which GB is most appropriate will depend on the proposed protection objective. In the case of this research, the estimated values for the CGB correspond to P75 and produced much lower values when compared to the estimated values of the EGB, equivalent to P98. CGB values are generally preferred by environmental agencies, in assessments carried out to verify the suitability and degree of chemical contamination of surface waters. However, it should be noted that if CGB values are adopted as indicators of the EGB, as they are lower values, there will be a greater number of water bodies that will need more detailed investigations to assess possible environmental contamination and this required high investments. It is noteworthy, however, that due to the fact that the natural conditions of the IRW have been altered by anthropic activities in a large part of the studied subbasin, the levels of CBG assumed may have contributions of anthropogenic origin. It is almost impossible not to have some influence from diffuse anthropogenic sources, such as aerial deposition, and, therefore, it is believed that any surface geochemical concentration, even if predominantly of natural or geogenic origin, may contain a small load of anthropogenic contribution. On the other hand, the estimated values of EGB, mainly considering the influence of natural factors on the composition of surface waters, as well as the contribution of diffuse agents of anthropogenic contamination, are more recommended for representing the baseline of chemical elements in surface waters of regions with interference from human activities, such as the subbasin studied. It is worth mentioning that the upper geochemical threshold only highlights the areas or microbasins that present unusual concentrations of the analyzed elements and, regardless of the technique used for its calculation, it does not allow to distinguish clearly contaminated areas in environmental terms from areas with enrichment of chemical elements to be from natural or geogenic sources, considered as potential targets for mineral exploration. Distinguishing these two types of areas requires expert judgment (Reimann et al. 2017).

7.5 Conclusions

This study revealed that the concentrations of most of the physical–chemical parameters and chemical elements analyzed in surface water of the MISB were within the limit of CONAMA 357/2005 – class II water quality standards. The parameters such as DO, turbidity, Fe, and Mn presented some disagreement. The distribution maps revealed significant seasonal variation in water quality, with the highest levels of most chemical elements especially Fe, Mn, Cu, Sr, and Ba, and turbidity in the rainy season is mainly linked with the geological substrate and the conditions of land use and land cover of the subbasin. No geochemical evidence was identified in the subbasin of widespread anthropic contamination, but some microbasins presented values that did not comply with current legislation and should be monitored to assess the persistence of contamination and identify its possible sources.

The geochemical baseline values were calculated as CGB and EGB, respectively, while the latter is recommended for representing the baseline of chemical elements in surface waters of regions with interference from diffuse sources. In general, both CGB and EGB values varied considerably between the two seasons and the sampling stations but were mostly below the threshold values recommended by CONAMA legislation, except in the case of Mn. The univariate and multivariate statistical treatment allowed us to infer that the water in the MISB rivers/streams is mainly influenced by the weathering of catchment rocks and, to a lesser extent, by atmospheric precipitation. It was also verified that the waters are enriched in Fe and Mn and are approximately neutral to slightly alkaline, regardless of the rainfall volume. In general, there were no significant alterations in the natural characteristics of surface waters in most of the subbasin of the middle Itacaiúnas River, but localized changes in a few microbasins, which apparently reflect particular geological scenarios or effects of local anthropogenic actions, which these watersheds are vulnerable. It is suggested that the microbasins that presented values in disagreement with CONAMA Resolution 357/05 be used as a reference point in the selection of areas that can be studied in greater detail.

Acknowledgments

The first author acknowledges the ITV, Belém, Brazil for providing an opportunity to carry out this research work, which was part of the ItacGMB project. This work was also finally supported by ITV and Vale (Gerência de Meio Ambiente Corredor Norte) and grants from Conselho Nacional de Desenvolvimento Científico e Tecnológico – CNPq to RD (proc. 402727/2018-5; proc. 304.648/2019-1). PKS would also like to thank the DST-FIST lab and DST-SERB Core research grant (CRG/2021/002567), both from India, for providing technical support to this work.

References

Alvares, C.A., Stape, J.L., Sentelhas, P.C. et al. (2013). Köppen's climate classification map for Brazil. *Meteorologische Zeitschrift* 22 (6): 711–728. https://doi.org/10.1127/0941-2948/2013/0507.

Ander, E.L., Johnson, C.C., Cave, M.R. et al. (2013). Methodology for the determination of normal background concentrations of contaminants in English soil. *Science of the Total Environment* 454–455: 604–618.

Arpine, H. and Gayane, S. (2016). Determination of background concentrations of hydrochemical parameters and water quality assessment in the akhuryan river basin. *Physics and Chemistry of the Earth* 94: 2–9. https://doi.org/10.1016/J.Pce.2016.03.011.

Brasil 2005, Conselho Nacional De Meio Ambiente (CONAMA). Resolução N. 357, De 17 De Março De 2005. Dispõe Sobre A Classificação Dos Corpos De Água E Diretrizes Ambientais Para O Seu Enquadramento, Bem Como Estabelece As Condições E Padrões De Lançamento De Efluentes, E Dá Outras Providências. Diário Oficial Da União: Brasília, Df, N. 053, P. 58–63 Mar. 2005.

BRASIL (2014). *Fundação Nacional da Saúde. Manual de controle da qualidade da água para técnicos que trabalham em ETAS*, 112. FUNASA: Brasília, DF.

Bundschuh, J., Prakash, J., Mushtaq, S. et al. (2016). Medical geology in the framework of the sustainable development goals. *Science of the Total Environment* 581: 87–104.

Chakrapani, G.J. (2005). Major and trace element geochemistry in upper Ganga river in the Himalayas, India. *Environmental Geology* 48: 189–201. https://doi.org/10.1007/S00254-005-1287-1.

da Conceição, F.T. et al. (2015). Influência sazonal no transporte específico de metais totais e dissolvidos nas águas fluviais da bacia do Alto Sorocaba (SP). *Geochimica Brasiliensis.* 29 (1): 23–34.

Dall'Agnol, R., Cunha, I.R.V., Guimarães, F.V. et al. (2017). Mineralogy, geochemistry, and petrology of Neoarchean ferroan to magnesian granites of Carajás Province, Amazonian Craton: the origin of hydrated granites associated with charnockites. *Lithos* 277: 3–32.

Dall'Agnol, R., Teixeira, N.P., Rämö, O.T. et al. (2005). Petrogenesis of the Paleoproterozoic rapakivi A-type granites of the Archean Carajás metallogenic province, Brazil. *Lithos* 80 (1–4): 101–129.

DOCEGEO 1988, Rio Doce Geologia E Mineração. Revisão Litoestratigráfica Da Província Mineral De Carajás – Litoestratigrafia E Principais Depósitos Minerais. In: Congresso Brasileiro De Geologia. 35., 1988, Belém. Resumo. . . Belém: Sbg, 1988. P. 11-54.

EPA (1993). *Environmental Protection Agency. Method 300.0: Determination of Inorganic Anions In Drinking Water By Ion Chromatography: Revision 12*. Cincinnati.

EPA (1999). Environmental Protection Agency. Method 300.1: Determination of Inorganic Anions in Drinking Water by Ion Chromatography: revision 12. Cincinnati: OH. https://www.epa.gov/sites/default/files/2015-06/documents/epa-300.1.pdf (accessed 31 October 2017).

EPA (2013). Environmental Protection Agency. Evaluation of Background Metal Soil Concentrations in Cuyahoga County – Cleveland Area. Summary Report. 2013. Columbus, Ohio: Division of Environmental Response and Revitalization, Voluntary Action Program. http://epa.ohio. gov/portals/30/vap/docs/Cleveland%20Background%20Summary%20Report%20%282%29.pdf (accessed 31 October 2017).

Foley, D., Mckenzie, G.D., and Utgard, R.O. (2009). *Investigations in Environmental Geology*, 13the. Pearson.

Fiorucci, A.R. and Benedetti, F.E. (2005). Oxigênio Dissolvido: Propriedades E Solubilidade. *Química Nova Na Escola* 22 (4): 10–16.

Gałuszka, A. (2007). Different approaches in using and understanding the term "geochemical background"—practical implications for environmental studies. *Polish Journal of Environmental Studies* 16 (3): 389–395.

Galuszka, A. and Migaszewski, Z.M. (2011). Geochemical background—an environmental perspective. *Mineralogia* 42: 7–17. https://doi.org/10.2478/V10002-011-0002-Y.

Gaillardet, J., Viers, J., and Dupré, B. (2003). Trace Elements in River Waters. *Treatise On Geochemistry* 5: 225–272. https://doi.org/10.1016/B0-08-043751-6/05165-3.

GAT (2007). Itacaiúnas: estudo de Gestão Ambiental Territorial da Bacia do Itacaiúnas. Caderno 3: Água. Relatório interno. Marabá.

Gloria, L.P., Horn, B.C., and Hilgemann, M. (2017). Avaliação Da Qualidade Da Água De Bacias Hidrográficas Através Da Ferramenta Do Índice De Qualidade Da Água - IQA. *Caderno Pedagógico* 14: 103–119. https://doi.org/10.22410/Issn.1983-0882.V14i1a2017.1421.

Gómez, R., Arce, M.I., Baldwin, D.S., and Dahm, C.N. (2017). Water physicochemistry in intermittent rivers and ephemeral streams. In: *Intermittent Rivers And Ephemeral Streams*, 109–134. Academic Press (Cap. 3, Subcap. 1). https://doi.org/10.1016/B978-0-12-803835-2.00005-X.

ICMBIO (2007). Reserva Biológica do Tapirapé - Plano De Manejo. Belém, PA: ICMBio, 109 p. (Encarte 3). https://www.gov.br/icmbio/pt-br/assuntos/biodiversidade/unidade-de-conservacao/unidades-de-biomas/amazonia/lista-de-ucs/flona-do-tapirape-aquiri/arquivos/flona_tapirape-aquiri.pdf (accessed 31 October 2017).

ICMBiO (2009). Plano de Manejo: Parque Nacional da Serra do Cipo. https://www.gov.br/icmbio/pt-br/assuntos/biodiversidade/unidade-de-conservacao/unidades-d e-biomas/cerrado/lista-de-ucs/parna-da-serra-do-cipo/arquivos/parna_serra_do_cipo _pm_encarte1e2.pdf (accessed 31 October 2017).

Ismaiel, I. A. et al. Establishment of background water quality conditions in the Great Zab River catchment : influence of geogenic and anthropogenic controls on developing a baseline for water quality assessment and resource management. *Environmental Earth Sciences*. v. 77. n. 50. p. 1–12. 2018. DOI: https://doi.org/10.1007/s12665-017-7190-8.

Kabata-Pendias, A. and Mukherjee, A.B. (2007). *Trace Elements from Soil to Human*, 3rde. Berlin: Springer Verlag 550 p. https://doi.org/10.1007/978-3-540-32714-1.

Keller, E A 2012, 'Introduction to Environmental Geology—Water Resources' 5. Pearson 550 p. [S.L.; S.N.]

De Laune, R.D. and Reddy, K.R. (2005). Redox potential. In: *Encyclopedia of Soils in the Environment* (ed. D. Hillel), 366–371. Amsterdam: Elsevier Ltd. https://doi.org/10.1016/B0-12-348530-4/00212-5.

Lepeltier, C.A. (1969). Simplified statistical treatment of geochemical data by graphical representation. *Economic Geology* 64: 538–550. https://doi.org/10.2113/Gsecongeo.64.5.538.

Levy, M.C., Lopes, A.V., Cohn, L.G., and Thompson, S.E. (2018). Land use change increases streamflow across the arc of deforestation in Brazil. *Geophysical Research Letters* 45: 1–11. https://doi.org/10.1002/2017gl076526.

Licht, O.A.B. and Bittencourt, A.V.L. (2013). Paisagens Geoquímicas - Naturais E Antrópicas - No Estado Do Paraná. *Revista Técnico-Científica Do Crea-Pr* 1: 1–27.

Matschullat, J., Ottenstein, R., and Reimann, C. (2000). Geochemical background—can we calculate it? *Environmental Geology* 39: 990–1000. https://doi.org/10.1007/S002549900084.

Manahan, S.E. (2010). *Environmental Chemistry*, 9the. Columbia: CRC Press; Taylor & Francis Group 786 p.

Moreto, C.P.N., Monteiro, L.V.S., Xavier, R.P. et al. (2015). Neoarchean and Paleoproterozoic iron oxide-copper-gold events at the Sossego deposit, Carajás Province, Brazil: Re-Os and U-Pb geochronological evidence. *Economic Geology* 110: 809, 809–835. https://doi.org/10.2113/econgeo.110.3.

Nakić, Z., Posavec, K., and Bačani, A. (2007). A visual basic spreadsheet macro for geochemical background analysis. *Ground Water* 45: 642–647.

Nóbrega, R.L.B. et al. (2018). A multiapproach and multi-scale study on water quantity and quality changes in the Tapajós River basin, Amazon. *Proceedings of the International Association of Hydrological Sciences* 377: 3–7.

Osté, L.A., Klein, J., and Zwolsman, G.J. (2011). Inventory and evaluation of methods to derive natural background concentrations of trace metals in surface water and application of two methods in a case study. Deltares Report 1206111.005, Utrecht.

Possas, H.M., Santos, A.A., and Sousa, L.S. (2016). "Recifrando" Acordes: Marabá Das Bordas. *Revista Do Gt De Literatura Oral E Popular Da Anpoll.* 21: 182–199.

Quantum Gis Development Team (2009). *Qgis Geographic Information System*. Open Source Geospatial Foundation Disponível Em: http://Qgis.Osgeo.Org.

Queiroz, M.M.A., Horbe, A.M.C., Seyler, P., and Moura, C.A.V. (2009). Hidroquímica Do Rio Solimões Na Região Entre Manacapuru E Alvarães: Amazonas – Brasil. *Acta Amazonica* 39: 943–952. https://doi.org/10.1590/S0044-59672009000400022.

R Core Team. R: A language and environment for statistical computing. R foundation for statistical computing. 2018. Disponível Em: https://www.R-Project.Org.

Rodrigues, A.S.D.L. and Nalini, H.A. Jr. (2009). Valores De Background Geoquímico E Suas Implicações Em Estudos Ambientais. *Rem: Revista Escola de Minas* 62 (2): 155–165. https://doi.org/10.1590/S0370-44672009000200006.

Rodrigues, A.S.L., Malafaia, G., Costa, A.T., and Júnior, H.A.N. (2013). Background values for chemical elements in sediments of the Gualaxo do norte river basin, MG, Brazil. *Revista De Ciências Ambienatis - Rca* 7 (2): 15–32.

Ramos, J., Gracia, J., Ortiz, S., and González-Chávez, J.L. (2018). Water quality in aguadas within a protected karstic rain forest: the role of the vegetation-soil-water interactions. *Ecological Engineering* 121: 2–11. https://doi.org/10.1016/J.Ecoleng.2017.12.006.

Reimann, C. and Garrett, R.G. (2005). Geochemical background - concept and reality. *Science of the Total Environmental* 350: 12–27. https://doi.org/10.1016/j.scitotenv.2005.01.047.

Reimann, C., Fabian, K., Birke, M. et al. (2018). GEMAS: establishing geochemical background and threshold for 53 chemical elements in european agricultural soil. *Applied Geochemistry* 88: 302–318. https://doi.org/10.1016/j.apgeochem.2017.01.021.

Reimann, C., Filzmoser, P., and Garrett, R.G. (2005). Background and threshold: critical comparison of methods of determination. *Science of the Total Environment* 346: 1–16. https://doi.org/10.1016/j.scitotenv.2004.11.023.

Reimann, C. and Caritat, P. (2017). Establishing geochemical background variation and threshold values for 59 elements in australian surface soil. *Science of the Total Environment* 578: 633–648. https://doi.org/10.1016/J.Scitotenv.2016.11.010.

Reimann, C. and Filzmoser, P. (2000). Normal and lognormal data distribution in geochemistry: death of a myth. consequences for the statistical treatment of geochemical and environmental data. *Environmental Geology* 39: 1001–1014.

Reimann, C., Filzmoser, P., Hron, K. et al. (2017). A new method for correlation analysis of compositional (environmental) data – a worked example. *Science of the Total Environment* 607–608: 965–971.

Salomão, G.N., Dall'Agnol, R., Sahoo, P.K. et al. (2018). Geochemical distribution and thresholds values determination of heavy metals in stream water in the sub-basins of Vermelho and Sororó rivers, Itacaiúnas river watershed, eastern Amazon, Brazil. *Geochimica Brasiliensis* 32: 179–197.

Sahoo, P.K., Tripathy, S., Panigraphi, M.K., and Equeenddin, S.M. (2016). Anthropogenic contamination and risk assessment of heavy metals in stream sediments influenced by acid mine drainage from a northeast coalfield, India. *Bulletin of Engineering Geology and the Environment* 76: 537–552.

Sahoo, P.K., Dall'Agnol, R., Salomao, G.N. et al. (2019). High resolution hydrogoechemical survey and estimation of baseline concentration of trace elements in surface water of the itacaiúnas river basin, southeastern amazonia: Implication for environmental studies. *Journal of Geochemical Exploration* 205: 106321.

Souza-Filho, P.W.M., Sahoo, P.K., Silva, M.S. et al. (2022). Methodological approach and general guidelines to geochemical mapping and background-baseline analysis for environmental assessment: a case study in the Carajás Mineral Province, Brazil. *Brazilian Journal of Geology* 52 (4): 1–17.

Salomão, G.N., Dall'Agnol, R., Sahoo, P.K. et al. (2021). Geochemical mapping in stream sediments of the Carajás Mineral Province: background values for the Itacaiúnas River watershed, Brazil. *Applied Geochemistry* 118: 104608.

Sahoo, P.K., Dall'Agnol, R., Salomão, G.N. et al. (2020). Regional-scale mapping for determining geochemical background values in soils of the Itacaiúnas River Basin, Brazil: The use of compositional data analysis (CoDA). *Geoderma* 376: 114504.

SMWW, 2012. APHA—American Public Health Association. Standard Methods for The Examination of Water and Wastewater, 22nd ed.; American Public Health Association: Washington, DC, USA; p. 1360.

Sondergaard, M. Redox potencial. Denmark. In book: Encyclopedia of Inland Waters (pp.852–859), Elsevier; University of Arthus. 2009. DOI:https://doi.org/10.1016/B978-012370626-3.00115-0.

Szikszay, M. (1993). Geoquímica das águas. Boletim IG-USP. n. 5. (Série Didática) DOI:https://doi.org/10.11606/issn.2316-896X.v0i5p1-166.

Quaresma, L.S., Silva, G.S., Sahoo, P.K. et al. (2022). Source apportionment of chemical elements and their geochemical baseline values in surface water of the Parauapebas River Basin, Southeast Amazon, Brazil. *Minerals* 12: 1579.

Sahoo, P.K., Tripathy, S., Equeenuddin, S.M., and Panigrahi, M.K. (2012). Geochemical characteristics of coal mine discharge vis-à-vis behavior of rare earth elements at Jaintia Hills coalfield, northeastern India. *Journal of Geochemical Exploration* 112: 235–243.

Schaefer, CE, LIma H, Teixeira, WG, et al. 2017, Solos Da Região Amazônica. In. Curi, N. João Carlos Ker, Roberto Ferreira Novais, Pablo Vidal-Torrado, Carlos Ernesto G R Schaefer

(Eds.). Pedologia: Solos Dos Biomas Brasileiros. Sociedade Brasileira de Ciência do Solo P. 111–176. (Capítulo 3).

Sedeme (2017). *Estudo De Impacto Ambiental. Diagnóstico Ambiental - Meio Físico: Recursos Hídricos. Espeleologia*. Belém: Áreas Ambientalmente Sensíveis.

SMEWW (2017). *Standard Methods for the Examination of Water and Wastewater*, 23e. Washington DC: American Public Health Association.

Silva, A.E.P., Angelis, C.F., Machado, L.A.T., and Waichaman, A.V. (2008). Influência Da Precipitação Na Qualidade Da Água Do Rio Purus. *Acta Amazonica* 38: 733–742.

Silva, M.S.R.S., Miranda, S.A.F., Domingos, R.N. et al. (2013). Classificação Dos Rios Da Amazônia : Uma Estratégia Para Preservação Desses Recursos. *Holos Environment* 13 (2): 163–174. https://doi.org/10.14295/Holos.V13i2.7344.

Silva, SC, Mariani, CF, & Pompêo, M 2015, 'Análise Crítica Da Resolução Conama N° 357 À Luz Da Diretiva Quadro Da Água Da União Europeia: Estudo De Caso (Represa Do Guarapiranga - São Paulo. Brasil). In: Ecologia De Reservatórios E Interfaces. São Paulo: Usp, P. 367–375. (Capítulo 24).

Silva Júnior, R.O. (2017). *Resposta Hidrológica Devido Às Mudanças No Uso Do Solo E Cobertura Vegetal Na Bacia Hidrográfica Do Rio Itacaiúnas (Bhri) - Amazônia Oriental*. Belém: 136 F. Tese (Doutorado) - Universidade Federal Do Pará, Instituto De Geociências, Programa De Pós-Graduação Em Ciências Ambientais.

Silva Junior, R., Queiroz, J.C.B., Ferreira, D. et al. (2017). Estimativa de precipitação e vazões médias para a bacia hidrográfica do rio Itacaiúnas (BHRI), Amazônia Oriental, Brasil. *Revista Brasileira de Geografia Física* 10 (5): 1638–1654. https://doi.org/10.26848/rbgf.v.10.5.p1638-1654.

Siqueira, G.W., Aprile, F., and Miguéis, A.M. (2012). Diagnóstico Da Qualidade Da Água Do Rio Parauapebas (Pará – Brasil). *Acta Amazonica* 42 (3): 413–422. https://doi.org/10.1590/S0044-59672012000300014.

Souza Filho, P.W.M., Souza, E.B., Silva Júnior, R.O. et al. (2016). Four decades of land-cover, land-use and hydroclimatology changes in the Itacaiúnas River watershed, southeastern Amazon. *Journal of Environmental Management* 167: 175–184. https://doi.org/10.1016/j.jenvman.2015.11.039.

Shapiro, S.S. and Wilk, M.B. (1965). An analysis of variance test for normality. *Biometrika* 52 (3): 591–611. https://doi.org/10.2307/2333709.

Schober, P., Boer, C., and Schwarte, L. (2018). Correlation coefficients: appropriate use and interpretation. *Anesthesia & Analgesia* 126 (5): 1763–1768. https://doi.org/10.1213/Ane.0000000000002864.

Teixeira, S.S. (2016). *Estudo Da Qualidade Das Águas Superficiais E Estimativa De Background Na Área De Influência Das Minas De Ferro Da Serra Norte, Carajás*. Belém: 128 F. Dissertação (Mestrado Profissional Em Uso Sustentável De Recursos Naturais Em Regiões Tropicais) - Instituto Tecnológico Vale.

Tukey, J.W. (1977). *Book Reviews: Exploratory Data Analysis*. Eua: Addison-Wesley, 688 P https://doi.org/10.1177/014662167800200115.

Urresti-Estala, B., Carrasco-Cantos, F., Vadillo-Pérez, I., and Jiménez-Gavilán, P. (2013). Determination of background levels on water quality of groundwater bodies: a methodological proposal applied to a Mediterranean river basin (Guadalhorce River. Málaga. Southern Spain). *Journal of Environmental Management* 117: 121–130.

Vasconcelos, F.D.M., Tundisi, J.G., and Tundisi, T.M. (2009). *Avaliação Da Qualidade De Água: Base Tecnológica Para A Gestão Ambiental*. Belo Horizonte: Smea 323 P.

Vasquez, L.V., Rosa-Costa, L.R., Silva, C.G. et al. (2008). *Geologia e recursos minerais do estado do Pará: Sistema de Informações Geográficas – SIG: texto explicativo dos mapas geológico e tectônico e de recursos minerais do estado do Pará*. Belém: CPRM.

Wilcoxon, F. (1945). Individual comparisons by ranking methods. *Biometrics Bulletin* 1 (6): 80–83. Disponível Em: Https://Sci2s.Ugr.Es/Keel/Pdf/Algorithm/Articulo/Wilcoxon1945.Pdf.

Zappi, C.D. (2017). *Paisagens E Plantas De Carajás = Landscapes And Plants Of Carajás*. Belém: Itv; Mpeg Disponível Em: http://www.Itv.Org/Wp-Content/Uploads/2017/12/Paisagem-E-Plantas-De-Caraj%C3%A1s-Vers%C3%A3o-Digital.Pdf.

Zucchetti, M 2007, Rochas Máficas Do Supergrupo Grão Pará E Sua Relação Com A Mineralização De Ferro Dos Depósitos N4 E N5. Carajás. Pa. 2007. 140 F. Tese (Doutorado) - Universidade Federal De Minas Gerais, Instituto De Geociências, Programa De Pós-Graduação Em Geologia. Belo Horizonte, 2007. Disponível Em: http://Hdl.Handle.Net/1843/Mpbb-77ekeb.

8

Identifying the Footprints of Meteorological, Tectonic, and Anthropogenic Parameters on Sediment Transport in the Indus River System: A Review

Prem Kumar[1]*, Jaya Rai*[1]*, and Chandra S. Dubey*[2]

[1] *Ram Lal Anand College, Department of Geology, University of Delhi, Delhi, India*
[2] *K.R. Manglam University, Gurugram, Haryana, India*

8.1 Introduction

The Indus River with its 27 tributaries is one of the oldest rivers documented in the history (Inam et al. 2007; Shrestha et al. 2019). With a water availability of 1329 m^3 per capita, the Indus Basin supports the lives and livelihoods of around 300 million people (Jianchu et al. 2009; Immerzeel et al. 2010; Rajbhandari et al. 2015; Shrestha et al. 2019). The Indus Basin with an annual runoff of 200 cubic kilometer and irrigation area of 144,900 ha serves to be of wide utility for various purposes like drinking, hydropower generation, irrigation, navigation, industrial purposes, and livelihood generation (Mirza et al. 2003; IPCC 2007; Kundzewicz et al. 2007; Sardar et al. 2008; Iqbal 2013; Khan and Hassan 2015). Thus, the IRB (Indus River Basin) plays crucial role in socioeconomic development of India and Pakistan (Aquastat 2011; Nepal and Shrestha 2015; Hasson et al. 2016).

The IRB exhibits wide hydrometeorological variability in terms of climate types, temperature, and precipitation (Aquastat 2011; Ali 2013; Hasson et al. 2017; ICIMOD 2017; Shrestha et al. 2019). However, in recent decades, due to the changing global climate the basin has become highly vulnerable and there has been increased concern amongst the researchers for the basin. The numerous hydrometeorological parameters like precipitation, temperature, evaporation, wind, and global circulation affect the hydrology and water resources of the Indus River. The ongoing climate change may result in occurrence of extreme events like draughts, floods, and GLOF which may cause spatiotemporal disturbances in the water resources (Khalid et al. 2013; Kiem et al. 2016; Thevakaran et al. 2019). Due to high rise in temperature and fluctuating precipitation, the Indus catchment has undergone significant changes in glacial cover, glacier, and snow melt and occurrences of seasonal extremes. Numerous studies have indicated toward shrinking of the glacial cover in the Himalayas, due to glacial ablation resulting from increase in temperature (Barnett et al. 2005; Immerzeel et al. 2010; Wulf et al. 2010; Bolch et al. 2012; Ka¨a¨b et al. 2012; Lutz et al. 2014; Nepal and Shrestha 2015; Lutz et al. 2016; Shrestha et al. 2019). This glacial

Weathering and Erosion Processes in the Natural Environment, First Edition. Edited by Virendra Bahadur Singh, Sughosh Madhav, Naresh Chandra Pant, and Ravi Shekhar.

ablation would lead to increase in meltwater initially but this would ultimately cause reduction in water availability in long-term end. The IPCC (Inter-Governmental Panel on Climate Change) panel report 2001 suggested significant fall in inflow by 2050, thus ranking Indus Basin among the top 10 vulnerable basins exposed to climate change (Rolland et al. 2001; Arfan et al. 2019).

Various attempts had been made to foresee and project the potential future climatic conditions and upcoming changes in temperature and precipitation. These studies have derived varying conclusions regarding the possible impacts of these fluctuations over the hydrological regimes and water availability in the Indus catchment. Consistent warming in glacierized mountainous terrain has been reported. These studies noted sharp and consistent rise in global temperature as well as Himalayan temperatures. However, these studies could not indicate toward any persistent and fixed pattern in precipitation in IRB. Studies have also reported about varying behavioral response of the Indus Basin. These studies have concluded about differential response of the Upper and Lower Indus Basins toward the erosional and climatic processes. These prominent uncertainties in the IRB make it difficult to analyze and project the climate change and resulting implications in the basin (Agro-Dev and International (Pvt) Ltd. 2000; Rolland et al. 2001; Hasson et al. 2016; Manzoor and Ahmad 2022; Zhou et al. 2022). This research work has been carried out to figure: The climate change in the basin in recent decades and projected changes in hydro-meteorological parameters in the basin. The review work attempts to figure out the impact of these changes on sedimentation processes operating in the basin.

8.2 Study Area

Indus is a trans-boundary river with unique catchment area and is one of the world's largest drainage basins (Ahmad et al. 2018; Tiwari et al. 2021). The Indus River with a length of ~3200 km originates in the mountain Kailash in the Himalayan orogenic belt in Tibet and finally drains into the Arabian Sea in the south (Ali & De Boer 2007; Jain et al. 2007; Nepal et al. 2015; Shrestha et al. 2019). The Indus Basin has spatial coverage of an area of 1,165,500 km^2. The drainage area of Indus River is 642,700 km^2 in Pakistan, 15,100 km^2 in Afghanistan and China, and 321,289 km^2 in India. The basin of Indus River and its tributaries with a length of 1114 km covers almost 9.8% of geographical area of India. In India the Indus Basin extends between 72°35′ E to 79°50′ E and 28°52′ N to 37°20′ N, in the states of Jammu and Kashmir, Ladakh, Himachal Pradesh, Punjab, Haryana, and Rajasthan (Immerzeel et al. 2010; Aquastat 2011; Laghari et al. 2012; Nepal and Shrestha 2015).

The Indus River with many subbasins is the world's 12th largest river in the world. The major subbasins of the IRB are Sutlej, Beas, Chenab, Jhelum, Ravi, and Kabul (Figure 8.1). According to the Indus Water Treaty 1960, the tributaries of the Indus River were divided into western and eastern rivers. The Jhelum, Chenab, and Kabul are the western rivers while the Ravi, Beas, and Sutlej are the eastern rivers, respectively (Argles et al. 2003; Ojeh 2006; Aquastat 2011). The Sutlej River originates from Tibet, the Kabul River originates from Afghanistan, the Jhelum River originates from Jammu and Kashmir, and the rivers Chenab, Ravi, and Beas originate from Himachal Pradesh in India (Khan 1999; Aquastat 2011).

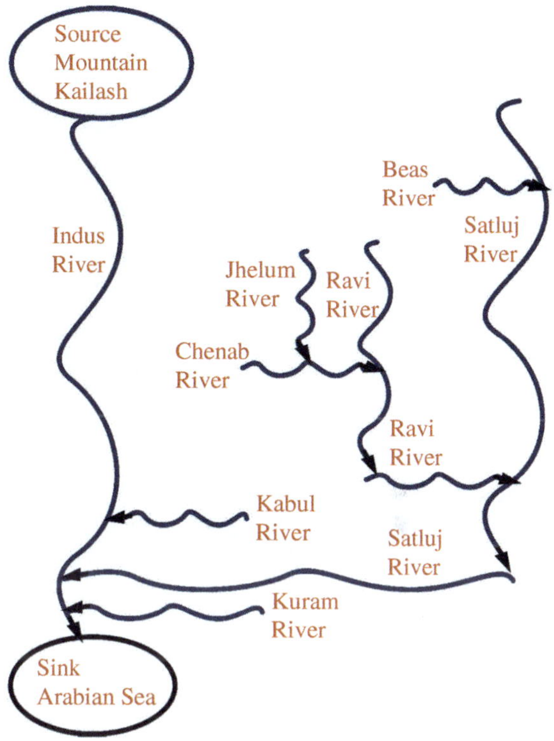

Figure 8.1 Flow diagram of Indus River (*Source:* Adapted from Jain et al. 2007; Ali 2013).

Apart from above classification based on the elevation and corresponding catchment area, the Indus Basin has been classified into the Upper and Lower Basins (Karim and Veizer 2002 and Sharma et al. 2017). In the UIB (Upper Indus Basin), the river traverses through the glacierized portions and rugged mountainous Himalayan terrain in Western Tibet, Ladakh, Central Crystalline Himalayas, Lesser Himalayas, Hindu Kush, and Himalayan Foothills (Mukhopadhyay and Dutta 2010; Sharma et al. 2017; Tiwari et al. 2021). On the other hand, the Lower Indus Basin with its vast expansion encompasses the fertile plains of Punjab and Pakistan. Thus, for countries like Pakistan and India, the Indus River with its vast catchment area proves to be highly vital for various purposes like irrigation and hydroelectric power generation (Rizwan et al. 2018; Adams 2019; Ashraf 2019). Also, the river catchment areas covered by glacial cover with their wider implications are extensively significant component of the planet's hydrological cycle (Thakur 1981; Owen et al. 2002; Paudyal et al. 2016; Tiwari et al. 2021).

8.3 Geological and Tectonic Settings

The major part of the IRB lies in the mountainous terrain and dry alluvial plains, with very significant spatial and temporal climatic variations (Aquastat 2011; Nepal and Shrestha 2015). The geological and tectonic settings of the Indus Basin are featured with very high variability and is shaped by the Indo-Eurasian collision occurred ~55 Mya

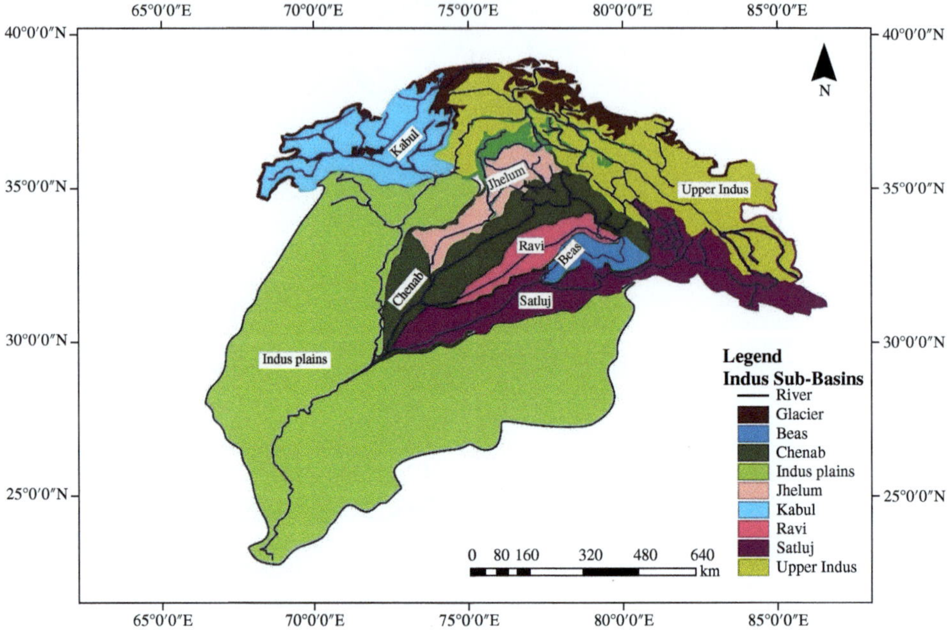

Figure 8.2 Indus catchment with subbasins and its tributaries (*Source:* Adapted from Shrestha et al. 2019).

(Clift and Jonell 2021) (Figure 8.2). The Indus River and its tributaries flow over and derive their runoff from the mountain ranges of Karakoram, Hindu Kush, Western Himalayas, and alluvial plains (Figure 8.3).

The Himalayan ranges in Pakistan are mostly constituted of unmetamorphosed strata, meta-sedimentary rocks, and fold nappes of high-grade metamorphic rocks (Badshah et al. 2000; Khan et al. 2021; Inam et al. 2022). The Himalayan front in Pakistan Himalayas is marked by the salt range and rock strata of Paleogene to Paleozoic time period (Warwick et al. 1998). West Pakistan ranges are made of large number of thrust belts of ophiolitic complexes with sedimentary rocks like limestone, conglomerate, and sandstone (Malkani 2022; Ullah et al. 2022).

The Western Himalayan ranges in India are characterized by the presence of Kohistan and Ladakh volcanic arcs formed during the continental collision (Rolland et al. 2001 and Sivaprabha et al. 2022). The Karakoram and Hindu Kush ranges are the part of Tran Himalayas, marking the northern most position of the Himalayan ranges composed of sedimentary successions characterized by low-grade metamorphic and poly phase deformational activities (Zanchi and Gaetani 2011; Palazzi et al. 2013; Khan et al. 2019; Akhter et al. 2022). These mountain ranges are tectono-morphic belts undergone crustal thickening, magmatic intrusions, transpression, and crustal melting activities. The Nanga Parbat massif comprising Pre-Cambrian Indian continental protoliths and Tethyan Himalayan lithofacies are made up of high-grade metamorphic and magmatic rocks. The Nanga Parbat is embarked with a very deep gorge created by the downcutting of the Indus River in its young stage. This zone is characterized by the presence of the Indus-Tsangpo Suture zone

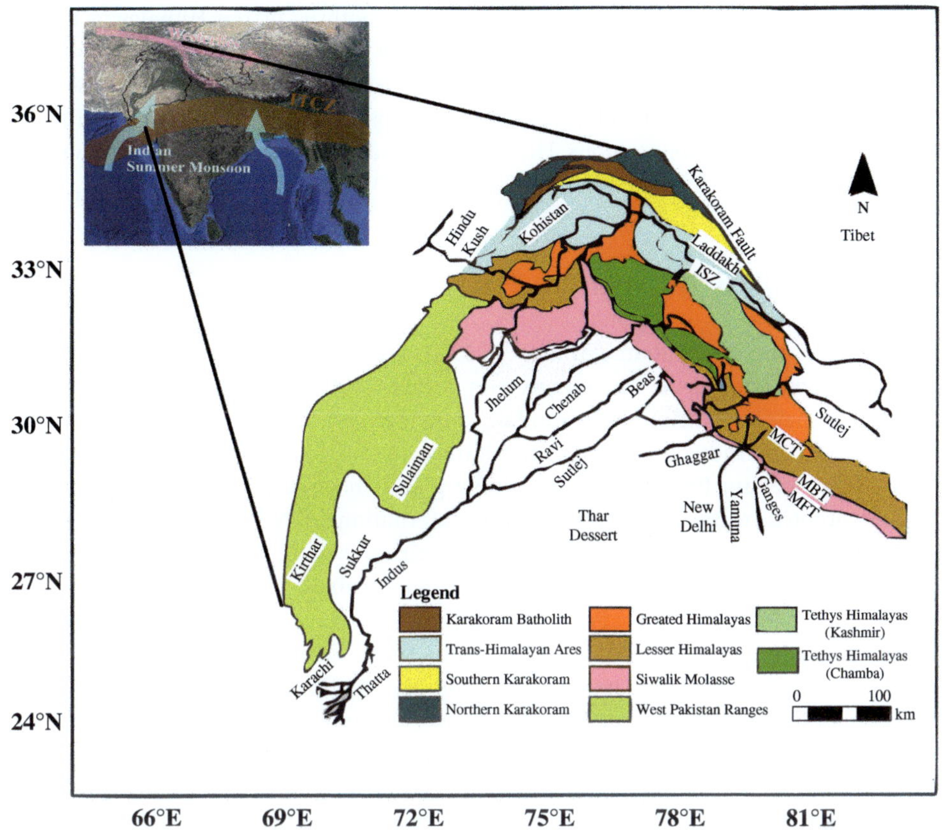

Figure 8.3 Geological map of Indus catchment with location map of the catchment (*Source:* Adapted from Clift and Jonell 2021).

which witnessed the complex collision between the Indian and Eurasian plate (Clift 2002; Henderson et al. 2011; Lone et al. 2020; Zhou et al. 2020; Inam et al. 2022). The upper river valley lies within this suture zone. Further to the Western Himalayan ranges, the tributaries of the Indus River traverse through the Trans-Himalayan arcs in Ladakh (Treloar et al. 1996). These Trans-Himalayan arcs are composed of ophiolitic sedimentary depositions. Along with this, the Himalayan ranges are distinguished by the unique occurrences of Indus molasses and Ladakh batholiths (Clift et al. 2001; Inam et al. 2007; Nag and Phartiyal 2015; Nag et al. 2016; Passang et al. 2022). The Trans-Himalayan sequences are underlain by the Tethys Himalayan which is made up of thick oceanic sedimentary sequence. Further in the south, the Greater Himalayan sequences lie, which are made up of thick sheet of high-grade metamorphic rocks. This zone is sandwiched between two large thrust faults, i.e. south Tibetan detachment zone and main central thrust. This area is also marked by the presence of a huge Zanskar fault. Toward south, the central Himalayas is followed by the lesser Himalayan sequences composed by low-grade meta-sedimentary rocks (Vannay et al. 2004). The Lesser Himalayan sequence is bounded by the large main boundary thrust in southern end. Further after crossing the Lesser Himalayan portions, the Indus River system traverses over the alluvial plains lying at the foothills of the Himalayan ranges (Pande et al. 1994; Karim and Veizer 2000; Henderson et al. 2010). The Himalayan foothills are marked by riverine depositional sequences of mudstone and limestone. Thus, we can say that the Indus River system and its runoff characteristics are influenced by two distinct geological regimes, i.e. mountainous Himalayan syntaxes and semiarid alluvial plains (Khan 1999; Nepal and Shrestha 2015; Shrestha et al. 2019; Inam et al. 2022).

8.4 Hydrologic Regime of the IRB

The Indus River originated at an elevation of 5500 m in the Himalayan range, the Nanga Parbat and is considered as the 12th largest river in the world (Ojeh 2006). The runoff in the IRB system is mainly contributed by rainfall, snowmelt, glacial melt, and surface runoff (Aquastat 2011). The Indus River and its tributaries are of perennial nature due to the continuous glacial feeding from the largest alpine glacial reserve with an aerial extent of 22 000 km^2 situated in the UIB (Agro-Dev and International (Pvt) Ltd. 2000). The entire Indus Basin forms a link between the glacial reservoirs present in the mountainous terrain and the riverine water reservoirs present in the alluvial plains.

The flow in the Indus Basin is affected by the prevailing seasons and is characterized by decreased flow in winters, increased flow in summers and flooded conditions in the monsoon (Bookhagen and Burbank 2006). Many scientific studies have indicated toward the interlinkage of the hydrological regimes of the basin and runoff with combined influential inputs of energy and precipitation (Archer 2003; Ahmad et al. 2012).

High-elevation zones like Karakoram and Hindu Kush catchments are characterized by glacial repositories and limited runoff due to prevalence of solid precipitation, low temperature, and diminished ablation (Yonus and Hassan 2022; Immerzeel et al. 2010; Nag et al. 2016; Forsythe et al. 2019; Shrestha et al. 2019). However, the mid-elevation zones lying in south to Karakoram ranges are characterized by both winter and summer precipitations. The winter precipitation at higher elevations governs the summer seasonal flow in these regions in the form of glacial melt. Although the seasonal runoff in the basin is determined by the winter precipitation, the overall annual runoff is highly influenced by the summer precipitation (Archer 2003, 2004). On contrary to the upper and mid-elevation zones, the foothill catchments of the IRB are mainly snow-free areas and are controlled by liquid form of precipitation. These areas experience liquid precipitation both in summers and winters and these areas of the basin are highly influenced by energy-controlled processes like evaporation and soil moisture content (Clift et al. 2001; Inam et al. 2007; Lone et al. 2020; Passang et al. 2022). Thus, the cryo-hydrological system of the IRB is highly susceptible to the changes in land surface temperature, planetary energy budget, and runoff-generating processes.

8.5 Climate Settings of the IRB

Owing to the changes in the geological settings and fluctuating altitudes, the IRB is characterized by varying and nonuniform climatic settings (Figure 8.4). The climate over the river basin varies with the changes in terrain from mountainous to alluvial plain provinces.

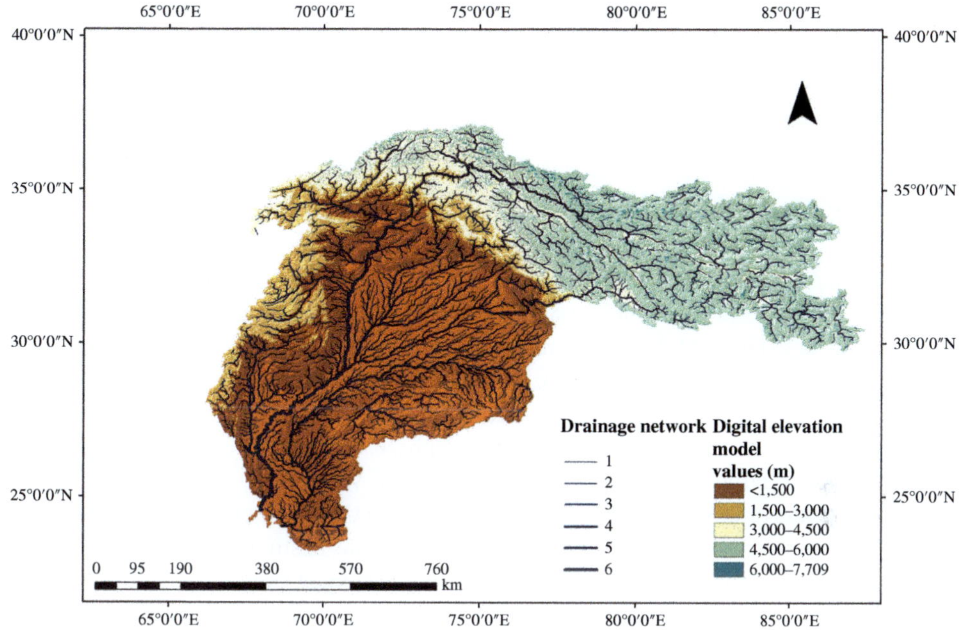

Figure 8.4 Overview of drainage network of Indus catchment projected on digital elevation model of the basin.

In mountainous regions, the river basin is characterized by high-elevation alpine climate settings while, in the plains, the river basin is characterized by temperate subhumid to semiarid and subtropical arid climate zone (Wulf et al. 2010; Aquastat 2011; Nepal and Shrestha 2015). The mountainous highland is bestowed by heavy rainfall of 2000 mm while the low-lying areas are characterized with annual precipitation of 150–500 mm.

Depending on the changes in the elevation, different portions of the IRB are characterized by distinct hydrometeorological regimes. The UIB situated in the high mountain region is characterized by large glacial cover. The mountainous terrain by limiting the intrusion of the monsoonal winds across the whole Himalayan belt leads to a decrease in the spatial influence of the monsoonal winds (Agro-Dev and International (Pvt) Ltd. 2000; Rolland et al. 2001; Archer 2004; Tiwari et al. 2021). In the upper IRB, the climate variables are influenced by altitudinal variations and here most of the precipitation is received during the winters. The Trans-Himalayan zones and Ladakh areas are characterized by arid climate with 100–200 mm annual precipitation (Forsythe et al. 2012a). The hydrology in these regions is mainly affected by the glacial melt runoff and snow distribution (Ali and De Boer 2007 and 2008; Ahmad et al. 2012; Adams 2019; Zhang et al. 2022a).

However, the mid-elevation zones like the north Western Himalayan ranges are characterized by small glacial cover and limited solid precipitation during the winters. The precipitation is mainly in form of liquid rainfall in these areas. On contrary, the low-lying Indus plains are characterized by arid to semiarid climatic regime. These areas are low precipitation zones with elevated temperature ranges. Due to the prevailing hot climate, these areas have semiarid characteristics (Agro-Dev and International (Pvt) Ltd. 2000;

Hasson et al. 2014). These areas are rain-fed basins bestowed with liquid form of precipitation. Thus, the complex interactions between the climate types and monsoon dynamics play crucial roles in determining the landscape of the IRB (Nepal and Shrestha 2015; Lutz et al. 2016; Kraaijenbrink et al. 2017).

8.6 Precipitation in the IRB

In the Himalayan orogens and the proximal regions, precipitation is chiefly controlled by two global atmospheric circulation systems – the ISM (Indian Summer Monsoon) and the winter Westerlies. The ISM derives its moisture content from the Bay of Bengal and sheds variable amounts of rainfall at different places ranging from ~1 m at mountain fronts to ~0.1 m at the northern portion of the mountain syntaxes (Wulf et al. 2010). However, the Westerlies winds arising in the northern hemisphere are stratiform types of monsoon winds possessing large spatial coverage (Dimri 2006; Wulf et al. 2010; Ali 2013). The Westerlies are less intensified and more prolonged while the summer monsoon is of highly intensified convective nature and short duration. The ISM bestows the frontal regions of the Himalayan mountain chains, whereas the northernmost portion of the Himalayan mountain chains receives their precipitation from Westerlies. The Westerlies get captured at high altitudes and pronounced relief zones of these orographic interiors, resulting in highly intense snowfall (Wulf et al. 2010; Nepal and Shrestha 2015; Arfan et al. 2019). Thus the non-glaciated catchments of the Himalayan orogens are dominantly monitored by the liquid form of summer precipitation, whereas the interior parts of the orogens are monitored by the solid form of precipitation (Wulf et al. 2010).

The precipitation in the IRB is spatially and temporally varied accounting to the altitudinal and seasonal variations as well. The annual precipitation in the IRB varies between ~1400 mm at the orographic fronts to <200 mm in the arid Himalayan zones (Bhutiyani et al. 2007; Barry 2008; Ateeq-Ur-Rehman et al. 2017; Forsythe et al. 2019; Shrestha et al. 2019). The southern foothills and other areas of the IRB like Chitral and Kabul receive a good amount of rainfall from the months of July–September. These rainfalls are orographic forced rain shedding occurring as a result of blocking of the pathway of the ISM winds (Forsythe et al. 2012b; Bookhagen et al. 2005; Kumar et al. 2021; Cendrero et al. 2022).

The ISM is characterized by heavy convective rainfall due to the presence of ample amount of moisture accumulated as a result of latent heat release and warming of the Asian landmass and the surrounding ocean (Webster and Chou 1980; Webster et al. 1998; Collins 1996; Clift et al. 2001; Chan et al. 2018). However, the path and moisture content of ISM winds are mainly controlled by the prevailing wind direction, speed of winds, and local topography; mainly the presence of topographic barriers and incised valleys is the principal influencer in moisture content of ISM winds (Lang and Barros 2002; Dimri 2006; Barry 2008). This interplay of moisture-laden winds and topography governs the amount and distribution patterns of precipitation (Barros and Lattenmaier 1994; Folland et al. 2001; Bookhagen et al. 2005). The river discharge and sediment transport in the IRB are mainly subject to the rainfall amounts of ISM thus affecting the sediment deposition and flooding processes in the alluvial plains of the IRB.

On contrary, the areas like Karakoram and Hindu Kush mainly receive their precipitation from the moisture-laden westerly disturbances. The Westerlies originating in the Mediterranean Sea are accounted for the snow form of precipitation received by the IRB. In association with the high-altitude topography, the winter monsoon governed by the Westerlies is the chief controlling factor for the complex glaciations pattern in the UIB (Galy and France-Lanord 2001; Fowler and Archer 2006; Clift and Jonell 2021).

8.7 Evaluation of Projections of Hydrometeorological Trends of the IRB

Numerous RCMs (regional climate models) and GCMs (general circulation models) have been developed and analyzed to deduce and predict the impact of climate changes undergone by IRB in the preceding decades and upcoming responses of the river system in future. The RCM and GCM are two major important sources of climate projection studies due to their integrated cumulative approach to study the interactions between the land surface processes and atmospheric phenomenon (Forsythe et al. 2019). However, due to primary dependency on the remotely sensed data, the resolution and fineness of the data are the major concerns while working on GCM and RCM (Huang et al. 2017; Shrestha et al. 2019). In order to provide better spatial resolution and refined results with reduced biased controls over the projections, the RCMs are generally coupled with GCM where due to coarse resolutions, GCMs are used as boundary conditions with RCM (Wilby and Dawson 2007; Lutz et al. 2016; Shrestha et al. 2019). In addition to all this, GCM data sets are also downscaled to improvise their spatial resolution and to reduce the inherent biasness (Rasmussen et al. 2011 and Chan et al. 2018).

8.7.1 Evaluation of Temperature Trends and Projections in the IRB

The IPCC in its report in 2007 has reported a warming rate of $0.128 \pm 0.026\,°C$ per year since last 59 years. However, in the recently released sixth report of IPCC in 2022, a rise of 1.5 °C in global temperature has been predicted. But if the current scenario persists for longer durations on account of high level of greenhouse gases emissions may lead to overshoot of global warming to greater than 1.5 °C by the year 2100 (IPCC AR 6 2022). However, owing to great altitudinal fluctuations, the IRB has been reported to exhibit nonuniform and inhomogeneous temperature trends (Nepal and Shrestha 2015 and Khan 2022b). The temperature trends between 1961 and 2000 were evaluated by Fowler and Archer, where consistent lowering of minimum and mean summer temperature was reported. On contrary, the same study also reported a consistent increase in winter mean temperature with less significant trend in winter minima temperature.

In a joint study carried out by PMD (Pakistan Meteorological Department) and global change impact studies center, increase in both maximum and minimum temperatures was predicted with positive trends in temperature in the UIB for the period 2011–2050 (Technical report No. PMD-22/2009). Similarly Chaudhry and Rasul 2007) observed an increasing trend in the annual mean temperature in the Himalayan regions of the UIB. The predicted trend is, however, characterized by intermediate periodic rise and fall in the temperature. Conversely, this portion of the basin is affected with seasonal temperature differences and studies have specifically pointed toward the high susceptibility of the basin toward these fluctuating seasonal temperature trends (Chaudhry and Rasul 2007; Khattak et al. 2011). By using the PRECIS system, based on the IPCC SRES AIB, a rise in seasonal as well as annual maximum and minimum temperatures was projected in high-elevation zones of the IRB. The winter temperatures particularly exhibit an increasing trend, while the annual maximum temperature projected maximum increase during 2071–2098 (Rajbhandari et al. 2015 and Shrestha et al. 2019). The three different simulations developed in the study projected an increase of 4 °C, 3.9 °C, and 4.6 °C in the annual maximum temperature during 2071–2098 and an increase of 4.3 °C, 3.9 °C, and 5.1 °C was projected in the annual minimum temperature.

Many studies have established consistent parallel behavior between the global temperature trends and the prevailing temperature in the north Himalayan mountainous regions in the last decade of the twentieth century (Chaudhry and Rasul 2007; Nepal and Shrestha 2015; Khan 2022a). Similarly, Bhutiyani et al. (2007) in their study had indicated about alarming warming rate of the Himalayan region at higher rates as compared to the global rates. With the help of an integrated implemental study by using a stochastic rainfall model and rainfall-conditioned generator, Forsythe et al. (2014) projected an increase in annual mean temperature by 4.8 °C during 2071–2100 in the UIB. Likewise, future temperature trends in a statistical study carried out by Kazmi et al. (2014) indicated toward significantly increasing trends in the minimum and maximum temperatures in the UIB during 2001–2030. Gebre and Ludwig (2014) predicted an increase in temperature ranging between 1 °C and 7 °C for the time periods 2030s and 2070s by the analysis of Coupled Model Inter Comparison Project Phase-5 (CMIP-5) global climate model.

Several authors have predicted elevation-dependent warming and high warming at high altitudinal zones in the IRB. They have also suggested several accounting factors like snow albedo, greenhouse gas emissions, change in atmospheric composition, and altered global circulations, which would lead to pronounced rise in temperature in the near future (Rajbhandari et al. 2015; Lutz et al. 2016; Kraaijenbrink et al. 2017; Shrestha et al. 2019; Manzoor and Ahanger 2022). Su et al. (2016) in their study based on CMIP-5 model predicted consistent increment in the annual mean temperature in the entire river basin and the results specifically highlighted the elevated increase in the UIB.

Similarly, Ali et al. (2015) and Rajbhandari et al. (2015) also highlighted projected temperature trends in the basin and proposed less temperature increase in the Lower Indus Basin as compared to the UIB and enhanced warming in the winters as compared to the other seasons. Most of the studies carried out in the IRB indicate toward the gradual rise of temperature in the basin. However, these changes are significantly affected by the seasonal temperature trends. Interestingly, these studies have not obtained systematic patterns in the temperature change and these studies are accompanied with certain uncertainties related to the projection studies (Nepal and Shrestha 2015; Shrestha et al. 2019).

8.7.2 Evaluation of Precipitation Trends and Projections in the IRB

The precipitation trends in the UIB were analyzed by Archer (2004) for the period 1895 to 2000. The study did not indicate toward any noteworthy long-term trends but periodic trends of upsurge in annual precipitation were reported at several stations for the period

1961–1999. The study indicated toward increasing contribution of winter rainfall in both southern and northern sides of the Himalayan divide, whereas summer rainfall exhibited statistical increase only on the northern side of the divide. However, this study could not indicate toward the cause of such uneven distribution of summer rainfall. It could be either the strong influence of Westerlies or the enhanced invasion of ISM winds resulting in such phenomenon. Similarly, in another study carried out by Khattak et al. (2011), precipitation exhibited inconsistent trend for the period from 1967 to 2005. These studies indicated toward persisting indefinite patterns of precipitation in the IRB since a long time period. There may be two probable reasons for such results – one may be the different stations considered, difference in the methods of analysis and different time periods of the data used. However, one of the most important contributing reasons may be due to division of the areas into windward and leeward sides accounting to the presence of orographic barrier (Barry 2008; Nepal and Shrestha 2015).

Chaudhary et al. (2009) projected the precipitation trends for the period 2011–2050. This study indicated toward decadal changes of +4.6 mm/decade, +2.9 mm/decade, and −1.3 mm/decade in precipitation for the study period. Similarly, Immerzeel et al. (2010) based on a study on five GCMs projected an increase of 25% in the mean precipitation in the Indus Basin during 2046–2065, whereas for the same time span Akhtar et al. (2008) based on a study on PRECIS RCM projected toward an increase up to 21% in mean precipitation.

A study conducted by PMD in collaboration with Global Change Impact Studies Center indicated toward nonuniform increasing and decreasing trends in precipitation in IRB portion situated in Pakistan. Similarly, Rajbhandari et al. (2015) in his study reported fluctuating precipitating pattern over the whole IRB with an increase in UIB and decrease in the Lower Indus Basin respectively. Hasson et al. (2016) in their study for 1961–2000 reported increase as well as decrease in mean precipitation. The study indicated toward ineffective signal capture by models. The study also pointed toward decreasing influence of westerly disturbances resulting in lowered winter precipitation. Ali et al. (2015) and Su et al. (2016) in their studies reported highly variable and erratic precipitation patterns. These studies also indicated toward highly significant increase in precipitation in the late twenty-first century. However, the projected trends exhibited disparity in seasonal and monsoonal precipitation.

It's very evident that the analysis and future climate projections are primarily subject to the global circulation models (GCMs) used in study. Although the majority of studies carried out in the IRB indicate toward an increase in annual and seasonal precipitation, these analyses could not indicate toward any significant and prominent temporal trends of precipitation. Unlike temperature, the proper projections in case of precipitation and significant results are absent. The major cause for such deficit may lie in research analysis methods and time periods adopted and involvement of complex orographic Himalayan barrier.

8.7.3 Sediment Yield and Related Factors in IRB

Sediment yield of any riverine system is mainly subject to two complimentary constantly ongoing processes. The two aforementioned processes are continental erosion and transfer of these products subsequently to the global ocean system (Chakrapani 2005; Zhang et al. 2022b). Few most important factors controlling the sediment yield in a river system are weathering and erosion, continental runoff, relief and geology of the catchment, basin area, temperature, precipitation, and anthropogenic influences (Chakrapani 2005; Clift and Jonell 2021; Cendrero et al. 2022; Rautela et al. 2022; Syvitski et al. 2022; Zhang et al. 2022b). Himalayan orogenic system with a mere cover of 4% land surface of earth, surprisingly, contributes to 25% of dissolved loads in the world ocean (Raymo and Ruddiman 1992 and Ahmad et al. 1998). The Indus River system drains through the Himalayan orogenic belts and dumps its total yield in the Arabian Sea. The river system feeds sediment to the second-largest submarine sedimentary body on our planet (Clift and Giosan 2012). The sediment yield and active erosional processes in the Indus catchment are subject to ongoing tectonic activities, rock uplift, and fluctuating meteorological parameters (Zhou et al. 2022). Modern and recent sediment flux in the ocean system has been estimated in various researches (Clift and Jonell 2021).

Pre-Industrial sedimentation rates in the IRB were proposed to be 250 mt/y and 300 mt/y by Milliman and Syvitski (1992) and Summerfield and Hulton (1994), respectively. However, due to the enhanced developmental activities like dam construction, sedimentation yield rates have shifted drastically (Syvitski and Kettner 2011). The best example illustrating this fact is decrease in sediment yield in the Lower Indus Basin at Kotri Barrage from 193 mt/y in 1940 to 13 mt/y in 2006 (Inam et al. 2007 and Clift and Jonell 2021). The Arabian Sea has faced abrupt decrease in sediment flux to about 10% or even lesser of the preindustrial levels. The probable reason for such exorbitant shift may be the limited water flow due to the damming activities (Clift and Jonell 2021). However, the erosional response and sediment transport dynamics of the UIB differ from the Lower Indus Basin. The causative factors for this may underlie amongst the various reasons such as lithology, glacial cover extent, debris content, subglacial erosion, supraglacial transport, mass movement, differential precipitation distribution, melt-water release, and alluvial Erosional activities (Ferguson 1984; Collins 1996; Ali and de Boer 2008). In the glacierized region of the UIB, large volumes of fine and coarse sediments are mainly released by debris transport in supraglacial zone and erosion in traction zone, respectively (Nag and Phartiyal 2015; Owen et al. 2003). However, the major portions of Indus catchment lying at higher relief with glacial cover exhibit reduced erosion rates as compared to the low-lying portions of the Indus catchment which are characterized by higher erosion rates. This behavioral difference exhibited by the Indus Basin may be accounted to the arid climate and low relief of Tibetan plateau region and high erosion activities in monsoon bestowed regions (Ali and de Boer 2008).

Tarar et al. (2018) in their study on sediment trends in the UIB deduced about the increasing trends of sedimentation in winter months and decreasing trends in the summer months. The study also indicated toward decreasing trends of sedimentation in the downstream regions of the IRB. Similarly, Wulf et al. (2012) in their study carried out in Sutlej River valley indicated toward more contribution of humid front to suspended sediment concentration as compared to the arid Tibetan region. The study dictated about the pronounced erosional activities in the Lower Indus Basin region lying at frontal regions, which are characterized by high precipitation activities. On contrary, sedimentation in the arid regions of the UIB was mainly found subject to occurrence of extreme events, mass movements, glacier melt, and GLOF activities. The study projected an increase in suspended sediment concentration in near future due to recent occurring climate change-induced catastrophes. Likewise, in a study carried out by Kumar and Srivastava (2017) rapid uplift, increased fluvial incision and high sediment supply in the upper reaches of Indus River in Ladakh Himalaya has been projected.

Rehman et al. (2018) in their study strongly pointed toward unconformity in future projections in suspended sediment load due to disproportional and fluctuating behavior of the IRB toward Erosional processes. Numerous studies have indicated toward fluctuating sedimentation patterns accounting to several controlling factors like fluvial activities, tectonic movements, extreme events, and glacial activities as dominant checking factors for the sedimentation patterns in the IRB (Leopold and Emmett 1976; Galy and France-Lanord 2001; Bookhagen et al. 2005; McCabe et al. 2007; Pratt-Sitaula et al. 2007; Gabet et al. 2008; Kumar et al. 2021).

8.8 Conclusion

The overall climate projections of the IRB indicated toward warming in the basin with significant temperature rise. In case of precipitation, no consistent trend has been projected yet. However, there are disparities in annual and seasonal trends and patterns of temperature and precipitation in the IRB. The significant findings of the study are summarized below:

- Majority of studies involving GCM and RCM indicate toward a common conclusion of increase in temperature, whereas there is very poor agreement on such common patterns for precipitation.
- Accounting to frequent variations in altitude, the IRB owes inhomogeneous temperature patterns. Majority of climate projection studies have indicated toward a rise of about 5 °C in annual minimum temperature and 4.5 °C in annual maximum temperature by the late twenty-first century (Bhutiyani et al. 2007; Chaudhry and Rasul 2007; Nepal and Shrestha 2015; IPCC AR 6 2022; Khan 2022b). For the end of the twentieth century, these studies have also substantiated a parallel behavior between the global temperature trends and temperature in the north Himalayan regions. This may result in glacial ablation and would lead to significant changes in water availability in the downstream regions.
- Numerous studies indicated toward an increase of 25% mean annual precipitation by the late twenty-first century (Immerzeel et al. 2010). However, these studies have not deduced any noteworthy long-term trends in precipitation, instead periodic upsurge have been reported for some decades.
- The Himalayan orogenic belt contributes to about 25% of the dissolutions in the ocean systems. However, the sediment yield in the IRB has decreased significantly as compared to the preindustrial sedimentation. The decrease has been from 193 mt/y in 1940 to 13 mt/y in 2006. This indicates toward a decrease up to 10% of the preindustrial sedimentation influx in the Arabian Sea (Milliman and Syvitski 1992; Summerfield and Hulton 1994; Zhang et al. 2022b; Zhou et al. 2022).
- However, the complex topography of the catchment, irregularities in precipitation and temperature, and no consideration of irrigation systems are the major limitations of the existing atmospheric and GCMs. Thus, incorporation of irrigation systems and orographic factors are the prerequisite requirements of such projection studies.

References

Adams, T.E. III (2019). Water Resources Forecasting within the Indus River Basin: a call for comprehensive modeling. In: *Indus River Basin* (ed. S.I. Khan and T.E. Adams III), 267–308. Elsevier.

Ahmad, I., Zhang, F., Tayyab, M. et al. (2018). Spatiotemporal analysis of precipitation variability in annual, seasonal and extreme values over upper Indus River basin. *Atmospheric Research* 213: 346–360.

Ahmad, Z., Hafeez, M., and Ahmad, I. (2012). Hydrology of mountainous areas in the upper Indus Basin, Northern Pakistan with the perspective of climate change. *Environmental Monitoring and Assessment* 184 (9): 5255–5274.

Ahmad, T., Khanna, P.P., Chakrapani, G.J., and Balakrishnan, S. (1998). Geochemical characteristics of water and sediment of the Indus River, Trans-Himalaya, India: constraints on weathering and erosion. *Journal of Asian Earth Sciences* 16 (2–3): 333–346.

Akhtar, M., Ahmad, N., and Booij, M.J. (2008). The impact of climate change on the water resources of Hindukush-Karakorum-Himalaya region under different glacier coverage scenarios. *Journal of Hydrology* 355: 148–163.

Akhter, G., Ge, Y., Hasan, M., and Shang, Y. (2022). Estimation of hydrogeological parameters by using pumping, laboratory data, surface resistivity and Thiessen technique in Lower Bari Doab (Indus Basin), Pakistan. *Applied Sciences* 12 (6): 3055.

Ali, S., Li, D., Congbin, F., and Khan, F. (2015). Twenty first century climatic and hydrological changes over upper Indus Basin of Himalayan region of Pakistan. *Environmental Research Letters* 10 (1): 1014007.

Ali, A. (2013). *Indus Basin Floods Mechanisms, Impacts, and Management*. Mandaluyong City: Asian Development Bank.

Ali, K.F. and de Boer, D.H. (2008). Factors controlling specific sediment yield in the upper Indus River basin, northern Pakistan. *Hydrological Processes: An International Journal* 22 (16): 3102–3114.

Ali, K.F. and De Boer, D.H. (2007). Spatial patterns and variation of suspended sediment yield in the Upper Indus River basin, Northern Pakistan. *Journal of Hydrology* 334: 368–387.

Aquastat, F.A.O. (2011). *FAO's information system on water and agriculture*. Rome, Italy: Food and Agriculture Organization of the United Nations (FAO).

Archer, D.R. (2004). Hydrological implications of spatial and altitudinal variation in temperature in the upper Indus basin. *Nordic Hydrology* 35: 213–227.

Archer, D. (2003). Contrasting hydrological regimes in the upper Indus Basin. *Journal of Hydrology* 274 (1–4): 198–210.

Arfan, M., Lund, J., Hassan, D. et al. (2019). Assessment of spatial and temporal flow variability of the Indus River. *Resources* 8 (2): 103.

Argles, T., Foster, G., Whittington, A. et al. (2003). Isotope studies reveal a complete Himalayan section in the Nanga Parbat syntaxis. *Geology* 31: 1109–1112.

Ashraf, A. (2019). Upstream and downstream response of water resource regimes to climate change in the Indus River basin. *Arabian Journal of Geosciences* 12 (16): 1–10.

Ateeq-Ur-Rehman, S., Bui, M.D., and Rutschmann, P. (2017). Variability and trend detection in the sediment load of the Upper Indus River. *Water* 10 (1): 16.

Badshah, M.S., Gnos, E., Jan, M.Q., and Afridi, M.I. (2000). Stratigraphy and tectonic evolution of the northwestern Indian plate and Kabul Block. *Special Publication. Geological Society of London* 170: 467–475.

Barnett, T.P., Adam, J.C., and Lettenmaier, D.P. (2005). Potential impacts of a warming climate on water availability in snow-dominated regions. *Nature* 438: 303–309. https://doi.org/10.1038/nature04141.

Barry, R.G. (2008). *Mountain weather and climate*. UK: Cambridge University Press.

Barros, A.P. and Lattenmaier, D.P. (1994). Dynamic modeling of orographically induced precipitation. *Reviews of Geophysics* 32 (3): 265–284.

Bhutiyani, M.R., Kale, V.S., and Pawar, N.J. (2007). Long-term trends in maximum, minimum and mean annual air temperatures across the Northwestern Himalaya during the twentieth century. *Climatic Change* 85: 159–177.

Bolch, T., Kulkarni, A., and Ka¨a¨b, A., Huggel, C., Paul, F. Cogley, J. G., . . . Stoffel, M. (2012). The state and fate of Himalayan glaciers. *Science* 336 (6079): 310–314.

Bookhagen, B. and Burbank, D.W. (2006). Topography, relief, and TRMM-derived rainfall variations along the Himalaya. *Geophysical Research Letters* 33: L08405. https://doi.org/10.1029/2006GL026944.

Bookhagen, B., Thiede, R.C., and Strecker, M.R. (2005). Late Quaternary intensified monsoon phases control landscape evolution in the northwest Himalaya. *Geology* 33: 149–152.

Cendrero, A., Remondo, J., Beylich, A. et al. (2022). Denudation and geomorphic change in the Anthropocene; a global overview. *Earth-Science Reviews* 233: 104186.

Chakrapani, G.J. (2005). Factors controlling variations in river sediment loads. *Current Science* 88 (4): 569–575.

Chan, S.C., Kendon, E.J., Roberts, N. et al. (2018). Large-scale predictors for extreme hourly precipitation events in convection-permitting climate simulations. *Journal of Climate* 31 (6): 2115–2131. https://doi.org/10.1175/JCLI-D-17-0404.1.

Chaudhry, Q.Z. and Rasul, G. (2007). *Development of a mesoscale convective system over the foothills of Himalaya into a severe storm*, vol. 10. Mountains Witnesses of Global Changes.

Chaudhry, Q.Z., Mahmood, A., Rasul, G., and Afzaal, M. (2009). A Report on Climate Indicators Pakistan. Pakistan Metrological Department. Technical Report No PMD-22.

Clift, P.D. and Jonell, T.N. (2021). Monsoon controls on sediment generation and transport: mass budget and provenance constraints from the Indus River catchment, delta and submarine fan over tectonic and multimillennial timescales. *Earth-Science Reviews* 220: 103682.

Clift, P. D., &Giosan, L. (2012). Sediment source-to-sink processes in the Indus River since the last glacial maximum. In EGU General Assembly Conference Abstracts (p. 1657) 22–27 April 2012 in Vienna, Austria.

Clift, P.D. (2002). A brief history of the Indus River. *Geological Society* 195 (1): 237–258.

Clift, P.D., Shimizu, N., Layne, G.D. et al. (2001). Development of the Indus Fan and its significance for the erosional history of the Western Himalaya and Karakoram. *Geological Society of America Bulletin* 113 (8): 1039–1051.

Collins, D.N. (1996). Sediment transport from glacierized basins in the Karakoram mountains. In: *Erosion and Sediment Yield: Global and Regional Perspective* (ed. D.E. Walling and B.W. Webb) IAHS Publication No. 236, 85–96. Wallingford: IAHS Press.

Dimri, A. (2006). Surface and upper air fields during extreme winter precipitation over the western Himalayas. *Pure and Applied Geophysics* 163: 1679–1698.

Ferguson, R.I. (1984). Sediment load of the Hunza River. In: *The International Karakoram Project* (ed. K.J. Miller), 580–598. Cambridge: Cambridge University Press.

Folland, C.K., Karl, T.R., Christy, J.R. et al. (2001). *Climate Change 2001: The Scientific Basis. Contribution of Working Group I to the Third Assessment Report of the Intergovernmental Panel on Climate Change*, 881. Cambridge, UK: Cambridge University Press.

Forsythe, N., Fowler, H.J., Blenkinsop, S. et al. (2014). Application of a stochastic weather generator to assess climate change impacts in a semi-arid climate: the Upper Indus Basin. *Journal of Hydrology* 517: 1019–1034.

Forsythe, N., Archer, D.R., Pritchard, D., and Fowler, H. (2019). A hydrological perspective on interpretation of available climate projections for the upper Indus Basin. In: *Indus River Basin* (ed. S.I. Khan and T.E. Adams III), 159–179. Elsevier.

Forsythe, N., Fowler, H.J., Kilsby, C.G., and Archer, D.R. (2012a). Opportunities from remote sensing for supporting water resources management in village/valley scale catchments in the Upper Indus Basin. *Water Resources Management* 26 (4): 845–871. https://doi.org/10.1007/s11269-011-9933-8.

Forsythe, N., Kilsby, C.G., Fowler, H.J., and Archer, D.R. (2012b). Assessment of runoff sensitivity in the upper Indus Basin to interannual climate variability and potential change using MODIS satellite data products. *Mountain Research and Development* 32 (1): 16–29. https://doi.org/10.1659/MRD-JOURNAL-D-11-00027.1.

Fowler, H. J., & Archer, D. R. (2006). Conflicting signals of climatic change in the upper Indus Basin. *Journal of Climate*, 19(17), 4276–4293. https://doi.org/10.1175/jcli3860.1

Gabet, E.J., Burbank, D.W., Pratt-Sitaula, B. et al. (2008). Modem erosion rates in the High Himalayas of Nepal. *Earth and Planetary Science Letters* 267: 482–494.

Galy, A. and France-Lanord, C. (2001). Higher erosion rates in the Himalaya: geochemical constraints on riverine fluxes. *Geology* 29: 23–26.

Gebre, S.L. and Ludwig, F. (2014). Spatial and temporal variation of impacts of climate change on the hydrometeorology of Indus River Basin using RCPs scenarios, South East Asia. *Journal of Earth Science and Climatic Change* 5 (10): 1–7.

Hasson, S., Lucarini, V., Khan, M.R. et al. (2014). Early 21st century snow cover state over the western river basins of the Indus River system. *Hydrology and Earth System Sciences* 18 (10): 4077–4100.

Hasson, S., Bohner, J., and Lucarini, V. (2017). Prevailing climatic trends and runoff response from Hindukush-Karakoram- Himalaya, upper Indus Basin. *Earth System Dynamics* 8: 337–355.

Hasson, S.U., Pascale, S., Lucarini, V., and Bohner, J. (2016). Seasonal cycle of precipitation over major river basins in south € and southeast Asia: a review of the CMIP5 climate models data for present climate and future climate projections. *Atmospheric Research* 180: 42–63.

Henderson, A.L., Najman, Y., Parrish, R. et al. (2011). Constraints to the timing of India–Eurasia collision; a re-evaluation of evidence from the Indus Basin sedimentary rocks of the Indus–Tsangpo Suture Zone, Ladakh, India. *Earth-Science Reviews* 106 (3-4): 265–292.

Henderson, A.L., Najman, Y., Parrish, R. et al. (2010). Geology of the Cenozoic Indus Basin sedimentary rocks: paleoenvironmental interpretation of sedimentation from the western Himalaya during the early phases of India-Eurasia collision. *Tectonics* 29 (6).

https://report.ipcc.ch/ar6/wg2/IPCC_AR6_WGII_FullReport.pdf.
https://www.pmd.gov.pk/CC%20Indicators.pdf.

Huang, J., Wang, Y., Fischer, T. et al. (2017). Simulation and projection of climatic changes in the Indus River Basin, using the regional climate model COSMO-CLM. *International Journal of Climatology* 37: 2545–2562.

ICIMOD (2017). Upper Indus Basin Network and Indus Forum Collaboration Meeting. Workshop Report, ICIMOD, Kathmandu.

Immerzeel, W.W., Droogers, P., de Jong, S.M., and Bierkens, M.F.P. (2010). Large-scale monitoring of snow cover and runoff simulation in Himalayan river basins using remote sensing. *Remote Sensing of Environment* 113: 40–49.

Inam, A., Clift, P.D., Giosan, L. et al. (2022). The geographic, geological, and oceanographic setting of the Indus River–an update. In: *Large Rivers: Geomorphology and Management*, 2nde (ed. A. Gupta), 488–520.

Inam, A., Clift, P.D., Giosan, L. et al. (2007). The geographic, geological and oceanographic setting of the Indus River. In: *Large Rivers: Geomorphology and Management* (ed. A. Gupta), 333–345.

IPCC (2007). Climate change 2007: impacts, adaptation and vulnerability: working group II contribution to the fourth assessment report of the IPCC intergovernmental panel on climate change. In: *Assessment* (ed. M.L. Parry, O.F. Canziani, J.P. Palutikof, et al.), 976. Cambridge: Cambridge University Press.

Iqbal, A.R., (2013). Environmental Issues of Indus River Basin: An Analysis. ISSRA Papers. Institute for Strategic Studies, Research & Analysis, National Defence University, Islamabad, pp. 89–112.

Jain, S.K., Agarwal, P.K., and Singh, V.P. (2007). *Hydrology and Water Resources of India*, vol. 57. Springer Science & Business Media.

Jianchu, X., Shrestha, A., & Eriksson, M. (2009). Climate change and its impacts on glaciers and water resource management in the Himalayan Region. *Assessment of Snow, Glaciers and Water Resources in Asia. International Hydrological Programme of UNESCO and Hydrology and Water Resources Programme of WMO. Koblenz, Germany*, *44*, 54.

Kääb, A., Berthier, E., Nuth, C. et al. (2012). Contrasting patterns of early twenty-first-century glacier mass change in the Himalayas. *Nature* 488: 495–498.

Karim, A. and Veizer, J. (2002). Water balance of Indus river basin and moisture source in the Karakoram and western Himalayas: implications from hydrogen and oxygen isotopes in river water. *Journal of Geophysical Research* 107: ACH 9-1–ACH 9-12. https://doi.org/10.1029/2000JD000253.

Karim, A. and Veizer, J. (2000). Weathering processes in the Indus River Basin: implications from riverine carbon, sulfur, oxygen, and strontium isotopes. *Chemical Geology* 170 (1–4): 153–177.

Kazmi, D.H., Li, J., Rasul, G. et al. (2014). Statistical downscaling and future scenario generation of temperatures for Pakistan region. *Theoretical and Applied Climatology* 120: 341–350.

Khalid, S., Qasim, M., and Farhan, D. (2013). Hydro-meteorological characteristics of Indus River Basin at extreme north of Pakistan. *Journal of Earth Science and Climatic Change* 5 (2).

Khan, F. (2022a). Water availability and response of Tarbela Reservoir under the changing climate in the Upper Indus Basin, Pakistan. *Scientific Reports* 12 (1): 1–15.

Khan, A.F. (2022b). Increasing Risk of Glacial Lake Outburst Floods (GLOFs) as a consequence of climate change in the Himalayan Region. In: *In 5th World Congress on Disaster Management*, 405–412. Routledge.

Khan, U., Janjuhah, H.T., Kontakiotis, G. et al. (2021). Natural processes and anthropogenic activity in the indus river sedimentary environment in Pakistan: a critical review. *Journal of Marine Science and Engineering* 9 (10): 1109.

Khan, A.A., ul Hassan, S.N., Baig, S. et al. (2019). The response of land surface temperature to the changing land-use land-cover in a mountainous landscape under the influence of urbanization: Gilgit City as a case study in the Hindu Kush Himalayan Region of Pakistan. *International Journal of Economic and Environmental Geology* 10 (3): 40–49.

Khan, H. and Hassan, S. (2015). Stochastic river flow modelling and forecasting of Upper Indus Basin. *Journal of Basic and Applied Sciences* 11: 630–636. https://doi.org/10.6000/1927-5129.2015.11.84.

Khan, A. R. (1999). An analysis of the surface water resources and water delivery systems in the Indus Basin.

Khattak, M., Babel, M., & Sharif, M. (2011). Hydro-meteorological trends in the upper Indus River basin in Pakistan. *Climate Research*, 46(2), 103–119. https://doi.org/10.3354/cr00957

Kiem, A.S., Austin, E.K., and Verdon-Kidd, D.C. (2016). Water resource management in a variable and changing climate: hypothetical case study to explore decision making under uncertainty. *Journal of Water and Climate Change.* 7 (2): 263–279.

Kraaijenbrink, P.D., Bierkens, M.F.P., Lutz, A.F., and Immerzeel, W.W. (2017). Impact of a global temperature rise of 1.5 degrees Celsius on Asia's glaciers. *Nature* 549 (7671): 257–260.

Kumar, P., Dubey, C.S., Kumar, O. et al. (2021). Deciphering the role of meteorological parameters controlling the sediment load and water discharge in the Sutlej basin, Western Himalaya. *Journal of Environmental Management* 298: 113413.

Kumar, A. and Srivastava, P. (2017). The role of climate and tectonics in aggradation and incision of the Indus River in the Ladakh Himalaya during the late Quaternary. *Quaternary Research* 87 (3): 363–385.

Kundzewicz, Z.W., Mata, L.J., Arnell, N.W., and Do¨ll, P., Kabat, P., Jime'nez, B., & . . . Scott, M. J. (2007). *Climate change 2007 – the physical science basis, science.* Cambridge: Cambridge University Press.

Laghari, A.N., Vanham, D., and Rauch, W. (2012). The Indus basin in the framework of current and future water resources management. *Hydrology and Earth System Sciences* 16: 1063–1083. https://doi.org/10.5194/hess-16-1063-2012.

Lang, T.J. and Barros, A.P. (2002). An investigation of the onsets of the 1999 and 2000 monsoons in central Nepal. *Monthly Weather Review* 130 (5): 1299–1316.

Leopold, L.B. and Emmett, W.W. (1976). Bedload measurements, East Fork River, Wyoming. *Proceedings of the National Academy of Sciences of the United States of America* 73: 1000–1004.

Lone, S.A., Jeelani, G., Mukherjee, A., and Coomar, P. (2020). Geogenic groundwater arsenic in high altitude bedrock aquifers of upper Indus river basin (UIRB), Ladakh. *Applied Geochemistry* 113: 104497.

Lutz, A.F., ter Maat, H.W., Biemans, H. et al. (2016). Selecting representative climate models for climate change impact studies: an advanced envelope-based selection approach. *International Journal of Climatology* 36 (12): 3988–4005.

Lutz, A.F., Immerzeel, W.W., Shrestha, A.B., and Bierkens, M.F.P. (2014). Consistent increase in High Asia's runoff due to increasing glacier melt and precipitation. *Nature Climate Change* 4: 587–592. https://doi.org/10.1038/nclimate2237.

Malkani, M.S. (2022). Plates boundary and structural geology of Balochistan and Indus basins through field observations on Chaman transform fault and Western Indus Suture (Pakistan): dinosaurs from Pakistan with attributed bones and key features: titanosaurs from India with updated assessment on jainosaurus. *Open Journal of Geology* 12 (11): 1032–1079.

Manzoor, S. and Ahmad Ahanger, M. (2022). Spatio-temporal trends in precipitation and temperature means/extremes in the Himalayan states of India. *Journal of Water and Climate Change* 13 (7): 2531–2558.

McCabe, G.J., Clark, M.P., and Hay, L.E. (2007). Rain-on-snow events in the western United States. *Bulletin of the American Meteorological Society* 88: 319–328.

Milliman, J.D. and Syvitski, J.P.M. (1992). Geomorphic/tectonic control of sediment discharge to the ocean; the importance of small mountainous rivers. *Journal of Geology* 100: 525–544.

Mirza, M.M.Q., Warrick, R.A., and Ericksen, N.J. (2003). The implications of climate change on floods of the ganges, Brahmaputra and Meghna rivers in Bangladesh. *Climatic Change* 57: 287–318. https://doi.org/10.1023/A:1022825915791.

Mukhopadhyay, B. and Dutta, A. (2010). A stream water availability model of upper Indus basin based on a topologic model and global climatic datasets. *Water Resources Management* 24: 4403–4443. https://doi.org/10.1007/s11269-010-9666-0.

Nag, D., Phartiyal, B., and Singh, D.S. (2016). Sedimentary characteristics of palaeolake deposits along the Indus River valley, Ladakh, Trans-Himalaya: implications for the depositional environment. *Sedimentology* 63 (6): 1765–1785.

Nag, D. and Phartiyal, B. (2015). Climatic variations and geomorphology of the Indus River valley, between Nimo and Batalik, Ladakh (NW trans Himalayas) during late quaternary. *Quaternary International* 371: 87–101.

Nepal, S. and Shrestha, A.B. (2015). Impact of climate change on the hydrological regime of the Indus, Ganges and Brahmaputra River basins: a review of the literature. *International Journal of Water Resources Development* 31 (2): 201–218.

Ojeh, E. (2006). Hydrology of the Indus Basin (Pakistan). GIS in water resources term project–CE K, 394.

Owen, L.A., Derbyshire, E., and Scott, C.H. (2003). Contemporary sediment production and transfer in high-altitude glaciers. *Sedimentary Geology* 155: 13–36.

Owen, L.A., Finkel, R.C., and Caffee, M.W. (2002). A note on the extent of glaciation throughout the Himalaya during the global last glacial maximum. *Quaternary Science Reviews* 21: 147–157. https://doi.org/10.1016/S0277-3791(01)00104-4.

Palazzi, E., Von Hardenberg, J., and Provenzale, A. (2013). Precipitation in the Hindu-Kush Karakoram Himalaya: observations and future scenarios. *Journal of Geophysical Research-Atmospheres* 118 (1): 85–100.

Pande, K., Sarin, M.M., Trivedi, J.R. et al. (1994). The Indus river system (India-Pakistan): major-ion chemistry, uranium and strontium isotopes. *Chemical Geology* 116 (3-4): 245–259.

Passang, S., Schmidt, S., and Nüsser, M. (2022). Topographical impact on snow cover distribution in the trans-Himalayan region of Ladakh, India. *Geosciences* 12 (8): 311.

Paudyal, R., Kang, S., and Sharma, C.M. (2016). Major ions and trace elements of two selected rivers near Everest region, southern Himalayas, Nepal. *Environmental Earth Sciences* 75: 46. https://doi.org/10.1007/s12665-015-4811-y.

Pratt-Sitaula, B., Garde, M., Burbank, D. et al. (2007). Bedload-to-suspended load ratio and rapid bedrock incision from Himalayan landslide-dam lake record. *Quaternary Research* 68: 111–120.

Rajbhandari, R., Shrestha, A.B., Kulkarni, A. et al. (2015). Projected changes in climate over the Indus river basin using a high resolution regional climate model (PRECIS). *Climate Dynamics* 44 (1): 339–357.

Rasmussen, R., Liu, C., Ikeda, K. et al. (2011). High-resolution coupled climate runoff simulations of seasonal snowfall over Colorado: a process study of current and warmer climate. *Journal of Climate* 24 (12): 3015–3048. https://doi.org/10.1175/2010JCLI3985.1.

Rautela, K.S., Kuniyal, J.C., Alam, M.A. et al. (2022). Assessment of daily streamflow, sediment fluxes, and erosion rate of a pro-glacial stream basin, Central Himalaya, Uttarakhand. *Water, Air, and Soil Pollution* 233 (4): 1–16.

Raymo, M.E. and Ruddiman, W.F. (1992). Tectonic forcing of late Cenozoic climate. *Nature* 359: 117–124.

Rehman Qaisar, F.U., Zhang, F., Pant, R.R. et al. (2018). Spatial variation, source identification, and quality assessment of surface water geochemical composition in the Indus River Basin, Pakistan. *Environmental Science and Pollution Research* 25: 12749–12763.

Rizwan, M., Bakhsh, A., Li, X. et al. (2018). Evaluation of the impact of water management technologies on water savings in the lower chenab canal command area, Indus River Basin. *Water* 10 (6): 681.

Sardar, M.S., Tahir, M.A., and Zafar, M.I. (2008). Poverty in riverine areas: vulnerabilities, social gaps and flood damages. *Pakistan Journal of Life and Social Sciences* 6: 25–31.

Sharma, A., Kumar, K., and Laskar, A. (2017). Oxygen, deuterium, and strontium isotope characteristics of the Indus River water system. *Geomorphology* 284: 5–16. https://doi.org/10.1016/j.geomorph.2016.12.014.

Shrestha, A.B., Wagle, N., and Rajbhandari, R. (2019). A review on the projected changes in climate over the Indus Basin. *Indus River Basin* 145–158.

Sivaprabha, S., Bhat, I.M., Ahmad, T. et al. (2022). Geochemical analysis of magmatic rocks from Shyok Suture Zone (SSZ) Trans-Himalaya, NW India: insights for geodynamic evolution of the terrane. *Lithos* 410: 106594.

Su, B., Huang, J., Gemmer, M. et al. (2016). Statistical downscaling of CMIP5 multi-model ensemble for projected changes of climate in the Indus River Basin. *Atmospheric Research* 178: 138–149.

Summerfield, M. and Hulton, N.J. (1994). Natural controls of fluvial denudaiton rates in major world drainage basins. *Journal of Geophysical Research* 99 (B7): 13871–13883.

Syvitski, J., Ángel, J.R., Saito, Y. et al. (2022). Earth's sediment cycle during the Anthropocene. *Nature Reviews Earth and Environment* 3 (3): 179–196.

Syvitski, J.P.M. and Kettner, A. (2011). Sediment flux and the Anthropocene. *Philosophical Transactions: Mathematical, Physical and Engineering Sciences* 369 (1938): 957–975.

Tarar, Z.R., Ahmad, S.R., Ahmad, I., and Majid, Z. (2018). Detection of sediment trends using wavelet transforms in the upper Indus River. *Water* 10 (7): 918.

Thakur, V.C. (1981). Regional framework and geodynamic evolution of the Indus Tsangpo Suture Zone in the Ladakh Himalaya. *Transactions of the Royal Society of Edinburgh* 72: 89–97. https://doi.org/10.1017/S0263593300009925.

Thevakaran, A., Suppiah, R., Sonnadara, U. (2019). Trends in extreme rainfall events in Sri Lanka, 1961-2010, *Journal of the National Science Foundation of Sri Lanka*, vol. 47, no. 3. Pg. 285–295.

Tiwari, A.K., Singh, A.K., Phartiyal, B., and Sharma, A. (2021). Hydrogeochemical characteristics of the Indus River water system. *Chemistry and Ecology* 37 (9-10): 780–808.

Treloar, P.J., Petterson, M.G., Jan, M.Q., and Sullivan, M.A. (1996). A re-evaluation of the stratigraphy and evolution of the Kohistan arc sequence, Pakistan Himalaya: implications for magmatic and tectonic arc-building processes. *Journal of the Geological Society (London)* 153: 681–693.

Ullah, S., Jan, I.U., Hanif, M. et al. (2022). Paleoenvironmental and Bio-Sequence Stratigraphic Analysis of the Cretaceous Pelagic Carbonates of Eastern Tethys, Sulaiman Range, Pakistan. *Minerals* 12 (8): 946.

Vannay, J.C., Grasemann, B., Rahn, M. et al. (2004). Miocene to Holocene exhumation of metamorphic crustal wedges in the NW Himalaya: evidence for tectonic extrusion coupled to fluvial erosion. *Tectonics* 23: TC1014.

Warwick, P.D., Johnson, E.A., and Khan, I.H. (1998). Collision-induced tectonism along the northwestern margin of the Indian subcontinent as recorded in the Upper Paleocene to Middle Eocene strata of central Pakistan (Kirthar and Suleiman Ranges). *Palaeogeography Palaeoclimatology Palaeoecology* 142: 201–216.

Webster, P.J., Magana, V.O., Palmer, T.N. et al. (1998). Monsoons: processes, predictability, and the prospects for prediction. *Journal of Geophysical Research, Oceans* 103 (C7): 14451–14510.

Webster, P.J. and Chou, L.C. (1980). Seasonal structure of a simple monsoon system. *Journal of the Atmospheric Sciences* 37 (2): 354–367.

Agro-Dev, A., International (Pvt) Ltd. (2000). Tarbela Dam and related aspects of the Indus River Basin Pakistan Final Report: November 2000 Prepared for the World Commission on Dams (WCD).

Wilby, R.L. and Dawson, C.W. (2007). *SDSM 4.2—A Decision Support Tool for the Assessment of Regional Climate Change Impacts, Version 4.2 User Manual*. Lancaster Univ. Lancaster/ Environment Agency Engl. Wales pp. 1–94. 1586. A REVIEW ON THE PROJECTED CHANGES IN CLIMATE OVER THE INDUS BASIN.

Wulf, H., Bookhagen, B., and Scherler, D. (2012). Climatic and geologic controls on suspended sediment flux in the Sutlej River Valley, western Himalaya. *Hydrology and Earth System Sciences* 16 (7): 2193–2217.

Wulf, H., Bookhagen, B., and Scherler, D. (2010). Seasonal precipitation gradients and their impact on fluvial sediment flux in the Northwest Himalaya. *Geomorphology* 118 (1–2): 13–21.

Rolland, Y., Pe^cher, A., and Picard, C. (2001). Middle Cretaceous back-arc formation and arc evolution along the Asian margin: the Shyok Suture Zone in northern Ladakh (NW Himalaya). *Tectonophysics* 325: 145–173.

Yonus, M. and Hassan, S.A. (2022). The impact of climate change on stochastic variations of the hydrology of the flow of the Indus River. *Slovak Journal of Civil Engineering* 30 (1): 33–41.

Zanchi, A. and Gaetani, M. (2011). The geology of the Karakoram range, Pakistan: the new 1: 100,000 geological map of Central-Western Karakoram. *Italian Journal of Geosciences* 130 (2): 161–262.

Zhang, F., Zeng, C., Wang, G. et al. (2022a). Runoff and sediment yield in relation to precipitation, temperature and glaciers on the Tibetan Plateau. *International Soil and Water Conservation Research* 10 (2): 197–207.

Zhang, L., Jiang, Y., Yang, M. et al. (2022b). Quantifying the impacts of land use and cover change (LUCC) and climate change on discharge and sediment load in the Hunhe River Basin, Liaoning Province, Northeast China. *Water* 14 (5): 737.

Zhou, P., Stockli, D.F., Ireland, T. et al. (2022). Zircon U-Pb age constraints on NW Himalayan exhumation from the Laxmi Basin, Arabian Sea. *Geochemistry, Geophysics, Geosystems* 23 (1): e2021GC010158.

Zhou, R., Aitchison, J.C., Lokho, K. et al. (2020). Unroofing the Ladakh Batholith: constraints from autochthonous molasse of the Indus Basin, NW Himalaya. *Journal of the Geological Society* 177 (4): 818–825.

9

An Implication of Enhanced Rock Weathering on the Groundwater Quality: A Case Study from Wardha Valley Coalfields, Central India

Priyadarshan S. Ganvir[1] *and Rajeeva Guhey*[2]

[1] *Department of Geology, M. G. Arts, Science and Late N. P. Commerce College Armori, Gadchiroli, MH, India*
[2] *Department of Geology, Govt. N. P. G. College of Science, Raipur, CG, India*

9.1 Introduction

The coalfields are crucial profitable setups of any country with an ample coal reserve. The productions from the coalfields not only suffice the energy requirements but also act as a column in economy. Despite such strategic play in the development of any country, they are consistently speculated for environmental pollutions. The interruption of hydrological conduits, adulteration of aquatic regime, and sinking of water table are the key quantifiable impacts of coal mining (Pulles et al. 1995; Younger and Wolkersdorfer 2004). Multiple endeavors manifested that most of the contaminants from the coal mines could be released into the periphery by the process of leaching (Krothe et al. 1980; Li 1988; Haigh 1995; Filcheva and Noustorova 2000). The Wardha valley coalfield of central India is no exception and the water quality issues have already been raised in terms of the hydro-geochemistry (Satapathy et al. 2009; Das 2013). The Wardha valley coalfield has been discovered long back in 1831 and around 5576 million tons of coal reserves have been estimated (Raja Rao 1982; Sarate 2017). The coal mining is the prime anthropogenic cause of the hydrogeological and geomorphological disruption in the Wardha valley coalfields (Ganvir and Guhey 2020a, 2020b). Such disruptions are potential enough to disturb the orderly system of groundwater regime in terms of quality and quantity. It is a well-established fact that the prevailing weathering process and respective groundwater quality in any locality are in the relationship of cause and effect. The type of weathering and the nature of rock together depict the picture of hydro-geochemistry; especially, where the rock-water interaction is dominant (Singh et al. 2020). The present endeavor is entirely motivated to understand the impact of enhanced weathering process due to anthropogenic activity on the water quality of the Wardha valley coalfields, whereas, the précised objectives can be enlisted as;

Weathering and Erosion Processes in the Natural Environment, First Edition. Edited by Virendra Bahadur Singh, Sughosh Madhav, Naresh Chandra Pant, and Ravi Shekhar.

- To assess the groundwater quality of the study area.
- To circumscribe the area of grave concern to groundwater quality of the study area.
- To elucidate the accountable temporal source in the study area for the prevailing groundwater quality.
- To comprehend the controlling mechanism for prevailing hydro-geochemistry of the study area.
- To examine and establish the causal relationship between the weathering process and the hydro-geochemistry of the study area.

9.2 Study Area

The study area is the major coal-producing region of central India and is traceable in the eastern part of Maharashtra state. The study area is divided by the Wardha River into two districts, viz. Chandrapur and Yavatmal; but the majority of it is in the Chandrapur district. It is stretched along NW-SW direction confining nearly 894.35 sq. km. area and is restricted by the latitude 19°47′to 20°10′ N and the longitude 78°57′ to 79°23′ E (Figure 9.1). The toposheet no. 55L/16, 55 P/4, 55 P/8, 56 I/13, 56M/1, and 56M/5 are sufficient enough to trace the entire study area. The topography of the study area is more or less plainer in nature and can be classified as pediment-pediplain complex originated by continuous fluvial denudation. The sub-dendritic drainage pattern prevailing in the study area justifies sedimentary terrain.

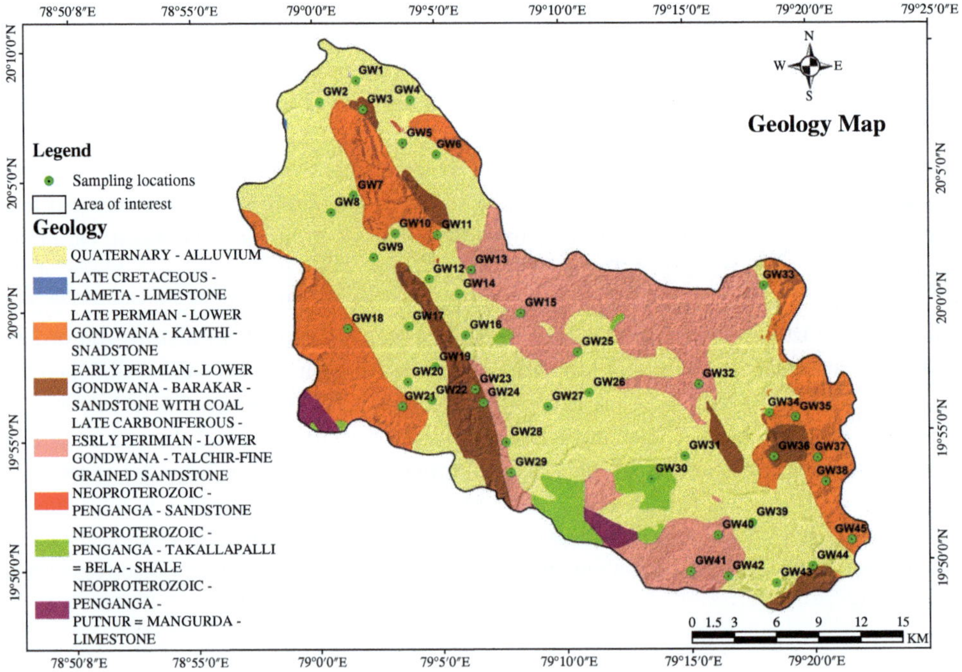

Figure 9.1 Geology map of the study area with groundwater sample locations (*Source:* Ganvir and Guhey 2021/Springer Nature).

9.3 Geology

The study area is well-familiar as the Wardha valley coalfield and structurally is a trough restricted to NW-SE trending faults in the Pranhita-Godavari sub-basin of the Gondwana basin (Hughes 1877; Raja Rao 1982). The Wardha valley coalfields are primarily dominated by the Gondwana supergroup rocks (Raja Rao 1982). Apart from the Archean basement, the Talchir formation is the lowermost sequence of the Gondwana supergroup. The striated pavement superimposed by the pebbly mudstone and the presence of the glacial varves elucidates the glacial origin of the Talchir formation (Fedden 1875; Ghosh 1962). The Barakar formation is overlain by the Talchir rocks and is the host of the one and only seam of the Wardha valley coalfields. In a wider range, the fining upward cycle of the sandstone, siltstone, shale, and coal can be seen in the Barakar formation, where the existence of the carbonaceous shale with the coal seams is certain (Singh et al. 1982; Kale and Soman 2012). The Kamthi formation is the top layer of the Gondwana supergroup acting as a blanket over the Barakars and mostly consists of ferruginous sandstone, variegated shale, and variegated sandstone (Raja Rao 1982). The few insignificant patches of the late Cretaceous Lameta limestone and rocks of the Pakhal supergroup are observable (Ramesh et al. 2018). The thick alluvium deposit mostly contains sand, silt and clays, which is widely spread over the study area and has been brought by the fluvial agencies (Figure 9.1).

9.4 Methodology

The pilot survey and the Geographical Information System (GIS) tools have aided in the selection of the sampling locations of the groundwater samples. Forty-five samples were collected separately in each season of 2019, i.e. pre-monsoon and post-monsoon, by the method of area sampling (a kind of cluster sampling); where, a due care has been taken to keep a uniform scattering along all the crucial aspects like residential, agricultural lands, and active and abandoned coalfields (Figure 9.1). The groundwater samples were done as per the guidelines of American Public Health Association (1998) which has assured the credibility of the samples. The groundwater samples were acidified with the conc. HNO_3 till the pH was below 2 for the stability of the metal ions after the spot analysis of pH, EC, and TDS. After the preparation of the samples, a cold bath of around 4 °C has been used for transportation and preservation.

The preserved samples were then subjected to the hydro-geochemical analysis by standard procedures in the pursuit of major anions and cations. The calcium and magnesium ions were measured by the titration procedures; whereas, the ions of sodium and potassium were done by the flame-photometer. Among anions, HCO_3^- and Cl^- are measured by titration; while SO_4^{2-} and NO_3^- were done by UV spectrophotometer. The metal ions i.e. Al, Cd, Cr, Cu, Fe, Ni, Pb, and Zn were assessed by the Atomic Absorption Spectrophotometer (AAS: GBC AA, Australia) with the standard procedures and prescribed wavelengths. The entire metal analysis has been done with the respective standard solutions prepared by the ultrapure base solution. The absorbance by the desired metal atom is measured in elemental state by subjecting it to a suitable flame. The observed absorbance is then compared with that of the standard sample's absorbance of same elemental species, ultimately giving the concentration in unknown as absorbance is proportional to its concentration. On results, the interpretation methods like the Piper plot, Gibbs plot, and PCA were applied to deduce reliable outcomes.

On opening results, the interpolation of the principal component done by XLSTAT and ArcGIS software on the base maps revealed the hotspots of the contaminations. These hotspots were visited for the rock sampling in the pursuit of the contamination sources. The petrography of sampled rocks suggested the accountability of the various rock samples for the observed hydro-geochemistry.

9.5 Characterization of the Groundwater

The groundwater of the Wardha valley has been sampled with a great precision to keep the results of utmost reliability. The 45 groundwater samples were adequate in terms of quantity and quality for the intended endeavor. The analyzed groundwater can be characterized by the following criterions;

9.5.1 Hydro-Geochemistry

The chemistry of the groundwater or hydro-geochemistry is a key aspect in contamination studies. It not only reveals the elemental distribution in the groundwater but also reflects the effect of the accompanying rocks over it. Hence, to comprehend the effect of the geology of Wardha valley coalfields on the groundwater system, particular hydro-geochemical parameters were evaluated. Those are pH, EC, TDS, major cations (Na^+, K^+, Ca^+ and Mg^+), and major anions (HCO_3^-, Cl^-, SO_4^{2-} and NO_3^-). In general, with respect to every hydro-geochemical parameter, the pre-monsoon samples were somewhat intense than the corresponding post-monsoon ones.

9.5.1.1 Hydro-Geochemical Parameters

The comprehensive results (Appendices 9.1 and 9.2) of the analyzed parameters are as follows;

pH

In spot analysis, at many locations, the pH of the groundwater samples was found to be lower than the prescribed Bureau of Indian Standards (2012) standard. In post-monsoon, the corresponding locations have represented higher pH values, but yet some were below the Bureau of Indian Standards (2012) standard. The lower pH can be justified by the presence of SO_4^{2-}, Cl^- and NO_3^- ions in the groundwater; whereas, the enhanced pH values in the post-monsoon are obvious and can be attributed to the dilution effect. The study area is an active coal mining and industrial region; hence, an anomalous pH in the groundwater is reasonable. The pH of the groundwater in some part of the Wardha valley coalfields have been referred to be of grave concern for human health (Das 2013).

EC

In pre-monsoon samples, the EC values were quite intense and around 53% of the observations were in the range of 1500–3000 μS/cm (not permissible).Whereas, in post-monsoon, only 7% were in the non-permissible range. The elevated values of EC in the groundwater samples can be attributed to the dissolved salts that obviously would be higher in the pre-monsoons.

TDS

Most of the groundwater samples range between the acceptable limit (500 mg/l) and permissible limit (2000 mg/l) of the TDS prescribed by the Bureau of Indian Standards (2012). The TDS values surpassing the acceptable limit, specifically in the pre-monsoon season, indicate the evaporative dominance, causing the precipitation of the salts.

Ca^+

It was a furthermost cation in the groundwater and nearly 11% were found to be surpassing the recommended permissible limit of 200 mg/l in pre-monsoons. Whereas, around 60% of the samples were traced between the acceptable limit (75 mg/l) and permissible limit (200 mg/l) prescribed by the Bureau of Indian Standards (2012). The rock-water interaction involving the minerals like feldspar, pyroxene, and limestone are the chief contributors of Ca^{2+}. In Wardha valley coalfields, the contribution of the Ca^{2+} from the plagioclase feldspar and the calcareous cementing material in the Barakar rocks are evident (Kale and Soman 2012). The limestone occurring in the Lameta formation could also be held accountable as a source of Ca^{2+}.

Mg^+

Around 11% of the sampled groundwater for the pre-monsoons were found exceeding the recommended permissible limit (100 mg/l) of the Bureau of Indian Standards (2012). While, nearly 75% of the observed samples were between the recommended acceptable limit (30 mg/l) and permissible limit (100 mg/l) of the Bureau of Indian Standards (2012). The minerals like orthoclase, microcline, and biotite in the lower Gondwana sediments are the possible source of Mg^{2+} concentration in the groundwater leached through rock-water interaction.

Na^+ and K^+

These are the typical ions in the groundwater and diverge a lot among all the samples. The Na^+ ion is the second most abundant among cations. The abundance of the K^+ was found the approximately ten times lesser than the Na^+ ions. Being next to the most abundant cation in the groundwater, the Na^+ can be predicted to be released from feldspar, clay minerals, etc., dominant in the lower Gondwana sediments. On the other hand, the K^+ ions would be released from the mineral-like orthoclase, microcline, and biotite from the lower Gondwana sediments.

HCO_3^-

The bicarbonate is quite dominant among the anions and approximately more than 250 mg/l is an average concentration in pre-monsoon and 200 mg/l in post-monsoon. The close companion of the bicarbonate i.e. CO_3^{2-} is not accountable for the alkalinity in the scenario of lower pH. The weathering of the limestone beds of the Lameta formation constituting $CaCO_3$ is accountable for HCO_3^- share in the groundwater. Also, the rocks with calcareous cementing material could also be held accountable for the same.

Cl^-

It was well within the permissible limit of 1000 mg/l (Bureau of Indian Standards 2012) in all the samples. Around 20% of the samples in the pre-monsoon were above the recommended acceptable limit (250 mg/l) of the Bureau of Indian Standards (2012). The dissolved ions form the NaCl existing in the rocks and soil contributes for higher Cl^- concentrations. Also, sewage effluents are the substantial cause for the same.

SO_4^{2-}

It is the leading anionic contributor in the groundwater samples in both the seasons. The average concentration in the pre-monsoons is more than 300 mg/l; whereas, in the post-monsoons, it is nearly 200 mg/l. More than 40% of the samples were above the permissible limit of 400 mg/l (Bureau of Indian Standards 2012) in the pre-monsoon and around 4% in the post-monsoon. The weathering of the metal sulfides viz. pyrite, pyrrhotite, chalcopyrite, galena, and sphalerite present in the rocks is the key mechanism for augmented SO_4^{2-} concentration in the groundwater. The presence of the coal and sulfide minerals in the Barakar formation is accountable for the raised concentration of SO_4^{2-} in the groundwater (Karanth 1997). The weathering of the sulfide phase usually produce AMD (acid mine drainage) when occurred in presence of water and ultimately lower the pH of the groundwater (Singh 1987).

NO_3^-

It is the least contributor among the anions and, in pre-monsoon, nearly 24% of samples was above the permissible limit of 45 mg/l (Bureau of Indian Standards 2012). The typical source of NO_3^- in the groundwater is the decaying organic matter, sewage wastes, and fertilizers. This can also be verified by the higher concentration of NO_3^- near the residential and industrial setups.

9.5.1.2 Hydro-Geochemical Interpretations

The percentage contributions of observed ions in the groundwater for both the seasons is in the order of $Ca^{2+} > Na^+ > Mg^{2+} > K^+$ and $SO_4^{2-} > HCO_3^- > Cl^- > NO_3^-$ with insignificant variations. The pre- and post-monsoon observation does not show any significant variation except the drop in ion concentrations in post-monsoon, evidently due to the dilution effect (Ganvir and Guhey 2022).

Piper Plots

The Piper plot for the pre-monsoon manifests 65% of the groundwater samples in mixed CaMgCl-type zone; whereas, around 31% of the samples in $CaHCO_3$ zone 1 and only around 4% in NaCl zone 2 (Figure 9.2a). In the cationic facies, about 74% of the groundwater samples are in the no dominance zone; whereas, approximately 22% are in the Ca^{2+} field and around 4% in the Na^+ field. In the anionic facies, the groundwater samples are dispersed into the no dominance field, HCO_3^-, and SO_4^{2-} fields. Where, around 49% are in the no dominance field, 33% in HCO_3^-, and 18% in SO_4^{2-} fields.

The Piper plot for the post-monsoon manifests 84% of the samples are of CaMgCl type; while, the rest are distributed between $CaHCO_3$ and CaCl zone (Figure 9.2b). In cations, 98% of the samples can be traced in the no dominating zone and remaining 2% in the Ca^{2+}

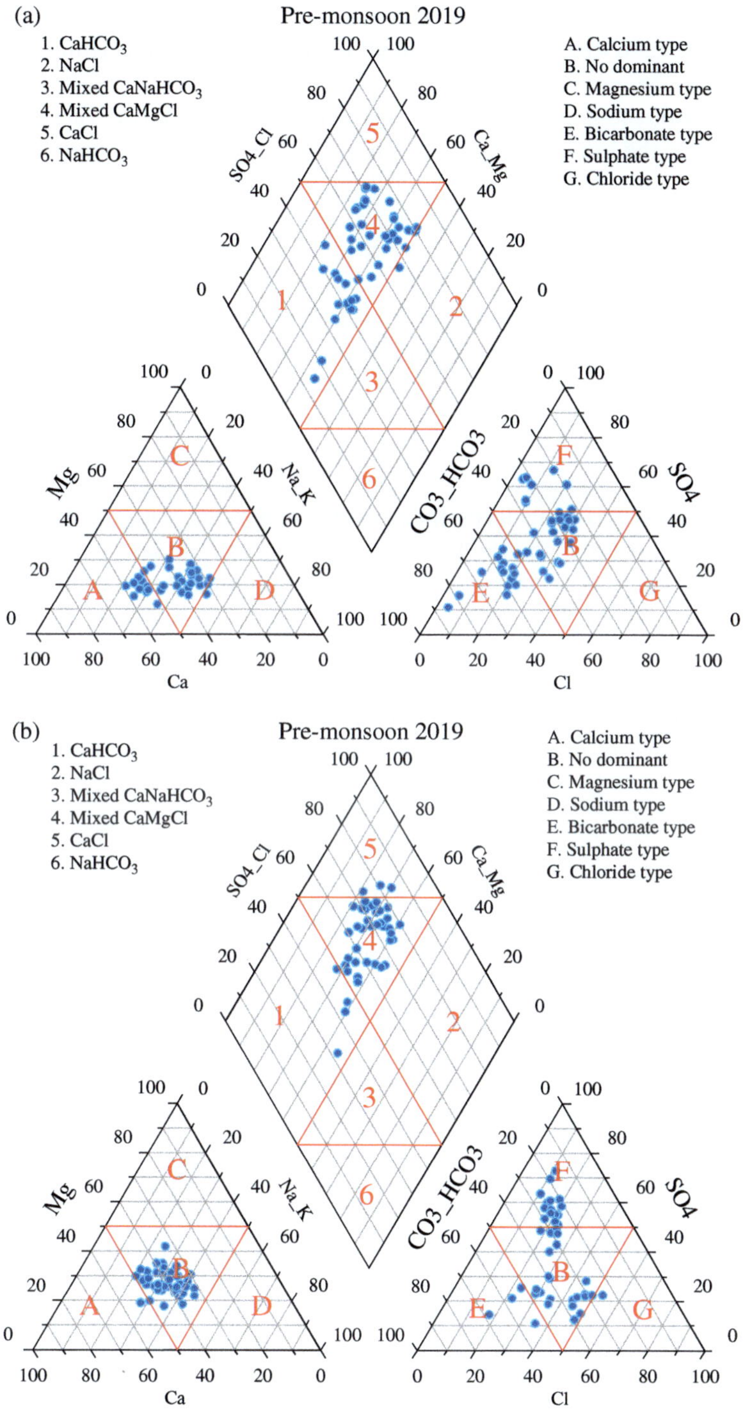

Figure 9.2 Piper plots for (a) Pre-monsoon and (b) Post-monsoon.

field. In the anionic facies, the samples are distributed into four fields; where, around 53% are in the no dominance field, 36% in SO_4^{2-} filed, and 9% in HCO_3^- field. An insignificant share of only around 2% can be traced in the Cl^- field.

The hydro-geochemistry of the groundwater not only discloses it's chemistry but also manifests the implication of the local geology on it. On hydro-geochemical evaluation of the groundwater, multifaceted outcomes were derived, of them the ascendency of the strong acids (SO_4^{2-}-Cl^-- NO_3^-) over the weak acids (HCO_3^-) is the promising one. In anionic facies, the dominance of SO_4^{2-} signifies the distinguished influx from the surrounding rocks. Such protruding influx of SO_4^{2-} clearly indicates the accelerated weathering of the pyrite mineral on exposure to the atmospheric agencies during coalfield excavations, which otherwise stays concealed. The pyrite mineral is abundant in form of disseminations, framboidal clusters, and cavity fillings in the coal of the Wardha valley coalfields (Singh et al. 1983; Sarate 2004, 2010, 2017).

Gibbs Plots

The plot suggested by Gibbs (1970) is symbolic for the process of controlling the groundwater chemistry and its association with the aquifer. The Gibbs plot for both the seasons clearly specifies the rock-water interaction as a prime process for the existing groundwater chemistry (Figures 9.3x,y). The dominance of the rock-water interaction-induced chemistry clearly indicates the ascendency of weathering process in the constituting process of groundwater chemistry. In the regions of active mining along with the tropical weather, the usual weathering rate can be observed to be in an accelerated pace due to dynamic terrestrial setting and the joint action of entangled weathering parameters, viz. temperature disparity and humidity.

9.5.2 Metal Chemistry

The average percentage contributions of the metals in the groundwater for both the seasons were derived in the order of Zn > Fe > Cu > Al > Cr > Ni = Pb > Cd for pre-monsoon and Fe > Zn > Cu > Al > Cr > Ni = Pb > Cd for post-monsoon. The inclusive results (Appendices 9.1 and 9.2) of the analzsed metals are as follows;

9.5.2.1 Al

In pre-monsoon, around 91% of the samples were between the range of acceptable and permissible limit suggested by Bureau of Indian Standards (2012). The sample GW9 is at the maximum concentration of 180 μg/l; whereas, the sample GW21 exhibited the least concentration of 20 μg/l. In post-monsoon, around 75% of the samples were between the range of acceptable and permissible limit suggested by Bureau of Indian Standards (2012). The highest was in the sample GW42 with the value of 143 μg/l; whereas, the least concentration of 10 μg/l was in the sample GW19. The presence of Al in the groundwater is obviously justified by its dominance in the rocks of the sedimentary origin in form of feldspar and its weathered product.

9.5.2.2 Cd

In pre-monsoon, around 48% of the samples were surpassing the acceptable limit of 3 μg/l prescribed by Bureau of Indian Standards (2012) (Figure 9.4a). The sample GW11 has

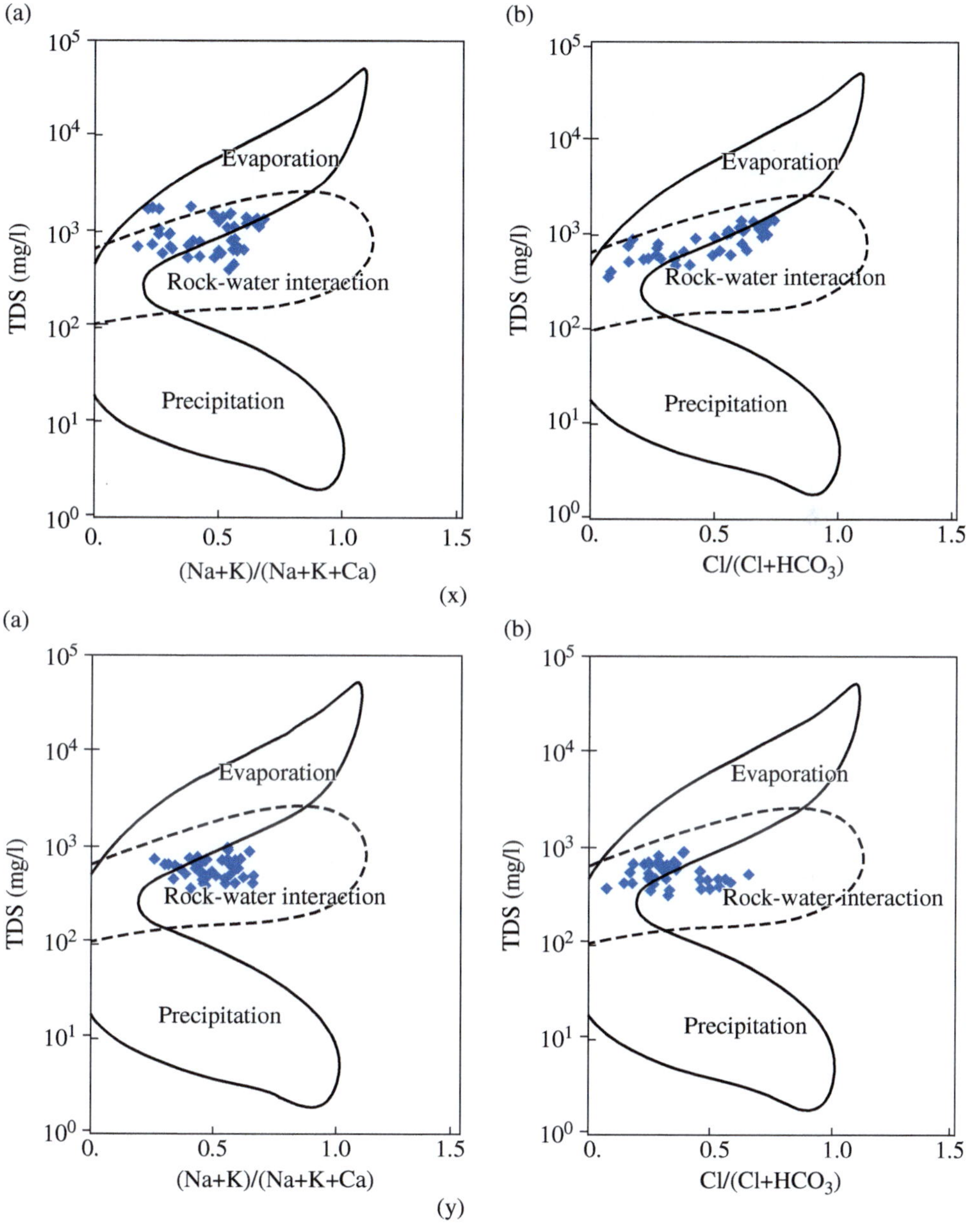

Figure 9.3 Gibbs plots for **x.** Pre-monsoon and **y.** Post-monsoon.

shown the maximum concentration of 52 µg/l. In post-monsoon, the surpassing samples magnitude was reduced to 42%. The sample GW11 had the utmost concentration of Cd with the value of 41 µg/l. The Barakar sandstone of the lower Gondwana have sulfide mineralization associated with the coal. The occurrence of sphalerite in the fractures of the coal and associated pyrite nodules is evident (Cobb et al. 1980).

9.5.2.3 Cr

In pre-monsoon, only samples GW7, GW32, and GW43 were found to be surpassing the acceptable limit of 50 μg/l (Bureau of Indian Standards 2012). The highest concentration was found in sample GW7 with an observed value of 66 μg/l. In post-monsoon, no samples surpassed the acceptable limit. While, the samples GW32, GW36, and GW43 were with the highest Cr concentration of 50 μg/l. The Cr concentration in the groundwater can be attributed to coal ash and igneous rocks (Satapathy et al. 2009).

9.5.2.4 Cu

In pre-monsoon, except samples GW9, GW14, GW22, and GW24, all were under the acceptable limit of 500 μg/l (Bureau of Indian Standards 2012). The sample GW14 has shown the maximum concentration of 710 μg/l; whereas, in post-monsoon, only GW14 was above the acceptable limit and the highest was evidently in GW14. The usual sulfide minerals associated with the coal along with pyrite (FeS_2) is chalcopyrite ($CuFeS_2$) (Finkelman 1981). Hence, on weathering, it could be certainly the source of the groundwater.

9.5.2.5 Fe

In pre-monsoon, around 44% of the samples were above and found to be surpassing the acceptable limit (300 μg/l) of Bureau of Indian Standards (2012) (Figure 9.4b). The uppermost concentration was 624 μg/l in the sample GW22. The magnitude of sample surpassing the acceptable limit was reduced to 42% in post-monsoon with the apex concentration in GW11. The oxidative weathering of the pyrite mineral in the coal and associated sediments is the prime source of Fe discharge in the groundwater (Chandra 1992). Along with this, the ferruginous cementing material is also accountable of the same.

9.5.2.6 Ni

In pre-monsoon, around 49% of the samples were surpassing the acceptable limit of 20 μg/l (Bureau of Indian Standards 2012) (Figure 9.4c). The sample GW33 was the most contaminated sample with 72 μg/l. In post-monsoon, the quantity of surpassing samples reduced to the 40% of total samples. The highest concentration was found in sample GW33 with 70 μg/l of concentration. A significant concentration of the pyrite-incorporated Ni has been observed in the Wardha valley coalfields (Mukherjee et al. 1988). This could be the possible source of Ni in the groundwater.

9.5.2.7 Pb

In pre-monsoon, almost 22 samples were above the permissible limit of 10 μg/l suggested by Bureau of Indian Standards (2012) with the maximum concentration of 10 μg/l was found in the sample GW11 (Figure 9.4d). Whereas, around 29% of samples were undetectable in concern to Pb concentration. In post-monsoon, around 33% samples were blank with Pb concentration and 21 samples were above the 10 μg/l with the highest of 74 μg/l in GW42. The sulfate of Pb is common in form of galena Pb and is much more observable in the carboniferous rocks (Satapathy *et al.* 2009). Around Ghugus coalfield of the Wardha Valley Coalfields, the coal has disclosed the concentration of Pb ranges from 5 ppm to 25 ppm (Singh *et al.* 1983).

9.5.2.8 Zn

In both the seasons, the concentration of the Zn was insignificant with respect to the prescribed limit of the Bureau of Indian Standards (2012) i.e. 5000 µg/l. In pre-monsoon, the highest was of 1620 µg/l in the sample GW12; whereas, in the same sample, the post-monsoon high was of 870 µg/l. The presence of the Zn is justifiable by the common sources of sedimentary rock, viz. sandstone, shale, and carbonates (Satapathy *et al.* 2009).

In general, the intensities of the contaminants were observed to be gradually descending in the post-monsoon season. Excepting the metals, viz. Cd, Fe, Ni, and Pb, all others in the sampled groundwater were well within the acceptable/permissible limit suggested by Bureau of Indian Standards (2012). In concern to the contaminants, around 18% of the samples were of grave concern in pre-monsoon that declines to only 9% in post-monsoon.

The enhanced weathering of disclosed rock strata comprising sulfide phases lowers the pH of water by adding SO_4^{2-}, which is always favored for the acid dissolving metals. Such chemistry is also a probable cause for higher metal dissolution and subsequent migration to groundwater. Therefore, it is evident to speculate the weathering of the pyrite or sulfide phase existing in the coal-associated strata is fairly responsible for the metal contaminations in the groundwater.

(a)

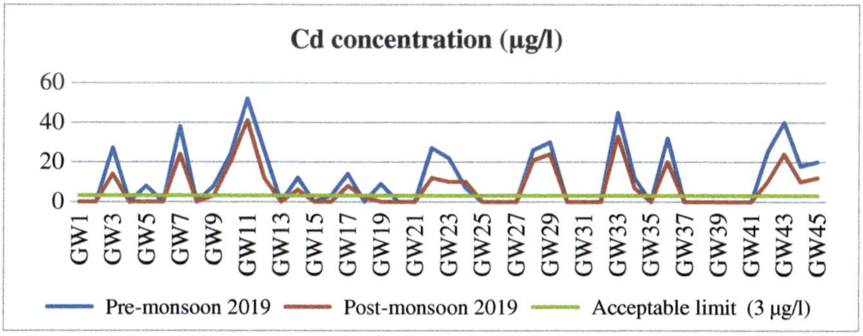

(b)

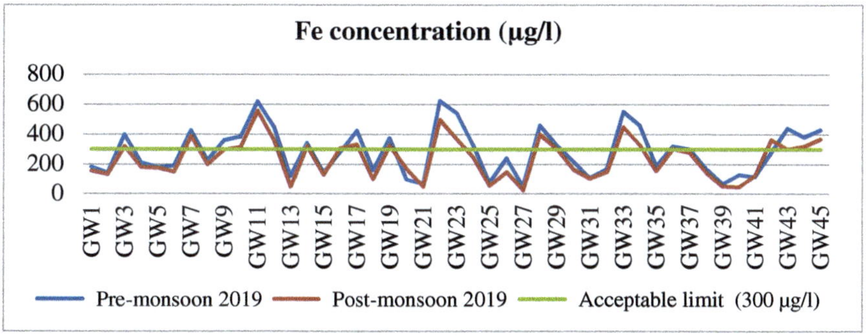

Figure 9.4 Line graphs for (a) Cd, (b) Fe, (c) Ni and (d) Pb.

(c)

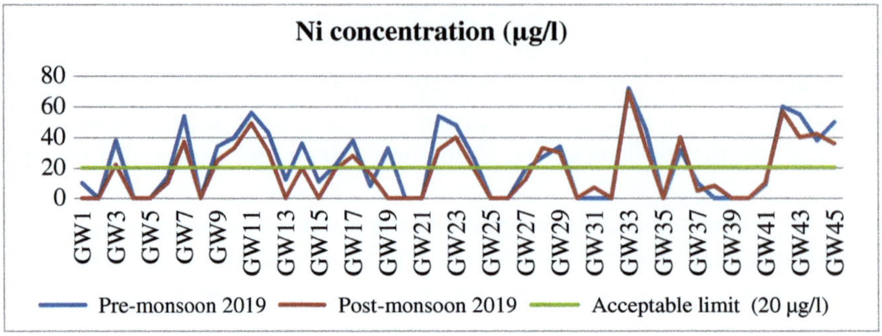

(d)

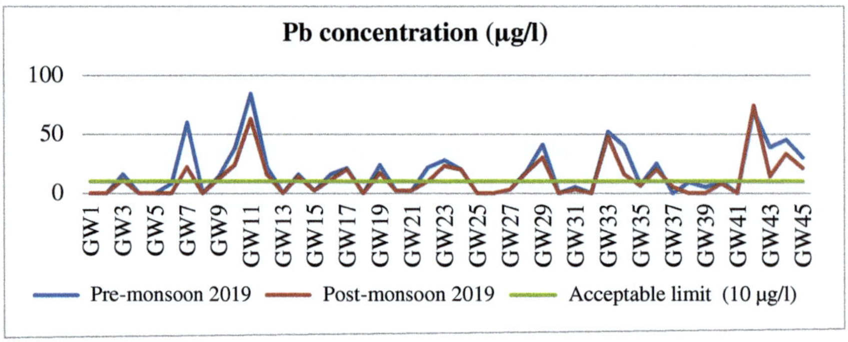

Figure 9.4 (Continued)

9.6 Spatial Source Approximation

The primary results from the hydro-geochemistry (including the metals' evaluation too) of the sampled groundwater demonstrating the contaminations of certain metals along with the associated significance of groundwater chemistry and arouses the venture of source approximation in the study area. The process of spatial source approximation pre-requisites a methodical tool that will sort out the hotspots over the study area in concern to the desired parameters. Hence, the PCA has been chosen to approximate the observed metals over the study area with an aid of spatial interpolation.

9.6.1 PCA

The PCA is a multivariate statistical technique often engaged for spatial source approximations and associated dispersal (Facchinelli *et al.* 2001). The PCA dedicated to the source finding is broadly practiced in the geological disciplines (Hwang *et al.* 2001). The PCA converts the subjected variables into the components from maximum variance to the least where the first factor/component (PC1) exhibits maximum. The individual PC can also be bifurcated with the factor loadings of the respective variables. The magnitude of factor loading of a specific variable in the specific PC represents the variance of the variable in the PC. On factorizing the obtained results from the groundwater analysis in concern to the metals, the PC1 specified the controlling variance of the contaminants viz., Cd, Fe, Ni, and Pb (Tables 9.1a and 9.1b).

Table 9.1a The factor loading and the respective percentage contributions of the first three principal components for the variables of the pre-monsoon 2019.

Variables	PC1	PC1's contribution in %	PC2	PC2's contribution in %	PC3	PC3's contribution in %
Al	0.648	8.852	0.543	19.468	−0.143	3.373
Cd	0.846	15.113	−0.402	10.710	−0.185	5.639
Cr	0.693	10.133	−0.298	5.889	0.603	59.815
Cu	0.614	7.959	0.605	24.201	0.356	20.811
Fe	0.929	18.232	−0.006	0.003	−0.114	2.149
Ni	0.956	19.273	−0.144	1.367	−0.102	1.726
Pb	0.828	14.485	−0.438	12.699	−0.118	2.303
Zn	0.531	5.953	0.623	25.663	−0.160	4.184
Eigen value	4.738	100.00	1.513	100.00	0.608	100.00
% of variance	59.225		18.908		7.602	

Table 9.1b The factor loading and the respective percentage contributions of the first three principal components for the variables of the post-monsoon 2019.

Variables	PC1	PC1's contribution in %	PC2	PC2's contribution in %	PC3	PC3's contribution in %
Al	0.647	9.594	0.414	12.259	0.407	25.415
Cd	0.799	14.630	−0.425	12.892	−0.195	5.795
Cr	0.669	10.235	−0.180	2.320	0.576	50.712
Cu	0.480	5.274	0.675	32.551	−0.195	5.827
Fe	0.888	18.059	−0.100	0.715	−0.131	2.640
Ni	0.938	20.128	−0.173	2.137	−0.043	0.287
Pb	0.835	15.969	−0.271	5.241	−0.205	6.416
Zn	0.517	6.112	0.668	31.886	−0.138	2.908
Eigen value	4.367	100.00	1.400	100.00	0.653	100.00
% of variance	54.585		17.502		8.164	

9.6.2 Interpolations of the Factor Scores

In the course of the analysis, the factor scores of the observations are the requisites derivative, which are the correlation coefficients between the PC scores and the observations. The positive and negative values of the factor scores signify the direct and inverse relationship between the observations and the respective principal component. The greater the score value will be, the stronger will be the relationship. The observations with analogous composition will have like scores with a tendency to group together and are practicable to be spatially interpolated. Such analogies among observations have potential to be interpreted for probable sources (Einax *et al.* 1997). On knowing the dominance of the contaminants over the PC1, the factor scores of the PC1 were utilized for the interpolation by the ordinary kriging method (Table 9.2). The ordinary kriging technique yields the fair linear approximations of the subjected variables in the un-sampled sites and is also employed for the

Table 9.2 The factor scores and the respective percentage contributions of the first principal component for the observations of the pre- and post-monsoon 2019.

Observations	PC1 Pre-monsoon	Contribution in %	PC1 Post-monsoon	Contribution in %
GW1	−2.123	2.114	−2.062	2.164
GW2	−2.057	1.985	−1.999	2.034
GW3	1.169	0.640	0.717	0.262
GW4	−1.672	1.312	−2.102	2.248
GW5	−2.181	2.231	−2.156	2.366
GW6	−1.640	1.261	−1.714	1.495
GW7	2.974	4.149	1.881	1.800
GW8	−1.432	0.961	−1.505	1.152
GW9	1.651	1.278	1.118	0.636
GW10	1.393	0.910	1.531	1.193
GW11	4.122	7.968	4.504	10.324
GW12	3.258	4.977	2.578	3.382
GW13	−2.278	2.433	−1.861	1.762
GW14	1.819	1.551	1.549	1.221
GW15	−2.322	2.530	−2.241	2.555
GW16	0.511	0.122	−0.424	0.092
GW17	1.704	1.362	2.035	2.107
GW18	−2.009	1.894	−1.468	1.097
GW19	0.649	0.197	−0.699	0.248
GW20	−2.060	1.990	−1.515	1.168
GW21	−2.875	3.877	−2.211	2.487
GW22	3.947	7.306	2.628	3.516
GW23	3.023	4.285	2.114	2.274
GW24	1.513	1.074	1.860	1.760
GW25	−1.670	1.308	−1.592	1.289
GW26	−1.699	1.354	−1.785	1.622

Table 9.2 (Continued)

Observations	PC1 Pre-monsoon	Contribution in %	PC1 Post-monsoon	Contribution in %
GW27	−2.097	2.062	−1.974	1.983
GW28	1.299	0.792	1.363	0.946
GW29	1.323	0.821	1.971	1.978
GW30	−1.918	1.726	−2.136	2.321
GW31	−2.004	1.883	−1.926	1.887
GW32	−1.476	1.022	−1.217	0.754
GW33	3.579	6.006	4.257	9.220
GW34	2.112	2.093	1.263	0.812
GW35	−1.938	1.762	−1.778	1.608
GW36	1.010	0.478	2.411	2.959
GW37	−1.831	1.573	−1.697	1.466
GW38	−1.627	1.242	−1.271	0.823
GW39	−2.566	3.087	−2.641	3.550
GW40	−2.194	2.257	−2.020	2.076
GW41	−2.209	2.289	−1.511	1.163
GW42	2.871	3.867	4.035	8.287
GW43	2.675	3.357	2.261	2.603
GW44	1.323	0.821	1.152	0.675
GW45	1.955	1.793	2.276	2.637

probabilistic recreations of ambiguity about the anonymous, by estimated predicted values (Deutsch and Journel 1998).

The interpolation of the factor scores over the base map overlain by the geology and coal mining region will reveal the hotspots of the subjected metals especially those with high magnitude of the variance over the study area. After the interpolation of the derived factor scores of PC1 of both the seasons, the regions of probable source were distinctly visible (Figures 9.5a and 9.5b). The correlation of the hotspots with the geology and the coal mining region can be done with a great ease. This correlation has disclosed the exact regions where a detailed petrography can be attempted for source confirmation.

9.7 Temporal Approximation

The PCA and interpolation of factor scores have efficiently pictured the source locations of the contaminants over the study area. These hotspots or probable sources were the most favored locations for the rock sampling to gather the direct field evidences. After the rigorous field visits in the hotspots some of the promising rocks strata were sampled by the

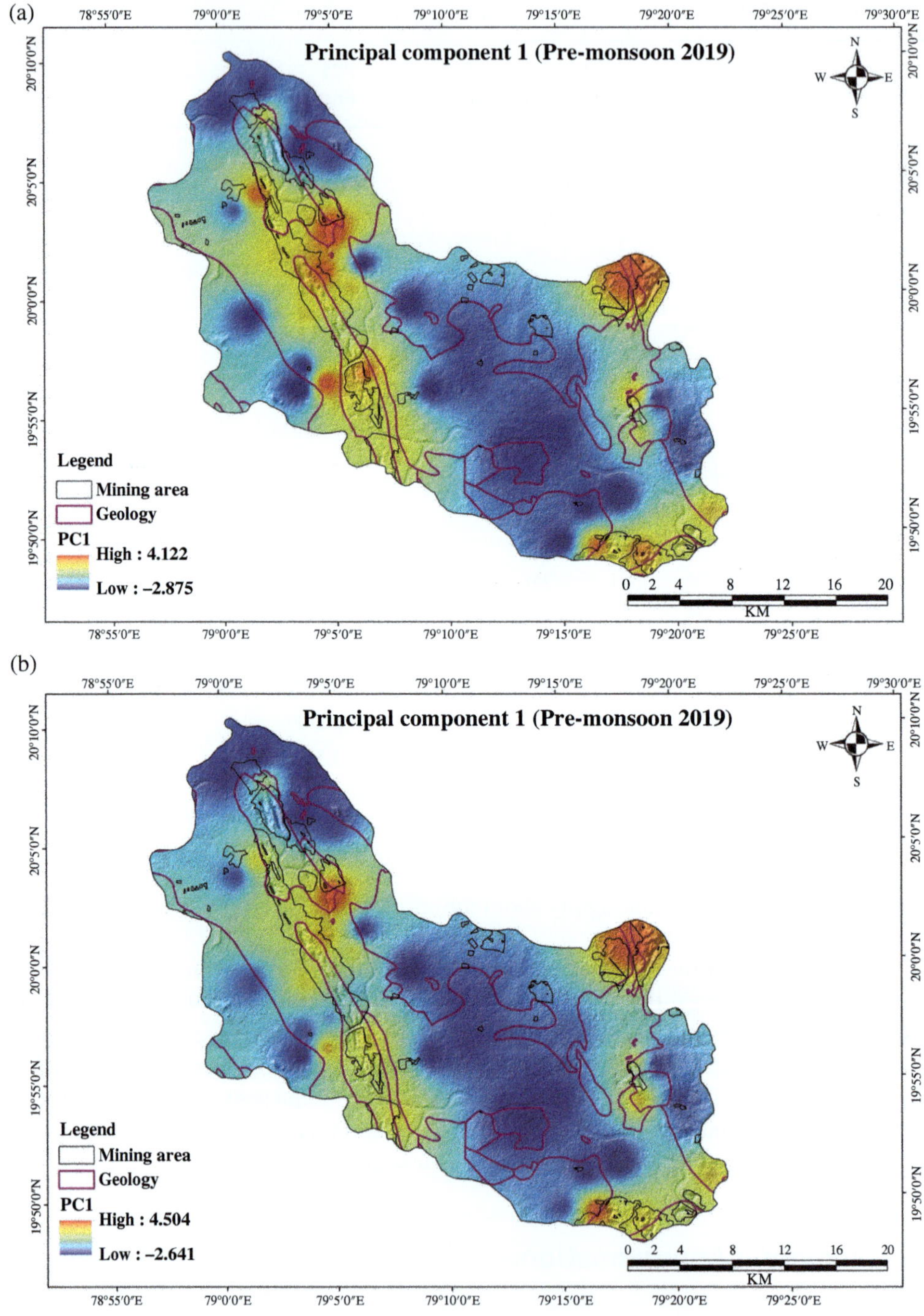

Figure 9.5 Interpolation maps of the PC1 demonstrating the enrichments of most contributing variables in the (a) pre-monsoon 2019 and (b) post-monsoon 2019.

standard procedures and subjected for petrographic study to approximate temporal sources. Following are some of the petrographic outcomes:

9.7.1 Gray Sandstone (P1)

The sample belongs to the Barakar formation, occurring in junction with the coal bed. The optical observations done in the PPL and CXN recommended the quartz in majority followed by alkali feldspar grains (Figure 9.6a,b). The characteristic existence of the gray-white feldspathic coarse grained sandstones is observable in the Barakar formation of the Wardha valley coalfields (Sarate 2010). The framework grains of the quartz and the feldspar are found to be floating on the sulfide constituting cement. Such dominating existence of the sulfide mineralization around the grains is slights unusual. On prima facie, the sulfide part appears to be the pyritic mineralization. The field observation revealed that the dominance of the pyrite in the cementation moderates away from the junction with the coal, suggesting the associated coal as the sulfide enricher. The sulfide minerals usually host multiple trace elements, in which Cd, Fe, Ni, and Pb are included (Pires *et al.* 1997).

9.7.2 Carbonaceous-Micaceous Siltstone (P2)

The P2 was sampled from the immediacy of the main coal seam. It was interbedded with the carbonaceous shale. The optical reflections disclosed the occurrence of fine grains floating over the carbonaceous clay matrix (Figure 9.7a,b). The high interference order of certain grains justified the existence of the muscovite (Figure 9.7b). The muscovite was quite plentiful after the quartz grains. The supremacy of the carbonaceous clay matrix and the muscovite grains is sufficient enough to claim it as carbonaceous-micaceous siltstone. The Barakar formation of the Wardha valley coalfield constitutes interposing argillaceous rocks with carbonaceous matter, along with the coal seam (Singh *et al.* 1983 and Sarate 2017).

(a) (b)

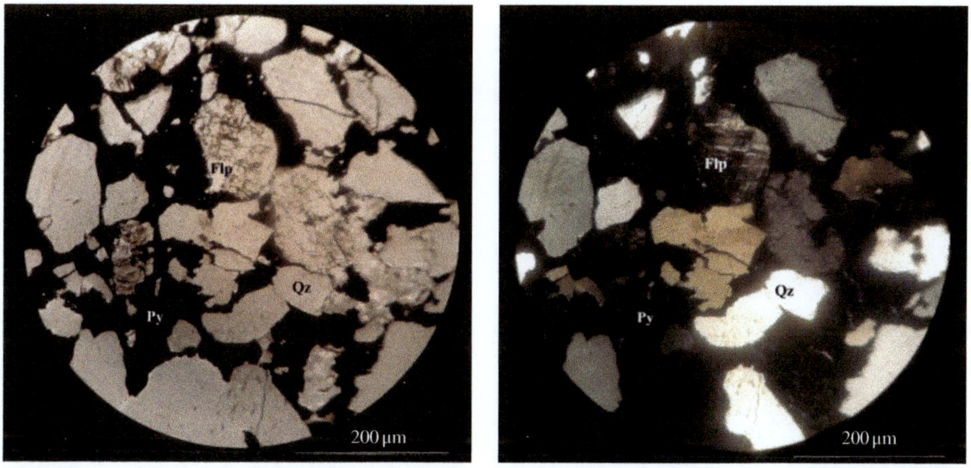

Figure 9.6 P1's microphotograph in (a) PPL (Qz- quartz, Flp- feldspar and Py- pyrite) and (b) CXN.

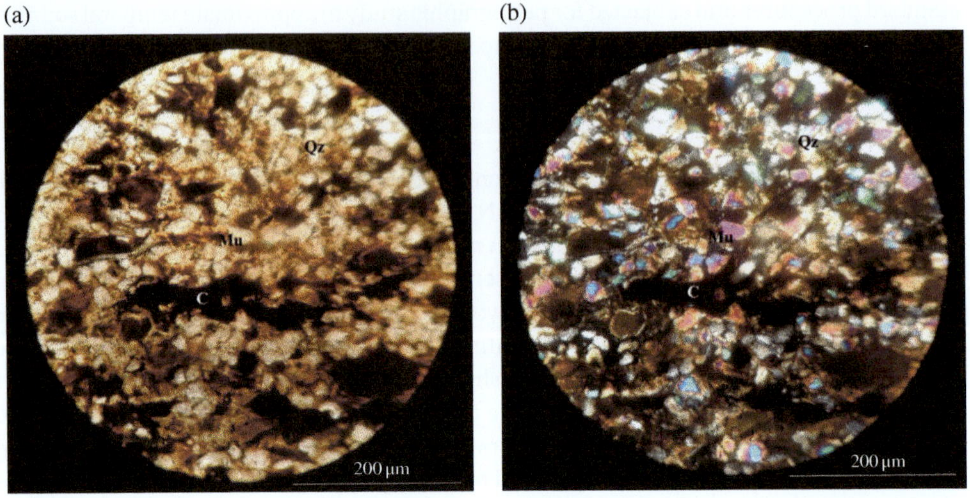

Figure 9.7 P2's microphotograph in (a) PPL (Qz- quartz, Mu- muscovite and C- carbonaceous matter) and (b) CXN.

9.7.3 Carbonaceous Shale (P3)

The P3 was sampled from the nearby coal bed and was interbedded with the carbonaceous-micaceous siltstone. The optical observations disclosed the occurrence of high proportion of the carbonaceous matrix with the fine grains of quartz and feldspar (Figure 9.8a,b). The ascendancy of the carbonaceous material with the constituents of fine size and indistinct mineralogy designates the section as of carbonaceous shale. The occurrence of the carbonaceous shale with the principal coal seams of the Wardha valley coalfield is quite certain (Singh *et al.* 1982). The carbonaceous part of the shale has also been traced with sulfur phase (Ganvir and Guhey 2021).

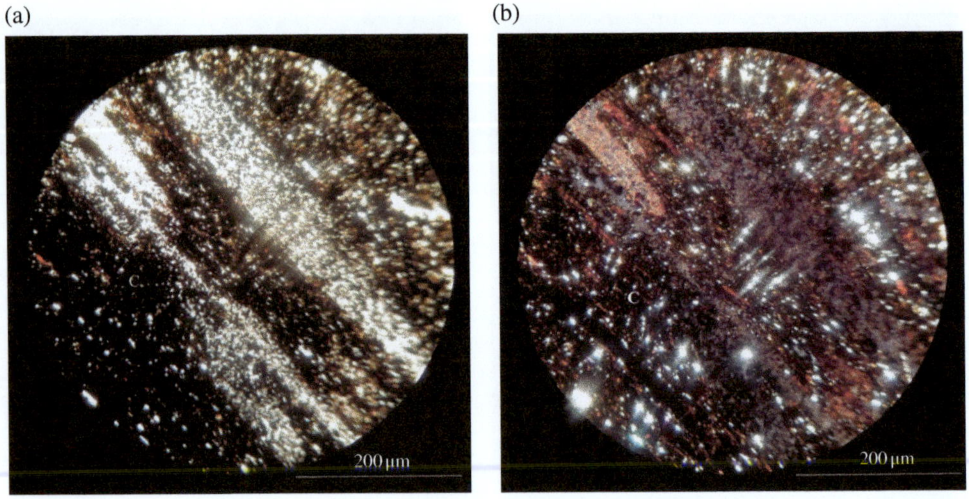

Figure 9.8 P3's microphotograph in (a) PPL (C- carbonaceous matter) and (b) CXN.

9.7.4 Pink-Colored Ferruginous Sandstone (P4)

The sample was collected from the Kamthi formation, which is overlying to the Barakar formation. The pink color is obviously attributable to the dominance of the pink feldspar. The thin section exposed the occurrence of quartz in abundance followed by the feldspar (Figure 9.9a,b). The grains of quartz and feldspar are cemented by the dark brown ferruginous material. Hence, the rock under concern was predicted as ferruginous sandstone.

9.7.5 Yellow-Colored Ferruginous Sandstone (P5)

The sample P5 was sampled from the yellow-colored sandstone basically belonging to the Kamthi formation. The yellow coloration may be due to the iron hydroxide staining to kaolin. The grain sizes are relatively coarser than the grains of sample P4 (Figures 9.9 and 9.10). The brownish-yellow sandstone containing ferruginous bands that are common in the Kamthi formation (Raja Rao 1982).

The petrography has revealed the enrichment of the sulfide phase in the rock strata of the Barakar formation. The primary observations imply the carbonaceous matter's (coal and associated carbonaceous strata) accountability for the sulfide phase enrichment in the associated strata. The customary sulfide minerals associated with the coal are pyrite (FeS_2), sphalerite ([Zn,Fe]S), galena (PbS), chalcopyrite ($CuFeS_2$), and pyrrhotite (FeS) (Finkelman 1981). Hence, the sulfide phase observed in sample P1 and the carbonaceous matter of sample P2 and sample P3 are sensibly possible hosts for the many metal sulfides, particularly which are sorted as contaminants in the groundwater analysis. The coal-associated pyrite is very common and is a noticeable cause for many trace and heavy metals' leaching (Saikia *et al.* 2009). The high iron percentage in the cementation of the Kamthi sandstones is an obvious source of Fe in the groundwater (Ganvir and Guhey 2021). On weathering, the scavenging influence of iron oxyhydroxide on the trace and heavy metals controls their concentration in the groundwater (Schurch *et al.* 2004).

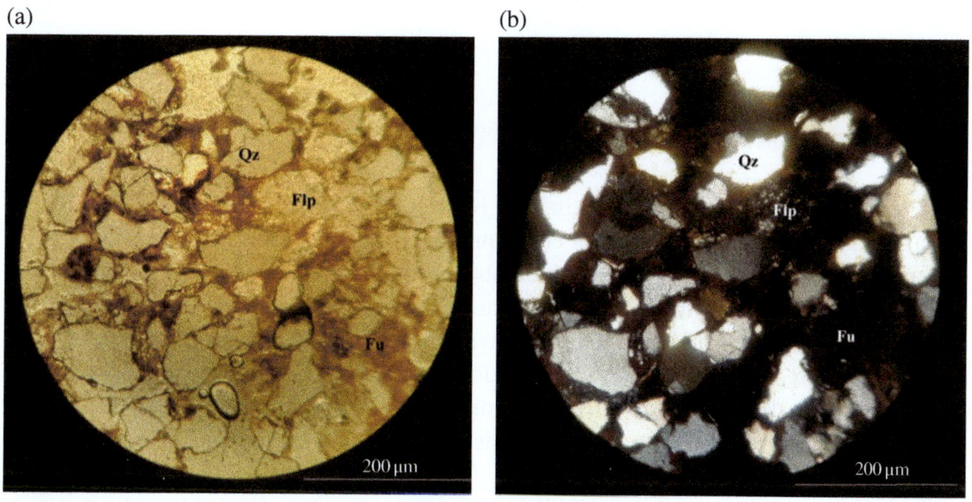

Figure 9.9 P4's microphotograph in (a) PPL (Qz- quartz, Flp- feldspar and Fu- ferruginous matter) and (b) CXN.

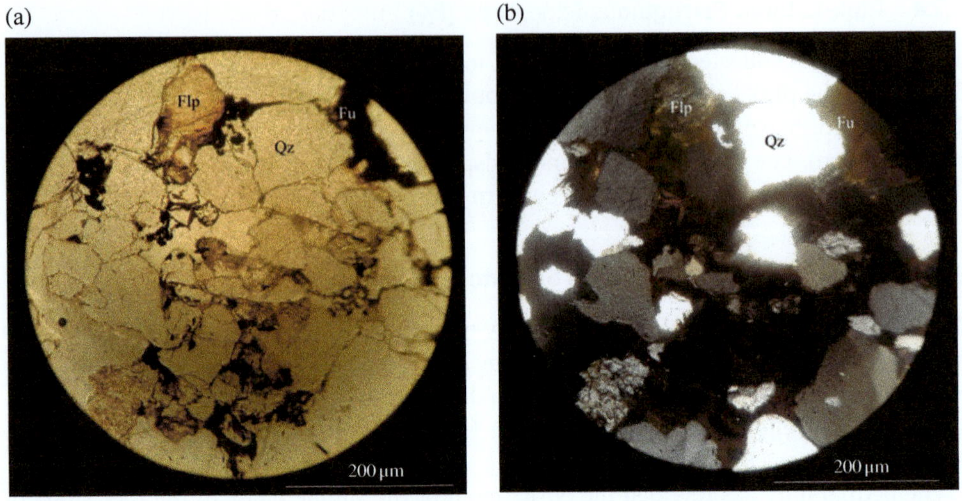

Figure 9.10 P5's microphotograph in (a) PPL (Qz- quartz, Flp- feldspar and Fu- ferruginous matter) and (b) CXN.

As in usual conditions, the rock strata encompassing the sulfide phase lies well within the Earths surface, far away from the zone of weathering. The enhanced weathering of these sulfide phases must be due to an act which provides exposure of air and water to the otherwise concealed sulfide phase (Ganvir and Guhey 2022). The PCA and interpolation studies indicated that the contaminants hotspots were in the proximity of the coal mines. The rock samples gathered from these hotspots have also reflected the presence of the sulfide phases. Hence, it can be assertively supposed that the excavation processes during the coal mining activity consequence into the exposures of the buried strata encompassing the sulfide phase, ultimately enhancing the rate of the weathering over it. This boosted weathering process augments the formation of AMD and eventually migration of the acid-dissolving metals to the local groundwater regime (Zipper *et al.* 2014).

9.8 Conclusion

The groundwater regime in the study area has been found to be vulnerable due to the dynamic nature of the terrain. As a consequence, the ionic assemblage in the sampled groundwater is somewhat atypical, which also partially depicts the entire mechanism of hydro-geochemical amendments. In concern to the metal chemistry, around 18% of the sampled groundwater was found to be in grave contamination with the metals viz. Cd, Fe, Ni, and Pb in pre-monsoon. The same is reduced to around 9% in the post-monsoon and is surely due to the dilution effect. Such protruding metal concentrations in the groundwater point toward the prevailing mechanism in the study area causing metal leaching from the host rocks strata.

The hydro-geochemistry of the sampled groundwater clearly indicated the dominance of the CaMgCl facies. The dominance of the strong acids over the weak ones is the factual clue leading to the probable cause of the metal contamination in the groundwater and is in affirmation with the obtained metal chemistry. The dominance of SO_4^{2-} itself specifies its boosted incursion in the groundwater from the nearby rock strata. Such influx is surely attributed to the presence of sulfide phase in the neighboring rock strata. The weathering of sulfide phases not only releases SO_4^{2-} ions but also the associated metals into the groundwater regime causing contamination.

The PCA and interpolation attempt has sorted the contamination hotspots, which lies over and around the coal mines of the study area. This encircles the source of the contamination in the proximity of coal mines of the study area. The petrography has clearly inferred the presence of sulfide phase mostly pyrite in the strata of the Barakar formation, sampled from the derived hotspots. This pyrite mineralization undoubtedly is the prime cause behind the release of SO_4^{2-} and creating a metal dissolving environment on the exposure to the weathering. Hence, the AMD produced due to the influx of SO_4^{2-} is reasonably responsible for the dissolution of the metals and concerned contaminations. The production of AMD is evidently been augmented by the enhanced weathering caused due to the exposures of the concealed rock strata to weathering agents in the due course of the mining excavations and is also verifiable from the high SO_4^{2-} concentration and contamination hotspots, distinctly found around the quarried portion of the coalfield in the study area.

The profundity attained after the entire study is the accelerated weathering of the otherwise concealed strata exposed during the coal mining has caused an alteration of usual ionic influx and consequently favored the environment for metal dissolution from the otherwise stable setup of rock strata and migration to the local groundwater regime.

Appendix 9.1

Sample	Locations		pH	EC	TDS	Ca^{2+}	Mg^{2+}	Na^{+}	K^{+}	HCO_3^-	Cl^-	SO_4^{2-}	NO_3^-	Al	Cd	Cr	Cu	Fe	Ni	Pb	Zn
GW1	20°08′46″N	79°01′48″E	7.2	979	627	98	23	61	8.41	248	91	113	13.22	22	ND	30	62	180	10	ND	110
GW2	20°07′58″N	79°00′16″E	6.4	1321	845	144	38	59	6.33	249	155	189	12.1	31	ND	30	190	143	ND	ND	34
GW3	20°07′39″N	79°02′04″E	6	1690	1082	146	79	64	8.36	288	76	442	42.97	66	27	45	255	398	38	16	220
GW4	20°07′59″N	79°03′58″E	6.9	1145	733	73	44	88	5.27	288	56	159	12.08	51	ND	40	60	209	ND	ND	95
GW5	20°06′20″N	79°03′38″E	7	999	639	66	52	49	5.17	198	89	143	8	44	8	22	41	180	ND	ND	47
GW6	20°05′52″N	79°05′00″E	6.8	1268	812	119	45	62	6	234	178	122	19.61	36	ND	33	114	188	14	8	112
GW7	20°04′21″N	79°01′37″E	6.2	1598	1023	129	74	64	8.32	248	32	476	7.88	48	38	66	140	426	54	60	180
GW8	20°03′42″N	79°00′40″E	6.9	1079	691	94	46	44	4.23	297	104	98	11	36	ND	43	208	225	ND	ND	40
GW9	20°01′57″N	79°02′23″E	4.9	2688	1720	177	109	187	24.29	329	387	423	23.12	180	8	44	620	360	34	14	186
GW10	20°02′51″N	79°03′16″E	6	1671	1069	146	74	64	8.31	281	76	412	8.61	44	24	42	228	384	40	38	400
GW11	20°02′47″N	79°04′59″E	5.7	1878	1202	177	74	78	9.19	279	42	568	4.05	55	52	48	205	618	56	84	194
GW12	20°01′05″N	79°04′37″E	4.8	2579	1651	177	94	182	24.2	389	331	436	15.71	144	26	38	455	450	43	22	1620
GW13	20°01′25″N	79°06′20″E	5.6	1923	1231	142	64	149	13.11	329	158	342	61.43	48	ND	ND	100	120	12	ND	300
GW14	20°00′29″N	79°05′50″E	4.2	2970	1901	299	99	126	21.9	321	399	532	85.87	120	12	49	710	340	36	16	320
GW15	19°59′43″N	79°08′17″E	6.1	1276	817	113	42	77	3.31	233	142	183	16.49	33	ND	14	165	145	11	2	159
GW16	19°58′54″N	79°06′04″E	4	3077	1969	275	102	169	34.28	287	379	578	99	108	3	38	332	280	22	16	893
GW17	19°59′16″N	79°03′45″E	4.9	2478	1586	189	78	187	17.75	266	290	501	10	150	14	40	330	424	38	21	515
GW18	19°59′13″N	79°01′17″E	5.5	1678	1074	164	52	88	4.26	251	223	199	39.32	89	ND	ND	55	160	8	ND	400
GW19	19°57′42″N	79°04′48″E	5	2391	1530	179	87	169	12.4	322	240	467	23.12	160	9	30	210	373	33	24	123
GW20	19°57′08″N	79°03′40″E	5.7	1411	903	117	49	84	4.91	222	207	174	34.91	49	ND	18	178	98	ND	2	465
GW21	19°56′12″N	79°03′26″E	6	1180	755	71	54	80	6.18	322	90	140	23.15	20	ND	10	ND	66	ND	2	328
GW22	19°56′25″N	79°04′38″E	4.4	3046	1949	288	125	121	8.79	365	321	588	59.03	168	27	41	540	624	54	22	1050
GW23	19°56′49″N	79° 06′25″E	4.4	2988	1912	301	111	119	9.27	344	351	547	62	177	22	32	440	540	48	28	830

GW24	19°56′18″N	79°06′44″E	4.8	2677	1713	201	98	177	15.1	272	300	532	42.17	160	7	41	530	330	27	20	665
GW25	19°58′11″N	79°10′33″E	6.5	1270	813	78	53	90	10.7	304	78	142	21.66	24	ND	39	240	76	ND	ND	415
GW26	19°56′37″N	79°10′59″E	7.2	741	474	55	22	59	4.31	336	19	43	9.63	41	ND	25	145	240	ND	ND	330
GW27	19°56′06″N	79°09′19″E	5.9	1440	922	98	51	114	13	280	188	162	49.96	40	ND	22	200	50	19	3	128
GW28	19°54′45″N	79°07′37″E	4.7	2191	1402	169	69	160	13.4	288	211	432	49.77	149	26	28	334	460	27	18	220
GW29	19°53′35″N	79°07′47″E	5.8	1699	1087	136	80	69	8.34	211	54	412	33.1	60	30	42	240	328	34	41	243
GW30	19°53′15″N	79°13′25″E	6.7	1075	688	72	42	69	7.17	316	43	122	13.54	29	ND	34	65	220	ND	ND	98
GW31	19°54′07″N	79°14′47″E	6.2	1161	743	61	51	78	7.1	262	56	167	31.09	45	ND	40	45	110	ND	5	30
GW32	19°56′53″N	79°15′24″E	6.9	831	532	54	31	61	5.5	330	23	67	16.4	40	ND	54	81	168	ND	ND	74
GW33	20°00′38″N	79°18′07″E	5.9	1981	1268	112	77	168	17.16	177	111	573	13.5	98	45	40	200	552	72	52	230
GW34	19°55′44″N	79°18′14″E	5.2	2341	1498	134	89	200	19.8	273	241	501	9	165	12	45	110	460	45	40	515
GW35	19°55′33″N	79°19′18″E	6	972	622	66	39	61	3.24	276	99	72	16	75	ND	25	47	188	ND	7	40
GW36	19°54′02″N	79°18′24″E	5.4	1995	1277	112	99	131	7.97	210	240	389	42.99	110	32	42	75	320	32	25	220
GW37	19°53′59″N	79°20′10″E	5.9	1260	806	132	43	43	5.17	332	116	166	47.08	66	ND	20	66	300	10	ND	33
GW38	19°53′03″N	79°20′29″E	5.9	1198	767	112	38	56	10.13	300	108	122	40.1	85	ND	35	60	170	ND	9	20
GW39	19°51′31″N	79°17′28″E	5.9	1401	897	99	66	77	4.25	343	142	123	22	49	ND	20	70	70	ND	5	20
GW40	19°51′03″N	79°16′04″E	6	1490	954	89	67	99	12	343	89	177	47.11	75	ND	18	63	130	ND	10	39
GW41	19°49′40″N	79°14′56″E	6.2	1145	733	79	44	74	7.25	247	87	163	19.65	120	ND	14	40	115	9	ND	68
GW42	19°49′27″N	79°16′27″E	5.2	2088	1336	127	76	170	13	181	221	412	49.53	102	25	48	320	280	60	70	395
GW43	19°49′10″N	79°18′24″E	5	2356	1508	122	98	185	31.07	178	198	577	12.19	85	40	52	105	440	55	39	400
GW44	19°49′48″N	79°19′53″E	4.9	2422	1550	141	101	189	13.66	288	261	478	41.01	120	18	40	130	382	38	45	120
GW45	19°50′48″N	79°21′29″E	5.1	2212	1416	136	69	202	16.57	299	231	380	54.73	140	20	42	220	430	50	30	270
Minimum			4	741	474	54	22	43	3.24	177	19	43	4.05	20	ND	ND	ND	50	ND	ND	20
Maximum			7.2	3077	1969	301	125	202	34.28	389	399	588	99	180	52	66	710	624	72	84	1620
Average			5.73	1761	1127.3	134.87	67.067	109	11.11	280.8	167.4	312.09	30.55	83.48	11.66	33.75	200.53	283.33	22.86	17.64	293.2

The results obtained for the hydro-geochemical parameters and metals concentrations from the groundwater samples of pre-monsoon 2019. (ND - not determined).

Appendix 9.2

Sample	Locations	pH	EC	TDS	Ca^{2+}	Mg^{2+}	Na^{+}	K^{+}	HCO_3^{-}	Cl^{-}	SO_4^{2-}	NO_3^{-}	Al	Cd	Cr	Cu	Fe	Ni	Pb	Zn
GW1	20⁰08′46″N 79⁰01′48″E	7	998	639	62	48	51	3.14	201	91	123	16.24	20	ND	24	48	154	ND	ND	71
GW2	20⁰07′58″N 79⁰00′16″E	7.2	979	627	87	21	78	4.45	254	101	107	7.22	30	ND	22	140	132	ND	ND	40
GW3	20⁰07′39″N 79⁰02′04″E	6.7	1199	767	97	44	71	10.14	222	98	212	26.67	62	14	35	201	318	22	11	200
GW4	20⁰07′59″N 79⁰03′58″E	7.4	779	499	66	23	53	7.43	190	91	75	13.54	50	ND	ND	42	180	ND	ND	56
GW5	20⁰06′20″N 79⁰03′38″E	7.3	755	483	57	28	49	3.45	185	67	77	7.77	24	ND	20	30	175	ND	ND	33
GW6	20⁰05′52″N 79⁰05′00″E	6.8	1173	751	109	38	67	4	228	164	106	17	28	ND	24	120	150	10	ND	102
GW7	20⁰04′21″N 79⁰01′37″E	6.9	1221	781	112	41	63	3.52	218	93	278	13.1	42	24	41	122	388	37	22	166
GW8	20⁰03′42″N 79⁰00′40″E	7.1	961	615	68	36	66	5.41	247	88	98	24.64	20	ND	32	180	200	ND	ND	42
GW9	20⁰01′57″N 79⁰02′23″E	5.6	1379	883	98	54	81	10.48	219	114	249	23.08	98	3	40	480	298	25	12	144
GW10	20⁰02′51″N 79⁰03′16″E	6.7	1376	881	132	49	61	4.26	219	71	259	17.33	38	20	34	214	315	33	24	314
GW11	20⁰02′47″N 79⁰04′59″E	6.5	1466	938	110	58	88	5.13	191	97	349	8	50	41	35	188	555	49	63	164
GW12	20⁰01′05″N 79⁰04′37″E	6.2	1191	762	102	45	61	4.32	186	98	256	38.23	123	12	22	444	360	31	16	870
GW13	20⁰01′25″N 79⁰06′20″E	6.9	1001	641	98	24	67	4.57	224	102	134	10.22	56	ND	ND	221	47	ND	ND	302
GW14	20⁰00′29″N 79⁰05′50″E	5.4	1222	782	81	43	92	8.1	171	89	271	64.09	126	6	25	661	322	20	14	300
GW15	19⁰59′43″N 79⁰08′17″E	7.4	691	442	59	21	42	4.14	173	81	58	8	20	ND	ND	118	130	ND	2	168
GW16	19⁰58′54″N 79⁰06′04″E	5.4	1361	871	104	38	112	5.28	165	77	298	52.21	44	ND	19	229	305	19	11	210
GW17	19⁰59′16″N 79⁰03′45″E	6	1231	788	111	48	54	8.18	176	85	280	12.11	134	8	39	279	332	28	20	415
GW18	19⁰59′13″N 79⁰01′17″E	6.2	989	633	71	41	57	6.11	141	136	119	6.63	89	2	12	16	98	16	ND	325
GW19	19⁰57′42″N 79⁰04′48″E	6.1	1357	868	109	51	78	7	169	60	291	14.71	10	ND	24	321	324	ND	17	120
GW20	19⁰57′08″N 79⁰03′40″E	6.3	832	532	61	33	51	4.19	143	129	87	3.33	81	ND	26	ND	168	ND	2	60
GW21	19⁰56′12″N 79⁰03′26″E	6.2	991	634	78	31	61	9	169	151	99	11.42	38	ND	ND	ND	48	ND	2	400
GW22	19⁰56′25″N 79⁰04′38″E	5.7	1433	917	112	61	81	4.78	211	74	309	39.23	102	12	32	364	498	32	10	680
GW23	19⁰56′49″N 79⁰ 06′25″E	5.9	1114	713	99	39	61	6.16	122	43	359	12.87	75	10	41	167	367	40	23	490

GW24	19°56′18″N	79°06′44″E	5.8	1021	653	81	32	78	3.17	147	51	276	14.05	112	10	42	460	247	20	20	441
GW25	19°58′11″N	79°10′33″E	6.9	956	612	58	29	91	4.73	171	133	89	9.6	25	ND	22	300	55	ND	ND	365
GW26	19°56′37″N	79°10′59″E	6.8	1010	646	68	32	81	5.17	249	91	109	25	33	ND	17	105	148	ND	ND	290
GW27	19°56′06″N	79°09′19″E	7.2	888	568	91	20	51	6.43	266	71	107	9.20	62	ND	12	241	24	12	3	55
GW28	19°54′45″N	79°07′37″E	5.9	1137	728	78	31	100	6.13	151	38	302	11.99	70	21	32	122	400	33	17	40
GW29	19°53′35″N	79°07′47″E	6.2	1411	903	97	52	107	9	175	69	365	11.92	62	24	40	232	304	30	30	190
GW30	19°53′15″N	79°13′25″E	6.8	896	573	57	31	77	6.19	267	61	81	7.04	22	ND	21	49	166	ND	ND	38
GW31	19°54′07″N	79°14′47″E	6.6	871	557	64	24	70	5.31	256	98	43	9	30	ND	32	30	105	7	3	21
GW32	19°56′53″N	79°15′24″E	7.1	790	506	51	21	77	6.5	311	47	54	11.1	32	ND	50	66	150	ND	ND	55
GW33	20°00′38″N	79°18′07″E	6.4	1722	1102	112	71	124	10	191	110	429	12.42	90	33	30	195	449	70	48	201
GW34	19°55′44″N	79°18′14″E	6	1319	844	101	48	81	12	180	74	312	15.06	110	7	33	120	330	33	16	325
GW35	19°55′33″N	79°19′18″E	6.2	941	602	69	32	66	4.4	138	143	91	8.53	55	ND	21	45	155	ND	6	55
GW36	19°54′02″N	79°18′24″E	5.9	1378	882	91	42	119	9	192	63	331	7.73	124	20	50	72	304	40	20	225
GW37	19°53′59″N	79°20′10″E	6	899	575	73	28	63	5.37	138	143	91	8.04	40	ND	15	68	280	5	5	24
GW38	19°53′03″N	79°20′29″E	6.4	778	498	57	24	62	5	169	122	44	5.18	90	ND	30	71	144	8	ND	31
GW39	19°51′31″N	79°17′28″E	6.4	801	513	60	18	71	8.23	161	131	55	32.22	30	ND	15	58	50	ND	ND	12
GW40	19°51′03″N	79°16′04″E	6.3	963	616	73	25	84	6.7	170	122	68	29.41	60	ND	22	50	47	ND	8	41
GW41	19°49′40″N	79°14′56″E	6	1099	703	91	31	69	6.16	132	169	100	12.7	105	ND	18	40	125	10	ND	60
GW42	19°49′27″N	79°16′27″E	5.7	1386	887	111	42	85	6.75	107	45	398	27	143	10	35	224	365	57	74	262
GW43	19°49′10″N	79°18′24″E	4.8	1828	1170	112	71	124	10	191	110	429	71.51	72	24	50	110	300	40	14	387
GW44	19°49′48″N	79°19′53″E	5	1676	1073	99	63	143	13	210	86	381	46.77	80	10	22	108	322	42	33	106
GW45	19°50′48″N	79°21′29″E	5.6	1391	890	98	37	118	11.30	147	79	362	19	130	12	45	163	368	36	21	241
Minimum			4.8	691	442	51	18	42	3.14	107	38	43	3.33	10	ND	ND	ND	24	ND	ND	12
Maximum			7.4	1828	1170	132	71	143	13	311	169	429	71.51	143	41	50	661	555	70	74	870
Average			6.33	1130.22	723.28	86.11	38.2	77.46	6.52	189.84	94.57	200.24	18.93	65.26	7.17	26.688	166.97	237.82	17.88	12.6	203.04

The results obtained for the hydro-geochemical parameters and heavy metals concentrations from the groundwater samples of post-monsoon 2019 (ND – not determined).

References

American Public Health Association (1998). *Standard Methods for the Examination of Water and Waste Water*, 20the, 10–161. Washington: Water Pollution Control Federation.

Bureau of Indian Standards (2012). *Indian Specifications for Drinking Water. IS: 10500*, 1–3. New Delhi: Bureau of Indian Standards.

Chandra, D. (1992). Mineral resources of India, 5: Jharia coalfields. *Journal Geological Society of India, Bangalore* 5: 1–149.

Cobb, J.C., Steele, J.D., Treworgy, C.G., and Ashby, J.F. (1980). The abundance of zinc and cadmium in sphalerite-bearing coals in Illinois. *Illinois minerals* no. 074.

Das, N.C. (2013). Physico-chemical characteristics of selected ground water samples of Ballarpur City of Chandrapur District, Maharashtra, India. *International Research Journal of Environment Sciences* 2 (11): 96–100.

Deutsch, C.V. and Journel, A.G. (1998). *GSLIB: Geostatistical Software Library and User's Guide*, 2nde. New York: Oxford University Press.

Einax, J.W., Zwanziger, H.W., and Geiss, S. (1997). *Chemometrics in Environmental Analysis.* Weinheim: Wiley-VCH.

Facchinelli, A., Sacchi, E., and Mallen, L. (2001). Multivariate statistical and GIS-based approach to identify heavy metal sources in soils. *Environmental Pollution* 114: 245–276.

Fedden, F. (1875). On the evidences of "ground-ice" in tropical India during the Talchir period. *Geological Survey of India Rec.* 8 (1): 16–18.

Filcheva, E. and Noustorova, M. (2000). Organic accumulation and microbial action in surface coal-mine spoils, Pernik, Bulgaria. *Ecological Engineering* 15: 1–15.

Finkelman, R.B. (1981). Modes of occurrence of trace elements in coal. U.S. Geological Survey Open-file Report, Series 81–99, pp. 312.

Ganvir, P.S. and Guhey, R. (2020a). Assessment of some heavy metals toxicity and its probable remediation in groundwater around telwasa and ghugus area of Wardha valley coalfileds, Maharashtra. *J Indian Asso. of Sedimentologists* 37 (01): 33–43.

Ganvir, P.S. and Guhey, R. (2020b). Groundwater quality assessment with reference to some heavy metals toxicity and its probable remediation around Ballarpur area of Wardha valley coalfields, Maharashtra. *IOP Conference Series: Earth and Environmental Science* 597 (1): 012001.

Ganvir, P.S. and Guhey, R. (2021). Geochemical studies of some heavy metals' toxicity in groundwater with their plausible sources around gondwana supergroup, Wardha valley coalfields, Maharashtra. *Journal Geological Society of India* 97: 1415–1421.

Ganvir, P.S. and Guhey, R. (2022). Hydro-geochemical elucidation and its implications in the Wardha valley coalfields of central India. *IOP Conference Series: Earth and Environmental Science* 1032: 012015.

Ghosh, P.K. (1962). Recurrence of glaciation in the Talchir series, North Karanpura coalfield, Bihar. *Geological Survey of India Rec.* 87 (4): 751–756.

Gibbs, R.J. (1970). Mechanisms controlling world water chemistry. *Science* 170: 1088–1090.

Haigh, M.J. (1995). Soil quality standards for reclaimed coalmine disturbed lands: a discussion paper. *International Journal of Surface Mineral Reclamation Environment* 9: 187–202.

Hughes, W.T.H. (1877). The Wardha Valley Coalfield. In: *Memoir of the Geological Survey of India*, vol. 13, 1–154.

Hwang, C.K., Cha, J.M., Kim, K.W., and Lee, H.K. (2001). Application of multivariate statistical analysis and geographical information system to trace element contamination in the Chungnam coal mine area, Korea. *Applied Geochemistry* 16: 1455–1464.

Kale, M.G. and Soman, G.R. (2012). Petrography of Barakar sediments from Ghugus Area, Pranhita-Godavari Basin, Maharashtra. *Gondwana Geological Magazine* 27 (1): 67–78.

Karanth, K.R. (1997). *Groundwater Assessment, Development and Management*. New Delhi: Tata McGrraw Hill pp. 720.

Krothe, N.C., Edkins, J.E., and Schubert, J.P. (1980). Leaching of metals and trace elements from sulfide-bearing coal waste in southwestern Illinois. In: *Proceedings of the Symposium on Surface Hydrology, Sedimentology and Reclamation* (ed. D.H. Graves), 455–463. Lexington: University of Kentucky O.E.S. Publications.

Li, F.C. (1988). Environmental effect of coal mine spoil and general method for against pollution. *Chongqing Environment Science* 10: 17–21.

Mukherjee, K.N., Datta, N.R., Chandra, A.N. et al. (1988). A statistical approach to the study of the distribution of trace elements and their organic/inorganic affinity in lower Gondwana coals of India. *International Journal of Coal Geology* 10: 99–108.

Pires, M., Fielder, H., and Teixeira, E.C. (1997). Geochemical distribution of trace elements in coal: modelling and environmental aspects. *Fuel* 76: 1425–1437.

Pulles, W., Howie, D., Otto, D., and Easton, J. (1995). A manual on mine water treatment and management in South Africa. Water Research Commission Rep. No. TT 80/96, Pretoria.

Ramesh, S., Sharan, P.S., and David, K. (2018). Tectono-stratigraphy of the Pakhal Basin, Warangal (Rural) district of Telangana State with special reference to the Bollapally and the Pattipally Area. *International Journal of Innovative Research in Engineering & Multidisciplinary Physical Sciences* 6 (5): 32–37.

Saikia, B.K., Goswamee, R.L., Baruah, B.P., and Baruah, R.K. (2009). Occurrence of some hazardous metals in Indian coals. *Coke and Chemistry* 52: 54–59.

Sarate, O.S. (2004). Biopetrological investigations on the coals of Kosar, Dongargaon and Mahadoli areas, Wardha Valley Coalfield, Maharashtra. *Journal Geological Society of India* 63: 406–414.

Sarate, O.S. (2010). Study of petrological characteristics of coal deposits from Marki-Jhari-Jamni area, Wardha Valley Coalfield, Maharashtra. *Journal Geological Society of India* 76: 361–368.

Sarate, O.S. (2017). Petrography of coal seams from Wilson area, Chandrapur District, Maharashtra and their depositional history. *Geophytology* 47: 193–201.

Satapathy, D.R., Salve, P.R., and Katpatal, Y.B. (2009). Spatial distribution of metals in ground/surface water in the Chandrapur district (Central India) and their plausible sources. *Environmental Geology* 56: 323–1352.

Schurch, M., Edmunds, W.M., and Buckley, D. (2004). Three-dimensional flow and trace metal mobility in shallow Chalk groundwater, Dorset, United Kingdom. *Journal of Hydrology* 292 (1–4): 229–248.

Singh, G. (1987). Mine water quality deterioration due to due to acid mine drainage. *International Journal of Mine Water* 6: 49–61.

Singh, M.P., Singh, R.M., Chandra, D., and Somsuddin, A.K. (1982). Petrology, sedimentology and depositional environment of Ghugus coals, Wardha Valley, district Chandrapur and

Yeotmal, Maharashtra, India. *Quaternary Journal of Geology, Mining and Metallurgical Society of India* 54: 1–17.

Singh, R.M., Singh, M.P., and Chandra, D. (1983). Occurrence, distribution and probable source of trace elements in Ghugus coals, Wardha valley, dist- Chandrapur and Yeotmal, Maharastra, India. *International Journal of Coal Geology* 2: 371–381.

Singh, S., Tignath, S., Deolia, D.K. et al. (2020). Geochemical assessment of groundwater quality for its suitability for drinking and irrigation purpose in Jabalpur, Madhya Pradesh, India. *International Journal of Research in Engineering, Science and Management* 3 (08): 130–134.

Raja Rao, C.S. (1982). Coal resources of Tamil Nadu, Andhra Pradesh, Orissa and Maharashtra. *Bulletin of the Geological Survey of India, coalfields of Maharashtra, Wardha Valley Coalfield; Series A.* 45: 85–87.

Younger, P.L. and Wolkersdorfer, C. (2004). Mining impacts on the fresh water environments: Technical and managerial guidelines for catchment scale management. *Mine Water and the Environment* 23: S2–S80.

Zipper, C., Skousen, J., and Jage, C. (2014). *Passive Treatment of Acid Mine Drainage*, 1–13. Virginia: Virginia Polytechnic Institute and State University.

10

Soil Loss Rates in Trans-Himalayan Region: Case Study of Shyok Suture Zone, Ladakh, India

Rohit Kumar[1]*, Rahul Devrani*[2]*, Shailendra Pundir*[3,4]*, Ihsan U. Lone*[1]*, Vikas Adlakha*[3]*, Kiran Sathunuri*[1]*, Benidhar Deshmukh*[1]*, and Anil Kumar*[3]

[1] *Discipline of Geology, School of Sciences, Indira Gandhi National Open University, New Delhi, India*
[2] *Jindal School of Environment and Sustainability, O.P. Jindal Global University, Sonipat, Haryana, India*
[3] *Sedimentology and Quaternary geology and Structure & Tectonic Department Division, Wadia Institute of Himalayan Geology, Dehradun, India*
[4] *Department of Geology, Centre of Advanced Study, Kumaun University, Nainital, India*

10.1 Introduction

Soil erosion, a geomorphic process, is a pressing global environmental issue that poses a threat to agriculture and the environment, particularly in nations like India, where population growth and the consequent rise in land demand are happening at an accelerated rate (Kouli et al. 2009; Lambin et al. 2013; Kumar et al. 2022a). Soil erosion can contribute to various land degradation issues, including a decreased soil mass and water quality and increased river sedimentation (Ping et al. 2020; Sidi Almouctar et al. 2021). On a global scale, high erosion affects 1094 million hectares, while severe erosion affects 224 million hectares and is irreversible (Oldeman 1993), costing $400 billion each year, resulting in increased degraded land (Fletcher 2006). The global rate of erosion is estimated to be between 12 and 15 billion t/ha/year (Biggelaar et al. 2004), 75 billion t/ha/year (Pimentel et al. 1995; Borrelli et al. 2017), 35–35.9 billion t/ha/year (Borrelli et al. 2017), and 43 ± 9.2 peta (Borrelli et al. 2020). Given that 99% of the world's food come from crop cultivation and that India's agriculture system feeds more than 1.2 billion people, the productivity of the land is crucial to the success of all agricultural endeavors (Pimentel et al. 2004). Soil erosion affects 550 Mha of land in the Indian subcontinent (Saha 2003) and 3.975 Mha along ravine zones in Chambal and Sabarmati (Dwivedi and Ramana 2003), and 5334 million tons erode every year with a mean of 16.35 t/ha/yr (Narayana and Babu 1983). The rate of soil loss varies depending upon the climatic region: 80 mg/ha/yr (36 t/acre) in the regions of arid, 20–30 mg/ha/yr (9–13 t/acre) in coastal zones, 5–10 mg/ha/yr in Gangetic plains, and 8.43–13.44 t/ha/yr in Chambal river basin ravines (Mannering 1981; Kumar et al. 2022a, 2022c).

Weathering and Erosion Processes in the Natural Environment, First Edition. Edited by Virendra Bahadur Singh, Sughosh Madhav, Naresh Chandra Pant, and Ravi Shekhar.

In India, severe rates of erosion have been observed in Himalayan and arid climatic conditions (Kumar et al. 2014, 2022c). Soil erosion decreases the growth in hilly terrain by reducing the land productivity and livestock husbandry (Zimmerer 1993; Lal 2001; Alewell et al. 2019). Due to the scarcity of water and soil resources and the significant spatiotemporal unpredictability and torrentiality of the region's rainfall, soil erosion rates are higher in arid regions (Zhang et al. 2015). The combination of arid climate and mountainous topography exposes soil to surface flow, resulting in massive soil loss, e.g. China, India (Ladakh region), and Afghanistan (Canora et al. 2015; Pour et al. 2020). The eroded sediments accumulate in reservoirs, stream channels, irrigation canals, and other water conveyance facilities, reducing their ability to carry out their primary responsibilities and frequently necessitating expensive remediation (National Research Council 1993; Renard et al. 1997). The fragile and delicate eco-balance in the cold and dry region (Ladakh) is threatened by the inappropriate and unstructured exploitation of sparse natural resources with insufficient replenishment; this has gradually contributed the long-term damage to the environment causing severe soil erosion (Wani et al. 2008). Every 4–5 years, the cloud bursts in the Leh-Ladakh region cause flash floods that cause damage to wastelands, agricultural areas, and other properties, including human lives and accelerating soil erosion rate (Singh et al. 2019). To estimate the rate of soil erosion at regional scale, appropriate planning, management, and conservation measures are essential (Fistikoglu and Harmancioglu 2002).

Numerous soil and water management solutions are increasingly used to reduce the rate of soil erosion. However, the efficiency of these mitigating strategies is diminished by varying climatic variations, particularly seasonal and spatiotemporal variations (Ben Zaied et al. 2021; Kumar et al. 2022b). To effectively direct mitigation interventions that call for in-depth analysis of soil erosion at the sub-watershed level, policymakers and conservationists need complete information on soil erosion forces and their impact (Ben Zaied et al. 2021). Due to lack of systematic mapping and monitoring, the data exchange is frequently a limiting factor in dry and high topographic regions, as are weather conditions, vandalism, and technical issues (presence of dust). Remote sensing data have been used from Himalayas to coastal regions in India to reduce human resources and enhance accuracy (Devrani et al. 2022; Kumar et al. 2023). In this sense, numerous physical- and empirical-based models have been used, evolved, and used, depending upon spatial and temporal variability, climatic conditions, and topography. Because some models fail to accurately represent the temporal and geographical variability of hydro-meteorological factors, as well as hydrologic and sedimentation processes (Olivera and Maidment 1999; Ben Zaied et al. 2021). The Revised Universal Soil Loss Equation (RUSLE) model, developed from the Universal Soil Loss Equation, is the most effective one and exhibits most potential when employed in a variety of natural environmental situations (Renard et al. 1997; Stone and Hilborn 2000). Around the world, the RUSLE model has been used on a variety of ecosystems and scales (Kumar et al. 2014; Panagos et al. 2015; Ganasri and Ramesh 2016; Teng et al. 2016; Alewell et al. 2019). At basin scale analysis, computational expenses are significant obstacles. The problem gets worse as the dataset's resolution goes up. However, by utilising cloud computing systems, i.e. GEE (Google Earth Engine (GEE), can process vast amount of data in incredibly very short time (Kiran et al. 2022; Kumar et al. 2022a). The RUSLE model and geographic information system (GIS) integrated with GEE have been successfully used in many studies (Wang and Zhao 2020; Kumar et al. 2021, 2022a).

With this background, this study has been carried out in the upper Shyok river basin, where higher exhumation occurs due to rapid uplift and erosion. The climatic change in the Quaternary period is known to have increased streamflow intensity, periglacial and glacial activities, and surface runoff in the region. Tectonics has led to regional faulting, folding, and slope failures on a geological timescale. An amalgamation of exogenic and endogenic parameters on different temporal scales suggests a higher rate of soil loss. Therefore, the objectives of this study are to estimate the short-term rate of soil loss and compare it with the long-term erosion rates in the area.

10.2 Study Area

The Shyok valleys are located in the Trans-Himalaya, between the Karakoram mountains and the Ladakh range of northwestern Himalaya. The study area lies in the upper Shyok river basin, a part of Karakoram bounded between 77.283 and 79.443 longitude and 34.078 and 35.554 latitudes (Figure 10.1). The Shyok river, originating from the Rimo group of glaciers, forms a significant tributary of the Indus river and flows parallel to the Indus river for about 500 km in a northwesterly direction prior to their confluence. The Shyok river is 550 km long and a tributary of river Indus, originates from the Rimo group of Glaciers at the elevation of 4950 m, flows through northern Ladakh (India) and Pakistan, and merges with river Indus at Keris at the elevation of 2275 m (Kaul 1998). Glaciers dominantly feed the Shyok river and flow through prominently longitudinal valleys, separated by the Karakoram batholith in north and Ladakh batholith in south (Juyal 2014). The Shyok river basin contains valley glaciers ranging from 1 km to 70 km and provides water to several river channels. The upper Shyok river basin area is ~14 156 km^2 and covers Rimo, Muztagh glaciers, Saser Muztagh glaciers, and Depsang and Chang Chenmo valleys (Bhambri et al. 2013). The upper Shyok river basin is in the snow regime during winters and monsoons in summer (Wake 1989; Thayyen and Gergan 2010). The study area is rich in Quaternary deposits of glacial, lacustrine, eolian, and fluvial origin. The region is dominated by deep gorges, alluvial fans, *U*-shaped valleys, moraines, bars, and sand dunes, occurring all along the Shyok river (Dortch et al. 2010; Sant et al. 2011; Juyal 2014). The study area lies in the rain shadow side of the Himalayas and is a cold hyper-arid high altitude desert with scanty precipitation of ~150 mm (Bhutiyani 2000; Juyal 2014). This region is bisected by NW-SE striking, right-lateral strike-slip fault known as Karakoram Fault. The Karakoram fault system mostly follows the Shyok river valley and separates the Karakoram block unit from the ranges of Ladakh. Geologically, the Shyok–Nubra valleys are mainly constituted by the rocks of the Shyok Suture Zone (SSZ). The rocks of the SSZ occur as intensely deformed tectonic units between the Ladakh plutonic complex in south and the Karakoram batholith in north. The SSZ, referred to as the accretionary complex, comprises WNW-ESE trending thrust-bound formations of Khardung volcanics, Shyok volcanics, Saltoro Flysch, Saltoro Molasse, and Shyok Ophiolitic Melange (Rai 1983; Sinha and Upadhyay 1997; Chandra et al. 1999; Upadhyay et al. 1999; Weinberg et al. 2000; Robertson and Collins 2002; Rolland 2002; Lone et al. 2023a; 2023b).

The Karakoram batholith, exposed to the north of SSZ, comprises intermediate to felsic calc-alkaline magmatism (ranging from granodiorite to granite) of the Late Jurassic to Late

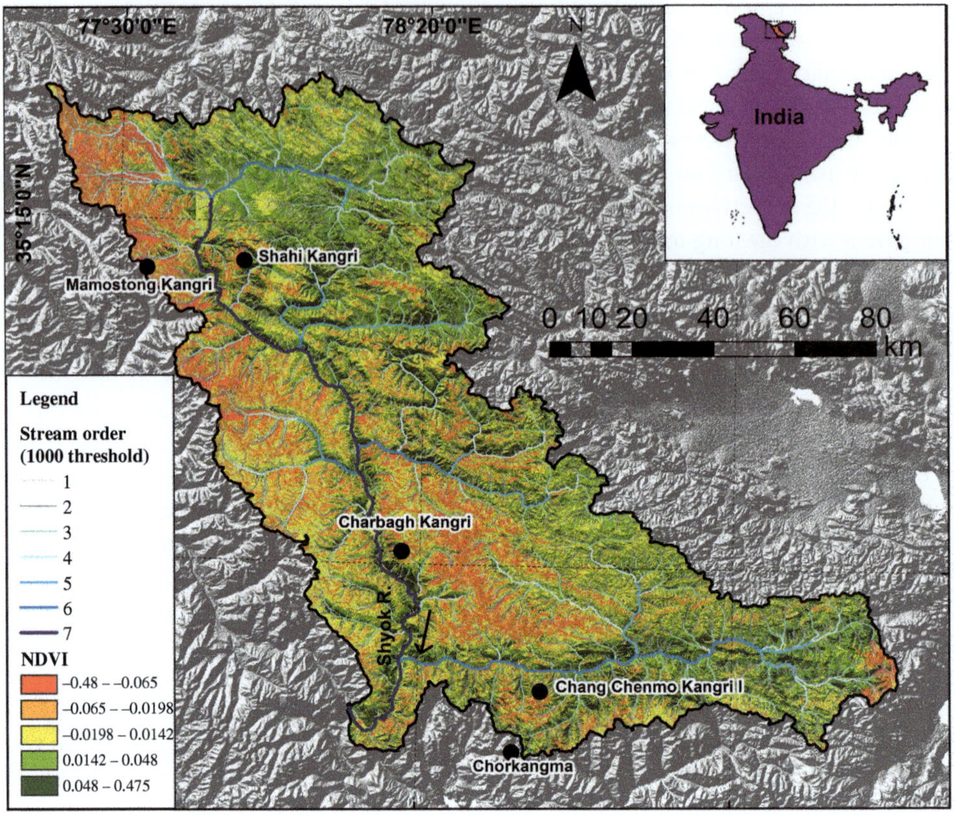

Figure 10.1 Study area located in the upper Shyok river basin. 90 m spatial resolution SRTM Digital Elevation Model (DEM) has been used for elevation in meters. The study area map is presented by mean Normalized Difference Vegetation Index (NDVI) value of 2020, draped over blue lines depicting streams and hillshade.

Cretaceous age (Borneman et al. 2015; Groppo et al. 2019; Pundir et al. 2020). The southern contact of the batholith is partially faulted against or intrusive into the SSZ. The NW-SE Karakoram fault separates the Karakoram batholith from the SSZ rocks (Weinberg et al. 2000). The Ladakh plutonic complex lies south of SSZ and forms part of the terrain extending from Pakistan (Kohistan arc) in the west through Ladakh to southern Tibet (Gangdese pluton). The Ladakh plutonic complex, representing an Andean-type calc-alkaline magmatism, is formed due to the northward subduction of Neo-Tethys oceanic crust beneath the Asian plate in the Early Cretaceous–Late Eocene (Allègre et al. 1984; Treloar et al. 1996; Rolland et al. 2002; Singh et al. 2007). This complex mainly consists of I-type lithologies, ranging from gabbro through diorites and granodiorites to granites, formed mostly between ca. 68 and 45 Ma, U—Pb dating of zircons (Schärer et al. 1984; Harrison et al. 1992; Rowley 1996; Singh et al. 2007; Ravikant et al. 2009; St-Onge et al. 2010; Bouilhol et al. 2013; Lakhan et al. 2020). Shyok volcanics defines the structurally lowest exposed units and is separated from Saltoro Flysch by an Unconrmity (Rai 1983; Borneman et al. 2015). These volcanics range in composition from basalt, basaltic andesite to andesite and have undergone low-grade metamorphism (Chandra et al. 1999; Sivaprabha et al. 2022). Shyok volcanics are overlain by Saltoro flysch, which mainly comprises slate, limestone, siltstone, sandstone, and phyllite of Aptian-Albian (Upadhyay 2014). The Saltoro Molasse overlies Shyok volcanics and Saltoro flysch and mainly consists of shales, conglomerates, sandstones, and gritstones (Rai 1983; Chandra et al. 1999; Borneman et al. 2015). The tectonically disrupted ophiolitic melange, which includes lenses of severely deformed Metapelits, pillow-basalts, meta-tuffs, basalts, and serpentinized ultramafics as well as limestones containing Orbitolina fossils from the Albian-Cenomanian period, crops out along the NW-SE trending Shyok and Nubra valleys (Rai 1983; Dunlap and Wysoczanski 2002; Thanh et al. 2012). Shyok ophiolite exposed at Bukdang in the western part of the Shyok valley comprises all the elements of the typical ophiolitic suite, i.e. ultramafics, gabbros, dolerite, diabase, pillow, and massive lavas (Srimal 1986; Lone et al. 2023a; 2023b).

10.3 Data and Methodology

The basin boundary of the upper Shyok river has been created from the 12.5 m spatial resolution DEM of ALOS-PALSAR procured from the Alaska Satellite facility (ASF) (https://asf.alaska.edu) using the tools of hydrology in a GIS environment (Kumar and Deshmukh 2018) (Table 10.1). A user-defined algorithm in GEE integrated with GIS has been used in this study to estimate the rate of soil loss using the RUSLE model. GIS and GEE have been used in many studies and different climatic and topographic conditions (Wang and Zhao 2020; Kumar et al. 2021, 2022a) (Figure 10.2). The RUSLE model is expressed as (Renard et al. 1991, 1997)

Table 10.1 Dataset used in this study.

Factor	Dataset	Source	Remarks
Rainfall Erosivity	Center for Hydrometeorology and Remote Sensing (CHRS) Rainfall	https://chrsdata.eng.uci.edu/	4 × 4 km resolution data produced by Center for Hydrometeorology and Remote Sensing at University of California
Soil Erodibility	NBSS and LUP soil texture	https://nbsslup.icar.gov.in	Soil texture map at 1 : 50000 scale paid data
Crop Management	Pre-post NDVI	USGS Earth Explorer https://earthexplorer.usgs.gov	10 m spatial resolution data procured
Support Practice	LULC	European Space Agency (ESA) https://worldcover2020.esa.int	Procured LULC map at 10 m spatial resolution
Slope length and steepness	ALOS-PALSAR	ASF https://asf.alaska.edu	12.5 m spatial resolution data procured

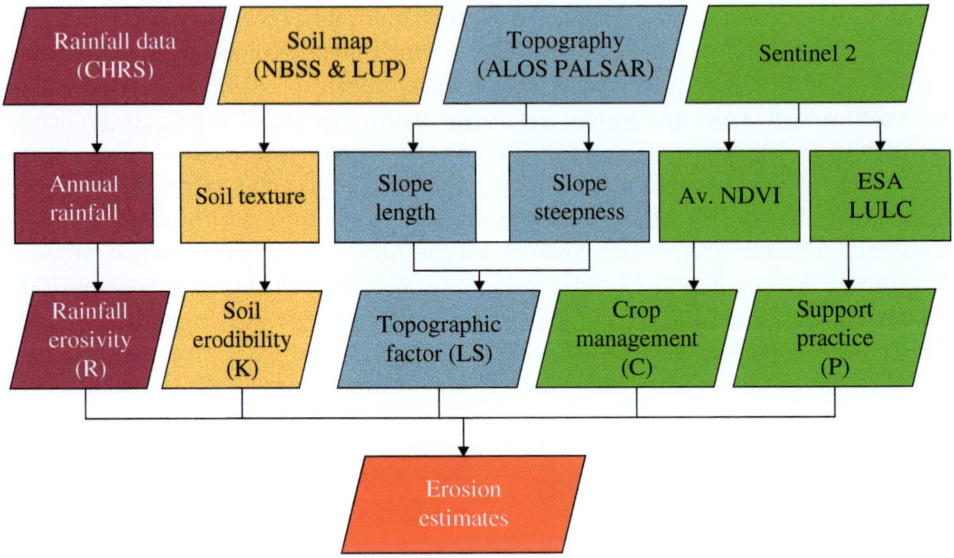

Figure 10.2 Overall methodology adopted in this study.

$$A = R \times K \times L \times S \times C \times P \tag{10.1}$$

where K is the soil erodibility (in Mg h/MJ/mm), R is the rainfall erosivity (in MJ mm/ha/hr/yr), L is the slope length (dimensionless factor), S is the slope steepness factor (dimensionless), C is the crop management (dimensionless), and P is the support practices factor (dimensionless).

10.3.1 Rainfall Erosivity (*R* Factor)

The *R* factor is an indicator of rainfall intensity, duration and size of raindrops to erode and transport sediments (Pandey et al. 2007). The average monthly rainfall data for 2021 at the 4×4 km spatial resolution from the CHRS has been procured. Monthly rainfall complied to create yearly rainfall and rescaled to 30 m spatial resolution using the inverse distance weighted method. The following equation developed for the Indian condition derived by Singh (1981) has been used for the calculation:

$$R = 79 + 0.363 \times \mathrm{AR} \tag{10.2}$$

where *R* is the rainfall erosivity, 79 and 0.363 are constant, and AR is the average rainfall.

10.3.2 Soil Erodibility (Factor K)

Soil erodibility reflects the sensitivity of soil characteristics to persist against erosion during raindrop strikes and surface runoff (Roslee and Sharir 2019). Soil properties, e.g. permeability, porosity, organic material, structure, and thickness, and hydraulic properties affect the soil erodibility (Wischmeier and Smith 1978). The calculation of soil erodibility and soil textures have been created from the soil map of the National Bureau of Soil Survey and Land Use Planning at 1 : 50000 scale. The vector files were converted into raster format using the K value assigned from the literature review (Figure 10.3b).

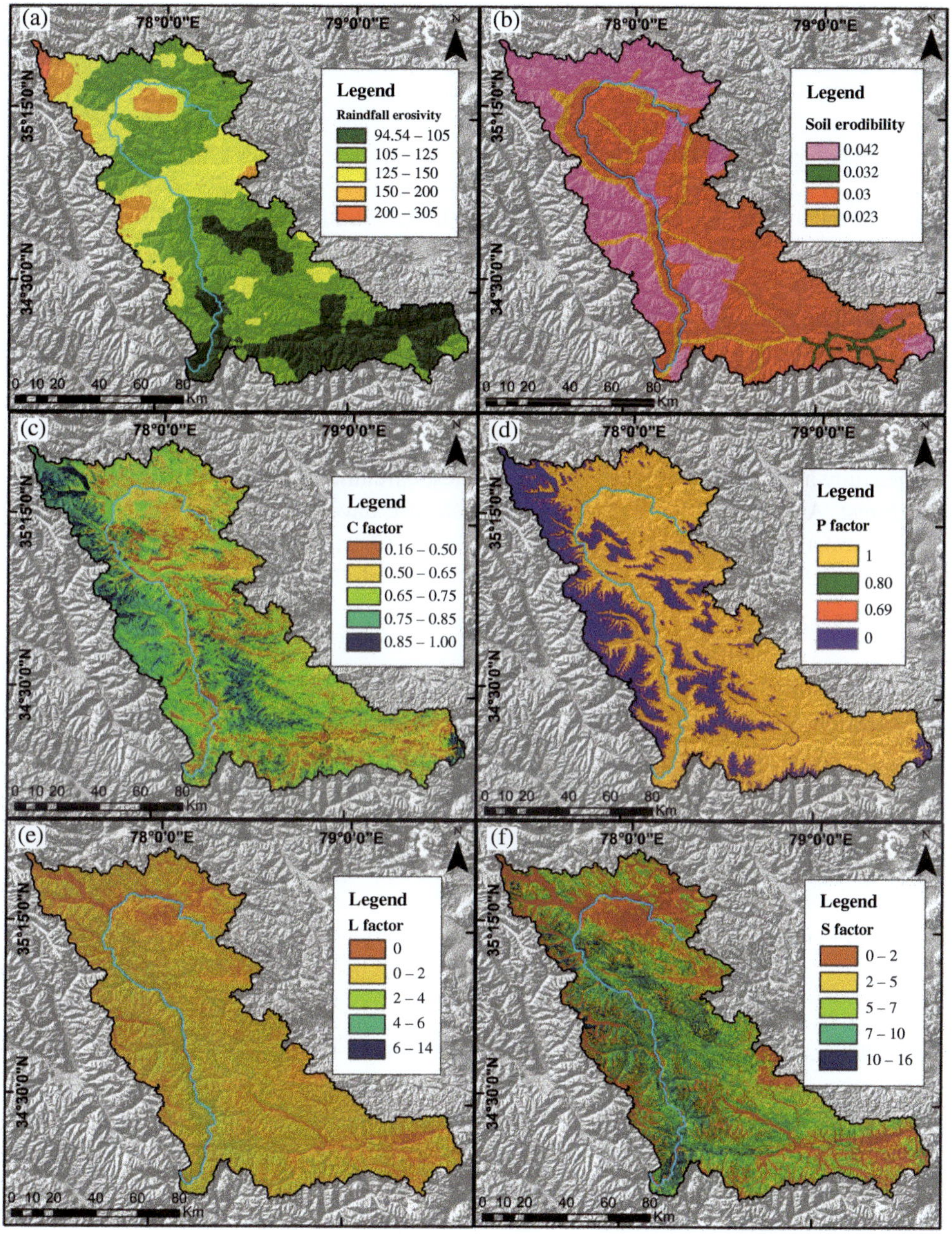

Figure 10.3 Factors used for the RUSLE model to calculate the rate of soil loss. (a) R factor, (b) K factor, (c) C factor, (d) P factor, (e) L factor, and (f) S factor.

10.3.3 Crop Management (C Factor) and Support Practice (P Factor)

This study used the pre-post rainfall NDVI of 2020 to calculate the C factor. NDVI has been calculated using 10 m spatial resolution Sentinel-2 surface reflectance imageries in GEE using the standard formula of NDVI, and then an average of the pre-post NDVI has been estimated. NDVI is generally expressed as:

$$\mathrm{NDVI} = (\mathrm{NIR} - \mathrm{RED})/(\mathrm{NIR} + \mathrm{RED})$$

For Sentinel imageries, NDVI can be expressed as:

$$(\mathrm{Band}\,5 - \mathrm{Band}\,4)/(\mathrm{Band}\,5 + \mathrm{Band}\,4)$$

For the calculation of the C factor from NDVI, the equation given by Zhou et al. (2008) and Kouli et al. (2009) have been followed. It expresses as follows:

$$C = \exp\left[-\alpha\left(\mathrm{NDVI}/\beta(\mathrm{NDVI})\right)\right]$$

where NDVI is the Normalized Difference Vegetation Index, $\alpha = 2$ (constant), $\beta = 1$ (constant) (Van der Knijff et al. 2000).

The LULC map was procured from the ESA for the P factor. ESA classifies seven classes in the study area, e.g. snow/glaciers, waterbodies, bareland, scrubland, grassland, builtup, and forest. Appropriate P values from the literature have been assigned to these LULC classes, and a P factor map has been created (Figure 10.3c, Table 10.2).

10.3.4 Topographic Factor (LS Factor)

Topographic factor consists of 'L' representing the length and 'S' representing the steepness of the topography. LS is directly proportional to soil erosion, as higher LS corresponds to a higher rate of erosion (Das et al. 2018). This current study uses ALOS-PALSAR 12.5 m spatial resolution DEM procured from ASF. First, the DEM has been reprojected to UTM 44 zone and further converted into ASCII format in a GIS environment. This format data has been used in the Arc Marco language developed tool by Hickey (2001) to calculate LS factor utilizing the equation proposed by Wischmeier and Smith (1978). The equation is as follows:

$$\mathrm{LS} = (l/72.6)^{m} \times (65.41\sin^2\beta + 4.56\sin\beta + 0.065)$$

Table 10.2 LULC classes *P* value have been assigned from the literature.

S. No	LULC classes	Area (Km^2)	Area (%)	*P* value	References
1.	Snow/glaciers	3433.32	24.25	0	Wanielista and Yousef (1993)
2.	Waterbodies	83.2481	0.59	0	Jain et al. (2001)
3.	Bareland	10 521.8	74.33	1	Wanielista and Yousef (1993)
4.	Scrubland	83.66	0.59	1	Jain et al. (2001)
5.	Grassland	31.71	0.22	0.7	Jain et al. (2001)
6.	Builtup	1.11	0.01	1	Wanielista and Yousef (1993)
7.	Forest	1.093	0.01	0.8	Pandey et al. (2007)
Total		14 156	100		

where l is the slope length, β represents the slope angle, m is the slope variable. If the slope is more than 2.86°, then m is 0.5. If the slope is between 1.72° and 2.86°, then m is 0.4; if between 0.57° and 1.72°, then m is 0.3; and if the slope is less than 0.57°, then the value of m is 0.2.

10.4 Result and Discussion

The RUSLE model has been used in this study to estimate the rate of soil loss. Therefore, all the factors are prepared and merged in a GIS environment. A maximum rainfall of 7493.91 mm and a minimum of 514.047 mm have been observed in 2020, having a mean value of 1288.49 mm with a standard deviation of 618.42 mm in the upper Shyok river basin. The yearly rainfall average ranges from 42.83 mm to 624.49 mm, with an average of 107.37 mm, and a standard deviation of 51.535 mm has been used to calculate rainfall erosivity (Figure 10.3a). The lowest rainfall erosivity has been observed in the southern part of the Shyok river basin, moderate erosivity in the middle, and the highest rainfall erosivity in the northern part of the basin (Figure 10.3a). Mostly, the study area is covered by sandy skeletal calcareous soil texture (53.33%), followed by sandy skeletal covering 35.85% of the area, calcareous covering 9.39%, and coarse loamy texture covering 1.43% of the area (Table 10.3). The pre-post mean NDVI ranges from −0.48 to 0.475. Less than 0.5 NDVI values indicate a lack of vegetation and more arid/barren hills in the area. Except for a few areas in the northern half, the higher NDVI values are primarily connected with the drainage. The C factor calculated from the NDVI ranges from 0.16 to 1 (Figure 10.3c). Seven LULC classes have been mapped in the study area. 74.33% area has been classified in the bare land or baren hill class, followed by 24.25% by Snow/Glaciers, and the rest 1.42% is scrubland, grassland, builtup, forest, and waterbodies. P factor has been calculated from the assigned values to the LULC classes from the literature (Figure 10.4). ALOS-PALSAR has been used to estimate relief and LS factor. The elevation ranges from 3558 m to 7601 m with a mean of 5418 m and a standard deviation of 440 m. LS factor has been calculated separately, and the L value ranges from 0 to 14 and the S value ranges from 0 to 16. Overall, the LS ranges from 0 to 169 having a mean value of 8 and a standard deviation of 10.8.

Table 10.3 Soil texture and assigned value.

S. no	Soil texture	Area (km^2)	Area (%)	K value	References
1.	Sandy skeletal calcareous	7549.01	53.33	0.03	Jain et al. (2001)
2.	Sandy skeletal	5074.97	35.85	0.042	Jain et al. (2001)
3.	Calcareous	1329.48	9.39	0.023	Jain et al. (2001)
4.	Coarse loamy	201.941	1.43	0.032	Jain et al. (2001)
Total		14156	100		

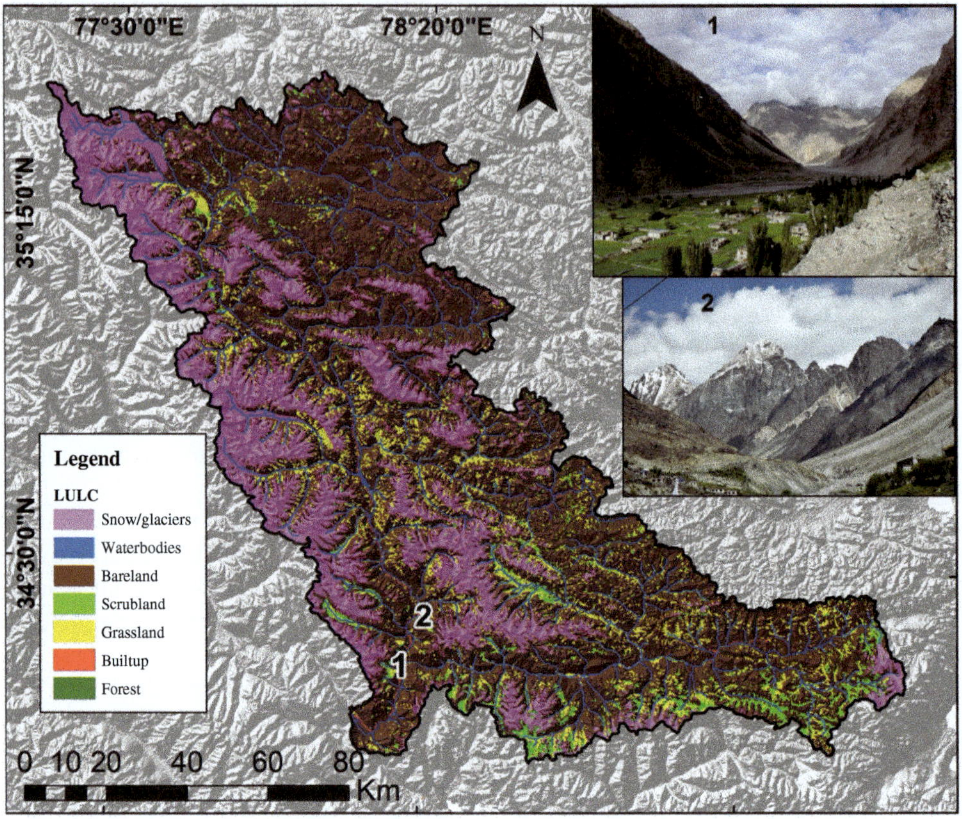

Figure 10.4 LULC map of the upper Shyok river basin, draped over hillshade. Field photos (1) and (2) depict a green region along the river's bank and glaciers, respectively.

10.4.1 Rate of Soil Loss and Spatial Distribution

The mean rate of soil loss is ~21.24 t/ha/yr. in the study area with a standard deviation of 37.74 t/ha/yr. (Figure 10.5). The highest soil loss rate (28.32 t/ha/yr) was observed in bareland class and the lowest in the forest (2.09 t/ha/yr) LULC classes. The rate of soil loss in the study area is above the tolerance limit set for India at ~16.35 t/ha/yr. The higher soil loss rate is due to the larger area covered by bareland (74.34% of the basin), as there are no vegetation covers that prevent soil erosion. The estimated soil loss has been categorized into seven classes based on the classification given by Singh et al. (1992) for the Indian subcontinent. Since the RUSLE model does not account for soil erosion caused by glaciers, snow, or water, 30% of the land is covered by the no erosion class (0 t/ha/yr). However, this does not imply any erosion in these locations. We acknowledge that soil erosion exists at these locations, but the RUSLE model cannot be used in these classes.

10.4.2 Rate of Soil Loss in Rainfall, LULC, and Soil Texture

The rate of erosion varies in different rainfall-classified regions. For the regions with rainfall ranging from 514 to 1088 mm (covering 6504.6 km^2), the erosion rate is 24.27 t/ha/yr (standard deviation 35.68 t/ha/yr). The regions where rainfall ranges from 1088 to 1581 mm

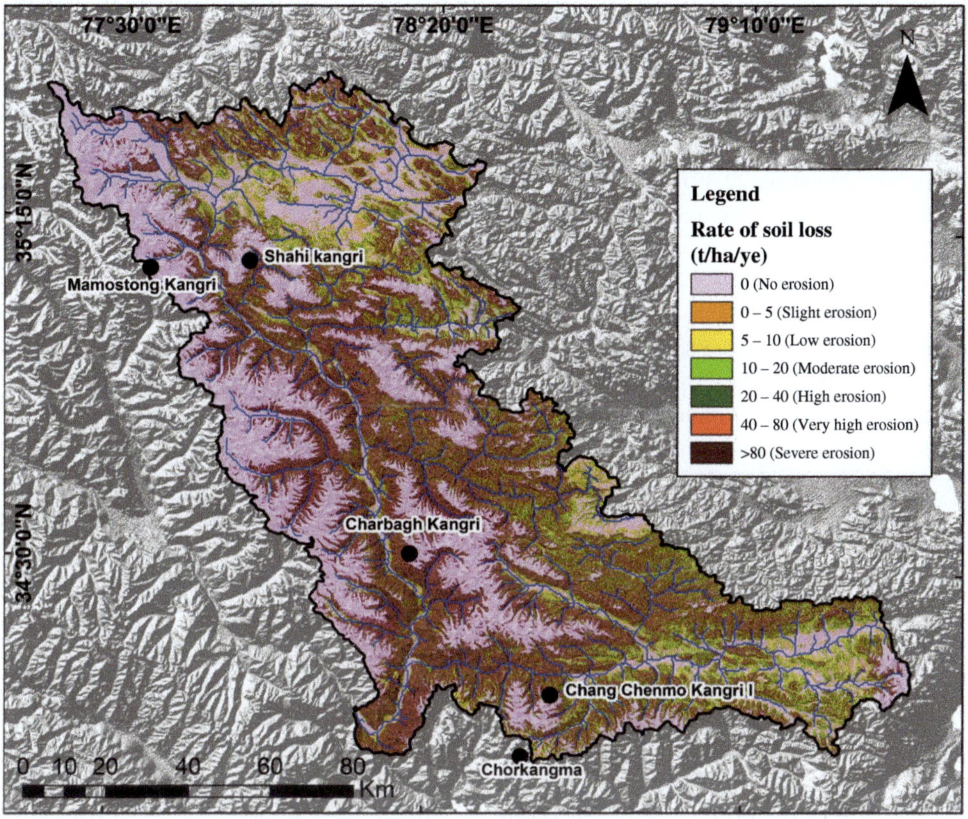

Soil loss class	Range	Area in sq. km	Percentage
No erosion	0-t/ha/yr	1683.15	11.89
Slight erosion	0–5-t/ha/yr	527.75	3.73
Low erosion	5–10-t/ha/yr	971.16	6.86
Moderate erosion	10–20-t/ha/yr	1827.75	12.91
High erosion	20–40-t/ha/yr	7593.73	53.64
Very high erosion	40–80-t/ha/yr	1155.50	8.16
Severe erosion	>80-t/ha/yr	397.29	2.81
Total		**14156**	**100**

Figure 10.5 Rate of soil loss in upper Shyok river basin and classified into seven classes. The classified map draped over hillshade and rivers in blue color.

(covering 4368.5 km^2) show 22.41 t/ha/yr (standard deviation 40.23 t/ha/yr), and for 1581–2320 mm (covering 2446.88 km^2) show 18.80 t/ha/yr (standard deviation 40.15 t/ha/yr). Further, the rainfall ranging from 2320 to 4209 mm (covering 761.2 km^2) shows 8.14 t/ha/yr (standard deviation 32.67 t/ha/yr) and rainfall ranging from 4209 to 7493 mm (covering 73.8 km^2) shows 0.88 t/ha/yr (standard deviation 16.44 t/ha/yr).

Soil loss in fluvial and snow LULC classes is 0 t/ha/yr, followed by 28.32 t/ha/yr (standard deviation 41.03 t/ha/yr) in bareland, 2.09 t/ha/yr (standard deviation 7.21 t/ha/yr) in the forest, 9.96 t/ha/yr in Scrub land (standard deviation 18.07 t/ha/yr), and 31.71 t/ha/yr in grassland (standard deviation 22.02 t/ha/yr). 22.19 t/ha/yr soil loss has been observed in sandy skeletal calcareous with a standard deviation of 32.98 t/ha/yr and 22.73 t/ha/yr in sandy skeletal having a standard deviation value of 46.47 t/ha/yr. In calcareous texture, a rate of 17.87 t/ha/yr (standard deviation 28.85 t/ha/yr) and, in coarse loamy, 14.13 t/ha/yr (standard deviation 22.30 t/ha/yr) have been observed. Therefore, the highest erosion rate is in sandy skeletal calcareous, and the lowest in coarse loamy has been observed.

10.4.3 Rate of Erosion in the Himalayas

Numerous RUSLE-based soil erosion estimation studies in the Himalayas have been carried out, demonstrating the various erosion rates and spatial distributions due to topography and climatic conditions. Erosion rates in Kashmir and Ladakh are on the higher side and it varies from 49.34 t/ha/yr in the Dudhganga watershed, a major tributary of the Jhelum river in the southwest of Kashmir valley (Ahmad et al. 2022), to 62.16 t/ha/yr in Narkara, a wetland catchment in the southwest of Kashmir valley (Rashid and Aneaus 2019). We observed the rate of soil erosion from 25.63 t/ha/yr in the Kangra region of Himachal Pradesh (Kumar et al. 2014) to 28.38 t/ha/yr in Tehri Garhwal (Gupta and Kumar 2017), 9.82 t/ha/yr in Rishiganga basin (Kumar et al. 2022c), 3.35 t/ha/yr in Assiganga (Kumar et al. 2022d). In Nepal, a study carried out by Koirala et al. (2019) estimates 15.6 t/ha/yr in the Mahakali river basin, 18 t/ha/yr in the Rapti river basin, 96.1 t/ha/yr in the Gandaki river basin, 79.7 t/ha/yr in the Kosi river basin, and 25 t/ha/yr in Nepal. Soil erosion rate of 3.53 t/ha/yr in the Tista river basin in northeastern Himalaya has been estimated (Chatterjee 2020). The Kosi river basin in the middle of the Himalayas has an estimated erosion rate of 40 t/ha/yr (Uddin et al. 2016). In the northeastern Himalayas, 36 t/ha/yr in Ri-Bhoi district, 1.38 t/ha/yr to 59.05 t/ha/yr in Tirap district, and 5.38 t/ha/yr in the Dzumah watershed of Nagaland have been estimated by RUSLE model (Das et al. 2018; Jena et al. 2018; Kichu and Dutta 2022).

According to the literature, areas with higher slopes, higher elevations, and more precipitation (Himalayas) experience higher rates of soil erosion than plain areas. For example, the Chambal river basin experiences 1.17 t/ha/yr, Parbati river basin experiences 1.2 t/ha/yr, Madhya Pradesh experiences 6.64 t/ha/yr, and Andipatti watershed experiences 6.73 t/ha/yr (Balasubramani et al. 2015; Kumar et al. 2021; 2022a). Similarly, our study calculates 28.32 t/ha/yr in Ladakh's upper Shyok river basin. This study is a new addition to soil erosion field and the use of conservation measures because no study relating to the RUSLE model has been conducted in the region.

10.4.4 Comparison of Long-Term and Short-Term Rate of Soil Erosion

In the north of SSZ, the Karakoram region has high relief and highly elevated topography as compared to the Ladakh region (Wallis et al. 2016; Pundir et al. 2021), and it is the most glaciated region on the Earth which are covered up to ~19,950 km^2 of glaciers (Dyurgerov and Meier 2005; Cogley 2011; Arendt et al. 2012; Bolch et al. 2012; Hewitt and Hewitt 2014). Studies suggest that unlike the Himalayas glaciers, the Karakoram glaciers are getting mass, termed as "Karakoram Anomaly" (Hewitt 2005; Bolch et al. 2012; Gardelle et al. 2012; Farinotti et al. 2020). There is evidence for the glacial advances in the Karakoram region compared to that of the Ladakh region, further suggesting the glacial cover variation in time and increasing trend of glaciation from southwest to northeast (Wallis et al. 2016).

If we see the uplift and exhumation history of the Trans-Himalayan, Ladakh, and Karakoram regions. In the middle Miocene, the southeast Karakoram range in the north of SSZ experienced fast exhumation rates of ~0.43 $mm\,yr^{-1}$ compared to the Ladakh range, where the rates are ~0.26 $mm\,yr^{-1}$ (Pundir et al. 2021). If we compare these millennial-scale exhumation rates to that of quaternary time, we see a similarity in the exhumation rates in the Karakoram region. Studies based on Quaternary glacial and river deposits landforms from the Trans-Himalayan region of Ladakh and Karakoram have been used to correlate long-term millennial scale erosion rates with the short-term erosion rates during Pleistocene and late Neogene and timing of glaciation in this region. Dortch et al. (2010) used 10Be dating on sediment of river channels flowing with multiple extents of glaciation and determined the long-term erosion rates over 105 years. Higher exhumation rates in the SSZ and Karakoram side as compared to that of the central and southern Ladakh block (approx. between 0.056 ± 0.012 mm/yr. and 0.074 ± 0.011 mm/yr. since 100 ka) are also observed. These rates are consistent with more glacial cover in the Karakoram region as compared to the Ladakh region. Evidence of more glaciation related to the Karakoram region's high uplift also comes from the study by Owen et al. (2002). They carried out 10Be and 26Al surface-exposure dating on the glacier moraines and associated landforms from the Karakoram and suggested that the glacier advances occurred at >60–9 ka. Similarly, other studies suggest the glacial stage that backs to ~311 ka that are preserved across the Trans-Himalayan region (e.g. Dortch et al. 2013). It is also considered that the older glacial stages in Pleistocene and late Neogene that are now erased may have occurred earlier in this region (Raymo and Ruddiman 1992; Zachos et al. 2001; Wallis et al. 2016).

Also, the geomorphic parameters (e.g. slope and relief) and long-term exhumation rates along the Trans-Himalayan region of India show a statistical correlation (Pundir et al. 2021). This also suggests that the million-year exhumation rate record can be correlated with present-day conditions in the Ladakh and Karakoram ranges.

10.5 Conclusion

A long-term average annual soil loss in the upper Shyok river basin (Ladakh) has been estimated using the RUSLE model, GEE, and GIS. RUSLE factors like rainfall erosivity have been calculated using rainfall data from CHRS, soil erodibility from the National Bureau of Soil Survey and Land Use Planning, slope length, and steepness using a 12.5 m spatial resolution of ALOS-PALSAR DEM, crop management using pre-post NDVI in the GEE, and support practice using LULC classes derived from the ESA. The results indicate that the mean rate of soil loss is ~21.24 t/ha/yr, highest (28.32 t/ha/yr) in bareland LULC classes and lowest in the forest (2.09 t/ha/yr) LULC classes, which is beyond the permissible limit of erosion in India. Due to snow cover and glaciers (11.89%), the RUSLE model has not computed erosion rate and requires a different model in such conditions. The geomorphic parameters (e.g. slope and relief) and long-term exhumation rates in Trans-Himalayan region of India show a statistical correlation. This also suggests that the million-year exhumation rate record can be correlated with present-day conditions in the Ladakh and Karakoram ranges. This study is an addition to the field of soil erosion and the use of conservation measures because no study relating to the RUSLE model has been conducted in the region.

Acknowledgments

RD is thankful to the O.P. Jindal Global University, Sonipat, Haryana, for providing the necessary facilities. RK and IL are grateful to IGNOU, New Delhi for providing financial support in the form of a fellowship. VA and SP would like to express sincere thanks to Wadia Institute of Himalayan Geology, Dehradun, India, for allowing carrying out this work. RK and IL thanks to Wadia Institute of Himalayan Geology and Ladakh University for organizing field training and 6th National Geo-Research meet in Ladakh to carried out field survey for this study.

References

Ahmad, W.S., Jamal, S., Taqi, M. et al. (2022). Estimation of soil erosion and sediment yield concentrations in Dudhganga watershed of Kashmir Valley using RUSLE & SDR model. *Environment, Development and Sustainability* 1–24. https://doi.org/10.1007/s10668-022-02705-9.

Alewell, C., Borrelli, P., Meusburger, K., and Panagos, P. (2019). Using the USLE: chances, challenges and limitations of soil erosion modelling. *International Soil and Water Conservation Research* 7: 203–225. https://doi.org/10.1016/j.iswcr.2019.05.004.

Allègre, C.J., Courtillot, V., Tapponnier, P. et al. (1984). Structure and evolution of the Himalaya-Tibet orogenic belt. *Nature* 307: 17–22. https://doi.org/10.1038/307017a0.

Arendt, A., Bolch, T., Cogley, J. G., Gardner, A., Hagen, J. O., Hock, R., & Zheltyhina, N. (2012). Randolph Glacier Inventory [v2. 0]: a dataset of global glacier outlines. Global Land Ice Measurements from Space, Boulder, Colorado, USA. Digital Media.

Balasubramani, K., Veena, M., Kumaraswamy, K., and Saravanabavan, V. (2015). Estimation of soil erosion in a semi-arid watershed of Tamil Nadu (India) using revised universal soil loss equation (rusle) model through GIS. *Modeling Earth Systems and Environment* 1 (3): 1–17. https://doi.org/10.1007/s40808-015-0015-4.

Ben Zaied, M., Jomaa, S., and Ouessar, M. (2021). Soil erosion estimates in arid region: a case study of the koutine catchment, southeastern Tunisia. *Applied Sciences* 11 (15): 6763. https://doi.org/10.3390/app11156763.

Bhambri, R., Bolch, T., Kawishwar, P. et al. (2013). Heterogeneity in glacier response in the upper Shyok valley, Northeast Karakoram. *The Cryosphere* 7 (5): 1385–1398. https://doi.org/10.5194/tc-7-1385-2013, 2013.

Bhutiyani, M.R. (2000). Sediment load characteristics of a proglacial stream of Siachen glacier and the erosion rate in Nubra valley in the Karakoram Himalayas, India. *Journal of Hydrology* 227: 84–92. https://doi.org/10.1016/S0022-1694(99)00174-2.

Biggelaar, C.D., Lal, R., Wiebe, K., and Breneman, V. (2004). The global impact of soil erosion on productivity. I. Absolute and relative erosion-induced yield losses. *Advances in Agronomy* 81: 1–48. https://doi.org/10.1016/S0065-2113(03)81001-5.

Bolch, T., Kulkarni, A., Kääb, A. et al. (2012). The state and fate of Himalayan glaciers. *Science* 336: 310–314. https://doi.org/10.1126/science.1215828.

Borneman, N.L., Hodges, K.V., Van Soest, M.C. et al. (2015). Age and structure of the Shyok suture in the Ladakh region of northwestern India: implications for slip on the Karakoram fault system. *Tectonics* 34 (10): 2011–2033. https://doi.org/10.1002/2015TC003933.

Borrelli, P., Robinson, D.A., Fleischer, L.R. et al. (2017). An assessment of the global impact of 21st century land use change on soil erosion. *Nature Communications* 8 (1): 2013. https://doi.org/10.1038/s41467-017-02142-7.

Borrelli, P., Robinson, D.A., Panagos, P. et al. (2020). Land use and climate change impacts on global soil erosion by water (2015-2070). *Proceedings of the National Academy of Sciences* 117 (36): 21994–22001. https://doi.org/10.1073/pnas.2001403117.

Bouilhol, P., Jagoutz, O., Hanchar, J.M., and Dudas, F.O. (2013). Dating the India-Eurasia collision through arc magmatic records. *Earth and Planetary Science Letters* 366: 163–175. https://doi.org/10.1016/j.epsl.2013.01.023.

Canora, F., D'Angella, A., and Aiello, A. (2015). Quantitative assessment of the sensitivity to desertification in the Bradano River basin (Basilicata, southern Italy). *Journal of Maps* 11 (5): 745–759. https://doi.org/10.1080/17445647.2014.980857.

Chandra, R., Upadhyay, R., and Sinha, A.K. (1999). Subduction and collision related magmatism in the Shyok suture and eastern Karakoram. *Paleobotanist* 48: 183–209. https://doi.org/10.54991/jop.1999.1303.

Chatterjee, N. (2020). Soil erosion assessment in a humid, eastern Himalayan watershed undergoing rapid land use changes, using RUSLE, GIS and high-resolution satellite imagery. *Modeling Earth Systems and Environment* 6 (1): 533–543. https://doi.org/10.1007/s40808-019-00700-0.

Cogley, J.G. (2011). Present and future states of Himalaya and Karakoram glaciers. *Annals of Glaciology* 52 (59): 69–73. https://doi.org/10.3189/172756411799096277.

Das, B., Paul, A., Bordoloi, R. et al. (2018). Soil erosion risk assessment of hilly terrain through integrated approach of RUSLE and geospatial technology: a case study of Tirap District, Arunachal Pradesh. *Modeling Earth Systems and Environment* 4 (1): 373–381. https://doi.org/10.1007/s40808-018-0435-z.

Devrani, R., Srivastava, P., Kumar, R., and Kasana, P. (2022). Characterisation and assessment of flood inundated areas of lower Brahmaputra River basin using multitemporal synthetic aperture radar data: a case study from NE India. *Geological Journal* 57 (2): 622–646. https://doi.org/10.1002/gj.4365.

Dortch, J.M., Owen, L.A., and Caffee, M.W. (2010). Quaternary glaciation in the Nubra and Shyok valley confluence, northernmost Ladakh, India. *Quaternary Research* 74: 132–144. https://doi.org/10.1016/j.yqres.2010.04.013.

Dortch, J.M., Owen, L.A., and Caffee, M.W. (2013). Timing and climatic drivers for glaciation across semi-arid western Himalayan–Tibetan orogen. *Quaternary Science Reviews* 78: 188–208. https://doi.org/10.1016/j.quascirev.2013.07.025.

Dunlap, W.J.N. and Wysoczanski, R. (2002). Thermal evidence for early cretaceous metamorphism in the Shyok suture zone and age of the Khardung volcanic rocks, Ladakh, India. *The Journal of Asian Earth Sciences* 20: 481–490. https://doi.org/10.1016/S1367-9120(01)00042-6.

Dwivedi, R.S. and Ramana, K.V. (2003). The delineation of reclamative groups of ravines in the Indo-Gangetic alluvial plains using IRS-1D LISS-III data. *International Journal of Remote Sensing* 24 (22): 4347–4355. https://doi.org/10.1080/0143116031000116994.

Dyurgerov, M.B. and Meier, M.F. (2005). *Glaciers and the Changing Earth System: A 2004 Snapshot*, vol. 58, 23. Boulder: Institute of Arctic and Alpine Research, University of Colorado.

Farinotti, D., Immerzeel, W.W., de Kok, R.J. et al. (2020). Manifestations and mechanisms of the Karakoram glacier anomaly. *Nature Geoscience* 13 (1): 8–16. https://doi.org/10.1038/s41561-019-0513-5.

Fistikoglu, O. and Harmancioglu, N.B. (2002). Integration of GIS with USLE in assessment of soil erosion. *Water Resources Management* 16: 447–467. https://doi.org/10.1023/A:1022282125760.

Fletcher, A. (2006). Soil erosion could devastate food sector. http://www.foodnavigator.com/Science-Nutrition/Soil-erosion-could-devastate-food-sector.

Ganasri, B.P. and Ramesh, H. (2016). Assessment of soil erosion by RUSLE model using remote sensing and GIS-a case study of Nethravathi Basin. *Geoscience Frontiers* 7 (6): 953–961. https://doi.org/10.1016/j.gsf.2015.10.007.

Gardelle, J., Berthier, E., and Arnaud, Y. (2012). Slight mass gain of Karakoram glaciers in the early twenty-first century. *Nature Geoscience* 5 (5): 322–325. https://doi.org/10.1038/ngeo1450.

Groppo, C., Rolfo, F., McClelland, W.C., and Coble, M.A. (2019). Pre-Cenozoic evolution of the Aghil range (western Tibetan plateau): a missing piece of the Tibet-PamirKarakorum geopuzzle. *Gondwana Research* 69: 122–143. https://doi.org/10.1016/j.gr.2018.12.006.

Gupta, S. and Kumar, S. (2017). Simulating climate change impact on soil erosion using RUSLE model—a case study in a watershed of mid-Himalayan landscape. *Journal of Earth System Science* 126 (3): 1–20. https://doi.org/10.1007/s12040-017-0823-1.

Harrison, T.M., Copeland, P., Kidd, W.S.F., and Yin, A.N. (1992). Raising Tibet. *Science* 255: 1663–1670.

Hewitt, K. (2005). The Karakoram anomaly? Glacier expansion and the 'elevation effect', Karakoram Himalaya. *Mountain Research and Development* 25 (4): 332–340. https://doi.org/10.1659/0276-4741(2005)025[0332:TKAGEA]2.0.CO;2.

Hewitt, K. and Hewitt, K. (2014). Glacier mass balance I: snowfall and glacier nourishment. In: *Glaciers of the Karakoram Himalaya: Glacial Environments, Processes, Hazards and Resources*, 87–116. https://doi.org/10.1007/978-94-007-6311-1_4.

Hickey, R. (2001). Slope angle and slope length solutions for GIS. *Cartography* 29 (1): 1–8. https://doi.org/10.1080/00690805.2000.9714334.

Jain, S.K., Kumar, S., and Varghese, J. (2001). Estimation of soil erosion for a Himalayan watershed using GIS technique. *Water Resources Management* 15 (1): 41–54. https://doi.org/10.1023/A:1012246029263.

Jena, R.K., Padua, S., Ray, P. et al. (2018). Assessment of soil erosion in sub-tropical ecosystem of Meghalaya, India using remote sensing, GIS and RUSLE. *Indian Journal of Soil Conservation* 46 (3): 273–282. Indian Associations of Soil & Water Conservationists.

Juyal, N. (2014). Ladakh: the high-altitude Indian cold desert. In: *Landscapes and Landforms of India, World Geomorphological Landscapes* (ed. V.S. Kale), 115–124. https://doi.org/10.1007/978-94-017-8029-2_10.

Kaul, H.N. (1998). *Rediscovery of Ladakh*. M. L. Gidwani, Indus Publishing Company 273.

Kichu, R. and Dutta, M. (2022). Assessment of soil erosion using GIS and remote sensing techniques in Dzumah watershed of upper Dhansiri, Nagaland. *The Pharma Innovation Journal* 11 (9): 2266–2273.

Kiran, S., Kumar, R., and Gogoi, K. (2022). Land use/land cover dynamics during 2001 and 2021 using google earth engine and GIS in Ramagundam coal mining area, a part of Pranhita Godavari Valley, southern India. *Journal of Scientific Research* 66 (1): https://doi.org/10.37398/JSR.2022.660107.

Van der Knijff, J. M., Jones, R. J. A., & Montanarella, L. (2000). Soil erosion risk assessment in Europe. European Soil Bureau Research Report EUR 19044 ENp, p. 34.

Koirala, P., Thakuri, S., Joshi, S., and Chauhan, R. (2019). Estimation of soil erosion in Nepal using a RUSLE modeling and geospatial tool. *Geosciences* 9 (4): 147. https://doi.org/10.3390/geosciences9040147.

Kouli, M., Soupios, P., and Vallianatos, F. (2009). Soil erosion prediction using the revised universal soil loss equation (RUSLE) in a GIS framework, Chania, Northwestern Crete, Greece. *Environmental Geology* 57 (3): 483–497. https://doi.org/10.1007/s00254-008-1318-9.

Kumar, A., Devi, M., and Deshmukh, B. (2014). Integrated remote sensing and geographic information system based RUSLE modelling for estimation of soil loss in western Himalaya, India. *Water Resources Management* 28 (10): 3307–3317. https://doi.org/10.1007/s11269-014-0680-5.

Kumar, R. and Deshmukh, B. (2018). Soil loss assessment using geospatial technology: challenges and opportunities with reference to the Chambal basin, Central India. In: *Environmental and Ecological Sustainability* (ed. B. Rupini, S. Baskar, S. Sahoo, and N.S. Kapur) Proceedings International Conference on Environmental and Ecological Sustainability. Engaging the Stakeholders, 304–316. New Delhi: IGNOU. ISBN: 978-93-87960-94-7.

Kumar, R., Deshmukh, B., & Sathunuri, K. (2021). RUSLE model based assessment of soil erosion in Parbati River Basin, central India using google earth engine and GIS. *EGU General Assembly Conference Abstracts* (pp. EGU21-1524). 10.5194/egusphere-egu21-1524

Kumar, R., Deshmukh, B., and Kumar, A. (2022a). Using Google earth engine and GIS for basin scale soil erosion risk assessment: a case study of Chambal river basin, Central India. *Journal of Earth System Science* 131 (4): 228. https://doi.org/10.1007/s12040-022-01977-z.

Kumar, R., Naqvi, H.R., Devrani, R. et al. (2022b). Sediment yield assessment, prioritisation and control practices in Chambal River basin employing SYI model. *Journal of the Geological Society of India* 98 (11): 1585–1594. https://doi.org/10.1007/s12594-022-2215-7.

Kumar, R., Devrani, R., and Deshmukh, B. (2022c). A review of remote sensing and GIS based soil loss models with a comparative study from the upper and marginal Ganga River basin. In: *Advances in Remote Sensing Technology and the Three Poles*, 321–339. https://doi.org/10.1002/9781119787754.ch22.

Kumar, R., Devrani, R., Dangwal, A. et al. (2022d). Extreme hydrological event-induced temporal variation in soil erosion of the Assiganga River basin, NW Himalaya. In: *Advances in Remote Sensing Technology and the Three Poles*, 230–246. https://doi.org/10.1002/9781119787754.ch15.

Kumar, R., Chatterjee, S., Kumar, R., and Deshmukh, B. (2023). Characteristics of tidal zone in Kalubhar Island, gulf of Kachchh, Gujarat and its geomorphological implications. *Journal of Geophysical Research* 8 (1): 58–62. 10.56153/g19088-022-0095-20.

Lakhan, N., Singh, A.K., Singh, B.P. et al. (2020). Zircon U–pb geochronology, mineral and whole-rock geochemistry of the Khardung volcanics, Ladakh Himalaya, India: implications for late cretaceous to Palaeogene continental arc magmatism. *Geological Journal* 55 (5): 3297–3320. https://doi.org/10.1002/gj.3594.

Lal, R. (2001). Soil degradation by erosion. *Land Degradation and Development* 12 (6): 519–539.

Lambin, E.F., Gibbs, H.K., Ferreira, L. et al. (2013). Estimating the world's potentially available cropland using a bottom-up approach. *Global Environmental Change* 23 (5): 892–901. https://doi.org/10.1016/j.gloenvcha.2013.05.005.

Lone, I.U., Mishra, M., and Chandra, R. (2023a). Petrological studies of Bukdang ophiolite, Shyok suture zone, trans-Himalaya. *Himalayan Geology* 44 (1): 130–146.

Lone, I.U., Mishra, M., Tiwary, S.K., and Chandra, R. (2023b). Recognition of Shyok ophiolites of NW Ladakh trans-Himalaya as a geoheritage: importance to Himalayan orogeny and remnant of Tethyan oceanic lithosphere. *Geoheritage* 15 (1): 3. https://doi.org/10.1007/s12371-022-00763-z.

Mannering, J.V. (1981). Use of soil loss tolerances as a strategy for soil conservation. In: *Soil Conservation Problems and Prospects: [Proceedings of Conservation 80, the International Conference on Soil Conservation, Held at the National College of Agricultural Engineering, Silsoe, Beford, UK, 21st-25th July, 1980]* (ed. R.P.C. Morgan), c1981. Chichester: Wiley.

Narayana, D.V. and Babu, R. (1983). Estimation of soil erosion in India. *Journal of Irrigation and Drainage Engineering* 109 (4): 419–434. https://doi.org/10.1061/(ASCE)0733-9437(1983).

National Research Council (1993). *Soil and Water Quality: An Agenda for Agriculture*. Washington: National Academies Press p. 2132, ISBN 978-0-309-04933-7.

Oldeman, L.R. (1993). An international methodology for an assessment of soil degradation and georeferenced soils and terrain database. *International Soil Reference and Information Center* https://doi.org/10.1016/0016-7061(93)90033-H.

Olivera, F. and Maidment, D. (1999). Geographic information systems (GIS)-based spatially distributed model for runoff routing. *Water Resources Research* 35 (4): 1155–1164.

Owen, L.A., Finkel, R.C., Caffee, M.W., and Gualtieri, L. (2002). Timing of multiple late quaternary glaciations in the Hunza Valley, Karakoram Mountains, northern Pakistan: defined by cosmogenic radionuclide dating of moraines. *Geological Society of America Bulletin* 114 (5): 593–604.

Panagos, P., Borrelli, J., Poesen, C. et al. (2015). The new assessment of soil loss by water erosion in Europe. *Environmental Science and Policy* 54: 438–447. https://doi.org/10.1016/j.envsci.2015.08.012.

Pandey, A., Chowdary, V.M., and Mal, B.C. (2007). Identification of critical erosion prone areas in the small agricultural watershed using USLE, GIS and remote sensing. *Water Resources Management* 21 (4): 729–746. https://doi.org/10.1007/s11269-006-9061-z.

Pimentel, D., Harvey, C., Resosudarmo, P. et al. (1995). Environmental and economic costs of soil erosion and conservation benefits. *Science* 267 (5201): 1117–1123. https://doi.org/10.1126/science.267.5201.1117.

Pimentel, D., Berger, B., Filiberto, D. et al. (2004). Water resources: agricultural and environmental issues. *BioScience* 54 (10): 909–918. https://doi.org/10.1641/0006-3568.

Ping, Z., Yajin, G.E., Jiang, Y. et al. (2020). Assessment of soil erosion by the RUSLE model using remote sensing and GIS: a case study of Jilin province of China. *Environmental Sciences* 1–22. preprints. https://doi.org/10.20944/preprints202011.0435.v1.

Pour, S.H., Abd Wahab, A.K., and Shahid, S. (2020). Spatiotemporal changes in aridity and the shift of drylands in Iran. *Atmospheric Research* 233: 104704. https://doi.org/10.1016/j.atmosres.2019.104704.

Pundir, S., Adlakha, V., Kumar, S., and Singhal, S. (2020). Closure of India–Asia collision margin along the Shyok suture zone in the eastern Karakoram: new geochemical and zircon U–pb geochronological observations. *Geological Magazine* 157 (9): 1451–1472. https://doi.org/10.1017/S0016756819001547.

Pundir, S., Adlakha, V., Devrani, R., and Kumar, S. (2021). Tectonic control over shallow crustal exhumation across the India-Asia convergent margin. *Tectonics* 40 (10): e2021TC006722. https://doi.org/10.1029/2021TC006722.

Rai, H. (1983). Geology of the Nubra valley and its significance on the evolution of the Ladakh Himalaya. In: *Geology of Indus Suture Zone of Ladakh* (ed. V.C. Thakur and K.K. Sharma), 79–91. Dehradun: Wadia Institute of Himalayan Geology.

Rashid, I. and Aneaus, S. (2019). High-resolution earth observation data for assessing the impact of land system changes on wetland health in Kashmir Himalaya, India. *Arabian Journal of Geosciences* 12 (15): 1–13. https://doi.org/10.1007/s12517-019-4649-9.

Ravikant, V., Wu, F.Y., and Ji, W.Q. (2009). Zircon U-pb and hf isotopic constraints on petrogenesis of the cretaceous-tertiary granites in eastern Karakoram and Ladakh, India. *Lithos* 110: 153–166. https://doi.org/10.1016/j.lithos.2008.12.013.

Raymo, M.E. and Ruddiman, W.F. (1992). Tectonic forcing of late Cenozoic climate. *Nature* 359 (6391): 117–122.

Renard, K.G., Foster, G.R., Weesies, G.A., and Porter, J.P. (1991). RUSLE: revised universal soil loss equation. *Journal of Soil and Water Conservation* 46 (1): 30–33.

Renard, K. G., Foster, G. R., Weesies, G. A., McCool, D. K., & Yoder, D. C. (1997). Predicting soil erosion by water: A guide to conservation planning with the Revised Universal Soil Loss Equation (RUSLE). Agriculture handbook, 1–703.

Robertson, A.H. and Collins, A.S. (2002). Shyok suture zone, N. Pakistan: late Mesozoic–tertiary evolution of a critical suture separating the oceanic Ladakh arc from the Asian continental margin. *Journal of Asian Earth Sciences* 20 (3): 309–351. https://doi.org/10.1016/S1367-9120(01)00041-4.

Rolland, J.P. (2002). The cross-cultural generalizability of the five-factor model of personality. In: *The Five-Factor Model of Personality Across Cultures*, 7–28. https://doi.org/10.1007/978-1-4615-0763-5_2.

Rolland, Y., Picard, C., Pecher, A. et al. (2002). The cretaceous Ladakh arc of NW himalaya—slab melting and melt–mantle interaction during fast northward drift of Indian plate. *Chemical Geology* 182 (2–4): 139–178. https://doi.org/10.1016/S0009-2541(01)00286-8.

Roslee, R. and Sharir, K. (2019, November). Soil erosion analysis using RUSLE model at the Minitod area, Penampang, Sabah, Malaysia. *Journal of Physics: Conference Series* 1358 (1): 012066. IOP Publishing.

Rowley, D.B. (1996). Age of initiation of collision between India and Asia: a review of stratigraphic data. *Earth and Planetary Science Letters* 145: 1–13.

Saha, S.K. (2003). Water and wind induced soil erosion assessment and monitoring using remote sensing and GIS. In: *Satellite Remote Sensing and GIS Applications in Agricultural Meteorology*, 315–330.

Sant, D.A., Wadhawan, S.K., Ganjoo, R.K. et al. (2011). Morphostratigraphy and palaeoclimate appraisal of the Leh valley, Ladakh Himalayas, India. *Journal of the Geological Society of India* 77: 499–510. https://doi.org/10.1007/s12594-011-0057-9.

Schärer, U., Xu, R.H., and Allègre, C.J. (1984). UPb geochronology of Gangdese (Transhimalaya) plutonism in the Lhasa-Xigaze region, Tibet. *Earth and Planetary Science Letters* 69 (2): 311–320. https://doi.org/10.1016/0012-821X(84)90190-0.

Sidi Almouctar, M.A., Wu, Y., Zhao, F., and Dossou, J.F. (2021). Soil erosion assessment using the RUSLE model and geospatial techniques (remote sensing and GIS) in south-Central Niger (Maradi region). *Water* 13 (24): 3511. https://doi.org/10.3390/w13243511.

Singh, G., Babu, R., Narain, P. et al. (1992). Soil erosion rates in India. *Journal of Soil and Water Conservation* 47 (1): 97–99.

Singh, G., Chandra, S. and Babu, R. (1981). Soil loss and prediction research in India. Central Soil and Water Conservation Research Training Institute Bull. No. T-12/D9.

Singh, R.K., Moharana, P.C., Goyal, R.K. et al. (2019). Soil erosion-status, causes and preventive measures in trans-Himalayan region of Leh-Ladakh in J&K. *Journal of Soil and Water Conservation* 18 (3): 235–240. https://doi.org/10.5958/2455-7145.2019.00033.X.

Singh, S., Kumar, R., Barley, M.E., and Jain, A.K. (2007). SHRIMP U–pb ages and depth of emplacement of Ladakh batholith, eastern Ladakh, India. *Journal of Asian Earth Sciences* 30: 490–503. https://doi.org/10.1016/j.jseaes.2006.12.003.

Sinha, A.K. and Upadhyay, R. (1997). Tectonics and sedimentation in the passive margin, trench, forearc and back-arc areas of the Indus suture zone in Ladakh and Karakoram: a review. *Geodynamica Acta* 10: 1–12. https://doi.org/10.1080/09853111.1997.11105289.

Sivaprabha, S., Bhat, I.M., Ahmad, T. et al. (2022). Geochemical analysis of magmatic rocks from Shyok Suture Zone (SSZ) Trans-Himalaya, NW India: Insights for geodynamic evolution of the terrane. *Lithos* 410: 106594. https://doi.org/10.1016/j.lithos.2022.106594.

Srimal, N. (1986). India-Asia collision: implications from the geology of eastern Karakoram. *Geology* 14: 523–527.

St-Onge, M.R., Rayner, N., and Searle, M.P. (2010). Zircon age determinations for the Ladakh batholith at Chumathang (Northwest India): implications for the age of the India–Asia collision in the Ladakh Himalaya. *Tectonophysics* 495 (3–4): 171–183. https://doi.org/10.1016/j.tecto.2010.09.010.

Stone R P and Hilborn D (2000). Universal soil loss equations fact sheet (USLE). Ministry of Agriculture, Food and Rural Affairs Version 2 (RUSLE2), 1–10. http://www.omafra.gov.on.ca/english/engineer/facts/00-001.htm.

Teng, H., Rossel, R.A.V., Shi, Z. et al. (2016). Assimilating satellite imagery and visible–near infrared spectroscopy to model and map soil loss by water erosion in Australia. *Environmental Modelling and Software* 77: 156–167.

Thanh, N.X., Rajesh, V.J., Itaya, T. et al. (2012). A Cretaceous forearc ophiolite in the Shyok suture zone, Ladakh, NW India: Implications for the tectonic evolution of the Northwest Himalaya. *Lithos* 155: 81–93. https://doi.org/10.1016/j.lithos.2012.08.016.

Thayyen, R.J. and Gergan, J.T. (2010). Role of glaciers in watershed hydrology: a preliminary study of a "Himalayan catchment". *The Cryosphere* 4 (1): 115–128. https://doi.org/10.5194/tc-4-115-2010.

Treloar, P.J., Petterson, M.G., Jan, M.Q., and Sullivan, M.A. (1996). A re-evaluation of the stratigraphy and evolution of the Kohistan arc sequence, Pakistan Himalaya: implications for magmatic and tectonic arc-building processes. *Journal of the Geological Society* 153: 681–693. https://doi.org/10.1144/gsjgs.153.5.0681.

Uddin, K., Murthy, M.S.R., Wahid, S.M., and Matin, M.A. (2016). Estimation of soil erosion dynamics in the Koshi basin using GIS and remote sensing to assess priority areas for conservation. *PLoS ONE* 11 (3): e0150494. https://doi.org/10.1371/journal.pone.0150494.

Upadhyay, R. (2014). Palaeogeographic significance of "Yasin-type" rudist and orbitolinid fauna of the Shyok suture zone, Saltoro Hills, northern Ladakh, India. *Current Science* 106: 223–228.

Upadhyay, R., Sinha, A.K., Chandra, R., and Rai, H. (1999). Tectonic and magmatic evolution of the eastern Karakoram, India. *Geodinamica Acta* 12 (6): 341–358.

Wake, C.P. (1989). Glaciochemical investigations as a tool for determining the spatial and seasonal variation of snow accumulation in the Central Karakoram, northern Pakistan. *Annals of Glaciology* 13: 279–284. https://doi.org/10.3189/S0260305500008053.

Wallis, D., Carter, A., Phillips, R.J. et al. (2016). Spatial variation in exhumation rates across Ladakh and the Karakoram: new apatite fission track data from the eastern Karakoram, NW India. *Tectonics* 35 (3): 704–721.

Wang, H. and Zhao, H. (2020). Dynamic changes of soil erosion in the taohe river basin using the RUSLE model and google earth engine. *Water* 12 (5): 1293. https://doi.org/10.3390/w12051293.

Wani, M.H., Baba, S.H., and Wani, S.A. (2008). Land degradation in cold arid eco-region of ladakh-extent, implications and strategies for sustainable land management. *Indian Journal of Agricultural Economics* 63 (3): 448.

Wanielista, M.P. and Yousef, Y.A. (1993). *Stormwater Management 21(2)*. Weinheim: Wiley https://doi.org/10.1002/aheh.19930210213.

Weinberg, R.F., Dunlap, W.J., and Whitehouse, M. (2000). New field, structural and geochronological data from the Shyok and Nubra valleys, northern Ladakh: linking Kohistan to Tibet. *Geological Society, London, Special Publications* 170 (1): 253–275. https://doi.org/10.1144/GSL.SP.2000.170.01.14.

Wischmeier, W.H. and Smith, D.D. (1978). *Predicting Rainfall Erosion Losses: A Guide to Conservation Planning (No. 537)*. Department of Agriculture, Science and Education Administration.

Zachos, J., Pagani, M., Sloan, L. et al. (2001). Trends, rhythms, and aberrations in global climate 65 ma to present. *Science* 292 (5517): 686–693.

Zhang, W., Zhou, J., Feng, G. et al. (2015). Characteristics of water erosion and conservation practice in arid regions of Central Asia: Xinjiang, China as an example. *International Soil and Water Conservation Research* 3 (2): 97–111. https://doi.org/10.1016/j.iswcr.2015.06.002.

Zhou, P., Luukkanen, O., Tokola, T., and Nieminen, J. (2008). Effect of vegetation cover on soil erosion in a mountainous watershed. *Catena* 75 (3): 319–325.

Zimmerer, K.S. (1993). Soil erosion and labor shortages in the Andes with special reference to Bolivia, 1953ndash; 1991: implications for "conservation-with-development". *World Development* 21 (10): 1659–1675.

11

Microbial Weathering of Rocks in Natural Habitat: Genetic Basis and Omics-Based Exploration

Mansi Podia, Prerna Yadav, Sunila Hooda, Prerna Diwan, and Rakesh K. Gupta

Department of Microbiology, Ram Lal Anand College, University of Delhi, New Delhi, India

11.1 Introduction

Mineral weathering is a process by which various nutrients are released from rocks and minerals. This process is dependent on both intrinsic and extrinsic parameters of the minerals (Brantley 2008), whereas bio-weathering is a process in which the weathering is carried out by living organisms, including microorganisms, through mechanical and chemical processes (Gadd 2007). Bio-weathering or microbial weathering contributes to pedogenesis (formation of soil) and facilitates the release of cations from phyllosilicates, thus considered as main buffer system in the ecosystem (White and Buss 2014). In bio-weathering, bacteria play a very crucial role in the biomining process in which extraction and recovery of metals takes place from the ores (Rawlings and Johnson 2007). This process also regulates the availability of nutrients, like Mg, K, and P, essential for plant growth and metabolism (Harley and Gilkes 2000). It has been reported that microorganisms accelerate the process of weathering through chelating reactions, formation of rock biofilms, and chemical dissolution (Finlay et al. 2009).

Microbial weathering has directly impacted the evolution of Earth's surface over geological time and also affected human life by influencing the water quality, development of soil affecting agriculture, structure of buildings, monuments, statues, etc. They have the ability to scavenge poorly bioavailable essential elements, such as phosphorus and iron (Ehrlich 1996). Microbial weathering is important for the agricultural industry as biofertilizers (Ribeiro et al. 2020) are used to remediate contaminated soils and increase soil fertility and crop productivity in arid lands. This also finds application as a bioactive agent which can degrade the building construction materials, such as asbestos (David and Geoffroy 2020), and in the maintenance of old monuments and stone artworks. Microbial integral proteins, such as bacterial membrane transport/channel proteins, are known to increase the absorption and transportation rate of minerals during the rock dissolution

Weathering and Erosion Processes in the Natural Environment, First Edition. Edited by Virendra Bahadur Singh, Sughosh Madhav, Naresh Chandra Pant, and Ravi Shekhar.

process (Chen et al. 2008). Contrarily, in rocky areas, these microorganisms are known to replenish the nutrients and minerals by regulating expression of certain genes that control the synthesis of proteins which are responsible for rock solubilization (Alahari and Apte 2004).

Though some microbes can grow on rock surfaces, the conditions for growth are not always favorable. In such cases, they produce siderophores, osmoprotectants, and chelating agents, forming complexes with metal ions like Fe, Cu, Cd, Pd, and Zn which in free state can be toxic to microbial cells (Kirtzel et al. 2020). Microorganisms produce organic acids, like citric acids and lactic acids, and inorganic acids, like sulfuric and nitric acids, which decrease the pH and enhance microbial weathering. The acidification process in bacteria plays a crucial role in the weathering process. The carbonic anhydrase (CA) enzyme of microbes which is involved in producing carbonic acid has been shown to dissolve feldspar and silicate rocks (Xiao et al. 2015). Microbes, like *Pseudomonas* and *Bacillus*, possess genes, like *ppq*, *gdh*, and others, to decrease the pH (Miller et al. 2010). Genes, such as *poxB*, *pta*, *mbaA*, are expressed by microbes to produce siderophores for chelating metal ions decreasing their toxicity and increasing their solubilization (Wang et al. 2018; Picard et al. 2022). Different metabolic pathways with different substrates regulate different physiological processes, for example, the addition of glucose and nitrogen substrates modify the relative abundance of microbial weathering bacteria together with increasing the phosphatase activity involved in organic phosphorus mobilization (Brucker et al. 2020). Microbes involved in weathering have been shown to rely on the concentration of C and N substrates present in the soil for their effective functioning (Uroz et al. 2015; Nicolitch et al. 2019). Glucose (C-source) confers the highest weathering effectiveness as compared to other derivatives and carbohydrates (Uroz et al. 2015), whereas NO_3 and NH_{4+} provide N source for improved growth and metabolic activities. The uptake of ammonium can induce dissolution of minerals by the process of proton extrusion (Nautiyal et al. 2000). Flagella for motility and biofilm formation are induced by the expression of genes, like *fliF*, *adh*, *fimA*, and *pilO*, found in *Pseudomonas* and *Rhizobium* to mineralize biotite (Wang et al. 2020b).

Microbes living in extreme conditions, like thermophiles, barophiles, and alkaliphiles, also have the tendency of "biomineralization." This biologically induced mineralization is likely unintentional and unregulated result of metabolic activities (Frankel and Bazylinski 2003), resulting in formation of minerals by the action of microorganisms (Lowenstam and Weiner 1989). It can take place in soils, rocks, sediments, seawater, lakes, etc. through the oxidation and reduction processes carried by bacteria, fungi, protozoa, and algae present in extreme conditions (Ghiorse and Ehrlich 1992). There are more than 60 different types of minerals known in our environment that are formed by microorganisms (Lowenstam and Weiner 1989). The organisms release one or more metabolic products that interact with ions and compounds in the environment, resulting in eventual deposition of mineral particles.

Thus, microbes are involved in various ways in the weathering process, and knowledge of diversity of microbes or microbiomes will enable the understanding of bioavailability of essential nutrients. This can also be applied to "Microbial Resource Managements" strategy to enhance the natural functional ability of microorganisms to promote an environment-friendly solution (Cohen and President 2007) for increasing the soil fertility.

11.2 Microbial Diversity of Extreme Habitats

Microbial weathering of rocks is a key component of the global carbon cycle and various other elements on the earth's near-surface and subsurface (Ehrlich 1998). Several strains of bacteria, fungi, lichens, etc. belonging to the diverse genera have been reported with different mineral-weathering abilities. Lithobiontic microorganisms play a vital role in bio-weathering through both enzymatic and nonenzymatic processes (Berthelin 1988). The enzymatic processes are based on the oxidation and reduction which leads to rock dissolution and is called redoxyolysis. The nonenzymatic processes given below are mediated by various extracellular metabolites released by microorganisms (Birch and Bachofen 1990).

1) Organic and inorganic metabolites solubilize rocks by acidolysis.
2) Organic compounds with functional groups result in the complexation of elements and lead to the complexolysis of rocks.
3) Redox-active compounds solubilize rocks by redoxyolysis.
4) Surface active metabolites modify the surface of minerals and thus facilitate weathering.

Hence, the rocks get modified by microbial activities leading to the mobilization of their components into the surrounding environment.

Microorganisms can adapt themselves and even grow in harsh conditions. Studying the microorganisms present in extreme habitats (Table 11.1) will provide information on their unique processes and interactions that allow them to keep their metabolic pathway active even under extreme conditions, like high temperature, pressure, dryness, low temperature, etc. Among extremophiles, rock-eating bacteria or lithotrophs are efficient in utilizing the inorganic compounds and facilitating renewal of lithosphere providing nutrients from the rocky lands and converting hard rocks into minerals (Enders et al. 2006). It is generally believed that life on Earth or any rocky planetary body would have been very limited if mineral weathering would not have taken place in the biosphere (Southam 2015), and there is a significant role of bio-weathering microorganisms in terms of Earth's evolution. Examples of micro-organisms involved in bio-weathering in different geological regions and their prevailing conditions present in extreme habitats have been compiled in Table 11.1, and they are being reviewed below.

11.2.1 Low-Temperature Regions

The microbial ecology of ice sheets and glaciers has been recently garnering significant scientific attention. These environments are recognized as the unique biome on Earth's surface and show diversity of prokaryotic and eukaryotic microorganisms. In a study done by Sanyal et al. (2020), 86 yeast strains were reported from blue ice and cryoconite hole environments from East Antarctica and cryoconite holes from a glacier in Western Himalayas. About 20–90% of the yeasts were producing extracellular enzymes, effectively degrading proteins, carbohydrates, pectin, esters, cellulose, tannin, and lignin. 10–29% of these yeasts exhibited the ability to solubilize rock minerals, like silicates and phosphates (Sanyal et al. 2020). Lichens have also been found in the regions of Antarctica which has bio-weathering properties. Lichen colonization results in the accumulation of organic matter, which along with cell excretions support heterotrophic bacterial population, thus

Table 11.1 Microorganisms present in extreme environments.

Environmental conditions	Geographical regions	Microorganisms involved	References
Low temperature	Arctic	Bacteria: *Cyanobacteria (Nostoc sp., Loriellopsis sp., Cyanothece sp.), α-Proteobacteria (Rhizobiales, Rubellimicrobium sp., Caulobacter sp., Sphingomonas sp), ß-Proteobacteria (Delftia sp., Rubrivivax sp.), Actinobacteria (Rubrobacter sp., Patulibacter sp.), Chroococcidiopsis sp.* Fungi: *Verrucaria sp., Thalidium sp., and BaglSiettoa sp.* Algae: *Trebouxia sp. and Trichosarcina sp.*	Ziolkowski et al. (2013)
	Arctic moraines	Bacteria: *Acidobacteria, Bacteroidetes, Sphingobacteria, Cyanobacteria, α-Proteobacteria (Rhodospirillales, Acetobacteraceae, Acidiphilium), β- Proteobacteria (Burkholderiales, Comamonadaceae, Rhodocyclales), γ- Proteobacteria (Xanthomonadales, Xanthomonadaceae, Pseudomonadales), δ-Proteobacteria (Mixococcales, Polyangiaceae, Cystobacteraceae), Actinobacteria (Acidimicrobiales, Acidimicrobiaceae, Microbacteriaceae), Planctomycetacia, Chloroflexi, Gemmatimonadetes, Clostridia, Verrucomicrobiae, Streptophyta, Chlorophyta*	Mapelli et al. (2011)
	Antarctic	Fungi: *Phoma sp., Dothideomycetidae, Friedmanniomyces endolithicus, F. simplex, Cryomyces antarcticus, C. minteri, Rachicladosporium antarcticum, R. mcmurdoi, Extremus antarcticus, Meristemomyces frigidus, Vermiconia Antarctica, Oleoguttula mirabilis, Glaciozyma watsonii, Phenoliferia sp., Mrakia robertii, Rhodotorula mucilaginosa*	Ziolkowski et al. (2013); Coleine et al. (2020); Sanyal et al. (2020)
	Damma Glacier (Switzerland)	Bacteria: *Arthrobacter sp., Frigoribacter sp., Paenibacillus sp., Janthinobacterium sp., Leifsonia sp., Oxalobacter sp., Paucibacter sp., Pedobacter steynii, Polaromonas sp., Pseudomonas sp., Rhodococcus, Erythropolis, Variovorax sp.* Algae: *Chlorella sp*	Frey et al. (2010)
	Sutri Dhaka glacier in Western Himalaya (India)	Yeast: *Mrakia robertii, Glaciozyma watsonii, Goffeauzyma gilvescens,* and *Rhodotorula mucilaginosa*	Sanyal et al. (2020)
	Guil River, glacial basin, Mount Viso (Italy)	Bacteria: *Verrucomicrobia, Proteobacteria, Planctomycetes, Firmicutes, Chloroflexi, Bacteroidetes, Actinobacteria, Acidobacteria*	Mahaney et al. (2016)
	Baelo Claudia (Spain)	Fungi: *Caloplaca velana, Lecanora albescens, Aspicilia contorta ssp. hoffmanniana, Caloplaca variabilis, Vemlcaria macrostoma, Collema crispum var. metzleri, Xanthoria calcicola,* and *Lecania turicensis*.	Ariño et al. (1995)

High temperature	Atacama Desert (Chile)	Bacteria: *Actinobacteria, Rubrobacterales, Actinomycetales, Acidimicrobiales, Gemmatimonadetes, Bacteroidetes, Thermoleophilia, Chloroflexi*	Crits-Christoph et al. (2013)
	Tsauchab River (Namibia)	Bacteria: *Arthrobacter sp., Kocuria sp., Lechevalieria sp., Microbacterium sp., Streptomyces sp., Bacillus sp., Massilia sp., Microvirga sp.*	Genderjahn et al. (2021)
	Desert oasis of Cuatro Ciénegas Basin (Mexico)	Bacteria: *Proteobacteria, Actinobacteria, Bacteroidetes, Bacillus, Clostridium, Chlorobi, Spirochaetes, Geobacillus,* and *Anoxybacillus*	Souza et al. (2018)
	Soil crust (Chile)	Algae: *Bracteacoccus, Chlamydomonas sp., Chlorella sp., Mymecia sp., Interfilum sp.*	Baumann et al. (2018)
		Cyanobacteria: *Chroococcidopsis sp., Leptolyngbya sp., Microcoleus vaginatus, Nostoc sp., Scytonema sp., Trichocoleus sp*	
	Rehai and Ruidian hot springs (China)	Bacteria: *Proteobacteria, Crenarchaeota, Firmicutes, Deinococcus-thermus, Crenarchaeota, Desulfurococcales (Sulfophobococcus* and *Aeropyrum), Caldimicrobium*	Li et al. (2021)
	Semiarid landscape in Torrollones De Gabarda Huesca (Spain)	Fungi: *Diploschistes actinostomus, Collema flaccidum, Grimmia orbicularis, Diploschistes*	Souza-Egipsy et al. (2004)
	Negev Desert (Israel)	Bacteria: *Proteobacteria, Actinobacteria, Deinococci (Deinococcus thermus), Chloroflexi, Bacteroidetes, Cyanobacteria, Acidobacteria, Firmicutes, and Gemmatimonadetes*	Wieler et al. (2019)
		Archaea: *Euryarchaeota, Thaumarchaeota, Candidatus Bat- hyarchaeota, Crenarchaeota, Diapherotrites, Parvarchaeota, Aenigmarchaeota, Nan- oarchaeota, Nanohaloarchaeota, Thaumarchaeota, Aigarchaeota, Crenarchaeota, and Korarchaeota*	Nir et al. (2022)
	Sahara Desert (Egypt) and Gibson Desert (Australia)	Bacteria: *Bacillus, Brevibacillus, Brevibacterium, Glutamicibacter, Micrococcus, Paracoccus, Pseudochrobactrum, Rhodococcus, Sphingobacterium (S. mizutaii), Streptomyces, Artrobacter (A. agilis), Bacillus, Leucobacter, Massilia, Micrococcus, Pseudarthrobacter, Stenotrophomonas*	Belov et al. (2018)

(Continued)

Table 11.1 (Continued)

Environmental conditions	Geographical regions	Microorganisms involved	References
Mines	Shuimuchong copper tailing impoundment (China)	Bacteria: *Proteobacteria, Actinobacteria, Euryarchaeota, Firmicutes, Nitrospirae, Rubrobacter, Acidithiobacillus, Sulfobacillus, and Leptospirillum*	Liu et al. (2014)
	Lubin Mine (Poland)	Bacteria: *Microbacterium sp., Microbacterium sp., Acinetobacter sp., Bacillus sp., Pseudomonas sp.*	Matlakowska and Sklodowska (2011)
	San Vittore mine and Settarme-Chassant mine (Italy)	Bacteria: *Verticillium leptobactrum,* Fungi: *Paecilomyces lilacinus, Aspergillus fumigatus, Mortierella spp., Myrothecium spp., Penicillium citreoviride, Paecilomyces farinosus, Epicoccum purpurescens*	Daghino et al. (2008)
	Eurasian and Philippine Sea plate (Taiwan)	Bacteria: *Proteobacteria, Patescibacteria, Gemmatimonadetes, Acidobacteria, Actinobacteria, Firmicutes, Cyanobacteria, Planctomycetes, Verrucomicrobia, Chloroflexi, and Bacteroidete*	Koner et al. (2022)
	Donghai County (China)	Bacteria: *Bacteroidetes, Proteobacteria, Cyanobacteria, Actinobacteria, Firmicutes, Chloroflexi, Acidobacteria, Gemmatimonadetes, Verrucomicrobia, Saccharibacteria, Planctomycetes* Fungi: *Ascomycota, Zygomycota, Rozellomycota,* and *Chytridiomycota*	Lu et al. (2022)
	Duluth Complex in Minnesota (USA)	Bacteria: *Neisseriaceae (Leeia spp.), Betaproteobacteria, Meiothermus spp., Sulfuriferula spp., Thiomonas spp., Acidobacteria, Chloroflexi (Ktedonobacteria, Thermomicrobia), Thiobacillus*	Jones et al. (2017)

imparting synergistic effect on weathering process (Ariño et al. 1995). Decomposition of lichens (especially nitrogen-fixing species) leads to the production of humic acid and fulvic acid which form strong complexes with iron and aluminum and thus accelerates microbial weathering of colonized rocks (Jones et al. 1981). Analysis of postglacial rocks and sediments of the Mount Viso area in Western Alps revealed that iron-oxidizing and iron-reducing bacteria accounted for the early stages of rock weathering (Mahaney et al. 2016).

Microbial weathering is primed by microbial iron oxidizers in the arctic glacier moraines, which increases fertility gradients and thus accelerates soil development and plant establishment. 15 bacterial classes were identified from these arctic moraines, dominating classes being Alphaproteobacteria and Cyanobacteria. At low pH, Acidobacteriacea and Acetobacteracea were predominant, and bacterial communities showed high iron-oxidizing activity (Mapelli et al. 2011).

Mergelov et al. (2018) analyzed bare bedrock surfaces in coastal oases of East Antarctica for its endolithic system (endolithic organisms like cyanobacteria, fungi, chlorophyta, lichens, and heterotrophic bacteria) and demonstrated transformation of silicate rocks by endolithic organisms, thus explaining the beginning of soils on the Earth's surface (Mergelov et al. 2018). Although much work has been done on polar regions and endolithic microorganisms over the last several decades, still there is much more to explore.

11.2.2 High-Temperature Regions

In dry environments, the biological soil crust of lichens, mosses, and cyanobacteria plays a vital role as agents for microbial weathering of natural rock surfaces (Cooke and Warren 1993). In a study conducted by Souza-Epigsy et al. (2004) in the Monegros region of NE Spain (arid region), it was reported that microbial weathering takes place in two steps. The first step being chemical dissolution process by lichen thallus, and the second is the disintegration of rocks because of hyphae penetration and lichen thallus expansion and contraction during wetting and drying cycles (Souza-Egipsy et al. 2004). Cryptogamic lithobiontic communities consisting of cyanobacteria, lichens, green algae, and fungi together with other heterotrophic organisms resulted in microbial weathering in the Atacama Desert (Jung et al. 2020). Chemical weathering was also observed in the lithomatrix with alkalinization, acidification, and enzyme activity of these communities in the lithomatrix. Pedogenesis and enrichment of elemental P (apatite and phytate-P, as well as P within sesquioxides) were observed during the mineralization process (Baumann et al. 2018). The microbial weathering process has also been recorded at petroglyph sites in the Negev Desert, Israel (Nir et al. 2022). The petroglyphs are rich assemblage of symbols, abstract elements, anthropomorphic and zoomorphic figures, and inscriptions written in a Semitic language (Eisenberg-Degen and Rosen 2013). Microbial weathering is causing deterioration of the petroglyphs which have historical and cultural value, with Cyanobacteria and Rhodophyta, the most abundant microorganisms in the samples collected (Eisenberg-Degen and Rosen 2013). 28 isolates belonging to three major phyla – Proteobacteria, Actinobacteria, and *Bacillus* – were found in limestone, quartz-rich shale, and quartz-rich sandstone categories of rocks from the Tsauchab river (Namibia) (Genderjahn et al. 2021). The potentially active lithic microorganisms were resulting in rock and mineral weathering with sustaining harsh climatic conditions (Genderjahn et al. 2021). Thus, understanding the weathering causing microbes and plants in arid lands would certainly contribute in designing and developing new environmental technologies, to enhance soil formation in high-temperature zones and counteract desertification.

11.2.3 Mines

Mines are reported to harbor diverse microorganisms that may aid in bio-weathering. The consortium of lithobiontic, neutrophilic, heterotrophic microorganisms was shown to cause weathering of the copper mine in a study conducted at a depth of 600 m, in the underground Lubin mine (Matlakowska and Sklodowska 2011). In this study, highly mineralized organic-rich Kupferschiefer black shale (SW Poland) was collected for identifying the factors responsible for microbial weathering of copper minerals. Eight bacterial strains were isolated from the Lubin mine and their consortium was used to check the bio-weathering of black shale in the laboratory. 415 cellular proteins were identified to be involved in rock degradation. The active compounds released by bacterial communities resulted in the mobilization of copper, iron, and sulfur (Włodarczyk et al. 2016). The samples from Olkusz mining area in Poland famous for the exploitation and processing of zinc-lead ore (Cabala et al. 2004) were exposed to the native microorganisms and sterile soil, with bioaugmentation by a single bacterial strain i.e. *Acidithiobacillus thiooxidans,* and sterile acidic growth medium containing different concentration of root exudates and water. The root exudates and *A. thiooxidans* showed significant influence on metallic elements (up to 44% for Zn, 78% for Cd, and 29% for Pb), whereas water played a minor role (0.13% for Zn, 1.23% for Cd, and 0.27% for Pb). Experimentally, it was proved that the dissolution of minerals, like smithsonite and hemimorphite, can take place, and their analogic dissolution can also happen under natural weathering conditions (Swęd et al. 2021). In another study, two asbestos mines, the Settarme Chassant mine and San Vittore mine, in Italy, were investigated. Three dominant species of fungi: *Paecilomyces lilacinus, Verticillium leptobactrum,* and *Aspergillus fumigatus* were found to actively remove the iron with *V. leptobactrum* being the most effective, as it produced a large number of chelating molecules, thereby increasing its iron-extracting ability. These studies suggest that fungal weathering can also be a possible tool for bioremediation of asbestos and fungal communities inhabiting ophiolitic sites can be useful for bioremediation applications (Daghino et al. 2008). Refer to Table 11.2 for the microorganisms identified and their specific solubilized rocks in various studies.

Table 11.2 Microorganisms involved in specific solubilization of rocks.

Microorganisms	Source of the solubilized compounds	Solubilized compounds	References
α-Proteobacteria			
Agrobacterium	Phlogopite (rock containing phosphate and mica)	Phosphate	Leyval and Berthelin (1989)
Azospirillium	Soluble phosphate	Phosphate (production of gluconic acid)	Rodriguez et al. (2004)

Table 11.2 (Continued)

Microorganisms	Source of the solubilized compounds	Solubilized compounds	References
Aminobacter	Soil	Biotite	Uroz et al. (2009)
Rhizobium	Soybean rhizosphere	Phosphate	Kim et al. (1998)
Labrys	Soil	Biotite	Uroz et al. (2009)
Sphingomonas	Forest soil	Biotite, phosphate, and iron	Uroz et al. (2007)
Rhanella	Soybean rhizosphere	Hydroxyapatite (production of gluconic acid)	Kim et al. (1998)
β-Proteobacteria			
Achromobacter	Conventional phosphate fertilizers and rock phosphate	Phosphate	Vassilev et al. (2006)
Ferribacterium	Sandstone rock	Iron [reduction of Fe(III)]	Potysz and Bartz (2022)
Burkholderia	Iceland spar	Biotite, phosphate (production of gluconic acid), iron, granite, and calcite	Jacobson and Wu (2009)
Janthinobacterium	Sterile biotite particles	Biotite	Uroz et al. (2009)
Collimonas	Forest soil	Biotite, phosphate, and iron	Uroz et al. (2007)
δ-Proteobacteria			
Acinetobacter	Conventional phosphate fertilizers and rock phosphate	Phosphate	Vassilev et al. (2006)
Geobacter	Sandstone rock	Iron [reduction of Fe(III)]	Potysz and Bartz (2022)
Azotobacter	Soil containing iron	Pyrite, glauconite, olivine, goethite, limonite, and hematite (production of dihydroxybenzoic acid and siderophore)	Page and Huyer (1984)
γ-Proteobacteria			
Acidithiobacillus	Pyrite	Pyrite and crystal	Mustin et al. (1992)
Citrobacter	Igneous rock	Phosphate	Puente et al. (2004)

(*Continued*)

Table 11.2 (Continued)

Microorganisms	Source of the solubilized compounds	Solubilized compounds	References
Shewanella	Sandstone rock	Iron [reduction of Fe(III)], smectite, calcium, and dolomite	Potysz and Bartz (2022)
Dyella	Soil	Biotite	Uroz et al. (2009)
Pseudomonas	Forest soil, compost, igneous rock	Biotite, phosphate (production of gluconic acid), iron, and rock	Uroz et al. (2007)
Enterobacter	Rock phosphate, compost	Phosphate (production of gluconic acid)	Hameeda et al. (2006)
Serratia	Rock phosphate, compost	Phosphate (production of gluconic acid)	Hameeda et al. (2006)
Frateuria	Soil	Biotite	Uroz et al. (2009)
Cyanobacteria			
Anabaena	Crystalline basalt and rhyolite	Basalt and rhyolite	Olsson-Francis et al. (2012)
Nostoc	Granite blocks	Granite, K-feldspar, and plagioclase	Seiffert et al. (2016)
Stigonema	Porphyroblastic granitic gneiss	Granite	Gaylarde et al. (2018)
Phormidium	Dolomite and limestone rocks	Dolomite and limestone	Tang et al. (2012)
Scytonema	Porphyroblastic granitic gneiss	Granite	Gaylarde et al. (2018)
Gram positive bacteria			
Arthrobacter	Hornblende (silicate mineral)	Hornblende (production of organic acids and siderophore)	Kalinowski et al. (2000)
Staphylococcus	Soil	Biotite	Uroz et al. (2009)
Bacillus	Rock phosphate, compost	Phosphate, rock, granite, and calcite	Hameeda et al. (2006)
Brevibacterium	Calcite and calcareous sodic soils	Calcite	Tamilselvi et al. (2016)
Streptomyces	Hornblende	Hornblende (production of siderophore)	Liermann et al. (2000)
Mycobacterium	Soil	Biotite	Uroz et al. (2009)
Paenebacillus	Soil	Biotite	Uroz et al. (2009)

Table 11.2 (Continued)

Microorganisms	Source of the solubilized compounds	Solubilized compounds	References
Lichens			
Diploicia canescens	Calcareous litharenite stones	Quartz and calcite	Ascasoa et al. (1998)
Lecanora fuscobrunnea	Rocks	Sandstone, granite, and dolerite	Selbmann et al. (2017)
Lecaniarabenhorstii	Calcareous litharenite stones	Quartz and calcite	Ascasoa et al. (1998)
Buellia frigida	Rocks	Sandstone, granite, and dolerite	Selbmann et al. (2017)
Caloplaca	Calcareous litharenite stones	Quartz and calcite	Ascasoa et al. (1998)
Verrucaria rubrocincta	Caliche plates	Calcite	Garvie et al. (2008)
Xanthoria	Calcareous litharenite stones	Quartz and calcite	(Ascasoa et al. (1998)

11.3 Factors Affecting Bio-Weathering

Soil is a rich habitat for microbial growth. Minerals in soil or rocks support microbial growth in their free state as well as in the consortia state. The physical and chemical factors play a crucial role in the bio-weathering process as it determines the type of microbes that will grow and establish interactions. This section reviews these factors affecting the weathering of minerals in the context of microbial weathering.

11.3.1 Physical Factors

Microbial weathering is observed in all the natural habitats involving extreme conditions (as discussed in the above section). The surface of minerals are exposed to diverse environmental conditions, like temperature, salt, pH, light, UV irradiation, oxygen, and water, which determine the kind of microbes that will exist in the environment (Wu et al. 2021). The attachment of microbes to the rock depends on the surface chemistry of microbes, rock surface, pH, and the presence of various organic and inorganic components. Water brings the microbial load in contact with the surface of rocks. Water presence can lead to hydrolysis, hydration, solution, and mineral formation, like clay, oxides, and hydroxides. Water is an important growth substituent that helps in preventing the desiccation of microbes or microbial colonies (Palmer and Hirsch 1991). Both water and wind lead to the formation of pits in the rocks where microbes can attach and form an association to survive (McNamara and Mitchell 2005). pH is another factor that influences the weathering of rocks. pH is affected by salt, acids, and components like iron, phosphorous, and other elements present (refer to Tables 11.1 and 11.3 showing microbial diversity with respect to environmental

Table 11.3 Recent studies using omics approaches to determine microbial diversity in different rock surfaces.

Technique	Weathering surface	Environmental factors	Microorganisms identified	Reference
Bacterial 16S rRNA gene (337F and 850R), Archaeal 16S rRNA gene (344F and 806R), and using specific primers for *mcrA* and *hydA* genes	Serpentine rock, Khalilovsky	pH 8.98 to 9.06	*Actinobacteria, Chloroflexi,* and *Proteobacteria*	Khilyas et al. (2019)
454 Pyrosequencing using a Roche 454 GS FLX Titanium	Waste rock, Antamine Mine	Rich in Cu and Zn, 80% mean annual precipitation, pH 8.2 to 8.6, Sulfur (0.53 to 1.4%)	*Acidiferrobacter, Ferritrophicum, Thiobacillus,* and *Acidithiobacillus*	Blackmore et al. (2018)
V4 region amplified by PCR and then Illumina MiSeq sequencing and rRNA transcript by RT-PCR	Sulfide-bearing rocks, Duluth complex	pH 4 to 6, rich in sulfur	*Meithermus, Sulfuriferula, Chloroflexi, Acidobacteria, Betaproteobacteria,* and *Sulfuriferula*	Jones et al. (2017)
Bacterial and archaeal (16S rRNA) and fungi (ITS) genes amplified by T-RFLP and then Illumina MiSeq sequencing	Soil from plots having sandstone or siltstone	pH 6.4 to 6.9	*Bacteria: Proteobacteria, Planctomyces, Acidobacteria,* and *Verrucomicrobia* *Fungi: Russula* and *Acer saccharum*	Yarwood et al. (2015)
V3-V4 regions 16S rRNA amplification by HotStart PCR and Illumina MiSeq	Serpentine rocks from Eurasian plate-tectonic plain and Philippines sea plate	Heavy metals like Ni, Cr, Mn, Ca, Mg	*Proteobactereia, Actinobacteria, Cyanobacteria, Planctomycetes,* and *Chloroflexi*	Koner et al. (2022)
V6-V9 16S rRNA amplification then barcode pyrosequencing to compare eubacterial 16S rRNA gene profiles	Different samples of Postglacial Rocks and Sediments, Mount Viso area, Western Alps	Chlorite, pH 4.8 to 6.3	*Verrucomicrobia, Proteobacteria,* and *Bacteroidetes*	Mahaney et al. (2016)
Amplified 16S rRNA genes by RFLP	Chaalis Abbey, France	Monument is subjected to wind, temperature, rain, relative humidity, condensation, and air pollution	*Sphingomonas, Rubrobacter,* and *Arthrobacter*	Mihajlovski et al. (2017)

factors, like low and high temperature and pH variations). Microbial weathering is also enhanced by the physical weathering of rocks. Cracks formed by physical factors become the habitat of microbes where they get attached to the surface of rocks and further solubilize the minerals. Physical and microbial weathering can be said to be correlated (Gorbushina and Broughton 2009).

11.3.2 Chemical Factors

Factors, like nutrient availability, acids, extracellular polymeric substance (EPS), chelators, and siderophore production by microbes, regulate the microbial weathering processes. The presence of nutrients, especially carbon and nitrogen, influences the growth of microorganisms. Glucose being a simple sugar is more effective in the process, as microbes can easily metabolize it as compared to the other complex sugars, like cellobiose and galactose. Glucose also enhances gluconic acid production and glucose dehydrogenase activity with an increase in the release of phosphorous, therefore solubilization of phosphorous was more effective in the presence of glucose (Nautiyal et al. 2000; Hameeda et al. 2006; Uroz et al. 2009). Ammonium can enhance the proton exchange mechanism resulting in the production of inorganic acids and with glucose can increase silicate weathering (Nautiyal et al. 2000; Wu et al. 2008). Solubilization of phosphorous depends on the availability of carbon and nitrogen sources, the presence of phosphorous decreases phosphorous solubilizing bacteria (PSBs) (Widdig et al. 2019; Brucker et al. 2020). Also, the presence of potassium, calcium and magnesium ions, and etheylenediaminetetraacetic acid (EDTA), along with nitrogen and carbon sources, intensifies the process of mineral weathering (Nautiyal et al. 2000; Nicolitch et al. 2019). EPS produced by a mixed population of microbes mostly in stress conditions also enhances the bio-weathering process. The weathering of rocks or dissolution of minerals depends on the microbial attachment to the rock surface that helps to build the microenvironment in close proximity to the microbes. This enhances the effects of EPS, chelating agents, and acids produced by microbes on the surface leading to porosity and dissolution of the minerals (Bennett et al. 1996; Ahmed and Holmström 2015). The microbes can attach to the surface in two ways. One can be the bonding between negatively charged residues of the cell wall and positively charged ions in the minerals, such as iron and aluminum ions. The second possible way can be bonding between negatively charged cell wall and negatively charged surface-like of silicates, but the microbial cells do not adhere to the surface and in such cases surface roughness, presence of pits and cracks come into play. This suggests that mineral composition (presence of positively charged ions) and surface property (presence of cracks) influence the microbial weathering of rocks (Roberts 2004).

Organic and inorganic acids produced by microbes as by-products also plays a major role in the weathering process. Organic acids (oxalic acid and citric acid) and siderophore production influence the dissolution of minerals (Buss et al. 2007). Acid production creates an acidic environment, some fungi decrease the pH by respiratory CO_2 but this is not the case for bacteria. Siderophores, on the other hand, are more effective compared to the organic acids produced because they can effectively and stably bind to Fe (III), helping in the dissolution of iron which depends on the concentration of siderophores being produced by the bacteria (Buss et al. 2007; Wu et al. 2008; Duckworth et al. 2014).

11.4 Genes and Microbial Pathways

Microbial weathering involves the breakdown of rocks and stones into simpler compounds. This requires specific genes and metabolic pathways (some of the genes identified to be involved in microbial weathering are listed in Table 11.4). These genes regulate the expression of extracellular compounds, proteins, and enzymes which further affect metabolic pathways required for the dissolution of minerals.

Table 11.4 Microbial genes/enzymes and metabolic pathways involved in weathering of different rocks.

Rock/mineral tested	Microorganisms	Genes/enzymes	Function	References
Acidification reaction/metabolic reactions				
K-bearing minerals	*Aspergillus fumigatus*	Aldehyde dehydrogenase	Acid production (aldehyde oxidized to acids)	Xiao et al. (2012)
		CA	Carbonic acid production	
Biotite	*Pseudomonas azotoformans F77*	*gcd* (glucose dehydrogenase for gluconic acid production)	Gluconic acid metabolism	Wang et al. (2020b)
		gad (gluconate 2-dehydrogenase for gluconic acid to ketogluconate)		
		glk (glucokinase for glucose to glucose 6-phosphate)		
		gntK (glucokinase for gluconic acid to glucose 6-phosphate)		
Biotite	*Enterobacter cloacae S71*	*ackA*	Acetic acid production, bacterial growth, mobilization of elements	Wang et al. (2018)
Mineral-Phosphate	*H.seropedicae Z67*	*Pqq*	Pyrroloquinoline quinone (PQQ) secretion and gluconic acid production result in mineral phosphate solubility	Wagh et al. (2014)
Soil inorganic phosphate	*Pseudomonas sp.*	*pqq, gcd*	Effects inorganic phosphate solubilization	Miller et al. (2010)

Table 11.4 (Continued)

Rock/mineral tested	Microorganisms	Genes/enzymes	Function	References
Calcite weathering	*Bacillus mucilaginosus*	*CA gene*	CA (promote calcite dissolution)	Xiao et al. (2014)
Inorganic phosphate, calcium phosphate	*Pseudomonas fluorescence*	*gcd* *pqqE* *pqqB* *pqq* *gad*	solubilization	Miller et al. (2010)
Phosphate solubilization	*Burkholderia cenocepacia strain 71-2*	*Eno*	Produce enolase enzyme for phosphate solubilization	Liu et al. (2019)
Tris-Cl rock phosphate	*Pseudomonas fluorescens strains*	*gltA1* (citrate synthase) *citC* (citrate transporter)	Citric and gluconic acid secretion, Mineral phosphate solubilization	Adhikary et al. (2014)
Redox reactions				
Biotite	*Enterobacter cloacae S71*	*poxB,* *pta*	Fe and Al release, siderophore	Wang et al. (2018)
Jurassic Lias rock		*amoA*	Ammonia monooxygenase gene for ammonia oxidizing	Cockell et al. (2011)
Serpentine bedrock		Chromium efflux pump, nickel/cobalt, iron transporter genes	Heavy metal resistance	Koner et al. (2022)
Siltstone		*ureC* *AOB amoA*	Ammonia oxidation	Yarwood et al. (2015)
Potassium feldspar	*Aspergillus niger*	Na^+, K^+-ATPase, cystathionine beta-synthase, cysteine synthase, glutathione synthase	Acquisition of potassium nutrition, transfer of the energy released by ATP hydrolysis	Wang et al. (2015)
Basalt, Torfajokul volcano, Iceland	*Cupriavidus metallidurans CH34*	*czc, cnr, hmy,* and *cop*	Efflux system and detox system for Resistance to Cd, Zn, and Co ions	Olsson-Francis et al. (2010)
K-bearing minerals, Fuquan in Guizhou Province	*Aspergillus fumigates*	Glutathione S transferase, cysteine synthase, thiamine biosynthesis protein	Oxidoreduction reaction	Xiao et al. (2012)

(*Continued*)

Table 11.4 (Continued)

Rock/mineral tested	Microorganisms	Genes/enzymes	Function	References
Basalt, Torfajokul volcano, Iceland	*Cupriavidus metallidurans CH34*	*czc, cnr, hmy,* and *cop*	Efflux system and detox system for Resistance to Cd, Zn, and Co ions	Olsson-Francis et al. (2010)
Chelation reaction				
Biotite	*Enterobacter cloacae S71*	*poxB* and *pta*	Siderophore production	Wang et al. (2018)
Hematite	*Collimonas pratensis PMB3(1)*	*mbaA* and *mbaB*	Siderophore production	Picard et al. (2022)
Iron and manganese oxides	*Pseudomonas chlororaphis PCL1391*	*phzB*	Phenazine production (functions as electron shuttle rather than iron chelator)	Hernandez et al. (2004)
Nontronites	*Pseudomonas aeruginosa*	*pvdA* *pvdC,* *pchEF*	Poverdine production for Iron bio-dissolution	Parrello et al. (2016)
Motility/biofilm formation				
Biotite	*Rhizobium pusense S41*	*fliF*	Flagellar protein synthesis, biofilm production	Wang et al. (2020a)
Biotite	*Pseudomonas azotoformans F77*	*adh* (adhesin for adhesion) fimA (type I fimbrial protein) *pilO* and *ecpD* (encode pilus assembly protein)	Pilus formation	Wang et al. (2020b)

Some of the major pathways (Figure 11.1) involved in microbial weathering are reviewed below:

a) *Acidification*: It is the process in which microorganisms produce acids, which decrease pH of the surroundings and hence result in the dissolution of minerals. Mineralization of rock occurs in the acidic microenvironment created by the microbes. Here, organic acid chelate metal ions on rock, which increases the mobilization of these ions (Zhu et al. 2014). Study shows that fungi and bacteria can produce organic acids in sandstone causing weathering of the building (Palmer et al. 1991). Microbes can produce organic acids, like acetic, formic, citric, gluconic, pyruvic, lactic, oxalic, and succinic acids. Inorganic acids produced by microbes are sulfuric acid by sulfur-oxidizing bacteria and nitric acid by nitrifying bacteria. Glucose and other substrates present in the environment are used by bacteria as their energy source, the main metabolic pathways involved

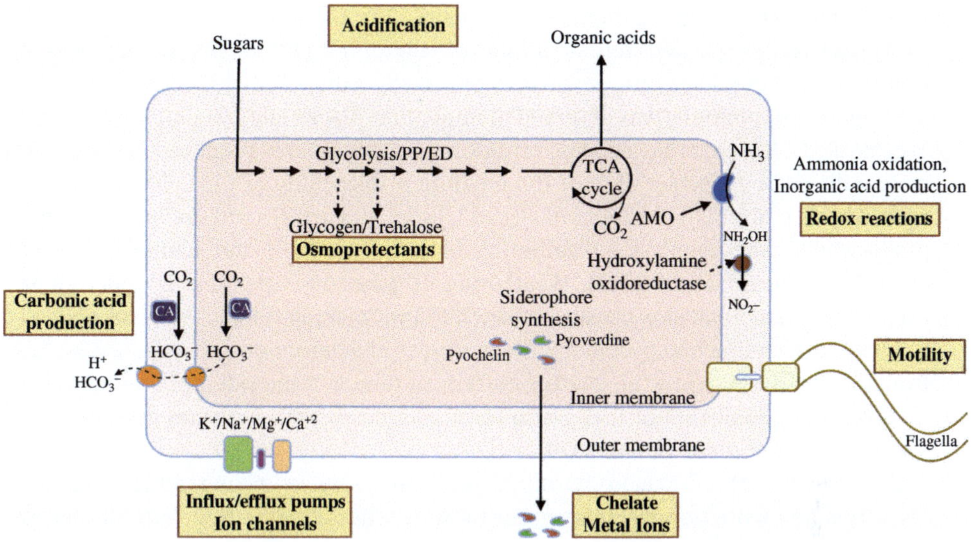

Figure 11.1 Major Metabolic pathways involved in microbial weathering of rocks. The acidification process decreases pH by secreting the organic acids produced during the breakdown of sugars to generate energy. Carbonic acid production decreases the pH. Different pumps/channels are present on the cell membrane mediating the influx and efflux of the elements. Siderophore production, such as Pyochelin and Pyoverdine, can chelate metal ions present in the rock. Ammonium oxidation reaction represents the example of redox reactions taking place. For the motility of bacterial cells, flagella is formed. Abbreviation: CA, carbonic anhydrase; TCA, tricarboxylic acid; AMO, ammonium monooxygenase; PP, pentose phosphate pathway; ED, Entner–Doudoroff pathway. (Created with biorender.com).

are glycolysis or Entner–Doudoroff and direct oxidative pathways and tricarboxylic acid cycle. These pathways also contribute to acidification as organic acids are produced during these processes (Uroz et al. 2022). Gluconic acid is produced by glucose dehydrogenase enzyme using PQQ as a cofactor. *Pseudomonas* is observed to express *pqq* gene and *gdh* gene, resulting in gluconic acid production that helps in inorganic phosphate solubilization (Oteino et al. 2015).

CA enzyme is produced by microbes in harsh conditions and low nutrient availability. Like, *Aspergillus fumigatus* and *Aspergillus niger* increase the expression of CA when potassium can be obtained only from available potassium feldspar (Xiao et al. 2012; Sun et al. 2013). CA hydrates CO_2 and forms carbonic acid (H_2CO_3), which results in silicate mineralization (Xiao et al. 2015). *Bacillus mucilaginosus* can effectively mineralize wollastonite by expressing CA genes, and CA gene expression has been determined to be negatively correlated with CO_2 concentration (Xiao et al. 2015). Hence, CA can produce carbonic acid and decrease the pH of surrounding which works effectively in low concentration of CO_2 and in harsh conditions where element mineralization is needed.

b) *Chelation reaction*: The chelation process helps to remove metals and plays a role in weathering in combined effects with organic acids and siderophores (Reichard et al. 2007). Siderophores are the main chelating agents that are related to Fe ions but they can also bind with metals, like Cd, Cu, Pb, and Zn, resulting in the dissolution of these elements by removing them (Kirtzel et al. 2020). *Rhodococcus* from dry serpentine rocks has been shown to secrete siderophores, siderophore-interacting protein, and two non-ribosomal peptide synthase clusters producing siderophore (Khilyas et al. 2021). *Schizophyllum commune* was observed to mineralize black slate by combined effects of organic acids, like ascorbic acid and tartaric acid, siderophores produced in the process were indicated by discoloration of the medium or bleaching of blue chrome azurol S agar (Kirtzel et al. 2020). *Pseudomonas* (*pvdA, pvdC,* and *pchEF*) and *Collimonas* (*mbaA*) produce chelating agents for weathering of chromazurol S and hematite (Parrello et al. 2016; Picard et al. 2022). Weathering of asbestos is possible by siderophore-producing microorganisms because it is rich in iron making it toxic. *Pseudomonas* also produces pyoverdine and pyochelin, the two types of siderophores that can extract iron from the surface (David et al. 2020). Desferrioxamine B is the siderophore that results in chromium mobilization; it can also form complexes with uranium and other rare earth elements (Kraemer et al. 2019).

c) *Redox reaction*: Metals act as terminal electron acceptor in anaerobic conditions. Oxidation and reduction reactions occur for the metals like iron that cause its dissolution (Štyriaková et al. 2012; Ribeiro et al. 2020), and the presence of such (like iron-oxidizing) bacteria determines the acidity and weathering process (Borin et al. 2010). In the sulfur and iron-rich area, acidophilic bacteria, such as *Acidithiobacillus ferrooxidans, A. thiooxidans, Leptospirillium, Sulfobacillus,* and *Thiobacillus,* are present that can oxidize sulfur and iron and increase their dissolution (Blackmore et al. 2018). A study on ultramafic rocks shows that oxides of Fe, Mn, Ni, and Co solubilization increase by bacterial reduction (Quantin et al. 2002). Ammonia-oxidizing archaea on monuments have an ammonia oxygenase gene (*amoA*) that helps in nitrogen transformation, and iron-reducing bacteria can transform iron oxides (F. Y. Bai et al. 2021).

d) *Mobility and biofilm formation*: Presence of fimbrial adhesin protein has been shown to affect the release of elements from the biotite by *Pseudomonas azotoformans* F77 as it helps in attaching bacterial cells to the surface (Wang et al. 2020b). EPSs produced by microbes help in the weathering process by applying mechanical pressure on the surface. *Schizophyllum commune* produces schizophyllan, EPS that forms a connection with the surface, covers hyphae and forms cracks in black slate (Kirtzel et al. 2020). EPS also acts as an osmoprotectant and nutrient reservoir and retains water as observed for endolithic microorganisms living in dolomite rocks (Horath and Bachofen 2009). Genes like Epsc, flagellin, FliS, FlaG and FlgN are responsible for cell-rock attachment and their aggregation (Wu et al. 2022). For mobility, *Rhizobium pusense* was studied and the *fliF* gene expresses flagellar protein responsible for biotite weathering (Wang et al. 2020a).

11.5 Microbial Interactions in Bio-Weathering

Microbes are present everywhere, and weathering of rocks is possible in every extreme condition, including the surface of rocks (epilithic), in pits and cracks (chasmolithic), or inside the interstitial spaces of rocks (endolithic) (De Belie et al. 2014). Microbes interact with the rock surface and among themselves enhancing the weathering process. Microbes can form microbial films and mats that enable cells to tolerate stress conditions. Some microbes have fimbriae helping in motility and reaching the depth of rocks (De Belie et al. 2014), and others have a symbiotic relationship like cyanobacteria (mosses and lichens), plant-microbe interaction that increases the weathering rate (F. Y. Bai et al. 2021). This section discusses two main microbial interactions: biofilms and symbiotic interactions.

11.5.1 Biofilms in Microbial Weathering

Biofilm is the association of microbial cells on the surface that is enclosed in the EPS matrix (Rehman et al. 2020). EPS has various functions protecting the microbes from environmental conditions, including erosion, antibiotic presence, and desiccation serving as the reservoir for energy storage and nutrients. Microbes can form permanent association with the rock surface, assisted by EPSs and flagella. EPS is involved in the attachment of cells to the surface and helps in the bioleaching process. *Thiobacillus ferrooxidans* are acidophilic and iron(II) ion oxidizing bacteria, found on the pyrite and sulfur attached by EPS (Gehrke et al. 1998). The main component of EPS is lipopolysaccharide, needed for the attachment initially onto the solid substrate. The type of surface on which biofilm formation takes place influences the synthesis of EPS-related genes. Sulfur-grown cells do not attach to the charged particles, like pyrite, whereas such substances need positively charged exopolymer-complexed Fe(III) ions. For corrosion, it is necessary to keep the iron in oxidized state, and this role is primarily played by the microbes. Therefore, iron-oxidizing bacteria can regenerate iron ions and can be used for their own growth (Gehrke et al. 1998). *In-situ* experiments have shown that biofilm production is important for mineralization, which is enhanced by bacterial acid production with mechanical processes (Papida et al. 2000). The biofilms tend to absorb water which results in the swelling and shrinking of EPS layer causing cracks in the stone (McNamara and Mitchell 2005). For example, quartz gets mineralized at higher pH (around 9) (Brehm et al. 2005). An environment with such conditions is rare; the presence of photosynthetic bacteria that consumes CO_2 increases the pH which can lead to quartz mineralization. The process is enhanced by biofilm formation because it creates a microenvironment where pH can go up to 9 (Brehm et al. 2005). Biofilms generally develop in the monsoon season so that it doesn't undergo desiccation. Study shows that biofilms consisting of CO_2, Al_2O_3, and FeO develop on lateritic bauxite before and after monsoon season, resulting in the weathering of surface (Boeva et al. 2021).

11.5.2 Symbiotic Interactions in Microbial Weathering

In-situ studies have shown that mixed microbial population and released EPS enhance the weathering process of rocks through enhanced porosity (Gómez-Alarcón et al. 1995b; Papida et al. 2000). Engravings on glass and quartz have been reported by the mixed cultures of diatoms and heterotrophic bacteria. The heterotrophic bacteria help in ion transport and mineralization process (Brehm et al. 2005). The laboratory experiments with *Knufia* and *Nostoc* have shown that mineral dissolution was significantly enhanced when mixed cultures of *Nostoc* and *Knufia* cells were used as compared to *Knufia* cells alone. In this association, *Nostoc* cells consume CO_2 and fungi produce organic acids to create acidic conditions facilitating the weathering process (Seiffert et al. 2014).

Plant–microbe interaction establishing the symbiotic relationship plays a major role in soil formation by breaking the rocks. *In-situ* experiments with ectomycorrhizal fungi grown with scots pine seeds on the biotite surface degraded it by penetrating up to 300 nm surface by extension of lamella (Bonneville et al. 2009). Plant roots get surrounded by symbiotically associated microbes and cause weathering of goethite and hematite (Boeva et al. 2021). Mosses show a symbiotic relationship with fungi and bacteria, which become the crucial space for these weathering microbes. Mosses have shown symbiotic relationship with fungi belonging to Ascomycota and Basidiomycota. The symbiotic interaction between mosses and bacteria drives weathering process as observed on the monolithic statue of Buddha in China (F. Y. Bai et al. 2021). Microbes present with mosses supply nutrients to it from the barren rocks. Bacterial communities of *Granulicella, Pseudarthrobacter, Rhizomicrobium, Rhodanobacter,* and *Paraburkholderia* are generally present which are plant-promoting and, rhizosphere bacteria; *Dyella* and *Ralstonia* are also found in mosses that show highest weathering activity. The microbial communities get influenced by growth rate and senescence of mosses as the latter supplies plant-derived amino acids as carbon, nitrogen, and sulfur sources. These microbes regulate the nitrogen cycle and, in turn, help in the growth of mosses, which supports the fact that ammonia-oxidizing bacteria are abundantly found in these habitats. Recent studies have shown that mosses inhabiting bare rocks can increase the weathering process by building favorable environment for better growth of microbes, by protecting them from UV and providing enough water to microbial cells in the harsh rock environment (Bai et al. 2021; Wang et al. 2021a).

Lichen is the association between fungi and algae, the latter mostly being cyanobacteria (McNamara and Mitchell 2005). Cyanobacteria being phototrophic provide nutrients to the fungi, whereas fungi protect the algae and helps in absorbing water from the surroundings. The N_2-fixing bacteria (cyanobacteria) enhance the weathering process by increasing the production of organic acids. Bio-weathering by lichen occurs during the adhesion of cells and hyphal penetration, leading to the fragmentation of rocks. It also involves dissolution process by organic acid production and formation of new minerals (Adamo and Violante 2000). Different lichens have been studied for the way they penetrate in surface of rocks, by forming pits, ridges, furrows, and splits, depending on the type of surface they inhabit. *Perusaria corallina* on labradorite grains show lamellar intergrowth by forming a furrow pattern (Wilson et al. 1981), *Xanthoria ectaneoides* inhabit calcium-rich plagioclase feldspar and *Stereocaulon vesuvianum* on leucite mineral, both form deep etch pits (Adamo et al. 1993). The splits and twists breaking patterns are shown by *Lecanora atra* on serpentinite rocks (Wilson et al. 1981). Also, the bacteria attached to the mycelial mats induce biofilm formation which, in turn, enhances the solubilization process. These heterotrophic nitrogen fixers present with green algae in lichens, such as *Placopsis gelidia*, increase the weathering process (Seneviratne and Indrasena 2006).

11.6 Importance of Bio-Weathering

Bio-weathering is a key process for the evolution of Earth's surface and has great impact on human life by enhancing the water quality, development of soil, in agriculture as well as in the preservation of valuable statues, precious paintings, and monuments. Microbes can forage various essential elements which are poorly bioavailable (Fe and P) in the environment (Ehrlich 1996). Microbial weathering has the potential to solve issues of contaminated soils and can increase the crop production and soil fertility, especially in arid lands where production is very less, thus have applications to provide environment-friendly practices, agriculture, and industries (Mapelli et al. 2012).

11.6.1 Soil Fertility and Plant Growth Promotion

The increased cost of commercial fertilizers is a serious issue faced by small farmers and is a threat to food security in the food production industry and also leads to environmental pollution. The world consumption of fertilizers was about 188 million tons from 2018 to 2019 (Randive et al. 2021). Several alternatives have been adopted to increase soil fertility, including the use of new technological fertilizers, exploration of beneficial plant–microbe interaction, and the development of genetically modified crops (Campbell et al. 2014). The use of rock powders is also one of the alternatives to manufacture fertilizers which have shown promising results in the recovery of degraded lands, rejuvenation of minerals in the soil and also increase in the availability of essential nutrients. Crushed rocks are comparatively cheaper than conventional fertilizers, and they are less susceptible to nutrient leaching (Fyfe et al. 2006). Microbial weathering organisms cause mineral dissolution, hydration, disaggregation, and formation of secondary minerals directly or indirectly (Jackson 2015).

Rhizosphere bacteria have the ability to directly solubilize minerals from the soil, thus increasing plant growth and its fitness (Figure 11.2). Diverse energy-consuming process results in the release of metabolic by-products that leads to mineral solubilization (Banfield

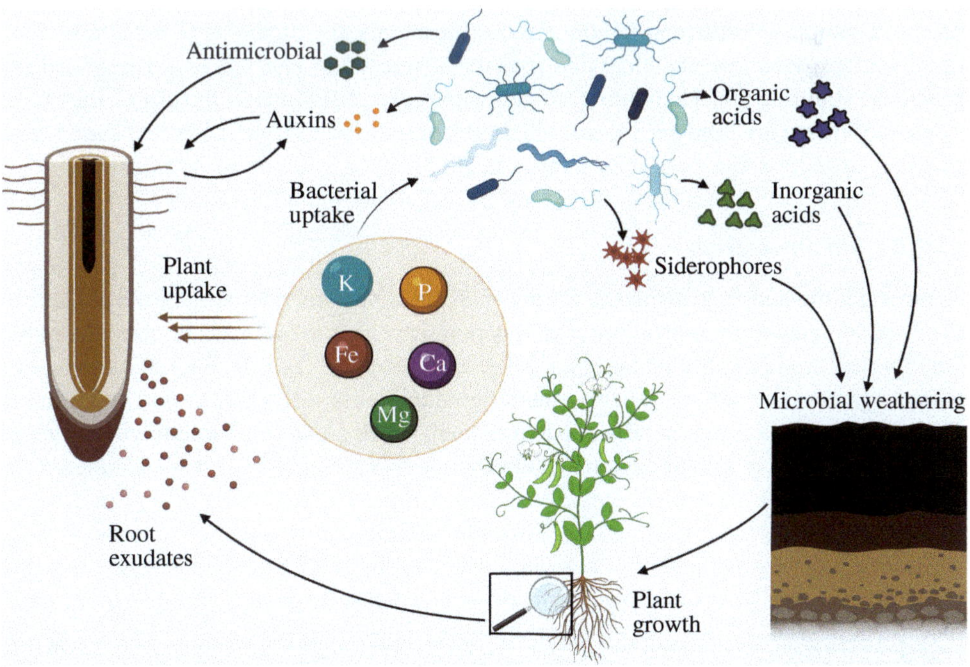

Figure 11.2 Mechanism of weathering of rocks and minerals by soil bacteria enhancing plant growth (created with biorender.com).

et al. 1999). Microbial weathering is induced by various mechanisms, like change in pH surrounding the mineral particles and proton-promoted dissolution, chelation of elements that are present in the mineral matrix and redox reactions (Samuels et al. 2020). N, P, and K are the most important nutrients for increased metabolism, growth, and development of plants (Amtmann and Armengaud 2009). Phosphorus (P) is the second most important nutrient which is required for maximum agricultural production. In the soil, P is present in both organic and inorganic forms but only a very small percentage of soluble P is available for plant uptake (Syers et al. 2008). PSBs include strains from the genera *Bacillus, Rhizobium, Pantoea, Methylobacterium, Bradyrhizobium, Enterobacter, Azospirillum, Gluconacetobacter, Paenibacillus, Microbacterium, Pseudomonas, Burkholderia,* and *Streptomyces* which have the capacity of solubilizing inorganic P (Ribeiro et al. 2020).

Plant-growth-promoting organisms and PSB bacteria act synergistically to increase the plant nutrition and enhance plant fitness (Yu et al. 2012). For inoculation of legumes, N-fixing symbiotic bacteria are combined with PSB, which is the most frequently used strategy. The co-inoculation of *Pseudomonas* PSB strains with *Bradyrhizobium* on soybean has been shown to increase the nodulation, grain yield, plant growth, and phosphorus uptake by plants and reduce the chemical P fertilizer input (Argaw 2012). Potassium is the most abundant essential element present in soil but only 0.1–2% of the potassium is available for plant uptake. K is directly available to plants as it is water soluble, K is exchangeable on the surface of organic matter and clay minerals, and non-exchangeable potassium is mainly present on the silicate surface and released very slowly. While insoluble rocks and minerals have more than 98% of potassium but they are not available for immediate use. Their availability is possible by a very slow and long-term soil weathering process (Reitemeier 1951). Potash (potassium-rich salt) produced from the mined salts is a great source of soluble K-fertilizers. As compared to the chemically produced N and P fertilizers, these sources need less chemical and industrial processing and are promising fertilizer products (Manning and Theodoro 2020). Silicate rocks are used as K fertilizers, they have low solubility but are used with K-solubilizing bacteria which very efficiently release K and provide nutrients to the plants (Etesami et al. 2017).

11.6.2 Bioremediation

Toxic pollutants are destroying the natural habitat of various microorganisms present in the soil; bioremediation is an alternative way that scavenges these toxic pollutants from the environment using natural biological activity. Bioremediation may be defined as the use of living organisms to degrade environmental contaminants into non- or less-toxic forms. Fungi and bacteria are generally used to detoxify or degrade the hazardous substances which are harmful to human health and environment (Varjani 2017). Bioremediation techniques are comparatively more economical as compared to the thermal and physiochemical remediation (Cho et al. 2000). Multiple microorganisms can be indigenous to contaminated sites, or they may be isolated from a different site and then brought to the contaminated site. These then process the contaminant compounds through their metabolic processes.

Removal of metals from industrial wastes and their by-products, metal-bearing minerals, and low grade ores takes place by solubilization, which is relevant to the bioremediation of soil wastes and soil matrices, and metal recovery and recycling (Burgstaller and Schinner 1993). Soil bioremediation is a combination of several factors, including the site location, type of contamination, cost-effectiveness, expected efficiency, duration of the remediation process, and its public acceptability. Microorganisms and plants break down the hydrocarbons present in soil by various methods into greener products, or they are converted by other microbial species into inorganic compounds with the release of water and CO_2 (Banerjee et al. 2016). In a study conducted by Elenga-Wilson (2021), 60 strains were isolated that were capable of degrading different hydrocarbons by the release of biosurfactants (Elenga-Wilson et al. 2021). The isolates were screened and used for biodegradation. It was observed that the consortium of *Bacillus* species was capable of decreasing the hydrocarbon content (Elenga-Wilson et al. 2021). Bioaugmentation and biostimulation are the most commonly used *in-situ* bioremediation techniques. Bioaugmentation is the process of addition of microorganisms to an environment to enhance the biological activity (remediation of soil), whereas biostimulation involves modification in the surrounding environment of microbes to stimulate bacteria that are capable of bioremediation. Among *ex-situ* systems of bioremediation, bioreactors are the most common ones. *Ex-situ* has advantage that it allows the treatment of contaminants in a controlled way, but it is costlier than the *in-situ* processes as it needs additional excavation and also includes transport expenses (Azubuike et al. 2016). Thus, bioremediation effectiveness can be increased by the combination of multiple techniques and by reducing the individual disadvantages of each (da Silva et al. 2020).

11.6.3 Biorestoration

Microbial weathering has beneficial effects in terms of preserving the monuments and artworks, one among them being biorestoration. It's a new approach that has been recently studied and applied. It involves the use of beneficial microorganisms which are safe for human and environmental health called "restorative agents" (Soffritti et al. 2019). The enzymatic and metabolic activities of these microorganisms help in the recovery and conservation of artworks and monuments. Microorganisms isolated from the natural habitats and environmental matrices are screened and selected for their ability to biocalcify stoneworks and monuments, removal of deposits of inorganic and organic substances, consolidation of artworks and act as a biological control (as potential decontaminating agents) (Fernandes 2006). Sulfate and nitrate-reducing bacteria, like *Desulfuromonas sulfuricans, D. vulgaris, Pseudomonas stutzeri,* and other enzymatically active bacteria are used for the removal of organic and inorganic residues. Biological controls, like Green biocides, biopesticides, biofungicides, and *Bacillus*-based bioactive compounds, are used to eradicate or prevent biodeteriogens contamination. Biocalcifying bacteria, like *Bacillus pastuerii, B. subtilis, Acinetobacter, Pseudomonas spp.*, and *Myxococcus xanthus,* are used for the stone consolidation and reinforcement (Soffritti et al. 2019). In a case study entitled, "The Etruscan tomb in Tarquinia: contribution of microorganisms to deterioration and restoration", five strains – DFVB 6 (*Staphylococcus haemoliticus*), TPBF 11 (*Cellulomonas cellulans*), DPBS 7 (*Actinomycetaceae*), TPBF 10, and TPID 1 (*Cellulosimicrobium cellulans*) – were shown to exhibit the ability to solubilize carbonates. The tomb was excavated in "macco," a calcareous rock, thus this ability to solubilize calcium carbonate is very interesting in relation to this environment. The strains were considered as potential degrader of unwanted depositions (Sprocati et al. 2008). Another example is of *Pseudomonas stutzeri* strain A29, which has biocleaning potential and was used to clean nearly 400 years old wall paintings at the Pisa Cathedral Cupola and the Vatican Museums in Italy. For the cleaning and removal of black crust from the Cathedral of Florence, microbes were the superior choice as compared to chemical and laser methods (Bosch-Roig et al. 2015). Microbes have also shown promising results in graffiti removal (Cattò et al. 2021).

11.7 Omics to Explore Microbial Weathering of Rocks

Conventionally for studying the texture, microstructure and geochemical composition of rocks, the techniques like scanning electron microscopy and energy-dispersive X-ray spectrometry (EDS) are used. Culture-based studies have identified bacteria, including *Pseudomonas, Actinobacteria, Bacillus, Staphylococcus,* and *Salmonella*, from different stone heritage (McNamara and Mitchell 2005). However, though microorganisms are present everywhere, only less than 2% of bacteria can be cultured in the laboratories (Wade 2002). Also, weathering by microbes occurs in extreme conditions, like desserts, mines, and artic regions (Cooke and Warren 1993; Daghino et al. 2008; Mapelli et al. 2011), and it is not easy to maintain or replicate these conditions in the lab. This is where omics (Table 11.4) based approaches, like 16S rRNA, whole genome sequencing, transcriptomics, and proteomics (Beata 2020), can bridge the gap in the knowledge of microbes and their role in weathering of rocks in different environmental conditions. These techniques enable an in-depth understanding of the dominant microbes and genes involved, in biorestoration, bio-weathering, bioremediation, deterioration, and corrosion in all natural conditions (Schröer et al. 2021; Picard et al. 2022). The information derived from these current approaches are summarized:

1) *Knowledge of Microbial diversity*: Omics techniques help to study the distribution of microbes present in the rocks in relation to depth. A study (Ke et al. 2021), using MiSeq high-throughput sequencing platform analyzed 16S rRNA sequences, reported the Proteobacteria, Actinobacteria, Chloroflexi, Bacteroidetes, Gemmatimonadetes, Planctomycetes, Acidobacteria, and Sachharibacteria as dominant phyla from the bauxite dissolution region. Serpentine rock has been studied to find the diversity of microbes present through 16S rRNA sequencing, showed that there was no difference at the phyla level with increase in the depthness of the rock, but bioweathering was enhanced by orders Actinomycetales, Gaiellales, Solirubrobacterales, Rhizobiales, and Burkholderiales showing positive correlation with depth (Khilyas et al. 2019).
2) *Identification of microbial interactions*: Omics approach can help to study microbe–plant interactions and microbes present in the vicinity of plants. Mosses and other microbes living in symbiotic associations have also been examined using high-throughput sequencing and quantitative PCR techniques, for analyzing the metabolic pathways and related genes for microbial weathering (F.-Y. Bai et al. 2021). This approach sheds light on microbes involved in the weathering of soil and thereby increasing soil fertility. Besides, the microbes involved, the omics approach can map genetic structure to the functions using metatranscriptomics and metabolomics (Bell et al. 2014).

3) *Functional and metabolic analysis*: A combination of genomics and transcriptomics approaches has been used by Wang et al. (2021b) in *Pseudomonas azotoformans* F77 strain, to identify the weathering gene clusters e.g. gene in gluconic acid metabolism, flagellar assembly and pilus biosynthesis, and acid tolerance-related gene clusters (Wang et al. 2021b). Kyoto Encyclopedia of Genes and Genomes (KEGG) pathways of flagellar assembly, pilus biosynthesis, and gluconic acid metabolism are active in the presence of biotite. Most recently, a study based on whole genome sequencing for the rock-dissolving bacterium was conducted followed by genome-wide transcriptome analysis, and for validation, RT-qPCR was conducted (Wu et al. 2022) to improve external soil spray seeding technology for the silicate rock mining area. They found that *Pseudomonas sp.* NLX-4 strain could efficiently dissolve silicate rocks. Further, RNA-sequencing techniques helped in the identification of regulatory genes involved in siderophore transport, amino acid synthesis, and organic acid metabolism.
4) *Comparative studies*: High-throughput techniques can be used for comparative studies for understanding the distribution of microbial communities at various locations where factors, like sunlight, temperature, moisture, and salt concentration, vary. Wu et al. (2021) showed the difference in microbial diversity through 16S rRNA sequencing when samples were collected from inside and outside the cave. Alpha diversity was significantly higher for outside samples as compared to the inside samples (Wu et al. 2021).
5) *Restoration of buildings and monuments*: Microbial weathering causes deterioration, discoloration, and dissolution of building stones and monuments (Gómez-Alarcón et al. 1995a; Sterflinger and Piñar 2013). The microbes resulting in the degradation remain largely unexplored; however, with the metagenomics approach, these microbes can also be identified. Biocolonization of microbes on the monuments can be studied in different seasons and to study the temporal variations in microbial composition to protect heritage (Schröer et al. 2021).

The limitation of 16S rRNA sequencing-based studies is that it only identifies the diversity of microbes present but lacks information about the genes and metabolic pathways involved as only 16S rRNA regions are amplified (Johnson et al. 2019). To know the presence of specific genes specific primers have to be used to amplify the gene of interest, and for functional studies, recombinant DNA technology is applied (Wu et al. 2017; Khilyas et al. 2019). The better approach is to use whole metagenome sequencing, in which all genomes are extracted from the sample and sequenced. This approach is more reliable because novel species can also be identified and as the entire genome is sequenced, all the genes and proteins can be predicted and can be identified by homology search.

11.8 Conclusion and Future Directions

Research in the area of bio-weathering can provide quantitative predictions and mechanistic explanations of best practices for promoting agriculture. By determining the nutrient release rate and creating rock preparations with appropriate microorganisms, this system can be effectively utilized in agricultural and forested lands. Such practices will bring benefits to the environment and ensure productivity, security, and sustainability.

Microbial weathering or bio-weathering is the process by which bacteria, fungi, lichens, and other microorganisms contribute to the weathering of rocks and minerals. Studying microbial weathering offers a comprehensive understanding of the geochemical processes involved in rock and mineral dissolution and formation of soil. This process is driven by both chemical and biological reactions and has far-reaching effects on the overall functioning and diversity of microbes in soil systems. Factors, such as temperature, climate, soil pH, soil solution composition, plant activity, changes in rhizosphere pH and redox, and the presence of various organic acids, play a role in determining the rate of rock and mineral dissolution by microorganisms. Additionally, the weathering of silicate, phosphate, and other minerals can be accelerated in the rhizosphere, leading to an increase in the availability of nutrients for plant uptake. This is due to the increased concentration of H-ions and the exudation of complex organic acids, which play a key role in mineral dissolution and nutrient uptake.

Though the field of geomicrobiology has seen significant progress in recent years, there remains a shortage of experimental evidence connecting the various mechanisms as it's difficult to replicate the natural conditions due to the interplay between biotic and abiotic factors. Hence, the integration of omics technologies with conventional tools and methods in geomicrobiology can provide a solution. Besides getting information on microbial diversity through 16S-based studies, proteomics, and metabolomics, can give an insight into how microbial communities interact with their geological surroundings and what pathways are involved in the bio-weathering of rocks and minerals. Additionally, metatranscriptomics can allow us to identify the specific genes that are hyperactive in microbial weathering, which can guide future studies through whole metagenome analysis.

References

Adamo, P., Marchetiello, A., and Violante, P. (1993). The weathering of Mafic Rocks by lichens. *The Lichenologist* 285–297. https://doi.org/10.1006/lich.1993.1033.

Adamo, P. and Violante, P. (2000). Weathering of rocks and neogenesis of minerals associated with lichen activity. *Applied Clay Science* 16 (5–6): 229–256. https://doi.org/10.1016/S0169-1317(99)00056-3.

Adhikary, H., Sanghavi, P.B., Macwan, S.R. et al. (2014). Artificial citrate operon confers mineral phosphate solubilization ability to diverse fluorescent pseudomonads. *PLoS ONE* 9 (9): https://doi.org/10.1371/journal.pone.0107554.

Ahmed, E. and Holmström, S.J.M. (2015). Microbe-mineral interactions: the impact of surface attachment on mineral weathering and element selectivity by microorganisms. *Chemical Geology* 403: 13–23. https://doi.org/10.1016/j.chemgeo.2015.03.009.

Alahari, A. and Apte, S.K. (2004). A novel potassium deficiency-induced stimulon in Anabaena torulosa. *Journal of Biosciences* 29 (2): 153–161. https://doi.org/10.1007/BF02703413.

Amtmann, A. and Armengaud, P. (2009). Effects of N, P, K and S on metabolism: new knowledge gained from multi-level analysis. *Current Opinion in Plant Biology* 12 (3): 275–283. https://doi.org/10.1016/J.PBI.2009.04.014.

Argaw, A. (2012). Evaluation of co-inoculation of *Bradyrhizobium japonicum* and phosphate solubilizing *Pseudomonas* spp. effect on soybean (Glycine max *L. Merr.*) in assossa area. *Journal of Agricultural Science and Technology* 14 (1): 213–224.

Ariño, X. Ortega-Calvo, J.J., Gomez-Bolea, A., Saiz-Jimenez, C. (1995) Lichen colonization of the Roman pavement at Baelo Claudia (Cadiz, Spain): biodeterioration vs. bioprotection.

Ascasoa, C., Wierzchosb, J., and Castelloa, R. (1998). Study of the biogenic weathering of calcareous litharenite stones caused by lichen and endolithic microorganisms. *International Biodeterioration and Biodegradation* 42 (1): 29–38. https://doi.org/10.1016/S0964-8305(98)00043-2.

Azubuike, C.C., Chikere, C.B., and Okpokwasili, G.C. (2016). Bioremediation techniques–classification based on site of application: principles, advantages, limitations and prospects. *World Journal of Microbiology and Biotechnology* 32 (11): 1–18. https://doi.org/10.1007/S11274-016-2137-X.

Bai, F.-Y., Chen, X.-P., Huang, J.-Z. et al. (2021). Microbial biofilms on a giant monolithic statue of Buddha: the symbiosis of microorganisms and mosses and implications for bioweathering. *International Biodeterioration and Biodegradation* 156: 105106. https://doi.org/10.1016/j.ibiod.2020.105106.

Banerjee, A., Roy, A., Dutta, S., and Mondal, S. (2016). Bioremediation of hydrocarbon-a review. *International Journal of Advanced Research* 4 (6): 1303–1313. https://doi.org/10.21474/IJAR01.

Banfield, J.F., Barker, W.W., Welch, S.A., and Taunton, A. (1999). Biological impact on mineral dissolution: application of the lichen model to understanding mineral weathering in the rhizosphere. *Proceedings of the National Academy of Sciences* 96 (7): 3404–3411. https://doi.org/10.1073/PNAS.96.7.3404.

Baumann, K., Jung, P., Samolov, E. et al. (2018). Biological soil crusts along a climatic gradient in Chile: richness and imprints of phototrophic microorganisms in phosphorus biogeochemical cycling. *Soil Biology and Biochemistry* 127: 286–300. https://doi.org/10.1016/J.SOILBIO.2018.09.035.

Beata, G. (2020). The use of -omics tools for assessing biodeterioration of cultural heritage: a review. *Journal of Cultural Heritage* 45: 351–361. https://doi.org/10.1016/j.culher.2020.03.006.

Bell, T.H., Joly, S., Pitre, F.E., and Yergeau, E. (2014). Increasing phytoremediation efficiency and reliability using novel omics approaches. *Trends in Biotechnology* 32 (5): 271–280. https://doi.org/10.1016/j.tibtech.2014.02.008.

Belov, A.A., Cheptsov, V.S., and Vorobyova, E.A. (2018). Soil bacterial communities of Sahara and Gibson deserts: physiological and taxonomical characteristics. *AIMS Microbiology* 4 (4): 685. https://doi.org/10.3934/MICROBIOL.2018.4.685.

Bennett, P.C., Hiebert, F.K., and Choi, W.J. (1996). Microbial colonization and weathering of silicates in a petroleum-contaminated groundwater. *Chemical Geology* 132 (1–4): 45–53. https://doi.org/10.1016/s0009-2541(96)00040-x.

Berthelin, J. (1988). Microbial weathering processes in natural environments. *Physical and Chemical Weathering in Geochemical Cycles* 33–59. https://doi.org/10.1007/978-94-009-3071-1_3.

Birch, L. and Bachofen, R. (1990). Complexing agents from microorganisms. *Experientia* 46 (8): 827–834. https://doi.org/10.1007/BF01935533.

Blackmore, S., Vriens, B., Sorensen, M. et al. (2018). Microbial and geochemical controls on waste rock weathering and drainage quality. *Science of the Total Environment* 640–641: 1004–1014. doi: 10.1016/j.scitotenv.2018.05.374.

Boeva, N., Bortnikov, N., Slukin, A. et al. (2021). Biofilms and biominerals in the lateritic weathering crust as exemplified by the central bauxite deposit (Siberian platform, Russia). *Minerals* 11 (11): https://doi.org/10.3390/min11111184.

Bonneville, S., Smits, M.M., Brown, A. et al. (2009). Plant-driven fungal weathering: early stages of mineral alteration at the nanometer scale. *Geology* 37 (7): 615–618. https://doi.org/10.1130/G25699A.1.

Borin, S., Ventura, S., Tambone, F. et al. (2010). Rock weathering creates oases of life in a High Arctic desert. *Environmental Microbiology* 12 (2): 293–303. https://doi.org/10.1111/j.1462-2920.2009.02059.x.

Bosch-Roig, P., Lustrato, G., Zanardini, E., and Ranalli, G. (2015). Biocleaning of cultural heritage stone surfaces and frescoes: which delivery system can be the most appropriate? *Annals of Microbiology* 65 (3): 1227–1241. https://doi.org/10.1007/S13213-014-0938-4/TABLES/2.

Brantley, S.L. (2008). Kinetics of mineral dissolution. In: *Kinetics of Water-Rock Interaction*, 151–210. New York: Springer https://doi.org/10.1007/978-0-387-73563-4_5/COVER.

Brehm, U., Gorbushina, A., and Mottershead, D. (2005). The role of microorganisms and biofilms in the breakdown and dissolution of quartz and glass. *Palaeogeography, Palaeoclimatology, Palaeoecology* 219 (1–2): 117–129. https://doi.org/10.1016/j.palaeo.2004.10.017.

Brucker, E., Kernchen, S., and Spohn, M. (2020). Release of phosphorus and silicon from minerals by soil microorganisms depends on the availability of organic carbon. *Soil Biology and Biochemistry* 143: 107737. https://doi.org/10.1016/J.SOILBIO.2020.107737.

Burgstaller, W. and Schinner, F. (1993). Leaching of metals with fungi. *Journal of Biotechnology* 27 (2): 91–116. https://doi.org/10.1016/0168-1656(93)90101-R.

Buss, H.L., Lüttge, A., and Brantley, S.L. (2007). Etch pit formation on iron silicate surfaces during siderophore-promoted dissolution. *Chemical Geology* 240 (3–4): 326–342. https://doi.org/10.1016/j.chemgeo.2007.03.003.

Cabala, J.E., Teper, E., Teper, L.E. et al. (2004). Mineral composition in rhizosphere of plants grown in the vicinity of a Zn-Pb ore flotation tailings pond. Preliminary study. *Acta Biologica Cracoviensia Series Botanica* 46: 65.

Campbell, B.M., Thornton, P., Zougmoré, R. et al. (2014). Sustainable intensification: what is its role in climate smart agriculture? *Current Opinion in Environmental Sustainability* 8: 39–43. https://doi.org/10.1016/J.COSUST.2014.07.002.

Cattò, C., Sanmartín, P., Gulotta, D. et al. (2021). Bioremoval of graffiti using novel commercial strains of bacteria. *Science of the Total Environment* 756: https://doi.org/10.1016/J.SCITOTENV.2020.144075.

Chen, S., Lian, B., and Liu, C. (2008). Effect of *Bacillus mucilaginosus* on weathering of phosphorite and a preliminary analysis of bacterial proteins. *Chinese Journal of Geochemistry* 27 (2): 209–216. https://doi.org/10.1007/S11631-008-0209-9.

Cho, Y.-G., Rhee, S.-K., and Lee, S.-T. (2000). Effect of soil moisture on bioremediation of chlorophenol-contaminated soil. *Biotechnology Letters* 22: 915–919.

Cockell, C.S., Pybus, D., Olsson-Francis, K. et al. (2011). Molecular characterization and geological microenvironment of a microbial community inhabiting weathered receding shale cliffs. *Microbial Ecology* 61 (1): 166–181. https://doi.org/10.1007/s00248-010-9730-6.

Cohen, Y. and President (2007). Microbial ecology and environmental biotechnology. *The ISME Journal* 1 (1): 4–8. https://doi.org/10.1038/ismej.2007.7.

Coleine, C., Stajich, J.E., de los Ríos, A., and Selbmann, L. (2020). Beyond the extremes: rocks as ultimate refuge for fungi in drylands. *Mycologia* 113 (1): 108–133. https://doi.org/10.1080/00275514.2020.1816761.

Cooke, R.U. and Warren, A. (1993). *Desert Geomorphology*. London: UCL Press.

Crits-Christoph, A., Robinson, C.K., Barnum, T. et al. (2013). Colonization patterns of soil microbial communities in the Atacama Desert. *Microbiome* 1 (1): 1–13. https://doi.org/10.1186/2049-2618-1-28/FIGURES/7.

Daghino, S., Martino, E., Vurro, E. et al. (2008). Bioweathering of chrysotile by fungi isolated in ophiolitic sites. *FEMS Microbiology Letters* 285 (2): 242–249. https://doi.org/10.1111/J.1574-6968.2008.01239.X.

da Silva, I.G.S., de Almeida, F.C.G., da Rocha e Silva, N.M.P. et al. (2020). Soil bioremediation: overview of technologies and trends. *Energies* 13 (18): https://doi.org/10.3390/EN13184664.

David, S.R. and Geoffroy, V.A. (2020). A review of asbestos bioweathering by siderophore-producing *Pseudomonas*: a potential strategy of bioremediation. *Microorganisms* 8: 1870. https://doi.org/10.3390/MICROORGANISMS8121870.

David, S.R., Ihiawakrim, D., Regis, R., and Geoffroy, V.A. (2020). Iron removal from raw asbestos by siderophores-producing *Pseudomonas*. *Journal of Hazardous Materials* 385: https://doi.org/10.1016/j.jhazmat.2019.121563.

De Belie, N., Wang, J., De Muynck, W. et al. (2014). Microbial interactions with mineral building materials. *Kuei Suan Jen Hsueh Pao/Journal of the Chinese Ceramic Society* 42 (5): 563–567. https://doi.org/10.7521/j.issn.0454-5648.2014.05.01.

Duckworth, O.W., Akafia, M.M., Andrews, M.Y., and Bargar, J.R. (2014). Siderophore-promoted dissolution of chromium from hydroxide minerals. *Environmental Science: Processes and Impacts* 16 (6): 1348–1359. https://doi.org/10.1039/c3em00717k.

Ehrlich, H.L. (1996). How microbes influence mineral growth and dissolution. *Chemical Geology* 132 (1–4): 5–9. https://doi.org/10.1016/S0009-2541(96)00035-6.

Ehrlich, H.L. (1998). Geomicrobiology: its significance for geology. *Earth-Science Reviews* 45 (1–2): 45–60. https://doi.org/10.1016/S0012-8252(98)00034-8.

Eisenberg-Degen, D. and Rosen, S. (2013). Chronological trends in negev rock art: the Har Michia Petroglyphs as a test case. *Arts* 2 (4): 225–252. https://doi.org/10.3390/ARTS2040225.

Elenga-Wilson, P.S., Kayath, C.A., Mokemiabeka, N.S. et al. (2021). Profiling of indigenous biosurfactant-producing *Bacillus* isolates in the bioremediation of soil contaminated by petroleum products and olive oil. *International Journal of Microbiology* https://doi.org/10.1155/2021/9565930.

Enders, M.S., Knickerbocker, C., Titley, S.R., and Southam, G. (2006). The role of bacteria in the supergene environment of the Morenci Porphyry copper deposit, Greenlee County, Arizona. *Economic Geology* 101 (1): 59–70. https://doi.org/10.2113/GSECONGEO.101.1.59.

Etesami, H., Emami, S., and Alikhani, H.A. (2017). Potassium solubilizing bacteria (KSB): mechanisms, promotion of plant growth, and future prospects a review. *Journal of*

Soil Science and Plant Nutrition 17 (4): 897–911. https://doi.org/10.4067/S0718-95162017000400005.

Fernandes, P. (2006). Applied microbiology and biotechnology in the conservation of stone cultural heritage materials. *Applied Microbiology and Biotechnology* 73 (2): 291–296. https://doi.org/10.1007/S00253-006-0599-8.

Finlay, R., Wallander, H., Smits, M. et al. (2009). The role of fungi in biogenic weathering in boreal forest soils. *Fungal Biology Reviews* 23 (4): 101–106. https://doi.org/10.1016/J.FBR.2010.03.002.

Frankel, R.B. and Bazylinski, D.A. (2003). Biologically induced mineralization by bacteria. *Reviews in Mineralogy and Geochemistry* 54 (1): 95–114. https://doi.org/10.2113/0540095.

Frey, B., Rieder, S.R., Brunner, I. et al. (2010). Weathering-associated bacteria from the damma glacier forefield: physiological capabilities and impact on granite dissolution. *Applied and Environmental Microbiology* 76 (14): 4788–4796. https://doi.org/10.1128/AEM.00657-10.

Fyfe, W.S., Leonardos, O.H., and Theodoro, S.H. (2006). Sustainable farming with native rocks: the transition without revolution. *Anais da Academia Brasileira de Ciências* 78 (4): 715–720. https://doi.org/10.1590/S0001-37652006000400007.

Gadd, G. M. (2007) 'Geomycology: biogeochemical transformations of rocks, minerals, metals and radionuclides by fungi, bioweathering and bioremediation', *Mycological Research*, 111(1), pp. 3–49. https://doi.org/10.1016/J.MYCRES.2006.12.001.

Garvie, L.A.J., Knauth, L.P., Bungartz, F. et al. (2008). Life in extreme environments: survival strategy of the endolithic desert lichen *Verrucaria rubrocincta*. *Naturwissenschaften* 95 (8): 705–712. https://doi.org/10.1007/S00114-008-0373-0/FIGURES/5.

Gaylarde, C., Baptista-Neto, J.A., Tabasco-Novelo, C., and Ortega-Morales, O. (2018). Weathering of granitic gneiss: a geochemical and microbiological study in the polluted sub-tropical city of Rio de Janeiro. *Science of the Total Environment* 644: 1641–1647. https://doi.org/10.1016/J.SCITOTENV.2018.07.303.

Gehrke, T., Telegdi, J., Thierry, D., and Sand, W. (1998). Importance of extracellular polymeric substances from Thiobacillus ferrooxidans for bioleaching. *Applied and Environmental Microbiology* 64 (7): 2743–2747. https://doi.org/10.1128/aem.64.7.2743-2747.1998.

Genderjahn, S., Lewin, S., Horn, F. et al. (2021). Living lithic and sublithic bacterial communities in Namibian Drylands', *Microorganisms 2021, Vol. 9, Page 235*. Multidisciplinary Digital Publishing Institute, 9(2), p. 235. *Microorganisms* 9: 235. https://doi.org/10.3390/MICROORGANISMS9020235.

Ghiorse, W.C. and Ehrlich, H.L. (1992). Microbial biomineralization of iron and manganese. *Catena* 21: 75–99.

Gómez-Alarcón, G., Cilleros, B., Flores, M., and Lorenzo, J. (1995a). Microbial communities and alteration processes in monuments at Alcala de Henares, Spain. *Science of the Total Environment* 167 (1–3): 231–239. https://doi.org/10.1016/0048-9697(95)04584-N.

Gómez-Alarcón, G., Muñoz, M., Ariño, X., and Ortega-Calvo, J.J. (1995b). Microbial communities in weathered sandstones: the case of Carrascosa del Campo church, Spain. *Science of the Total Environment* 167 (1–3): 249–254. https://doi.org/10.1016/0048-9697(95)04586-P.

Gorbushina, A.A. and Broughton, W.J. (2009). Microbiology of the atmosphere-rock interface: how biological interactions and physical stresses modulate a sophisticated microbial ecosystem. *Annual Review of Microbiology* 63: 431–450. https://doi.org/10.1146/annurev.micro.091208.073349.

Hameeda, B., Reddy, Y.H.K., Rupela, O.P. et al. (2006). Effect of carbon substrates on rock phosphate solubilization by bacteria from composts and macrofauna. *Current Microbiology* 53 (4): 298–302. https://doi.org/10.1007/S00284-006-0004-Y/TABLES/2.

Harley, A.D. and Gilkes, R.J. (2000). Factors influencing the release of plant nutrient elements from silicate rock powders: a geochemical overview. *Nutrient Cycling in Agroecosystems* 56 (1): 11–36. https://doi.org/10.1023/A:1009859309453.

Hernandez, M.E., Kappler, A., and Newman, D.K. (2004). Phenazines and other redox-active antibiotics promote microbial mineral reduction. *Applied and Environmental Microbiology* 70 (2): 921–928. https://doi.org/10.1128/AEM.70.2.921-928.2004.

Horath, T. and Bachofen, R. (2009). Molecular characterization of an endolithic microbial community in dolomite rock in the central Alps (switzerland). *Microbial Ecology* 58 (2): 290–306. https://doi.org/10.1007/s00248-008-9483-7.

Jackson, T.A. (2015). Weathering, secondary mineral genesis, and soil formation caused by lichens and mosses growing on granitic gneiss in a boreal forest environment. *Geoderma* 251–252: 78–91. https://doi.org/10.1016/J.GEODERMA.2015.03.012.

Jacobson, A.D. and Wu, L. (2009). Microbial dissolution of calcite at T = 28 °C and ambient pCO_2. *Geochimica et Cosmochimica Acta* 73 (8): 2314–2331. https://doi.org/10.1016/J.GCA.2009.01.020.

Johnson, J.S., Spakowicz, D.J., Hong, B.Y. et al. (2019). Evaluation of 16S rRNA gene sequencing for species and strain-level microbiome analysis. *Nature Communications* 10 (1): 1–11. https://doi.org/10.1038/s41467-019-13036-1.

Jones, D., Wilson, M.J., and McHardy, W.J. (1981). Lichen weathering of rock-forming minerals: application of scanning electron microscopy and microprobe analysis', *Journal of Microscopy*. *Journal of Microscopy* 124 (1): 95–104. https://doi.org/10.1111/J.1365-2818.1981.TB01309.X.

Jones, D.S., Lapakko, K.A., Wenz, Z.J. et al. (2017). Novel microbial assemblages dominate weathered sulfidebearing rock from copper-nickel deposits in the Duluth Complex, Minnesota, USA. *Applied and Environmental Microbiology* 83 (16): https://doi.org/10.1128/AEM.00909-17.

Jung, P., Baumann, K., Emrich, D. et al. (2020). Lichens bite the dust – a bioweathering scenario in the Atacama desert. *iScience* 23 (11): 101647. https://doi.org/10.1016/J.ISCI.2020.101647.

Kalinowski, B.E., Liermann, L.J., Brantley, S.L. et al. (2000). X-ray photoelectron evidence for bacteria-enhanced dissolution of hornblende. *Geochimica et Cosmochimica Acta* 64 (8): 1331–1343. https://doi.org/10.1016/S0016-7037(99)00371-3.

Ke, W., Zhang, X., Zhu, F. et al. (2021). Appropriate human intervention stimulates the development of microbial communities and soil formation at a long-term weathered bauxite residue disposal area. *Journal of Hazardous Materials* 405: 124689. https://doi.org/10.1016/j.jhazmat.2020.124689.

Khilyas, I.V., Sorokina, A.V., Elistratova, A.A. et al. (2019). Microbial diversity and mineral composition of weathered serpentine rock of the Khalilovsky massif. *PLoS ONE* 14 (12): 1–14. https://doi.org/10.1371/journal.pone.0225929.

Khilyas, I.V., Sorokina, A.V., Markelova, M.I. et al. (2021). Genomic and phenotypic analysis of siderophore-producing Rhodococcus qingshengii strain S10 isolated from an arid weathered serpentine rock environment. *Archives of Microbiology* 203 (2): 855–860. https://doi.org/10.1007/s00203-020-02057-w.

Kim, K.Y., Jordan, D., and Krishnan, H.B. (1998). Expression of genes from *Rahnella aquatilis* that are necessary for mineral phosphate solubilization in *Escherichia coli*. *FEMS Microbiology Letters* 159 (1): 121–127. https://doi.org/10.1111/J.1574-6968.1998.TB12850.X.

Kirtzel, J., Ueberschaar, N., Deckert-Gaudig, T. et al. (2020). Organic acids, siderophores, enzymes and mechanical pressure for black slate bioweathering with the basidiomycete Schizophyllum commune. *Environmental Microbiology* 22 (4): 1535–1546. https://doi.org/10.1111/1462-2920.14749.

Koner, S., Tsai, H.C., Chen, J.S. et al. (2022). Exploration of pristine plate-tectonic plains and mining exposure areas for indigenous microbial communities and its impact on the mineral-microbial geochemical weathering process in ultramafic setting. *Environmental Research* 214 (P2): 113802. https://doi.org/10.1016/j.envres.2022.113802.

Kraemer, D., Frei, R., Viehmann, S., and Bau, M. (2019). Mobilization and isotope fractionation of chromium during water-rock interaction in presence of siderophores. *Applied Geochemistry* 102: 44–54. https://doi.org/10.1016/j.apgeochem.2019.01.007.

Leyval, C. and Berthelin, J. (1989). Interactions between *Laccaria laccata*, Agrobacterium radiobacter and beech roots: influence on P, K, Mg, and Fe mobilization from minerals and plant growth. *Plant and Soil* 117 (1): 103–110. https://doi.org/10.1007/BF02206262.

Li, F., Wang, S., He, Q. et al. (2021). Minerals determined a special ecological niche and selectively enriched microbial species from bulk water communities in hot springs. *Microorganisms* 9: 1020. https://doi.org/10.3390/MICROORGANISMS9051020.

Liermann, L.J., Kalinowski, B.E., Brantley, S.L., and Ferry, J.G. (2000). Role of bacterial siderophores in dissolution of hornblende. *Geochimica et Cosmochimica Acta* 64 (4): 587–602. https://doi.org/10.1016/S0016-7037(99)00288-4.

Liu, C., Mou, L., Yi, J. et al. (2019). The eno gene of *Burkholderia cenocepacia* strain 71-2 is involved in phosphate solubilization. *Current Microbiology* 76 (4): 495–502. https://doi.org/10.1007/s00284-019-01642-7.

Liu, J., Hua, Z.S., Chen, L.X. et al. (2014). Correlating microbial diversity patterns with geochemistry in an extreme and heterogeneous environment of mine tailings. *Applied and Environmental Microbiology* 80 (12): 3677–3686. https://doi.org/10.1128/AEM.00294-14/SUPPL_FILE/ZAM999105423SO1.PDF.

Lowenstam, H.A. and Weiner, S. (1989). *On Biomineralization*. Oxford University Press on Demand.

Lu, M., Wang, X., Li, Y. et al. (2022). Soil microbial community structure and environmental effects of serpentine weathering under different vegetative covers in the serpentine mining area of Donghai County, China. *Science of the Total Environment* 835: 155452. https://doi.org/10.1016/J.SCITOTENV.2022.155452.

Mahaney, W.C., Somelar, P., Dirszowsky, R.W. et al. (2016). A microbial link to weathering of postglacial rocks and sediments, mount viso area, western alps, demonstrated through analysis of a soil/paleosol bio/chronosequence. *Journal of Geology* 124 (2): 149–169. https://doi.org/10.1086/684442.

Manning, D.A.C. and Theodoro, S.H. (2020). Enabling food security through use of local rocks and minerals. *The Extractive Industries and Society* 7 (2): 480–487. https://doi.org/10.1016/J.EXIS.2018.11.002.

Mapelli, F., Marasco, R., Rizzi, A. et al. (2011). Bacterial communities involved in soil formation and plant establishment triggered by pyrite bioweathering on arctic moraines. *Microbial Ecology* 61 (2): 438–447. https://doi.org/10.1007/S00248-010-9758-7.

Mapelli, F., Marasco, R., Balloi, A. et al. (2012). Mineral-microbe interactions: biotechnological potential of bioweathering. *Journal of Biotechnology* 157 (4): 473–481. https://doi.org/10.1016/j.jbiotec.2011.11.013.

Matlakowska, R. and Sklodowska, A. (2011). Biodegradation of Kupferschiefer black shale organic matter (Fore-Sudetic Monocline, Poland) by indigenous microorganisms. *Chemosphere* 83 (9): 1255–1261. https://doi.org/10.1016/j.chemosphere.2011.03.003.

McNamara, C.J. and Mitchell, R. (2005). Microbial deterioration of historic stone. *Frontiers in Ecology and the Environment* 3 (8): 445–451. https://doi.org/10.1890/1540-9295(2005)003[0445:MDOHS]2.0.CO;2.

Mergelov, N., Mueller, C.W., Prater, I. et al. (2018). Alteration of rocks by endolithic organisms is one of the pathways for the beginning of soils on Earth. *Scientific Reports* 8 (1): 1–15. https://doi.org/10.1038/s41598-018-21682-6.

Mihajlovski, A., Gabarre, A., Seyer, D. et al. (2017). Bacterial diversity on rock surface of the ruined part of a French historic monument: the Chaalis abbey. *International Biodeterioration and Biodegradation* 120: 161–169. https://doi.org/10.1016/j.ibiod.2017.02.019.

Miller, S.H., Browne, P., Prigent-Combaret, C. et al. (2010). Biochemical and genomic comparison of inorganic phosphate solubilization in *Pseudomonas* species. *Environmental Microbiology Reports* 2 (3): 403–411. https://doi.org/10.1111/j.1758-2229.2009.00105.x.

Mustin, C., Berthelin, J., Marion, P., and De Donato, P. (1992). Corrosion and Electrochemical Oxidation of a Pyrite by Thiobacillus ferrooxidans. *Applied and Environmental Microbiology* 58 (4): 1175–1182. https://doi.org/10.1128/AEM.58.4.1175-1182.1992.

Nautiyal, C.S., Bhadauria, S., Kumar, P. et al. (2000). Stress induced phosphate solubilization in bacteria isolated from alkaline soils. *FEMS Microbiology Letters* 182 (2): 291–296. https://doi.org/10.1016/S0378-1097(99)00605-9.

Nicolitch, O., Feucherolles, M., Churin, J.L. et al. (2019). A microcosm approach highlights the response of soil mineral weathering bacterial communities to an increase of K and Mg availability. *Scientific Reports* 9 (1): 1–13. https://doi.org/10.1038/s41598-019-50730-y.

Nir, I., Barak, H., Kramarsky-Winter, E. et al. (2022). Microscopic and biomolecular complementary approaches to characterize bioweathering processes at petroglyph sites from the Negev Desert, Israel. *Environmental Microbiology* 24 (2): 967–980. https://doi.org/10.1111/1462-2920.15635.

Olsson-Francis, K., Van Houdt, R., Mergeay, M. et al. (2010). Microarray analysis of a microbe-mineral interaction. *Geobiology* 8 (5): 446–456. https://doi.org/10.1111/j.1472-4669.2010.00253.x.

Olsson-Francis, K., Simpson, A.E., Wolff-Boenisch, D., and Cockell, C.S. (2012). The effect of rock composition on cyanobacterial weathering of crystalline basalt and rhyolite. *Geobiology* 10 (5): 434–444. https://doi.org/10.1111/J.1472-4669.2012.00333.X.

Oteino, N., Lally, R.D., Kiwanuka, S. et al. (2015). Plant growth promotion induced by phosphate solubilizing endophytic *Pseudomonas* isolates. *Frontiers in Microbiology* 6: 1–9. https://doi.org/10.3389/fmicb.2015.00745.

Page, W.J. and Huyer, M. (1984). Derepression of the *Azotobacter vinelandii* siderophore system, using iron-containing minerals to limit iron repletion. *Journal of Bacteriology* 158 (2): 496–502. https://doi.org/10.1128/JB.158.2.496-502.1984.

Palmer, R.J. and Hirsch, P. (1991). Photosynthesis based microbial communities on two churches in northern germany: weathering of granite and glazed brick. *Geomicrobiology Journal* 9 (2–3): 103–118. https://doi.org/10.1080/01490459109385992.

Palmer, R.J., Siebert, J., and Hirsch, P. (1991). Biomass and organic acids in sandstone of a weathering building: production by bacterial and fungal isolates. *Microbial Ecology* 21 (1): 253–266. https://doi.org/10.1007/BF02539157.

Papida, S., Murphy, W., and May, E. (2000). Enhancement of physical weathering of building stones by microbial populations. *International Biodeterioration and Biodegradation* 46 (4): 305–317. https://doi.org/10.1016/S0964-8305(00)00102-5.

Parrello, D., Zegeye, A., Mustin, C., and Billard, P. (2016). Siderophore-mediated iron dissolution from nontronites is controlled by mineral cristallochemistry. *Frontiers in Microbiology* 7 (MAR): 1–11. https://doi.org/10.3389/fmicb.2016.00423.

Picard, L., Paris, C., Dhalleine, T. et al. (2022). The mineral weathering ability of *Collimonas pratensis* PMB3(1) involves a Malleobactin-mediated iron acquisition system. *Environmental Microbiology* 24 (2): 784–802. https://doi.org/10.1111/1462-2920.15508.

Potysz, A. and Bartz, W. (2022). Bioweathering of minerals and dissolution assessment by experimental simulations—implications for sandstone rocks: a review. *Construction and Building Materials* 316: 125862. https://doi.org/10.1016/J.CONBUILDMAT.2021.125862.

Puente, M. E. Bashan, Y., Li, C.Y., Lebsky, V.K., (2004) Microbial populations and activities in the rhizoplane of rock-weathering desert plants. I. Root colonization and weathering of igneous rocks', *Plant Biology*. 6(5), pp. 629–642. https://doi.org/10.1055/S-2004-821100/ID/56.

Quantin, C., Becquer, T., Rouiller, J.H., and Berthelin, J. (2002). Redistribution of metals in a new caledonia ferralsol after microbial weathering. *Soil Science Society of America Journal* 66 (6): 1797–1804. https://doi.org/10.2136/sssaj2002.1797.

Randive, K., Raut, T., and Jawadand, S. (2021). An overview of the global fertilizer trends and India's position in 2020. *Mineral Economics* 34 (3): 371–384. https://doi.org/10.1007/S13563-020-00246-Z/TABLES/1.

Rawlings, D.E. and Johnson, D.B. (2007). The microbiology of biomining: development and optimization of mineral-oxidizing microbial consortia. *Microbiology (Reading)* 153 (Pt 2): 315–324. https://doi.org/10.1099/MIC.0.2006/001206-0.

Rehman, F., Thebo, K.H., Aamir, M., and Akhtar, J. (2020). Nanomembranes for water treatment. *Nanotechnology in the Beverage Industry: Fundamentals and Applications* 8 (9): 207–240. https://doi.org/10.1016/B978-0-12-819941-1.00008-0.

Reichard, P.U., Kretzschmar, R., and Kraemer, S.M. (2007). Dissolution mechanisms of goethite in the presence of siderophores and organic acids. *Geochimica et Cosmochimica Acta* 71 (23): 5635–5650. https://doi.org/10.1016/j.gca.2006.12.022.

Reitemeier, R.F. (1951). Soil potassium. *Advances in Agronomy* 3: 113–164. https://doi.org/10.1016/S0065-2113(08)60368-5.

Ribeiro, I.D.A., Volpiano, C.G., Vargas, L.K. et al. (2020). Use of mineral weathering bacteria to enhance nutrient availability in crops: a review. *Frontiers in Plant Science* 11: 1–20. https://doi.org/10.3389/fpls.2020.590774.

Roberts, J.A. (2004). Inhibition and enhancement of microbial surface colonization: the role of silicate composition. *Chemical Geology* 212 (3-4): 313–327. https://doi.org/10.1016/j.chemgeo.2004.08.021.

Rodriguez, H., Gonzalez, T., Goire, I., and Bashan, Y. (2004). Gluconic acid production and phosphate solubilization by the plant growth-promoting bacterium *Azospirillum* spp. *Naturwissenschaften* 91 (11): 552–555. https://doi.org/10.1007/S00114-004-0566-0/FIGURES/1.

Samuels, T., Bryce, C., Landenmark, H. et al. (2020). Microbial weathering of minerals and rocks in natural environments. *Biogeochemical Cycles: Ecological Drivers and Environmental Impact* 59–79. https://doi.org/10.1002/9781119413332.ch3.

Sanyal, A., Antony, R., Ganesan, P., and Thamban, M. (2020). Metabolic activity and bioweathering properties of yeasts isolated from different supraglacial environments of Antarctica and Himalaya. *Antonie van Leeuwenhoek International Journal of General and Molecular Microbiology* 113 (12): 2243–2258. https://doi.org/10.1007/S10482-020-01496-1/TABLES/3.

Schröer, L., Boon, N., De Kock, T., and Cnudde, V. (2021). The capabilities of bacteria and archaea to alter natural building stones – a review. *International Biodeterioration and Biodegradation* 165: https://doi.org/10.1016/j.ibiod.2021.105329.

Seiffert, F., Bandow, N., Bouchez, J. et al. (2014). Microbial colonization of bare rocks: laboratory biofilm enhances mineral weathering. *Procedia Earth and Planetary Science* 10: 123–129. https://doi.org/10.1016/j.proeps.2014.08.042.

Seiffert, F., Bandow, N., Kalbe, U. et al. (2016). Laboratory tools to quantify biogenic dissolution of rocks and minerals: a model rock biofilm growing in percolation columns. *Frontiers in Earth Science* 4: 31. https://doi.org/10.3389/FEART.2016.00031/XML/NLM.

Selbmann, L., Onofri, S., Coleine, C. et al. (2017). Effect of environmental parameters on biodiversity of the fungal component in lithic Antarctic communities. *Extremophiles* 21 (6): 1069–1080. https://doi.org/10.1007/S00792-017-0967-6/FIGURES/5.

Seneviratne, G. and Indrasena, I.K. (2006). Nitrogen fixation in lichens is important for improved rock weathering. *Journal of Biosciences* 31 (5): 639–643. https://doi.org/10.1007/BF02708416.

Soffritti, I., D'Accolti, M., Lanzoni, L. et al. (2019). The potential use of microorganisms as restorative agents: an update. *Sustainability* 11 (14): 3853. https://doi.org/10.3390/SU11143853.

Southam, G. (2015) Geology, Life, and Habitability. https://doi.org/10.1016/B978-0-444-53802-4.00175-5.

Souza, V., Moreno-Letelier, A., Travisano, M. et al. (2018). The lost world of cuatro ciénegas basin, a relictual bacterial niche in a desert oasis. *eLife* 7: https://doi.org/10.7554/ELIFE.38278.

Souza-Egipsy, V., Wierzchos, J., Sancho, C. et al. (2004). Role of biological soil crust cover in bioweathering and protection of sandstones in a semi-arid landscape (Torrollones de Gabarda, Huesca, Spain). *Earth Surface Processes and Landforms* 29 (13): 1651–1661. https://doi.org/10.1002/ESP.1118.

Sprocati, A. R. De Leo, F., Urzi, C., Alisi, C., Tasso, F., Vedovato, E., Barbabietola, N., Cremisini, C., (2008) 'A microbiological survey of the Etruscan Mercareccia tomb (Italy): contribution of microorganisms to deterioration and restoration'. *9th International Conference on NDT of Art, Jerusalem Isr. 25–30 May 2008, 9 pp.*

Sterflinger, K. and Piñar, G. (2013). Microbial deterioration of cultural heritage and works of art - tilting at windmills? *Applied Microbiology and Biotechnology* 97 (22): 9637–9646. https://doi.org/10.1007/s00253-013-5283-1.

Štyriaková, I., Štyriak, I., and Oberhänsli, H. (2012). Rock weathering by indigenous heterotrophic bacteria of *Bacillus* spp. at different temperature: a laboratory experiment. *Mineralogy and Petrology* 105 (3–4): 135–144. https://doi.org/10.1007/s00710-012-0201-2.

Sun, L.L., Xiao, L.L., Xiao, B. et al. (2013). Differences in the gene expressive quantities of carbonic anhydrase and cysteine synthase in the weathering of potassium-bearing minerals by *Aspergillus niger*. *Science China Earth Sciences* 56 (12): 2135–2140. https://doi.org/10.1007/s11430-013-4704-4.

Swęd, M., Potysz, A., Duczmal-Czernikiewicz, A. et al. (2021). Bioweathering of Zn–Pb-bearing rocks: experimental exposure to water, microorganisms, and root exudates. *Applied Geochemistry* 130: 104966. https://doi.org/10.1016/j.apgeochem.2021.104966.

Syers, J.K., Johnston, A.E., and Curtin, D. (2008). Efficiency of soil and fertilizer phosphorus use Reconciling changing concepts of soil phosphorus behaviour with agronomic information. *FAO Fertilizer and Plant Nutrition Bulletin* 18.

Tamilselvi, S.M., Thiyagarajan, C., and Uthandi, S. (2016). Calcite dissolution by brevibacterium sp. SOTI06: a futuristic approach for the reclamation of calcareous sodic soils. *Frontiers in Plant Science* 7: 1828. https://doi.org/10.3389/FPLS.2016.01828/XML/NLM.

Tang, Y., Lian, B., Dong, H. et al. (2012). Rocks from the Nanjiang Canyon in Guizhou Karst Area (China). *Geomicrobiology Journal* 29 (3): 213–225. https://doi.org/10.1080/01490451.2011.558560.

Uroz, S., Calvaruso, C., Turpault, M.P. et al. (2007). Effect of the mycorrhizosphere on the genotypic and metabolic diversity of the bacterial communities involved in mineral weathering in a forest soil. *Applied and Environmental Microbiology* 73 (9): 3019–3027. https://doi.org/10.1128/AEM.00121-07/ASSET/A2C9F8B6-299B-4C36-B6DF-62B29E1A4E05/ASSETS/GRAPHIC/ZAM0090777590004.JPEG.

Uroz, S., Calvaruso, C., Turpault, M.P., and Frey-Klett, P. (2009). Mineral weathering by bacteria: ecology, actors and mechanisms. *Trends in Microbiology* 17 (8): 378–387. https://doi.org/10.1016/J.TIM.2009.05.004.

Uroz, S., Kelly, L.C., Turpault, M.-P. et al. (2015). The mineralosphere concept: mineralogical control of the distribution and function of mineral-associated bacterial communities. *Trends in Microbiology* 23 (12): 751–762. https://doi.org/10.1016/j.tim.2015.10.004.

Uroz, S., Picard, L., and Turpault, M.P. (2022). Recent progress in understanding the ecology and molecular genetics of soil mineral weathering bacteria. *Trends in Microbiology* 1–16. https://doi.org/10.1016/j.tim.2022.01.019.

Varjani, S.J. (2017). Microbial degradation of petroleum hydrocarbons. *Bioresource Technology* 223 (1): 277–286. https://doi.org/10.1016/j.biortech.2016.10.037.

Vassilev, N., Vassileva, M., and Nikolaeva, I. (2006). Simultaneous P-solubilizing and biocontrol activity of microorganisms: potentials and future trends. *Applied Microbiology and Biotechnology* 71 (2): 137–144. https://doi.org/10.1007/S00253-006-0380-Z/TABLES/2.

Wade, W. (2002). Unculturable bacteria - the uncharacterized organisms that cause oral infections. *Journal of the Royal Society of Medicine* 95 (2): 81–83. https://doi.org/10.1258/jrsm.95.2.81.

Wagh, J., Shah, S., Bhandari, P. et al. (2014). Heterologous expression of pyrroloquinoline quinone (pqq) gene cluster confers mineral phosphate solubilization ability to *Herbaspirillum seropedicae* Z67. *Applied Microbiology and Biotechnology* 98 (11): 5117–5129. https://doi.org/10.1007/s00253-014-5610-1.

Wang, Q., Zhang, Z., Zhu, Y. et al. (2018). Impact of poxB, pta, and ackA genes on mineral-weathering of Enterobacter cloacae S71. *Journal of Basic Microbiology* 58 (7): 633–642. https://doi.org/10.1002/jobm.201800013.

Wang, Q., Cheng, C., Agathokleous, E. et al. (2021a). Enhanced diversity and rock-weathering potential of bacterial communities inhabiting potash trachyte surface beneath mosses and lichens —a case study in Nanjing, China. *Science of the Total Environment* 785: 147357. https://doi.org/10.1016/j.scitotenv.2021.147357.

Wang, W., Lian, B., and Pan, L. (2015). An RNA-sequencing study of the genes and metabolic pathways involved in *Aspergillus niger* weathering of potassium feldspar. *Geomicrobiology Journal* 32 (8): 689–700. https://doi.org/10.1080/01490451.2014.991812.

Wang, Y., Li, L., Sun, Y. et al. (2020a). Impact of the fliF gene on biotite weathering of *Rhizobium pusense* S41. *Geomicrobiology Journal* 37 (4): 308–314. https://doi.org/10.1080/01490451.2019.1697396.

Wang, Y.-L., Sun, L.-J., Xian, C.-M. et al. (2020b). Interactions between biotite and the mineral-weathering bacterium *Pseudomonas azotoformans* F77. *Applied and Environmental Microbiology* 86 (7): 1–17. https://doi.org/10.1128/AEM.02568-19.

Wang, Y.L., Dong, W., Xiang, K.X. et al. (2021b). A combination of genomics, transcriptomics, and genetics provides insights into the mineral weathering phenotype of pseudomonas azotoformans f77. *Applied and Environmental Microbiology* 87 (24): 1–18. https://doi.org/10.1128/AEM.01552-21.

White, A.F. and Buss, H.L. (2014). Natural weathering rates of silicate minerals. *Treatise on Geochemistry: Second Edition* 7: 115–155. https://doi.org/10.1016/B978-0-08-095975-7.00504-0.

Widdig, M., Schleuss, P.M., Weig, A.R. et al. (2019). Nitrogen and phosphorus additions alter the abundance of phosphorus-solubilizing bacteria and phosphatase activity in grassland soils. *Frontiers in Environmental Science* 7: 1–15. https://doi.org/10.3389/fenvs.2019.00185.

Wieler, N., Ginat, H., Gillor, O., and Angel, R. (2019). The origin and role of biological rock crusts in rocky desert weathering. *Biogeosciences* 16 (6): 1133–1145. https://doi.org/10.5194/BG-16-1133-2019.

Wilson, M.J., Jones, D., and McHardy, W.J. (1981). The weathering of serpentinite by lecanora atra. *The Lichenologist* 13 (2): 167–176. https://doi.org/10.1017/S0024282981000212.

Włodarczyk, A., Szymańska, A., Skłodowska, A., and Matlakowska, R. (2016). Determination of factors responsible for the bioweathering of copper minerals from organic-rich copper-bearing Kupferschiefer black shale. *Chemosphere* 148: 416–425. https://doi.org/10.1016/j.chemosphere.2016.01.062.

Wu, F., Zhang, Y., He, D. et al. (2021). Community structures of bacteria and archaea associated with the biodeterioration of sandstone sculptures at the Beishiku Temple. *International Biodeterioration and Biodegradation* 164 (June): 105290. https://doi.org/10.1016/j.ibiod.2021.105290.

Wu, L., Jacobson, A.D., and Hausner, M. (2008). Characterization of elemental release during microbe-granite interactions at T = 28 °C. *Geochimica et Cosmochimica Acta* 72 (4): 1076–1095. https://doi.org/10.1016/j.gca.2007.11.025.

Wu, Y.-W., Zhang, J.-C., Wang, L.-J., and Wang, Y.-X. (2017). A rock-weathering bacteria isolated from rock surface and its role in ecological restoration on exposed carbonate rocks. *Ecological Engineering* 101: 162–169. https://doi.org/10.1016/j.ecoleng.2017.01.023.

Wu, Y., Kameshwar, A.K.S., Zhang, B. et al. (2022). Genome and transcriptome analysis of rock-dissolving *Pseudomonas* sp. NLX-4 strain. *Bioresources and Bioprocessing* 9 (1): 1–16. https://doi.org/10.1186/s40643-022-00548-w.

Xiao, B., Lian, B., Sun, L., and Shao, W. (2012). Gene transcription response to weathering of K-bearing minerals by *Aspergillus fumigatus*. *Chemical Geology* 306–307: 1–9. https://doi.org/10.1016/j.chemgeo.2012.02.014.

Xiao, L., Hao, J., Wang, W. et al. (2014). The up-regulation of carbonic anhydrase genes of bacillus mucilaginosus under soluble Ca^{2+} deficiency and the heterologously expressed enzyme promotes calcite dissolution. *Geomicrobiology Journal* 31 (7): 632–641. https://doi.org/10.1080/01490451.2014.884195.

Xiao, L., Lian, B., Hao, J. et al. (2015). Effect of carbonic anhydrase on silicate weathering and carbonate formation at present day CO_2 concentrations compared to primordial values. *Scientific Reports* 5: 1–10. https://doi.org/10.1038/srep07733.

Yarwood, S., Wick, A., Williams, M., and Daniels, W.L. (2015). Parent material and vegetation influence soil microbial community structure following 30-years of rock weathering and pedogenesis. *Microbial Ecology* 69 (2): 383–394. https://doi.org/10.1007/s00248-014-0523-1.

Yu, X., Liu, X., Zhu, T.H. et al. (2012). Co-inoculation with phosphate-solubilzing and nitrogen-fixing bacteria on solubilization of rock phosphate and their effect on growth promotion and nutrient uptake by walnut. *European Journal of Soil Biology* 50: 112–117. https://doi.org/10.1016/J.EJSOBI.2012.01.004.

Zhu, Y.G., Duan, G.L., Chen, B.D. et al. (2014). Mineral weathering and element cycling in soil-microorganism-plant system. *Science China Earth Sciences* 57 (5): 888–896. https://doi.org/10.1007/s11430-014-4861-0.

Ziolkowski, L.A., Mykytczuk, N.C.S., Omelon, C.R. et al. (2013). Arctic gypsum endoliths: a biogeochemical characterization of a viable and active microbial community. *Biogeosciences* 10 (11): 7661–7675. https://doi.org/10.5194/BG-10-7661-2013.

12

Occurrence of Arsenic (As) in the Aquatic Environment Due to Weathering and Erosion

Shailesh K. Yadav[1]*, Alagappan Ramanathan*[2]*, and Rakesh K. Ranjan*[1]

[1] *Department of Geology, Sikkim University, Sikkim, India*
[2] *School of Environmental Sciences, Jawaharlal Nehru University, New Delhi, India*

12.1 Introduction

At the current time, it is estimated that the earth is a habitat of around 8 billion people and they are using the earth's resources for their daily life (Molden et al. 2007; Mannan et al. 2018; Kushawaha et al. 2021). According to Rees (2018), the rate of natural resource consumption by human beings is 1.7 faster than resource replenishment in our ecosystem. Water is an essential and valuable innate (natural) resource that plays a significant role in the socioeconomic development of a country (Mannan et al. 2018; Rees 2018; Kushawaha et al. 2021). The world population was nearly threefold during the nineteenth and twentieth centuries, despite a sixfold increase in the utilization of renewable water resources. With the increasing freshwater demand and overexploitation of water, humanity is currently experiencing a water quality deterioration and crisis (Mannan et al. 2018; Kushawaha et al. 2021). In many locations with hydric stress, the problems brought on by natural water pollution are getting worse due to the overexploitation of aquifers. Millions of people worldwide are affected by the amounts of fluoride, arsenic, and other chemical compounds that are dangerous to health, creating a very serious health problem (Bhattacharya et al. 2011; Shankar et al. 2014; Chakraborti et al. 2017; Shaji et al. 2021).

In recent days, arsenic received high attention due to its behavior, environmental fate, and toxicity level. In the earth's crust, arsenic is the 20th most common element. It has been found in the rocks and soil and can enter the air and water by wind and water runoff (Bhattacharya et al. 1997, 2002; Herath et al. 2016; Chakraborti et al. 2017; Yadav et al. 2020; Raju 2022). Naturally, As is derived from pyrite (FeS_2) and arsenopyrite (FeAsS) minerals present in the earth's crust. Sulfide minerals such as realgar, orpiment, enargite, etc. are the common As-bearing minerals that can release As into the water. Besides these, secondary minerals of some metal oxides may also contribute As to the soil and water (Akai et al. 2004; Nordstrom et al. 2014; Chakraborti et al. 2017). On the other hand, man-made sources such as agricultural runoff, industrial discharge, fossil fuel burning, as well as mining of the

Weathering and Erosion Processes in the Natural Environment, First Edition. Edited by Virendra Bahadur Singh, Sughosh Madhav, Naresh Chandra Pant, and Ravi Shekhar.

metal ores can also introduce As into soil and groundwater (Bhattacharya et al. 1997; Nordstrom and Archer 2003; Kumar 2015; Yadav et al. 2020; Shaji et al. 2021; Raju 2022). Due to the direct release of As from natural and anthropogenic sources, many groundwater systems around the globe significantly surplus the acceptable As guideline value of 10 μg/l, recommended by WHO (2004) (Németi and Gregus 2002; Rahman et al. 2006; Chakraborti et al. 2018; Yadav et al. 2020; Shaji et al. 2021; Raju 2022). As a result, the increased As content in groundwaters throughout the globe has drawn a lot of attention recently as a major concern to both human health and agricultural operations (Smith et al. 1998; Nordstrom 2002; Nordstrom and Archer 2003; Flora 2015).

Naturally, As combined with oxygen and sulfur forms several arsenic oxides and sulfides. Furthermore, most of the As species (oxides and sulfides) are highly mobile and soluble in an aqueous medium; hence As easily can enter into the food chain through contaminated food and water (Saha et al. 1999; Herath et al. 2016; Jain and Chandramani 2018). The uppermost layer comprises small amounts of As which can be released into the natural aquatic environment via weathering and erosion processes. This may subsequently alter the chemistry of the aquatic system (Bhattacharya et al. 1997, 2002; Landner 1998; Herath et al. 2016; Masuda 2018). The total estimated arsenic reservoir and fluxes in the earth's crust and earth's surface indicate that anthropogenic activities are a significant contributor to the spread of arsenic pollution into the water, soil, and atmosphere (Flora 2015; Masuda 2018).

Millions of people are affected by arsenic poisoning globally due to occupational and environmental exposure, mental illness, and eventually intentional suicide and homicide attempts (Bhattacharya et al. 2002; Shankar et al. 2014; Chakraborti et al. 2018; Yadav et al. 2020). As enters the human body through three major pathways: ingestion, inhalation, and adsorption. However, ingestion is identified as the primary source of arsenic exposure (Bhattacharya et al. 2002; Bundschuh and Garcia 2008; Flora 2015). The International Agency for Research on Cancer classifies arsenic as a class four carcinogen. Arsenic is highly toxic to humans and especially dangerous because it has no taste or odor, so people can be exposed to it without knowing it. Arsenic affects nearly all organ systems because it easily targets dispersed enzyme reactions and long-term As exposure caused exfoliative dermatitis (ATSDR 2011; Engwa et al. 2019). As poisoning can be classified into two types: acute and chronic. Acute As poisoning causes diarrhea, vomiting, and encephalopathy: however, chronic As poisoning causes several types of cancer and eventually multi-organ failure (Ratnaike 2003).

12.2 History and Extent of Arsenic Poisoning in an Aquatic System

The first incident of As contamination in groundwater was investigated in Germany in 1885 (Raju 2022) and the first As poisoning medical case due to the consumption of As contaminated water was reported from the state of Cordoba, Argentina, in 1917 (Mukherjee et al. 2006; Raju 2022). By the end of the twentieth century, a total of five Asian nations such as Taiwan, China, India, Bangladesh, and Thailand were acknowledged as having As pollution in groundwater (Shaji et al. 2021; Raju 2022). Due to growing knowledge of the toxicity and potentially, the WHO decreased the maximum permissible As content in

Table 12.1 List of the countries with the maximum allowed As in their drinking water.

Countries	Maximum permissible limit (μg/l)	Reference
Argentina	50	Smedley and Kinniburgh (2002)
Australia	7	Chakraborti et al. (2018)
Bangladesh	50	Ahmad et al. (2019)
China	50	Guo et al. (2001)
Chile	50	Nordstrom (2002)
India	10	Sarkar (2002)
Mexico	50	Ongley et al. (2001)
Nepal	50	Shrestha et al. (2003)
New Zealand	50	McLaren and Kim (1995)
Taiwan	10	Tseng et al. (2005)
USA	5	Chakraborti et al. (2018)
Vietnam	10	Berg et al. (2001)

Source: Adopted from Raju (2022).

drinking water to 10 μg/l (WHO 2004). Likewise, The Bureau of Indian Standards also reduced the acceptable limit of As in drinking water to 10 μg/l in 2012. The majority of poor nations continue to adhere to the outdated WHO norm of 50 μg/l. Developed nations such as the United States, Australia, and Canada also revised the maximum allowed As in groundwater and set their own As limits in drinking water, which are shown in Table 12.1.

According to Chakraborti et al. (2018), Asia and Europe are severely affected followed by Africa, North America, South America, and Australia, and currently, more than 108 countries have reported moderate to overhead As concentrations in drinking water worldwide (Figure 12.1). It has been estimated that around 100–230 million people around the globe, including 150 million from Asia, might be at risk due to ingestion of arsenic-contaminated water and food (Chakraborti et al. 2018; Erickson et al. 2021; Shaji et al. 2021). South Asian countries like Bangladesh, India, China, Nepal, Mongolia, Burma, Thailand, and Vietnam are severely affected by arsenic contamination and menace (Bhattacharya et al. 2002; Mukherjee et al. 2009, 2012). However, in India, around 20 states and 4 union territories are affected by As poisoning in groundwater (Acharyya and Shah 2007; Shankar et al. 2014; Saha and Sahu 2016; Yadav et al. 2020; Shaji et al. 2021). As poisoning is directly or indirectly related to the aquifer type and surrounding subsurface environment. Furthermore, the worst damaged groundwater comes from unconfined aquifers, especially those that are reported from the world's orogenic belts.

Currently, Bangladesh is the worst affected country in the world, and it was reported that the maximum As-contaminated areas as well as the highest number of cases were reported from Bangladesh (Figure 12.2). Before 1970, surface water was the major drinking source for Bangladeshi people and hence they easily got infected with water-borne diseases like diarrhea, cholera, etc. (Raju 2022). Therefore, using groundwater resources was a different way to deal with the issue of water-borne illnesses.

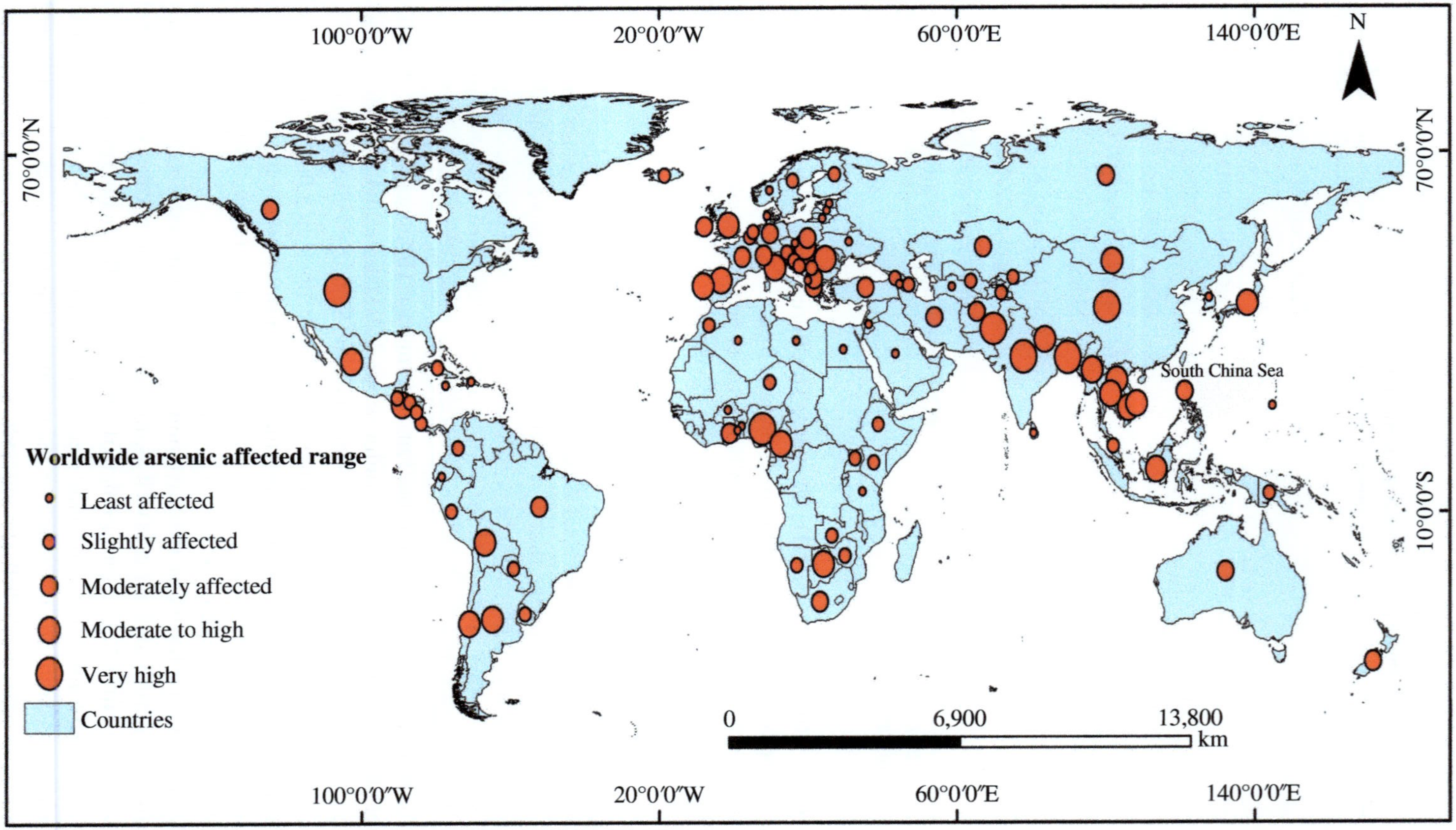

Figure 12.1 World map showing As affected countries with As intensity (range). The map is showing that south Asia, south central Europe, and North America are affected worstly by As. *Source:* Shaji et al. (2021) and references therein.

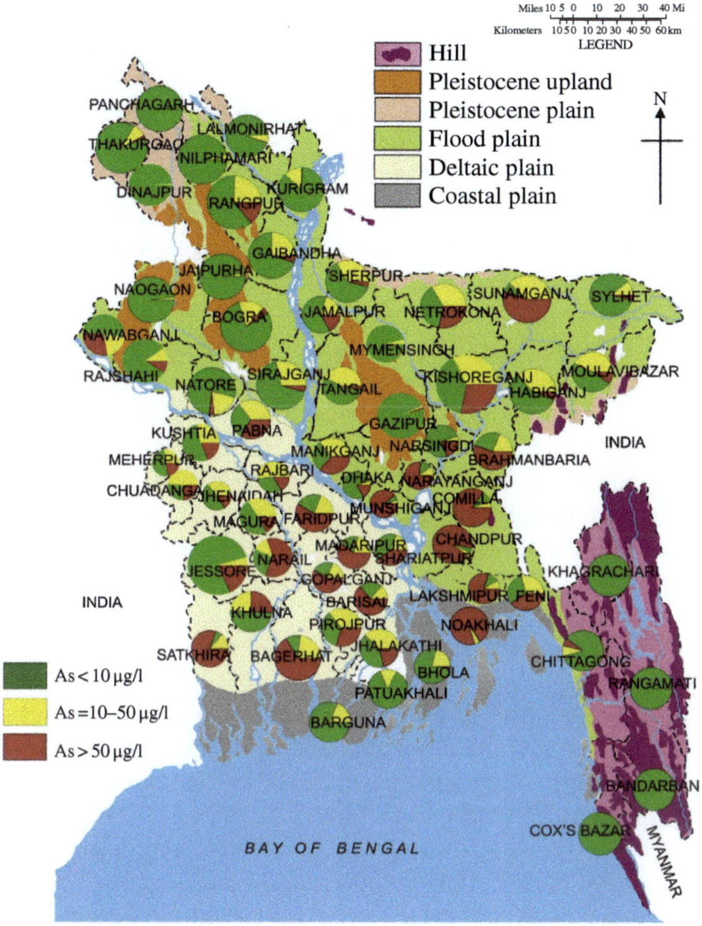

Figure 12.2 Bangladesh arsenic pollution scenarios. Distribution of arsenic concentration ranges throughout all geomorphological zones. *Source:* Adopted from Chakraborti et al. (2015).

After 1993, the Bangladesh government installed millions of shallow depth (20–75 m deeper) handpumps to resolve the surface water drinking problems with the collaboration and support of international cooperation. Unintentionally, these groundwater supplies led to the biggest As poisoning outbreak in human history (Smith et al. 2000). No one anticipated the massive scope of the poisoning in Bangladesh or the previous incident recorded in the West Bengal delta plains (Figure 12.2). At the same time, India was also facing As menace but did not receive enough media and research attention. Therefore, maximum As analysis and groundwater testing agencies (national and international) were implemented only in Bangladesh. Since 1993, numerous research carried out in Bangladesh and are well documented that arsenic extensively supply to the groundwater from naturally As bearing minerals in shallow aquifers. As dissolution or precipitation in subsurface aquifers are controlled by redox reactions (Smedley and Kinniburgh 2002; Anawar et al. 2003; Kinniburgh et al. 2003; McArthur et al. 2004; Harvey et al. 2006; Rahman et al. 2006; Mazumder et al. 2010; Olea et al. 2018; Ahmad et al. 2019). As of now, nearly 35 million Bangladeshis have been exposed to poisoning from these deposits, making it the largest known poisoning.

The As catastrophic situation generated in the Bengal delta plain has upraised the concerns of the research community and international health organizations in areas with similar geological conditions. Studies on water quality linked to As pollution in Vietnam and West Bengal were initiated since both the Bengal delta basin and the Vietnam delta contain alluvial soil with essentially the same geology (Das et al. 1996; Chowdhury et al. 1997; Berg et al. 2001, 2007). Following that, several researchers also reported that slow As poisoning in drinking water affects groups of individuals from many other nations.

The first occurrence of As contamination in groundwater in India was reported in 1976 from the Chandigarh region (Datta and Kaul 1976), while the second report on As contamination was reported from the West Bengal region in 1978 (Vorosmarty et al. 2000). After five years, the first incidence of As poisoning to human health was discovered in the state in 1983 (Saha 1995; Mukherjee et al. 2009; Chakraborti et al. 2018). It is estimated that drinking As-polluted water puts more than 50 million people at risk in India. Several research works have been carried out on As poisoning in groundwater and it was identified that As contamination in groundwater is densely occurring in the Gangetic basin (Chakraborti et al. 2018; Yadav et al. 2020; Shaji et al. 2021). The Gangetic fluvial region is a notably fertile part of India that makes up around 26% of the country's total area and provides food and shelter to more than 500 million people (Chakraborti et al. 2018; Shaji et al. 2021).

According to the CGWB (2018), 10 Indian states have reported As concentrations greater than the WHO prescribed limit in shallow groundwater in shallow aquifers. As poses a serious threat to the states of West Bengal, Bihar, Jharkhand, and Uttar Pradesh (Chakraborti et al. 2015; Olea et al. 2018; Yadav et al. 2020; Shaji et al. 2021). According to CGWB (2018), 20 Indian states and 4 union territories are now affected by As contamination (Figure 12.3), and it is believed that roughly more than 50 million people get skin lesions due to the ingestion of As-polluted water. Both the number impacted by the issue and the geographic area where it is present is frighteningly expanding annually. As entered into the food chain through vegetables and cereals cultivated in the Indo-genetic plain (Saha 1995; Saha and Sahu 2016).

There are two types of terrain in India that are associated with high arsenic in groundwater: alluvial terrain and hard rock terrain. More than 90% of As in groundwater is coming from the alluvial terrain while only 10% of the As in the groundwater may originate from the hard rock terrain (Figure 12.4). In Karnataka, As is released into the aquifer system from the oxidation of primary minerals like pyrite and arsenopyrite. Raichur and Yadgir districts are famous for gold mining. The availability of As in this region is limited. However, in Chhattisgarh, As mostly released from the Korti lineament.

12.3 Chemistry of Arsenic (Inorganic and Organic)

Arsenic is a metalloid that is rarely found in the natural environment. Arsenic occurs more frequently in the sulfide ores in the form of metal arsenides, e.g., nickel diarsenide, cobalt diarsenide, cobalt arsenide sulfide, iron diarsenide. Few metals (like Al, Ca, Mg, Fe, Mn,

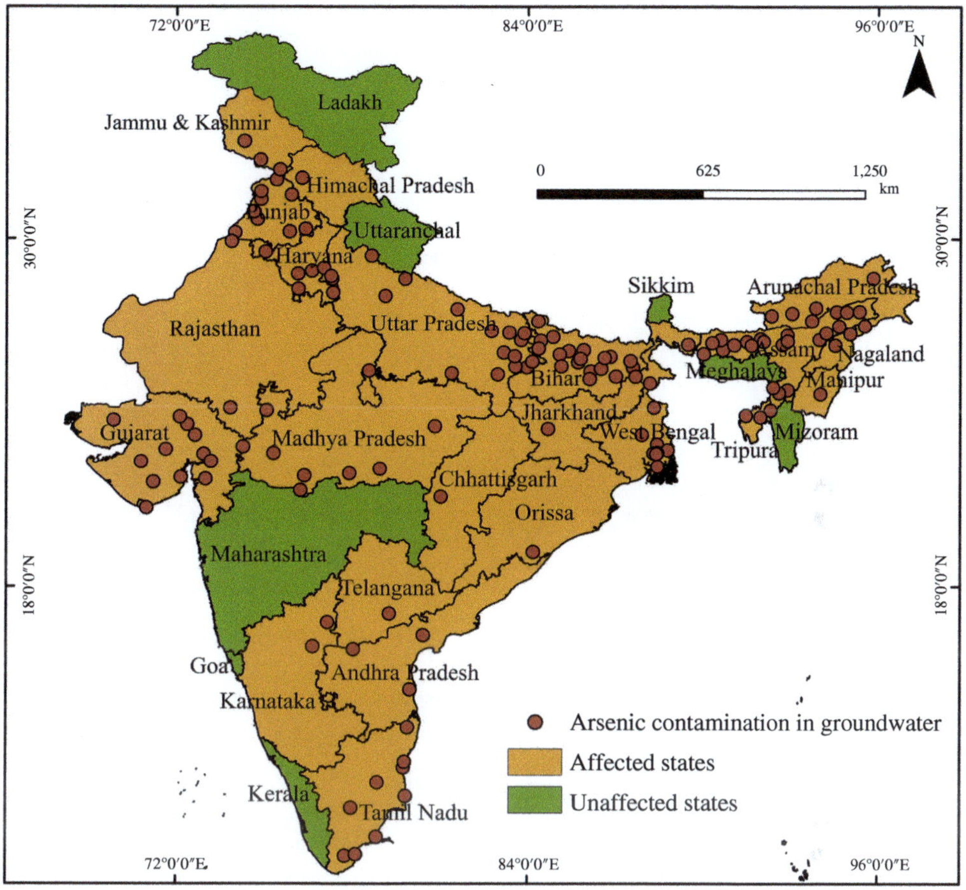

Figure 12.3 State map of India showing Arsenic-affected states and union territories. *Source:* Shaji et al. (2021) and CGWB (2018).

Ni, U, and Zn) are tending to form the complexes with arsenate and arsenite in the natural environment, which is a product of weathering of arsenides (Smedley and Kinniburgh 2002; Acharyya and Shah 2007; Chakraborti et al. 2015; Yadav et al. 2020; Shaji et al. 2021; Raju 2022). Realgar and orpiment, two important arsenic sulfide ores, occur in the natural environment. Both realgar and orpiment are formed in similar geological environments and have similar physical properties. Local arsenic concentrations are now significantly higher than background levels due to the commercial usage and manufacturing of inorganic and organic arsenic compounds (Bhattacharya et al. 1997; Flora 2015). In an aqueous environment, As exists in four oxidation states viz. −3, 0, +3 and +5 (Smedley and Kinniburgh 2002; Flora 2015; Raju 2022). Arsenic has two forms: inorganic and organic. Inorganic As compounds occur in soil, sediment, and water, while organic As compounds are mainly identified in fish and shellfish. Inorganic arsenic is the predominant form that is having trivalent and pentavalent inorganic arsenic compounds (As_2O_3, AsO_3^{4-}, $AsCl_3$, As_2O_5, AsO_4^{3-}), arsenous acid (H_3AsO_3 and $H_2AsO_3^-$), arsenic acids (H_3AsO_4, $H_2AsO_4^-$,

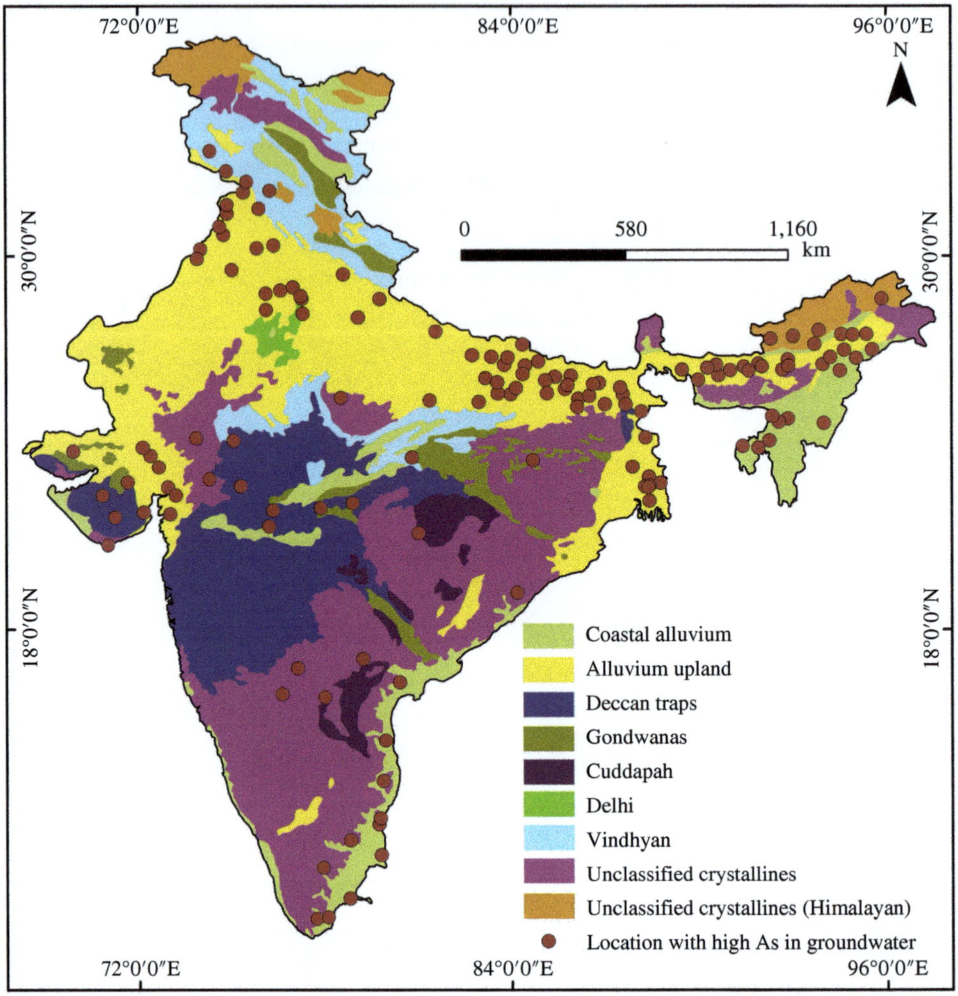

Figure 12.4 Geology map of India, superimposed with the As contamination in groundwater. The map showed that maximum As contamination is associated with alluvium aquifers. *Source:* Aquifer systems of India; CGWB (2017, 2018).

$HAsO_4^{2-}$), and arsine (AsH_3), while arsanilic acid, methylarsonic acid, dimethylarsinic acid (DMA), and arsenobetaine are organic forms of As species (Flora 2015). Two well-known oxidation states of inorganic arsenic are oxyanions of trivalent also known as arsenite (AsIII) and oxyanions of pentavalent also known as arsenate (AsV) and both are highly toxic in an aqueous environment (Nickson et al. 1998; Bhattacharya et al. 2002; Smedley and Kinniburgh 2002). As(III) acts like a hard acid and forms complexes with oxides and nitrogen more frequently than As(V), which behaves like a soft acid and forms complexes with sulfides (Bodek et al. 1988; Langmuir et al. 2003; Mohindru et al. 2017). The valency state affects sorption behavior and mobility in the aquatic environment in addition to other factors. In stable oxygen-rich aerobic groundwater, arsenate (AsV) is dominant while in reduced anaerobic aqueous groundwater arsenite (AsIII) is the dominant species (Bodek et al. 1988; Langmuir et al. 2003; Flora 2015). Arsenite (AsIII) is highly mobile and more toxic than arsenate (AsV). However, when we compared the toxicity level of organic and inorganic arsenic in groundwater, inorganic As is generally more toxic than organic As.

In surface water generally, arsenate (AsV) will be persistent since oxidizing conditions are dominant there. Arsenate exists in oxyanionic species such as H_3AsO_4, $H_2AsO_4^-$, $HAsO_4^{2-}$, and AsO_4^{3-}. These oxyanionic species are significantly influenced by pH and redox potential. At low pH (<6.9) and oxidizing environment, H_3AsO_4 and $H_2AsO_4^-$ are the dominant species (Figure 12.5). However, at higher pH (pH > 7), $HAsO_4^{2-}$ and AsO_4^{3-} are the dominant species.

In an extremely acidic environment, H_3AsO_4 is the dominant species, while in an alkaline environment AsO_4^{3-} is the dominant species. In reducing aquifers system, the non-ionic forms of H_3AsO_3 are more stable when pH is less than 9.2 (Zheng et al. 2004; Ghosh et al. 2006; Lu and Zhu 2011). Surface and river waters are mostly oxidizing in nature; therefore, the trivalent form of arsenic is commonly not reported. However, when the aquatic is reducing in nature, trivalent arsenic can exist in $As(OH)_2$ and H_3AsO_3 in an acidic solution and $H_2AsO_3^-$ and AsO_2^- in basic solutions, and these forms of arsenic are highly mobile in natural water due to their anionic nature (Moore et al. 1988; Flora 2015).

At low pH (5–6), the sorption of arsenate is higher than arsenite, although arsenite has a stronger affinity for the solids above pH 7–8. The sorption of arseniate and arsenate is

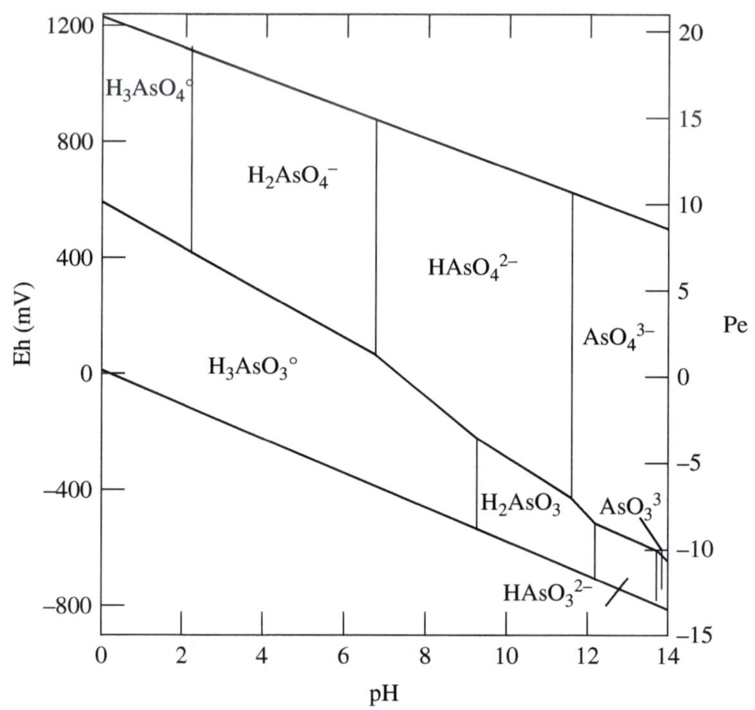

Figure 12.5 Pourbaix diagram for aqueous species of As at 25 °C and 1 bar total pressure. *Source:* Adapted from Smedley and Kinniburgh (2002)/with permission of Elsevier.

mainly pH-dependent (Figure 12.5) and pH is significantly affected by factors, viz. solid-to-solution ratio, solution type, and the specific surface area of the minerals (Bhattacharya et al. 2002; Lu and Zhu 2011). Sometimes when phosphate is present in the aqueous solution, it acts as a competitive ion for the sorption site and shifts the pH range to lower values (Sarkar 2002; Flora 2015). High bicarbonate in groundwater also enhanced arsenic mobilization by successfully releasing the arsenic from iron hydroxide-coated sediment (Anawar et al. 2004). Hence, As mobilization in groundwater can be dependent on several physico-chemical factors. In the natural environment, As can be released into the groundwater through the weathering and dissolution of the arsenopyrite minerals. Once As enters the aquatic environment, its solubility and mobility play a significant role in As mobilization. In an acidic environment, arsenate is highly mobile and soluble, while in neutral to basic pH, arsenate mobility gets reduced due to the co-precipitation of hydrous iron oxides and form scorodite (Shaji et al. 2021; Raju 2022). The overall following chemical processes can be written as:

$$4\text{FeAsS} + 13\text{O}_2 + 6\text{H}_2\text{O} = 4\text{FeSO}_4 + 4\text{H}_3\text{AsO}_4$$

$$\text{Fe}\left(\text{OH}\right)_3 + \text{H}_3\text{AsO}_4 = \text{FeAs}_4.2\text{H}_2\text{O}$$

The process of modification of As species by adding a methyl group is known as As methylation. Inorganic arsenic is metabolized mostly by a process called bio-methylation, which involves the alternating processes of two-electron reduction and oxidative addition of methyl group, commonly known as oxidative methylation (Flora 2015). In anaerobic conditions or with a pH lower than 2, arsenate reduction could be possible non-enzymatically. Arsenic is methylated by the methyl donor *S*-adenosylmethionine into monomethylarsonic acid and ultimately DMA. Methyltransferases and glutathione are used in this process as a catalyst and co-factor, respectively (Flora 2015; Bühl et al. 2017; Masuda 2018). It is observed that arsenic methylation is not dependent on the concentration of arsenic in water, while when it is present in its organic form, it is less ionizable and less assessable for chemical and biochemical reactions.

12.4 Source, Occurrence, and Distribution of Arsenic

Both man-made activities and natural geological processes allowed arsenic to infiltrate terrestrial and aquatic habitats. The typical amount of As in the earth's crust is between 1.5 and 5 mg/kg. Yet, As concentration is highly variable in rock, soil, and sediment. In the sediment, As ranges from 0.1 to 40 mg/kg (Sarkar 2002; Shankar et al. 2014). The concentration of As is highly variable in the type of rock. The lowest to moderate arsenic concentration is reported in undifferentiated igneous rock with an average of 1.5 mg/kg (Bowen et al. 1982). Volcanic rocks have an average As slightly greater than 6 mg/kg (Smedley and Kinniburgh 2002). The concentration of arsenic in sedimentary rocks is slightly higher than that found in igneous rocks and volcanic glass, with levels ranging from 5 to 10 mg/kg, and Pelitic metamorphic rocks, such as slate and phyllites, contain the highest amounts of arsenic, averaging around 18 mg/kg (Webster et al. 1999). Compared to igneous and metamorphic rocks, sedimentary rocks typically contain a higher proportion of minerals that have a greater capacity to absorb arsenic. However, among sedimentary rocks, argillaceous deposits tend to have a higher arsenic content, averaging around 13 mg/kg, compared to sandstone and limestone (Bowen et al. 1982). In the earth's crust, more than 200 minerals commonly have arsenic as a major constituent, in which pyrite (FeS_2) and arsenopyrite (FeAsS) are the most prevalent As ores (Nordstrom 2002); however, orpiment (As_2S_3) and realgar (As_4S_4) are also arsenic bearing sulfur minerals, but contains comparatively low arsenic. Scorodite ($FeAsO_4.2H_2O$) is a secondary mineral that forms by the oxidation of arsenic-bearing minerals. Few metals and minerals also have arsenic in trace amounts such as cobaltite (CoAsS), niccolite (NiAs), and enargite (CuAsS) (Smedley and Kinniburgh 2002; Masuda 2018). Arsenic-bearing sulfur minerals, i.e. orpiment (As_2S_3) and realgar (AsS), release arsenic into an aqueous solution at low pH and reducing conditions (Cullen and Reimer 1989; Sarkar 2002).

The concentration of As is dependent on free sulfide in water. If free sulfide is high in the water, the availability of As in water is limited (Moore et al. 1988; Masuda 2018) because both are having almost similar chemistry, and As can replace sulfur from the lattice of sulfides. In marine sediment, elevated concentration of arsenic is particularly associated with fine-grained sediments, organic matter, sulfide minerals, phosphate, and iron oxides (Mirlean et al. 2012; Saha et al. 2021). Ganga-Brahmaputra flood plains are formed by the quaternary alluvial deposits that are severely As contaminated in India (Yadav et al. 2020; Raju 2022). Yet, arsenic concentration is not uniformly distributed and it can be changed with time. Arsenic ranges from 1.2 to 1.5 mg/kg in the Gangetic plain, while in the Brahmaputra plain, it varies from 1.4 to 5.9 mg/kg. The availability of As in subsurface matrices depends on the amount of arsenic present in the soil and the nature of the chemical bonding between soil and metals (Flora 2015; Masuda 2018). Besides these, As concentration is also dependent on the oxidation reduction potential of the groundwater (Mukherjee et al. 2012). Arsenic speciation significantly determined the chemical nature of arsenic in the aqueous environment (Figure 12.5). Soil variables such as pH, redox condition, microbial activity, and organic matter play a significant role in As transformation in soil (Bhattacharya et al. 1997). The presence of As in soil is remarkably determined by the physical and chemical properties of the soil. In addition, physiological factors such as climate change, geomorphological features, and hydrogeological activities also have an impact on the availability and transportation of arsenic (Anawar et al. 2004; Mahanta et al. 2015; Saha et al. 2021).

The average concentration of arsenic in river and lake water can vary depending on various factors such as the geological characteristics of the region, human activities, and environmental conditions. On average, the concentration of arsenic in river water is typically ranges from 2 to 10 micrograms per liter (μg/l). However, certain regions with specific geological features or pollution sources may experience higher concentrations. It has been observed that the presence of elevated alkalinity and pH, combined with water flowing through local rocks containing arsenic, can lead to an enhanced concentration of arsenic in surface water (Ghosh et al. 2006). While the presence of As in the groundwater is dependent on several factors viz. source of arsenic, rock–water interaction, and retention time and geochemistry of the groundwater (Mukherjee et al. 2012; Verma et al. 2015). Due to extensive mineral water interactions and ideal circumstances for arsenic mobilization and accumulation, groundwater has the widest range and highest amounts of arsenic of all the sources and posing risks to human health and ecosystems. Government agencies, outside organizations, educational institutions, and business research and development facilities are working hard to locate rising arsenic concentrations in groundwater, food, and sediment across the world.

Since the start of the Industrial Revolution, human-generated sources such as industrial disposal; sewage and sludge discharge; landfill, mining, and agricultural inputs can introduce metals into the soil and water. Besides, the burning of fossil fuels can also contribute arsenic in the air because sulfide-rich coals contain a trace amount of pyrite and other primary minerals of arsenic. Despite this fact, there is a lot of published literature in which it is frequently mentioned that As is released into the atmosphere through fossil fuel burning, chemical manufacturing industries, metallurgical emissions, and cement kilns (Bhattacharya et al. 2002; Flora 2015; Masuda 2018). But none of the studies have been found in the Gangetic basin and Bengal delta plain based on the atmospheric deposition of As. The availability of As in the soil depends on physical, chemical, and biological properties as well as nutritional uptake from the soil to the ecosystem (Corkhill and Vaughan 2009; Jang et al. 2016; Bühl et al. 2017). Organic compounds of As like monomethylarsonous acid, dimethylarsonous acid, trimethyl arsine, and arsenobetaine are the As forms that can be bioaccumulated in plants and other living things. Furthermore, breakdown of these organic arsenic in the presence of bacteria, fungi, and other microbes can release inorganic As in sediment and water.

12.5 Geochemistry and Arsenic Mobilization

Several anthropogenic and natural processes influence the geochemistry of As in groundwater. Therefore, it is essential to understand how arsenic is dissolved and mobilized into drinking water between naturally occurring solids and aqueous solutions. Depending on the speciation or chemical form, the leaching of the As from the As-bearing minerals, ores, rocks, and sediment at a low temperature is known as mobilization. In groundwater, As is primarily found in three phases, i.e. sulfide and organic matter phases, oxides phases of Fe and Mn, and silicate phases (Anawar et al. 2003), and it is believed that a three-stage process led to the emergence of these stages (Acharyya and Shah 2007; Raju 2022). Several research had been documented on arsenic mobilization, and they have revealed their diverse point of view and the possibility of As mobilization mechanisms in arsenic-affected areas. Currently, two major and well-accepted mechanisms for the As mobilization in the groundwater: (i) pyrite dissolution, (ii) reductive dissolution of iron oxides (Bose and Sharma 2002; Zheng et al. 2004; Ghosh et al. 2006; Herath et al. 2016). Besides these two mechanisms, several other small mechanisms are also involved in As mobilization.

12.5.1 Arsenic Containing Pyrite Oxidation

Oxidation of As-bearing sulfide minerals like pyrite and arsenopyrite can lead to arsenic in the groundwater. Several well-documented studies have been carried out on the releasing mechanism of arsenic from the oxidative dissolution of pyrite under acidic conditions. It is a possible pathway for the remobilization of As under an oxic environment (Cheng et al. 2004; Corkhill and Vaughan 2009). Few previously documented studies have suggested that the availability of chemical constituents in the groundwater significantly influences the oxidation of pyrite and As adsorption on minerals (Ghosh et al. 2006; Mahanta et al. 2015; Zhang et al. 2017). However, the role of organic matter and bicarbonate on the release of As from pyrite oxidation remains unknown. The excessive groundwater pumping in alluvial aquifers causes the water-level dropdown and eventually, an oxidizing environment is created, which supports pyrite dissolution (Zheng et al. 2004). Another process such as excessive use of fertilizers and mining in an area causes subsequent exposure of pyrite to the atmosphere, which can also be mobilized As in an aqueous environment (Mandal and Suzuki 2002; Raju 2022). The dissolution of arsenopyrite is dependent on several factors. At circum-neutral pH, arsenopyrite followed a series of the oxidation reaction, where arsenic acid (AsIII) is formed initially which is further oxidized and forms iron oxyhydroxide and As(V):

$$4FeAsS + 9O_2 + 10H_2O \rightarrow 4Fe^{2+} + 4H_3AsO_3 + 4H_2SO_4^{2-}$$

$$4Fe^{2+} + O_2 + 10H_2O \rightarrow 4Fe(OH)_3 + 8H^+$$

$$2H_3AsO_3 + O_2 \rightarrow 2HAsO_4^{2-} + 4H^+$$

$$2H_3AsO_3 + O_2 \rightarrow 2H_2AsO_4^{2-} + 2H^+$$

Through the above reactions, Walker et al. (2006) suggested that there is no complete oxidation of iron and arsenic takes place in the reaction. The rate constant of the reaction at pH 6.5–6.9 is around 10^{-10} mol/m/s. Other researchers investigated the rate of the reaction of arsenopyrite in the presence of oxygen over a wide range of pH, temperature, and dissolved oxygen and suggested that the rate of arsenic release into the groundwater increases with increasing the concentration of dissolved oxygen as well as with rising temperature, and it was reported similar at pH 7 and 10 (Neil et al. 2014; Khatri and Tyagi 2015; Panagiotaras and Nikolopoulos 2015). The oxidation rate of As was observed minimum at pH 7–8, which indicates that the dominant oxidant species has changed from aqueous Fe(III) (Sarkar 2002; Flora 2015). Hence, it concludes that at higher pH and dissolved oxygen, the surface reaction would be the rate-determining step that needed large activation energy.

12.5.2 Reductive Dissolution of Metal Oxyhydroxides

The reductive dissolution of metal oxyhydroxide is one of the well-known and accepted theories to release arsenic in groundwater in an anoxic environment and at neutral pH (Zheng et al. 2004; Chakraborti et al. 2018). The microbial degradation of organic matter in anoxic conditions (low level of dissolved oxygen) ultimately reduced the ferric ions to soluble ferrous ions and released As(III) in the groundwater (Nickson et al. 1998). Those aquifers having a higher concentration of Fe and Mn, as well as comparatively low concentrations of NO_3, SO_4, and PO_4, can undergo reductive dissolution of metal oxides (Stüben et al. 2003). In the Gangetic plain and Brahmaputra-Meghna delta basin, the reductive dissolution of iron oxyhydroxide is one of the well-documented mechanisms for arsenic mobilization (Bhattacharya et al. 2002; Anawar et al. 2003; McArthur et al. 2004; Mukherjee et al. 2006; Chakraborti et al. 2015; Olea et al. 2018; Yadav et al. 2020). In alluvial-deposited deltaic regions, hydrous iron oxides act as secondary minerals with a distinctively large specific surface area for adsorption (600 m^2/g) (Smith 1999; Raghav et al. 2015). In an oxidizing environment, iron oxyhydroxide is incorporated with inorganic arsenic coating due to its high adsorption affinity toward As(V). In the central Gangetic plain, the extent of flooded land increased due to an expansion in agricultural land utilized for rice cultivation, which also blocked off oxygen to the aquifers, and created a redox state in the subsurface region. The reducing environmental circumstances are also brought on by the biodegradation of the peat present in the sediment. Arsenic is extensively distributed within the deltaic region, particularly in alluvial deposits and buried swamp lands. However, its release into groundwater occurs solely during the process of organic matter decomposition. The microbial reduction of organic matter-rich argillaceous sediment generates a reducing environment and facilitates Fe and As transformation and release As in the groundwater (Nickson et al. 1998; Harvey et al. 2006; Kumar et al. 2018; Saha et al. 2021; Raju 2022).

Arsenic also can be released from the soil or sediment through the competitive exchange of organic anions such as phosphate and sulfates (detailed explained in letter Sections) available in the aqueous medium; however, in that reaction redox processes probably play a significant role (Herath et al. 2016; Zhang et al. 2017). In an aqueous medium, pH also plays an important role in As desorption. At higher pH, the desorption rate of arsenic from the ferric oxyhydroxide surface has been increased, leading to a high level of As in the groundwater (Herath et al. 2016). As mobilization by reductive dissolution is mostly accomplished by sulfidation of ferrihydrite $Fe(III)OAs(OH)_2$. The $Fe(III)OAs(OH)_2$ can be reductively dissolved by sulfides, especially H_2S, deposited in sediments and releasing As(III) in groundwater. The presence of the minerals like hematite and goethite in the soil or sediment can create chemical circumstances that promote sulfate reduction thermodynamically in an aqueous medium (Verma et al. 2015; Kumar et al. 2018; Yadav et al. 2020).

12.5.3 Role of Physicochemical Factors on As Mobilization

In addition to the above-described factors, several more factors affect the As removal from the rocks, sediments, and minerals. For instance, the change in crystal structure from ferrihydrite to goethite lowers the adsorption sites and causes the release of arsenic into the groundwater. Appelo et al. (2002) developed a novel idea for the origin of high-As groundwater that relies on the displacement of As by dissolved bicarbonate as a distinct process. Bicarbonate is produced by weathering and the breakdown of sedimentary carbonates causes the leaching of As into groundwater (detailed explanation given in Section 12.5.3.4). The presence of nitrate and sulfate in the groundwater can influence the As concentration. According to an experiment carried out by the USEPA (2016), the addition of extra nitrate in the groundwater can reduce the As content in groundwater.

Arsenic and phosphate are having similar chemical behaviors; hence, phosphate displaced arsenic from their sorption site on sediment and releases it into the groundwater. The excessive use of phosphate-containing fertilizers and irrigation causes the leaching of phosphate down to the shallow aquifers resulted, anion exchange-induced arsenic mobilization from the surface reactive mineral surface (Acharyya 2002; Raju 2022). In the coarse aquifer sediment, ferrihydrite (FeIII) has a large number of surface adsorption sites compared to the other oxides hence both As and phosphate adsorbed on Fe(III). Therefore, reductive dissolution of ferrihydrite (FeIII) releases As and PO_4^{3-} into the groundwater (Anawar et al. 2003; McArthur et al. 2004; Mukherjee et al. 2009; Zhang et al. 2017; Yadav et al. 2020).

12.5.3.1 The Role of Nitrate in As Mobilization

It is widely known that nitrate may participate in a variety of microbe-mediated anoxic redox processes in soils and aquifers resulting in high As. In the subsurface environment, nitrate is reduced into nitrite and finally into ammonia. This reaction is coupled with the oxidation of numerous electron donors, including Fe(II) and As(III), acetate, lactate, pyruvate, sulfide, and other organic and inorganic compounds (Zhang et al. 2017; Jiang et al. 2019; Islam and Mostafa 2021). In the process of arsenic desorption from the FeO(OH) mineral, nitrate, nitrite, and ammonium have complex and controversial roles that are difficult to understand. The effectiveness of the functioning capacity of the nitrate group is also influenced by certain contributing components like organic matter, microbial population, soil profile, pH, and electrode potential. Ammonium and nitrate can build up in the subsoil as a result of nitrogen-rich fertilizer leaching due to numerous chemical reactions. The pooled concentration of each of these ions is highly associated with arsenic, similar to that of phosphate (Kurosawa et al. 2008, 2013; Islam and Mostafa 2021).

Recently, few studies have been carried out to identify the role of nitrate on the bacterial-catalyzed reductive dissolution of arsenic in contaminated aquifers soils, and it was found that nitrate/nitrite significantly reduces the microbial activity required for the reductive dissolution of As and Fe in arsenic-contaminated soils. It was identified that nitrate significantly resists the reductase protein in the genetic factor of the respiring bacteria cell (Chen et al. 2017; Zhang et al. 2017; Shakya and Ghosh 2018; Shakya et al. 2018; Hare et al. 2020). According to Zhu et al. (2020), the denitrifying bacteria facilitate the reductive dissolution of arsenic and iron through the indirect oxidation at the solid-phase surface of the iron oxyhydroxide in the arsenic-laden soil/sediments.

12.5.3.2 Role of Sulfate in As Mobilization

Similar to nitrate, sulfate behaves as a weak electron acceptor. Sulfate absorbed electrons from the organic material and creates sulfide and eventually, these electrons are added to Fe(II) and As(III). Resulted, all the sulfate is removed from the aqueous solution and co-precipitation as sulfide of iron and arsenic (Altun et al. 2014; Shakya and Ghosh 2018; Islam and Mostafa 2021). Similar to nitrate, sulfate also generates H^+ on reduction, hence acidity increases. Furthermore, high acidity retards the formation of arsenic-sulfide precipitation (Henke 2009). However, the contribution of sulfate in arsenic dissolution is very negligible.

12.5.3.3 Role of Phosphate in As Mobilization

Arsenic and phosphorus are thought to behave similarly in their environmental condition because of their analogous structure. However, some of the chemical properties are different from each other, hence they behave differentially in their biogeochemical characteristics (Strawn 2018). Arsenic is significantly influenced by redox potential; however, the chemistry of phosphorus is not influenced by redox conditions. In an aquatic environment, As exists in two inorganic oxyanions forms As(III) and As(V). In contrast, phosphorus occurs in the form of orthophosphate (PO_4^{-3}), which is the highest and most stable oxidation state of phosphate in an aquatic environment (McArthur et al. 2004; Mukherjee et al. 2009; Bonsor et al. 2017; Yadav et al. 2020; Raju 2022). In biogeochemical reactions, arsenic and phosphate behave as competitive ions, and this can be often observed in adsorption and desorption reactions, precipitation and dissolution reactions. According to O'reilly et al. (2001), only 35% of adsorbed arsenate can be desorbed from goethite in the presence of phosphate solution. This finding suggests that the arsenate bonds on iron oxide are stronger than the phosphate bonds. Arsenate may be hazardous to organisms when it is used as a substitute for phosphate during metabolism. In a competitive biogeochemical reaction of As and phosphate, the bioavailability and toxicity of As are more dominant over phosphate. However, redox reaction significantly creates a distinct change in the biogeochemical reactions of arsenic but that does not occur for phosphate. A watered rice paddy field is the best example of a reducing environment where arsenate changed into arsenite (Sanz et al. 2007; Strawn 2018). Wetland soil, where soils are preferentially formed with iron oxides, may show an association with As, while phosphate may be associated with both iron oxide and jarosite minerals. This happens because of the difference in their radius, which leads to providing different adsorption strengths to them. Few documents have demonstrated that phosphate amendments immobilized Pb in soil containing As as a co-contaminants but enhanced As availability through competitive adsorption processes (Sanz et al. 2007; Strawn 2018). In As-contaminated soils, biochar amendment reduced As bioavailability but not phosphate; this may be because the biochar encouraged biological processes that preferentially allowed phosphate absorption over arsenic (Strawn 2018).

12.5.3.4 Role of Bicarbonate and Alkaline Earth Metals (Ca and Mg) in As Mobilization

According to Anawar et al. (2004), carbonate and bicarbonate are very effective in extracting arsenic from arsenic-laden soil and sediments. The rate of the leaching of As from the soil or sediment raised by increasing the concentration of carbonate and bicarbonate (alkalinity) in the solution. The As leaching rate is also dependent on reaction time. Hence, it is concluded that over a geological period, carbonate and bicarbonate play a significant role in arsenic mobilization (Anawar et al. 2004). A group of researchers from Bangladesh documented that arsenic and bicarbonate are strongly correlated in groundwater (Nickson et al. 2000; Anawar et al. 2003). It was observed that the groundwater collected from the Ganga-Meghna delta is enriched with arsenic, iron and bicarbonate in reducing conditions. Kim et al. (2000) reported that the carbonation of arsenic sulfide minerals by bicarbonate ions resulted in the leaching of arsenic into groundwater. However, in the deltaic region of Bengal and Bangladesh, arsenic sulfide is not a prominent mineral for As contamination in soil and groundwater. Minerals of As such as pyrite and arsenopyrite are rarely found in the alluvial aquifer of the deltaic plain. Therefore, As must be released into the groundwater from secondary minerals such as clay, organic matter, and oxyhydroxides of Fe and Mn, which are responsible for As mobilization in the alluvial deposits (Takahashi et al. 2004; Saha 2009; Kumar 2015). At some locations of Ganga-Meghana fluvial deposits, pyrite and arsenopyrite are reported, which are formed as a result of diagenesis processes (Nickson et al. 2000). These minerals and biological materials are linked to arsenic in the sediments of the delta plains. Hence, As mobilization can be possible when bicarbonate replaces As from the surface site of these minerals, soil/sediment, and organic matter.

12.6 Variation in As with the Groundwater Depth

Numerous research has been carried out in the alluvial region to examine the spatial and vertical distribution of As. They have also well documented the major hydrogeochemical processes, which are significantly involved in As retention and liberation in groundwater. The chemistry of arsenic is also influenced by the local geomorphology and how the groundwater interacts with sediment (Mukherjee et al. 2009, 2012; Mahanta et al. 2015; Verma et al. 2015; Chakraborti et al. 2018; Kumar et al. 2018; Yadav et al. 2020). In groundwater, arsenic concentration considerably changed with vertical depth due to changes in redox potential, soil composition, and texture. Upper layers of soil consist of shallow Holocene aquifer with recently recharged. In recently deposited (newer alluvium) regions, sometimes the upper layer contained an enormous amount of organic matter. Furthermore, the breakdown of organic matter due to the bacterial reduction caused a reducing environment, which triggered As mobilization into the surrounding aquifer. According to Mihajlov et al. (2020), deeper aquifers (55–60 m) that are recharged recently within 10–15 years and have low As concentrations can also trigger As in the presence of clay layers that expels organic carbon as an indirect response to excessive groundwater pumping. Hence, the untouched deeper aquifers, which are recharged 60–70 years before, can only be safe from As contamination.

12.7 Role of Geomorphology and Geo-stratigraphy in As Mobilization

Several researchers around the globe have well documented that geomorphology and geo-stratigraphy play a significant role in groundwater arsenic contamination in fluvio-deltaic regions (McArthur et al. 2004; Saha and Sahu 2016; Kumar et al. 2018; Yadav et al. 2020; Das and Mondal 2021). The geomorphic features that are reported from the Gangetic plain such as sediment nature, elevation differences, soil texture, etc., indicate that the foreland basin of Ganga is formed at different times and climatic conditions. The sediment deposited in the Gangetic plain indicates that they were deposited during the Pleistocene-Holocene period (Acharyya and Shah 2004; Raju 2022). The influences of geomorphology and geologic constraints on As pollution in river channels and flood plains from the central and lower Ganga plain of the Indo-Gangetic area can be identified and explained in the study of quaternary geomorphic and geo-stratigraphic research. The geomorphic features available in the Gangetic plain are characterized as local sources of organic matter in the shallow aquifers, which eventually release As in anaerobic conditions (Postma et al. 2007; Saha 2009; Neumann et al. 2010). Other researchers also found that the high As concentration in the central Gangetic plain is associated with the abandoned meander scar and oxbow lakes (Nath et al. 2008; Das and Mondal 2021). The oxbow lakes and meander scar are formed by finned-grained sediment (mainly clay and silt) and were identified as an ample source of organic matter to the shallow aquifers, resulting in the release of arsenic in the shallow aquifer (Postma et al. 2007; Kumar et al. 2018). Therefore, high As in Gangetic plain is associated with the deposited organic matter. That is why As deposits (hotspots) are not consistently discovered but rather exist in isolated areas throughout the Ganga plain of several Indian states. The central Gangetic plain is further divided into three subdivisions, based on the height of the upland terrace surface from the river channel; the upland interfluve surface is also known as Pleistocene older alluvium, the river valley surface is also known as Holocene newer alluvium and the flood plains is also known as Holocene to recent active channels (Nickson et al. 2000; Mahanta et al. 2015; Verma et al. 2015; Saha and Sahu 2016; Kumar et al. 2018; Yadav et al. 2020). Active flood channels have the highest amount of organic matter followed by Holocene younger alluvium and Pleistocene older alluvium and the amount of organic matter is directly linked with As concentration (Chakraborti et al. 2018; Kumar et al. 2018).

12.8 Role of Clay Minerals on As Mobilization

The adsorption and oxidation/reduction reactions of As at clay mineral surfaces significantly assist the natural attenuation and transformation of As in groundwater systems. Aluminum oxide and silica oxides combined to form aluminosilicate clay minerals, which provide a large binding site for adsorbing metal ions (Herath et al. 2016). The aluminum ion present on the clay surface is bound with OH groups and acts as a proton acceptor, and it is responsible for the generation of anionic species of As. The adsorption of As occurred on clay minerals, hence clay minerals types significantly affect the oxidation rate and adsorption capacity (Lin and Puls 2000). The absorption of the arsenite on clay minerals is dependent on several parameters such as pH, ionic strength, and the oxidation state of As (Manning and Goldberg 1997; Mohapatra et al. 2007). The maximum As(V) adsorption by kaolinite, montmorillonite, illite, halloysite, and chlorite may happen up to pH – 7, and on increasing the pH (toward basic nature) the adsorption rate of As on clay minerals is reduced. Competing anions, such as orthophosphate, chromate, sulfate, and molybdate, which are present in the media, might affect the rate of adsorption due to their similar charge (Manning and Goldberg 1996; Wang and Mulligan 2006). The rate of arsenate adsorption on kaolinite, montmorillonite, and illite decreases to their respective maximum values at pH – 5, 6, and 6.5 in the presence of competing anions, such as orthophosphates (Herath et al. 2016). Another crucial aspect influencing the adsorption of As by clay particles is surface area. It was observed that montmorillonites had a larger surface area than kaolinite, hence it can absorb about twice as much As(III) and As(V) as the latter (Frost and Griffin 1977).

12.9 Conclusion

Arsenic is one of the highly toxic elements found in soil, water and food (cereal and vegetables) and causes a variety of cancer and dermal exposure to the human being. Even low concentrations of As may be hazardous to human beings. This study reviews the literature on the occurrence of Arsenic, its history, speciation, chemistry, and the most accepted mechanism for arsenic mobilization. The concentration of As in soil and water is highly variable. In groundwater, maximum As concentration was mainly derived from natural rock–water interaction except for limited anthropogenic As contamination sources. As speciation suggests that pH and redox potential play a significant role in speciation and species stability in an aqueous medium. But there are limits to the many scientific methods that have been created for the As estimate and speciation investigations. It is well studied and reported that As concentration was concentrated in alluvial regions of the globe. A large population from Bangladesh, West Bengal Vietnam, and Myanmar has been facing health-related issues due to the consumption of contaminated water and food. The mobilization processes of As in groundwater and soil have been governed by several processes. The mechanism of As release into the groundwater is still poorly known, but three well-accepted mechanisms such as the dissolution of arsenic-bearing sulfide minerals, reductive dissolution of metal oxyhydroxide, as well as competitive ion exchange between As and anions trigger the release of As from soil to the aquifer system. The scientific community is working to resolve the As poisoning and they developed several techniques to resolve the As issue on a small scale. Besides these, a few alternative options can be considered to get As free drinking water such as (i) installation of a deeper well (>200 m) rather than a shallow well, (ii) identification and drilling of the well in As safe aquifer regions, (iii) treated surface water can be used in the place of As contaminated groundwater. The areas with high As pollution are urgently required comprehensive water resource management.

Declaration of Competing Interest

The authors declare that they are clear of any financial or personal conflicts that may have seemed to affect the research described in this chapter.

References

Acharyya, S.K. (2002). Arsenic contamination in groundwater affecting major parts of southern West Bengal and parts of western Chhattisgarh: source and mobilization process. *Current Science* 82: 740–744.

Acharyya, S.K. and Shah, B.A. (2004). Risk of arsenic contamination in groundwater affecting the Ganga Alluvial Plain, India. *Environmental Health Perspectives* 112 (1): A19–A21.

Acharyya, S.K. and Shah, B.A. (2007). Groundwater arsenic contamination affecting different geologic domains in India – a review: influence of geological setting, fluvial geomorphology and Quaternary stratigraphy. *Journal of Environmental Science and Health, Part A* 42 (12): 1795–1805. https://doi.org/10.1080/10934520701566744.

Agency for Toxic Substances and Disease Registry (A.T.S.D.R.) (2011). Atlanta (GA): CCA-treated wood. Fact sheet [Internet]. https://www.atsdr.cdc.gov/CCATreated_Wood_Factsheet.pdf (accessed 6 December 2022).

Ahmad, A., Wal, A.V.N., Bhattacharya, P., and Genuchten, C.M.V. (2019). Characteristics of Fe and Mn bearing precipitates generated by Fe(II) and Mn(II) co-oxidation with O_2, MnO_4 and

HOCl in the presence of groundwater ions. *Water Research* 161: 505–516. https://doi.org/10.1016/j.watres.2019.06.036.

Akai, J., Izumi, K., Fukuhara, H. et al. (2004). Mineralogical and geomicrobiological investigations on groundwater arsenic enrichment in Bangladesh. *Applied Geochemistry* 19 (2): 215–230. https://doi.org/10.1016/j.apgeochem.2003.09.008.

Altun, M., Sahinkaya, E., Durukan, I. et al. (2014). Arsenic removal in a sulfidogenic fixed-bed column bioreactor. *Journal of Hazardous Materials* 269: 31–37. https://doi.org/10.1016/j.jhazmat.2013.11.047.

Anawar, H.M., Akai, J., Komaki, K. et al. (2003). Geochemical occurrence of arsenic in groundwater of Bangladesh: sources and mobilization processes. *Journal of Geochemical Exploration* 77 (2–3): 109–131. https://doi.org/10.1016/S0375-6742(02)00273-X.

Anawar, H.M., Akai, J., and Sakugawa, H. (2004). Mobilization of arsenic from subsurface sediments by effect of bicarbonate ions in groundwater. *Chemosphere* 54 (6): 753–762. https://doi.org/10.1016/j.chemosphere.2003.08.030.

Appelo, C.A.J., Van der Weiden, M.J.J., Yournassat, C., and Charlet, L. (2002). Surface complexation of ferrous iron and carbonate on ferrihydrite and the mobilization of arsenic. *Environmental Science & Technology* 36 (14): 3096–3103.

Berg, M., Tran, H.C., Nguyen, T.C. et al. (2001). Arsenic contamination of groundwater and drinking water in Vietnam: a human health threat. *Environmental Science & Technology* 35 (13): 2621–2626.

Berg, M., Stengel, C., Trang, P.T.K. et al. (2007). Magnitude of arsenic pollution in the Mekong and Red River Deltas – Cambodia and Vietnam. *Science of the Total Environment* 372 (2–3): 413–425. https://doi.org/10.1016/j.scitotenv.2006.09.010.

Bhattacharya, P., Chatterjee, D., and Jacks, G. (1997). Occurrence of arsenic-contaminated groundwater in alluvial aquifers from Delta Plains, Eastern India: options for safe drinking water supply. *International Journal of Water Resources Development* 13 (1): 79–92. https://doi.org/10.1080/07900629749944.

Bhattacharya, P., Frisbie, S.H., Smith, E. et al. (2002). Arsenic in the environment: a global perspective. In: *Handbook of Heavy Metals in the Environment* (ed. B. Sarkar), 147–215. New York: Marcell Dekker Inc.

Bhattacharya, P., Mukherjee, A., and Mukherjee, A.B. (2011). Arsenic in groundwater of India. In: *Encyclopedia of Environmental Health* (ed. J.O. Nriagu), 150–164. Elsevier. https://doi.org/10.1016/B978-0-444-52272-6.00345-7.

Bodek, I., Lyman, W.J., Reehl, W.F., and Rosenblatt, D.H. (1988). *Environmental Inorganic Chemistry: Properties, Processes, and Estimation Methods*, 1199. New York: Pergamon Press.

Bonsor, H.C., MacDonald, A.M., Ahmed, K.M. et al. (2017). Hydrogeological typologies of the Indo-Gangetic basin alluvial aquifer, South Asia. *Hydrogeology Journal* 25 (5): 1377–1406. https://doi.org/10.1007/s10040-017-1550-z.

Bose, P. and Sharma, A. (2002). Role of iron in controlling speciation and mobilization of arsenic in subsurface environment. *Water Research* 36 (19): 4916–4926. https://doi.org/10.1016/S0043-1354(02)00203-8.

Bowen, H.J.M., Ure, A.M., and Berrow, M.L. (1982). The elemental constituents of soils. In: *Environmental Chemistry*, vol. 2 (ed. H.J.M. Bowen), 94–204.

Bühl, V., Álvarez, M.C., Torre, M.H. et al. (2017). Biomonitoring of arsenic in woodworkers exposed to CCA and evaluation of other non-occupational sources in Uruguay. *International*

Journal of Occupational and Environmental Health 23 (1): 71–80. https://doi.org/10.1080/10773525.2018.1425654.

Bundschuh, J. and Garcia, M.E. (2008). Rural Latin America – a forgotten part of the global groundwater arsenic problem? In: *Groundwater for Sustainable Development* (ed. P. Bhattacharya, A. Ramanathan, A.B. Mukherjee, et al.), 335–346. CRC Press.

Bureau of Indian Standards (2012). *10500 Indian Standard Drinking Water-Specification, Second Revision*. New Delhi: Bureau of Indian Standards.

CGWB (2017). *Dynamic Ground Water Resources of India*, 1–39. Faridabad: CGWB (cited 6 December 2022).

CGWB (2018). *Groundwater Quality in Shallow Aquifers of India*, 18–21. Faridabad: CGWB (cited 6 December 2022).

Chakraborti, D., Rahman, M.M., Mukherjee, A. et al. (2015). Groundwater arsenic contamination in Bangladesh – 21 years of research. *Journal of Trace Elements in Medicine and Biology* 31: 237–248. https://doi.org/10.1016/j.jtemb.2015.01.003.

Chakraborti, D., Rahman, M.M., Das, B. et al. (2017). Groundwater arsenic contamination and its health effects in India. *Hydrogeology Journal* 25 (4): 1165–1181. https://doi.org/10.1007/s10040-017-1556-6.

Chakraborti, D., Singh, S.K., Rahman, M.M. et al. (2018). Groundwater arsenic contamination in the Ganga River Basin: a future health danger. *International Journal of Environmental Research and Public Health* 15 (2): 180. https://doi.org/10.3390/ijerph15020180.

Chen, Y., Han, Y.H., Cao, Y. et al. (2017). Arsenic transport in rice and biological solutions to reduce arsenic risk from rice. *Frontiers in Plant Science* 8: https://doi.org/10.3389/fpls.2017.00268.

Cheng, T., Barnett, M.O., Roden, E.E., and Zhuang, J. (2004). Effects of phosphate on uranium(VI) adsorption to goethite-coated sand. *Environmental Science & Technology* 38 (22): 6059–6065. https://doi.org/10.1021/es040388o.

Chowdhury, T.R., Mandal, B.K., Samanta, G. et al. (1997). Arsenic in groundwater in six districts of West Bengal, India: the biggest arsenic calamity in the world: the status report up to August, 1995. In: *Arsenic: Exposure and Health Effect* (ed. C.O. Abernathy, R.L. Calderon, and W.R. Chappell), 93–111. Washington, DC: US EPA, Springer Dordrecht.

Corkhill, C.L. and Vaughan, D.J. (2009). Arsenopyrite oxidation – a review. *Applied Geochemistry* 24 (12): 2342–2361. https://doi.org/10.1016/j.apgeochem.2009.09.008.

Cullen, W.R. and Reimer, K.J. (1989). Arsenic speciation in the environment. *Chemical Reviews* 89 (4): 713–764.

Das, A. and Mondal, S. (2021). Geomorphic controls on shallow groundwater arsenic contamination in Bengal basin, India. *Environmental Science and Pollution Research* 28 (31): 42177–42195. https://doi.org/10.1007/s11356-021-13761-5.

Das, D., Samanta, G., Mandal, B.K. et al. (1996). Arsenic in groundwater in six districts of West Bengal, India. *Environmental Geochemistry and Health* 18 (1): 5–15. https://doi.org/10.1007/BF01757214.

Datta, D.V. and Kaul, M.K. (1976). Arsenic content of drinking water in villages in northern India. A concept of arsenicosis. *The Journal of the Association of Physicians of India* 24 (9): 599–604. https://doi.org/10.1007/s10661-016-5507-9.

Engwa, G.A., Ferdinand, P.U., Nwalo, F.N., and Unachukwa, M.N. (2019). Mechanism and health effects of heavy metal toxicity in humans. In: *Poisoning in the Modern World-new Tricks for an Old Dog*, vol. 10 (ed. O. Karcioglu and B. Arslan), 70–90. IntechOpen.

Erickson, M.L., Elliott, S.M., Brown, C.J. et al. (2021). Machine-learning predictions of high arsenic and high manganese at drinking water depths of the glacial aquifer system, northern continental United States. *Environmental Science & Technology* 55 (9): 5791–5805. https://doi.org/10.1021/acs.est.0c06740.

Flora, S.J.S. (2015). Arsenic. In: *Handbook of Arsenic Toxicology*, 1–49. Elsevier. https://doi.org/10.1016/B978-0-12-418688-0.00001-0.

Frost, R.R. and Griffin, R.A. (1977). Effect of pH on adsorption of arsenic and selenium from landfill leachate by clay minerals. *Soil Science Society of America Journal* 41 (1): 53–57. https://doi.org/10.2136/sssaj1977.03615995004100010019x.

Ghosh, A., Sáez, A.E., and Ela, W. (2006). Effect of pH, competitive anions and NOM on the leaching of arsenic from solid residuals. *Science of the Total Environment* 363 (1–3): 46–59. https://doi.org/10.1016/j.scitotenv.2005.06.018.

Guo, X.J., Fujino, Y., Kaneko, S. et al. (2001). Arsenic contamination of groundwater and prevalence of arsenical dermatosis in the Hetao plain area, Inner Mongolia, China. *Molecular Mechanisms of Metal Toxicity and Carcinogenesis* 222: 137–140. https://doi.org/10.1007/978-1-4615-0793-2_16.

Hare, V., Chowdhary, P., and Singh, A.K. (2020). Arsenic toxicity: adverse effect and recent advance in microbes mediated bioremediation. In: *Microorganisms for Sustainable Environment and Health* (ed. J. Hayton and L. Sparks), 53–80. Elsevier. 10.1016/B978-0-12-819001-2.00004-8.

Harvey, C.F., Ashfaque, K.N., Yu, W. et al. (2006). Groundwater dynamics and arsenic contamination in Bangladesh. *Chemical Geology* 228 (1–3): 112–136. https://doi.org/10.1016/j.chemgeo.2005.11.025.

Henke, K. (2009). *Arsenic: Environmental Chemistry, Health Threats and Waste Treatment.* Wiley.

Herath, I., Vithanage, M., Bundschuh, J. et al. (2016). Natural arsenic in global groundwaters: distribution and geochemical triggers for mobilization. *Current Pollution Reports* 2 (1): 68–89. https://doi.org/10.1007/s40726-016-0028-2.

Islam, M.S. and Mostafa, M.G. (2021). Influence of chemical fertilizers on arsenic mobilization in the alluvial Bengal delta plain: a critical review. *Journal of Water Supply: Research and Technology-AQUA* 70 (7): 948–970. https://doi.org/10.2166/aqua.2021.043.

Jain, N. and Chandramani, S. (2018). Arsenic poisoning – an overview. *Indian Journal of Medical Specialities* 9 (3): 143–145. https://doi.org/10.1016/j.injms.2018.04.006.

Jang, Y.C., Somanna, Y., and Kim, H. (2016). Source, distribution, toxicity and remediation of arsenic in the environment – a review. *International Journal of Applied Environmental Sciences* 11 (2): 559–581.

Jiang, Z., Li, P., Wang, Y. et al. (2019). Arsenic mobilization in a high arsenic groundwater revealed by metagenomic and Geochip analyses. *Scientific Reports* 9 (1): 1–10. https://doi.org/10.1038/s41598-019-49365-w.

Khatri, N. and Tyagi, S. (2015). Influences of natural and anthropogenic factors on surface and groundwater quality in rural and urban areas. *Frontiers in Life Science* 8 (1): 23–39. https://doi.org/10.1080/21553769.2014.933716.

Kim, M.-J., Nriagu, J., and Haack, S. (2000). Carbonate ions and arsenic dissolution by groundwater. *Environmental Science & Technology* 34 (15): 3094–3100. https://doi.org/10.1021/es990949p.

Kinniburgh, D.G., Smedley, P.L., Davies, J. et al. (2003). The scale and causes of the groundwater arsenic problem in Bangladesh. In: *Arsenic in Ground Water* (ed. T. Appelo), 211–257. Springer. https://doi.org/10.1007/0-306-47956-7_8.

Kumar, C.P. (2015). Status and mitigation of arsenic contamination in groundwater in India. *The International Journal of Earth and Environmental Sciences* 1 (1): 1–10.

Kumar, M., Ramanatham, A.L., Mukherjee, A. et al. (2018). Hydrogeo-morphological influences for arsenic release and fate in the central Gangetic Basin, India. *Environmental Technology and Innovation* 12: 243–260. https://doi.org/10.1016/j.eti.2018.09.004.

Kurosawa, K., Egashira, K., Tani, M. et al. (2008). Variation in arsenic concentration relative to ammonium nitrogen and oxidation reduction potential in surface and groundwater. *Communications in Soil Science and Plant Analysis* 39 (9–10): 1467–1475. https://doi.org/10.1080/00103620802004318.

Kurosawa, K., Egashira, K., and Tani, M. (2013). Relationship of arsenic concentration with ammonium–nitrogen concentration, oxidation reduction potential and pH of groundwater in arsenic-contaminated areas in Asia. *Physics and Chemistry of the Earth, Parts A/B/C* 58: 85–88. https://doi.org/10.1016/j.pce.2013.04.016.

Kushawaha, J., Borra, S., and Kushwaha, A.K. (2021). Climate change and its impact on natural resources. In: *Water Conservation in the Era of Global Climate Change* (ed. B. Thokchom, P. Singh, P. Qiu, and P.K. Iyer), 333–346. Elsevier. https://doi.org/10.1016/B978-0-12-820200-5.00002-6.

Landner, L. (1998). Arsenic in the Aquatic Environment-speciation and Biological Effects. Technical Report, Sweden.

Langmuir, D., Charostowski, P., Vigneault, B., and Chaney, R. (2003). *Issue Paper on the Environmental Chemistry of Metals*, vol. 114. US Environmental Protection Agency.

Lin, Z. and Puls, R.W. (2000). Adsorption, desorption and oxidation of arsenic affected by clay minerals and aging process. *Environmental Geology* 39 (7): 753–759. https://doi.org/10.1007/s002540050490.

Lu, P. and Zhu, C. (2011). Arsenic Eh–pH diagrams at 25°C and 1 bar. *Environmental Earth Sciences* 62 (8): 1673–1683. https://doi.org/10.1007/s12665-010-0652-x.

Mahanta, C., Enmark, G., Nordborg, D. et al. (2015). Hydrogeochemical controls on mobilization of arsenic in groundwater of a part of Brahmaputra river floodplain, India. *Journal of Hydrology: Regional Studies* 4: 154–171. https://doi.org/10.1016/j.ejrh.2015.03.002.

Mandal, B.K. and Suzuki, K.T. (2002). Arsenic round the world: a review. *Talanta* 58 (1): 201–235. https://doi.org/10.1016/S0039-9140(02)00268-0.

Mannan, M., Al-Ansari, T., Mackey, H.R., and Al-Ghamdi, S.G. (2018). Quantifying the energy, water and food nexus: a review of the latest developments based on life-cycle assessment. *Journal of Cleaner Production* 193: 300–314. https://doi.org/10.1016/j.jclepro.2018.05.050.

Manning, B.A. and Goldberg, S. (1996). Modeling arsenate competitive adsorption on kaolinite, montmorillonite and illite. *Clays and Clay Minerals* 44 (5): 609–623. https://doi.org/10.1346/CCMN.1996.0440504.

Manning, B.A. and Goldberg, S. (1997). Adsorption and stability of arsenic(III) at the clay mineral–water interface. *Environmental Science & Technology* 31 (7): 2005–2011. https://doi.org/10.1021/es9608104.

Masuda, H. (2018). Arsenic cycling in the Earth's crust and hydrosphere: interaction between naturally occurring arsenic and human activities. *Progress in Earth and Planetary Science* 5 (1): 68. https://doi.org/10.1186/s40645-018-0224-3.

Mazumder, D.N.G., Ghosh, A., Majumdar, K.K. et al. (2010). Arsenic contamination of ground water and its health impact on population of district of Nadia, West Bengal, India. *Indian Journal of Community Medicine: Official Publication of Indian Association of Preventive & Social Medicine* 35 (2): 331. https://doi.org/10.4103/0970-0218.66897.

McArthur, J.M., Banerjee, D.M., Hudson-Edwards, K.A. et al. (2004). Natural organic matter in sedimentary basins and its relation to arsenic in anoxic ground water: the example of West Bengal and its worldwide implications. *Applied Geochemistry* 19 (8): 1255–1293. https://doi.org/10.1016/j.apgeochem.2004.02.001.

McLaren, S.J. and Kim, N.D. (1995). Evidence for a seasonal fluctuation of arsenic in New Zealand's longest river and the effect of treatment on concentrations in drinking water. *Environmental Pollution* 90 (1): 67–73. https://doi.org/10.1016/0269-7491(94)00092-R.

Mihajlov, I., Mozumder, M.R.H., Bostick, B.C. et al. (2020). Arsenic contamination of Bangladesh aquifers exacerbated by clay layers. *Nature Communications* 11 (1): 2244. https://doi.org/10.1038/s41467-020-16104-z.

Mirlean, N., Medeanic, S., Garcia, F.A., and Travassos, M.P. (2012). Arsenic enrichment in shelf and coastal sediment of the Brazilian subtropics. *Continental Shelf Research* 35: 129–136. https://doi.org/10.1016/j.csr.2012.01.006.

Mohapatra, D., Mishra, D., Chaudhury, G.R., and Das, R.P. (2007). Arsenic adsorption mechanism on clay minerals and its dependence on temperature. *Korean Journal of Chemical Engineering* 24 (3): 426–430. https://doi.org/10.1007/s11814-007-0073-z.

Mohindru, J.J., Garg, U.K., and Gupta, R. (2017). Coagulation-flocculation technologies for arsenic removal – a review. *Asian Journal of Research in Chemistry* 10 (3): 405–413. https://doi.org/10.5958/0974-4150.2017.00069.4.

Molden, D., Frenken, K., Barker, R. et al. (2007). Trends in water and agricultural development. In: *Water for Food, Water for Life: A Comprehensive Assessment of Water Management in Agriculture* (ed. S. Attapatu, M. Giordano, A. Inocencio, et al.), 57–89. Londres: Earthscan.

Moore, J.N., Ficklin, W.H., and Johns, C. (1988). Partitioning of arsenic and metals in reducing sulfidic sediments. *Environmental Science & Technology* 22 (4): 432–437. https://doi.org/10.1021/es00169a011.

Mukherjee, A., Sengupta, M.K., Hossain, M.A. et al. (2006). Arsenic contamination in groundwater: a global perspective with emphasis on the Asian scenario. *Journal of Health, Population and Nutrition* 24 (2): 142–163. www.jstor.org/stable/23499353.

Mukherjee, A., Fryar, A.E., and Thomas, W.A. (2009). Geologic, geomorphic and hydrologic framework and evolution of the Bengal basin, India and Bangladesh. *Journal of Asian Earth Sciences* 34 (3): 227–244. https://doi.org/10.1016/j.jseaes.2008.05.011.

Mukherjee, A., Scanlon, B.R., Fryar, A.E. et al. (2012). Solute chemistry and arsenic fate in aquifers between the Himalayan foothills and Indian craton (including central Gangetic plain): influence of geology and geomorphology. *Geochimica et Cosmochimica Acta* 90: 283–302. https://doi.org/10.1016/j.gca.2012.05.015.

Nath, B., Stuben, D., Mallik, S.B. et al. (2008). Mobility of arsenic in West Bengal aquifers conducting low and high groundwater arsenic. Part I: comparative hydrochemical and hydrogeological characteristics. *Applied Geochemistry* 23 (5): 977–995. https://doi.org/10.1016/j.apgeochem.2007.11.016.

Neil, C.W., Yang, J., Schupp, D., and Jun, Y.S. (2014). Water chemistry impacts on arsenic mobilization from arsenopyrite dissolution and secondary mineral precipitation: implications for managed aquifer recharge. *Environmental Science & Technology* 48 (8): 4395–4405. https://doi.org/10.1021/es405119q.

Németi, B. and Gregus, Z. (2002). Reduction of arsenate to arsenite in hepatic cytosol. *Toxicological Sciences* 70 (1): 4–12. https://doi.org/10.1093/toxsci/70.1.4.

Neumann, R.B., Ashfaque, K.N., Badruzzaman, A.B.M. et al. (2010). Anthropogenic influences on groundwater arsenic concentrations in Bangladesh. *Nature Geoscience* 3 (1): 46–52. https://doi.org/10.1038/ngeo685.

Nickson, R., McArthur, J., Burgess, W. et al. (1998). Arsenic poisoning of Bangladesh groundwater. *Nature* 395 (6700): 338–338. https://doi.org/10.1038/26387.

Nickson, R.T., McArthur, J.M., Ravenscroft, P. et al. (2000). Mechanism of arsenic release to groundwater, Bangladesh and West Bengal. *Applied Geochemistry* 15 (4): 403–413. https://doi.org/10.1016/S0883-2927(99)00086-4.

Nordstrom, D.K. (2002). Worldwide occurrences of arsenic in ground water. *Science* 296 (5576): 2143–2145. https://doi.org/10.1126/science.1072375.

Nordstrom, D.K. and Archer, D.G. (2003). Arsenic thermodynamic data and environmental geochemistry: an evaluation of thermodynamic data for modeling the aqueous environmental geochemistry of arsenic. In: *Arsenic in Ground Water*, 1–25. Boston, MA: Springer US https://doi.org/10.1007/0-306-47956-7_1.

Nordstrom, D.K., Majzlan, J., and Konigsberger, E. (2014). Thermodynamic properties for arsenic minerals and aqueous species. *Reviews in Mineralogy and Geochemistry* 79 (1): 217–255. https://doi.org/10.2138/rmg.2014.79.4.

O'reilly, S.E., Strawn, D.G., and Sparks, D.L. (2001). Residence time effects on arsenate adsorption/desorption mechanisms on goethite. *Soil Science Society of America Journal* 65 (1): 67–77. https://doi.org/10.2136/sssaj2001.65167x.

Olea, R.A., Raju, N.J., Egozcue, J.J. et al. (2018). Advancements in hydrochemistry mapping: methods and application to groundwater arsenic and iron concentrations in Varanasi, Uttar Pradesh, India. *Stochastic Environmental Research and Risk Assessment* 32 (1): 241–259. https://doi.org/10.1007/s00477-017-1390-3.

Ongley, L.K., Armienta, M.A., Heggeman, K. et al. (2001). Arsenic removal from contaminated water by the Soyatal Formation, Zimapán Mining District, Mexico – a potential low-cost low-tech remediation system. *Geochemistry: Exploration, Environment, Analysis* 1 (1): 23–31. https://doi.org/10.1144/geochem.1.1.23.

Panagiotaras, D. and Nikolopoulos, D. (2015). Arsenic occurrence and fate in the environment; a geochemical perspective. *Journal of Earth Science & Climatic Change* 6 (4): 1. https://doi.org/10.4172/2157-7617.1000269.

Postma, D., Larsen, F., Hue, N.T.M. et al. (2007). Arsenic in groundwater of the Red River floodplain, Vietnam: controlling geochemical processes and reactive transport modeling. *Geochimica et Cosmochimica Acta* 71 (21): 5054–5071. https://doi.org/10.1016/j.gca.2007.08.020.

Raghav, M., Sáez, A.E., and Ela, W.P. (2015). Understanding abiotic ferrihydrite re-mineralization by ferrous ions. *International Journal of Environmental Science and Technology* 12 (6): 1945–1956. https://doi.org/10.1007/s13762-014-0599-7.

Rahman, M., Vahter, M., Wahed, M.A. et al. (2006). Prevalence of arsenic exposure and skin lesions. A population based survey in Matlab, Bangladesh. *Journal of Epidemiology & Community Health* 60 (3): 242–248. https://doi.org/10.1136/jech.2005.040212.

Raju, N.J. (2022). Arsenic in the geo-environment: a review of sources, geochemical processes, toxicity and removal technologies. *Environmental Research* 203: 111782. https://doi.org/10.1016/j.envres.2021.111782.

Ratnaike, R.N. (2003). Acute and chronic arsenic toxicity. *Postgraduate Medical Journal* 79 (933): 391–396.

Rees, W. (2018). Ecological footprint. In: *Companion to Environmental Studies* (ed. N. Castree, M. Hulme, and J.D. Proctor), 43–48. Routledge.

Saha, K.C. (1995). Chronic arsenical dermatosis from tube-well water in West Bengal during 1983–1987. *Indian Journal of Dermatology* 40: 1–12.

Saha, D. (2009). Arsenic groundwater contamination in parts of middle Ganga plain, Bihar. *Current Science* 97 (6): 753–755.

Saha, D. and Sahu, S. (2016). A decade of investigations on groundwater arsenic contamination in Middle Ganga Plain, India. *Environmental Geochemistry and Health* 38 (2): 315–337. https://doi.org/10.1007/s10653-015-9730-z.

Saha, J.C., Dikshit, A.K., Bandyopdhayay, M., and Saha, K.C. (1999). A review of arsenic poisoning and its effects on human health. *Critical Reviews in Environmental Science and Technology* 29 (3): 281–313. https://doi.org/10.1080/10643389991259227.

Saha, S., Reza, A.H.M., and Roy, M.K. (2021). Arsenic geochemistry of the sediments of the shallow aquifer and its correlation with the groundwater, Rangpur, Bangladesh. *Applied Water Science* 11 (10): 1–11. https://doi.org/10.1007/s13201-021-01495-1.

Sanz, E., Muñoz-Olivas, R., Cámara, C. et al. (2007). Arsenic speciation in rice, straw, soil, hair and nails samples from the arsenic-affected areas of Middle and Lower Ganga plain. *Journal of Environmental Science and Health, Part A* 42 (12): 1695–1705. https://doi.org/10.1080/10934520701564178.

Sarkar, B. (2002). *Heavy Metals in the Environment*. CRC Press. https://doi.org/10.1201/9780203909300.

Shaji, E., Santosh, M., Sarath, K.V. et al. (2021). Arsenic contamination of groundwater: a global synopsis with focus on the Indian Peninsula. *Geoscience Frontiers* 12 (3): 101079. https://doi.org/10.1016/j.gsf.2020.08.015.

Shakya, A.K. and Ghosh, P.K. (2018). Simultaneous removal of arsenic, iron and nitrate in an attached growth bioreactor to meet drinking water standards: importance of sulphate and empty bed contact time. *Journal of Cleaner Production* 186: 1011–1020. https://doi.org/10.1016/j.jclepro.2018.03.176.

Shakya, A.K., Rajput, P., and Ghosh, P.K. (2018). Investigation on stability and leaching characteristics of mixtures of biogenic arsenosulphides and iron sulphides formed under reduced conditions. *Journal of Hazardous Materials* 353: 320–328. https://doi.org/10.1016/j.jhazmat.2018.04.031.

Shankar, S., Shanker, U., and Shikha (2014). Arsenic contamination of groundwater: a review of sources, prevalence, health risks, and strategies for mitigation. *The Scientific World Journal* 2014: 1–18. https://doi.org/10.1155/2014/304524.

Shrestha, R.R., Shrestha, M.P., Upadhyay, N.P. et al. (2003). Groundwater arsenic contamination, its health impact and mitigation program in Nepal. *Journal of*

Environmental Science and Health, Part A 38 (1): 185–200. https://doi.org/10.1081/ESE-120016888.

Smedley, P.L. and Kinniburgh, D.G. (2002). A review of the source, behaviour and distribution of arsenic in natural waters. *Applied Geochemistry* 17 (5): 517–568. https://doi.org/10.1016/S0883-2927(02)00018-5.

Smith, K.S. (1999). Metal sorption on mineral surfaces: an overview with examples relating to mineral deposits. *Reviews in Economic Geology* 6: 161–182.

Smith, E.R.G., Naidu, R., and Alston, A.M. (1998). Arsenic in the soil environment. PhD thesis. Academic Press, Cambridge, MA.

Smith, A.H., Lingas, E.O., and Rahman, M. (2000). Contamination of drinking-water by arsenic in Bangladesh: a public health emergency. *Bulletin of the World Health Organization* 78 (9): 1093–1103.

Strawn, D.G. (2018). Review of interactions between phosphorus and arsenic in soils from four case studies. *Geochemical Transactions* 19 (1): 10. https://doi.org/10.1186/s12932-018-0055-6.

Stüben, D., Berner, Z., Chandrasekharam, D., and Karmakar, J. (2003). Arsenic enrichment in groundwater of West Bengal, India: geochemical evidence for mobilization of As under reducing conditions. *Applied Geochemistry* 18 (9): 1417–1434. https://doi.org/10.1016/S0883-2927(03)00060-X.

Takahashi, Y., Minamikawa, R., Hattori, K.H. et al. (2004). Arsenic behavior in paddy fields during the cycle of flooded and non-flooded periods. *Environmental Science & Technology* 38 (4): 1038–1044. https://doi.org/10.1021/es034383n.

Tseng, C.H., Huang, Y.K., Huang, Y.L. et al. (2005). Arsenic exposure, urinary arsenic speciation, and peripheral vascular disease in blackfoot disease-hyperendemic villages in Taiwan. *Toxicology and Applied Pharmacology* 206 (3): 299–308. https://doi.org/10.1016/j.taap.2004.11.022.

United States Environmental Protection Agency (U.S.EPA) (2016). Overview of wood preservative chemicals [Internet]. https://www.epa.gov/ingredients-used-pesticideproducts/overview-wood-preservative-chemicals-0 (accessed 6 December 2022).

Verma, S., Mukherjee, A., Choudhury, R., and Mahanta, C. (2015). Brahmaputra river basin groundwater: solute distribution, chemical evolution and arsenic occurrences in different geomorphic settings. *Journal of Hydrology: Regional Studies* 4: 131–153. https://doi.org/10.1016/j.ejrh.2015.03.001.

Vorosmarty, C.J., Green, P., Salisbury, J., and Lammers, R.B. (2000). Global water resources: vulnerability from climate change and population growth. *Science* 289 (5477): 284–288.

Walker, F.P., Schreiber, M.E., and Rimstidt, J.D. (2006). Kinetics of arsenopyrite oxidative dissolution by oxygen. *Geochimica et Cosmochimica Acta* 70 (7): 1668–1676. https://doi.org/10.1016/j.gca.2005.12.010.

Wang, S. and Mulligan, C.N. (2006). Natural attenuation processes for remediation of arsenic contaminated soils and groundwater. *Journal of Hazardous Materials* 138 (3): 459–470. https://doi.org/10.1016/j.jhazmat.2006.09.048.

Webster, J.G., Marshall, C.P., and Fairbridge, R.W. (1999). *Encyclopaedia of Geochemistry*, 21–22. London: Chapman Hall.

World Health Organization (2004). *Guidelines for Drinking-Water Quality*, vol. 1. World Health Organization.

Yadav, S.K., Ramanathan, A., Kumar, M. et al. (2020). Assessment of arsenic and uranium co-occurrences in groundwater of central Gangetic Plain, Uttar Pradesh, India. *Environmental Earth Sciences* 79 (6): 154. https://doi.org/10.1007/s12665-020-8892-x.

Zhang, Z., Moon, H.S., Myneni, S.C.B., and Jeffe, P.R. (2017). Phosphate enhanced abiotic and biotic arsenic mobilization in the wetland rhizosphere. *Chemosphere* 187: 130–139. https://doi.org/10.1016/j.chemosphere.2017.08.096.

Zheng, Y., Stute, M., van Geen, A. et al. (2004). Redox control of arsenic mobilization in Bangladesh groundwater. *Applied Geochemistry* 19 (2): 201–214. https://doi.org/10.1016/j.apgeochem.2003.09.007.

Zhu, F., Yang, M., Luo, Z.X. et al. (2020). Bioaccumulation and biotransformation of arsenic in *Leptolyngbya boryana*. *Environmental Science and Pollution Research* 27 (24): 29993–30000. https://doi.org/10.1007/s11356-020-09294-y.

13

Atmospheric CO_2 Consumption Associated With Chemical Weathering in the Riverine Ecosystem

Sushil Kumar

Department of Environmental Studies, Siksha Bhavana (Institute of Science), Visva-Bharati, Santiniketan, West Bengal, India

13.1 Introduction

Atmospheric carbon dioxide (CO_2) is an important component of the biogeochemical cycle, climate change, pH balance in aquatic systems, and terrestrial and aquatic photosynthesis in the riverine environment (Liu et al. 2011; Moosdorf et al. 2011; Regnier et al. 2022; Wen et al. 2022). The rapid economic and developmental activities suddenly increased the atmospheric CO_2 level in the last few decades (Inal et al. 2022; Mohsin et al. 2022). With the industrial revolution and rapid growth of industrial clusters across the globe, land use patterns, CO_2 consumption, and emission pattern are severely affected (Zhonghua et al. 2022). In addition, coal mining and burning activities exponentially amplify the atmospheric CO_2 budget in the ambient atmosphere. The continuous rise of CO_2 in the atmosphere is responsible for global warming and climate change (Shabani et al. 2022). The long-term shifting of weather patterns across the globe is the result of elevated levels of CO_2 in the atmosphere (Helbig et al. 2022). The uncontrolled discharge of polluted water due to exponential industrial growth severely degrades the water quality and sediment, and soil chemistry of catchment areas of rivers across the world. The catchment area of the riverine environment becomes a receptor for various wastes, such as municipal and industrial, and degrades the river and soil sediment. In addition, the atmospheric deposition of various pollutants in the riverine environment is an additional source of degradation of the physicochemical environment in the river catchment area. Further, the exchange of atmospheric CO_2 between inland water, ocean, and atmosphere is important for biogeochemical and ecological processes in the environment. The global budget for atmospheric CO_2 is severely affected due to the exponential growth of industrialization, urbanization, and fossil fuel burning (Liu et al. 2018). The continuous circulation of CO_2 between the ocean, atmosphere, and land is affected due to intense human activities. The consumption of atmospheric CO_2 through chemical weathering (silicate and carbonate weathering) mainly regulates the global climate for a long time (Kump et al. 2000). The atmospheric dissolution of CO_2 and interaction with calcium silicate fixed CO_2 in the form of $CaCO_3$ in the aquatic

Weathering and Erosion Processes in the Natural Environment, First Edition. Edited by Virendra Bahadur Singh, Sughosh Madhav, Naresh Chandra Pant, and Ravi Shekhar.

system for a long time scale. In addition, the biological fixation of atmospheric CO_2 during photosynthesis in the plant is the sink of carbon. On death and decay of plants, their burial deposit on surface sediments in aquatic environments is a long-term sink of carbon. However, the uplifting and the rapid dissolution of carbonate minerals regulates the carbon flux. The balanced ratio of silicate and carbonate weathering over the earth is important for the long-term global balance of atmospheric CO_2. The movement of carbon from the atmosphere to the biological organism (photosynthetic organism) to the earth's crust through death and deposition in the aquatic system regulates carbon distribution.

The river is considered the lifeline of human civilization development, and in ancient times, major civilizations were located along the river. The availability of resources in terms of food, water, and shelter makes it favorable to settle various civilizations along the river. across the world. The long-term changes in the riverine ecosystem make it hospitable for human settlement through weathering and erosion processes. The deposited carbon in the aquatic system (ocean, river, pond, lake, etc.) is present in the form of dissolved organic carbon (DOC) and dissolved inorganic carbon (DIC) (Donnini et al. 2016). The dissolved carbon content in the river system moves toward the ocean. The rate of riverine transport of dissolved carbon to the ocean depends upon weathering rate, climate, and lithological factors of the catchment area. The global landscape (22% of the global area) is occupied by a tropical humid environment, and rivers originating in this region contribute 38% dissolved solute and 65% dissolved silica from the land to the ocean (Achard et al. 2002; Amrish et al. 2022). The current land use change pattern over the globe mainly affects carbon transport toward the ocean system. The earth's atmosphere and climate are ocean–atmosphere-coupled phenomena that regulate global temperature. The exchange of atmospheric carbon dioxide (CO_2) between the ocean atmosphere and land atmosphere also plays a key role in global temperature balance and climate regulation. The continuous transport of dissolved CO_2 from land to the ocean through rivers regulates the global carbon cycle. Further, the dissolution and transport of atmospheric CO_2 play a vital role in climate change, global warming, nutrient dynamics, rock weathering, and pH regulation in surface water and rainwater chemistry.

This chapter provides a detailed insight into the current role of atmospheric CO_2 in chemical weathering in the riverine environment. In addition, it also provides a detailed overview of the possible factors affecting CO_2-induced weathering, chemical weathering and ecosystem development, ocean–atmospheric interaction, emission flux in the atmosphere from combustion sources, global carbon cycle, and nutrient dynamics.

13.2 Weathering and Ecosystem

The fundamental concept of understanding the weathering processes is mainly the interaction of biotic with an abiotic component of the ecosystem. Figure 13.1 describes the interaction of biotic and abiotic components and their role in chemical weathering. However, various models are used to understand chemical weathering considering physical, chemical, and biological aspects. In addition, it covers laboratory-based observation to understand the weathering pattern compared with the field-based weathering rate (Sverdrup, 2009). Brantley et al. (2011) have explained the development of a hospitable

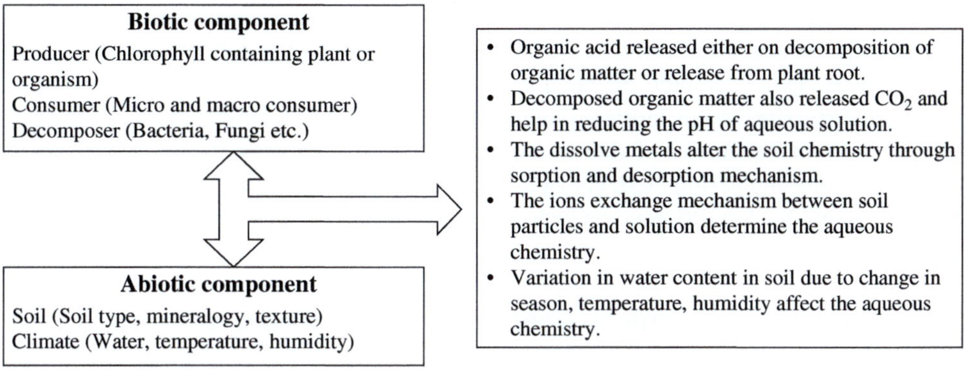

Figure 13.1 Graphical representation showing the role of biotic and abiotic interaction in the ecosystem and chemical weathering.

environment on the earth's surface using 12 hypotheses. They summarize the role of plants, and microorganic and dead organic matter on earth's surface development for a hospitable environment and nutrient dynamics. Chemical weathering is initiated by the physical and biological activity of various organisms including plants and microorganisms in the environment. The soil formation on the earth's crust is strongly associated with biotic components, and it determines the rate of chemical weathering and biological activity (Dontsova et al. 2020). The formation of aqueous soil solutions in the riverine environment is the result of interaction with soil minerals, decomposed organic matter, living organic matter, plants release matter, and water content. In addition, dissolved atmospheric CO_2 affects the chemistry of aqueous solution. The weathering processes start with the formation of a colloidal solution due to the dissolution of organic matters and minerals. The microorganism-induced degradation of biomass residue and dead organic matter favored the dissolution of gases in an aqueous medium. The plantation and soil amendment with powder of basalt rock potentially sequestered the atmospheric CO_2 for a longer time. The rate of atmospheric CO_2 sequestration and chemical weathering processes increased in silicate rich powder of basalt rock (Goll et al. 2021). Recently, soil amendment with basalt powder in agriculture improves the productivity of soil and sequestered the atmospheric CO_2 and quality of the soil. Beerling et al. (2020) have observed the enhanced silicate weathering to reduce atmospheric CO_2 using mining overburden material and concrete use in soil amendment practices. Balogh-Brunstad et al. (2008) performed a 20-year sandbox experiment to understand the minerals weathering pattern and its effect on nutrient uptake in the plant. Further, they suggested that highly chemical weather shows quite high soil-nutrient uptake (Ca^{2+}, Mg^{2+}, and K^{+}) and retention in plant biomass. Wen et al. (2022) have observed that the soil biotas have exported atmospheric CO_2 vertically while associated rivers and streams in the ecosystem latterly transport carbon in the form of DIC and DOC. Further, they quantify using the model and reported underwater stress conditions that 86% of inorganic carbon was emitted in the form of CO_2, while 14% was exported as lateral in the form of DIC.

However, the concept for dissolution of different chemical species (inorganic and organic) in the microenvironment is feasible, while in the macro environment, the interaction of vegetation (herb, shrub, and tree) with their plant root could be quite different.

The long roots of higher plants forcefully penetrate the soil and release various chemical species (organic acid and CO_2). The weathering pattern and development of nude area for ecological succession mainly regulated by the climatic variability and biotic component. The weathering trend using biotic and abiotic proxies data over Chilika Lagoon, India were studied and it shows the rate of chemical weathering continuously increases from the early to the late Holocene period Dash et al. (2023).The weathering regime pattern shifted from chemical to physical in different geological time scales Harzhauser et al. (2023).

13.3 Drivers of Chemical Weathering in the Riverine Ecosystem

The disintegration of parent rock into small fractions undergoes physical, chemical, and biological weathering in the environment (Dethier et al. 2022). In geological time scales, riverine environments are an active site for weathering processes due to the availability of various weathering agents (Moosdorf et al. 2011). The development of newly deposited sediment over the riverine environment is an active site for chemical weathering. The global transport of weathered sediment from the terrestrial environment through the river system is mainly affected by the land use change pattern (Kump et al. 2000). Dethier et al. (2022) have used the field-based measurement of 414 major river water and satellite-based investigation (1980 onward) and reported that 41% sediment transport are trapped in the human constructed dam while $41 \pm 7\%$ transport to the ocean due to change land use pattern over the tropical environment. The chemical weathering reaction in the riverine environment is mainly regulated by the weathering agent released from weak and strong acids. Among various agents, protons (H^+) released from weak and strong acids from multiple sources are a key component of the weathering process. However, mineral composition, regional geology, temperature, and water content play important roles in chemical weathering (Kump et al. 2000). The generation of the soluble product (e.g. HCO_3^-) is a product of carbonate weathering) and their concentrations totally depend upon the hydrogeochemistry of water. The availability of weathering minerals over the riverine environment also depends upon the process of formation. The deposited minerals are either fresh or old and deposited over the weathering environment. The freshly deposited minerals are sensitive to chemical weathering. In contrast, the old deposited minerals are poorly sensitive to weathering process. The uplifting of deposited sediment due to tectonic events and long-term deposition through geological time scale are also drivers of chemical weathering. Aquatic pH balance is maintained via the following chemical reactions (Eqs. 13.1–13.3):

$$CO_2 + H_2O \rightleftharpoons H_2CO_3 \tag{13.1}$$

$$H_2CO_3 \rightleftharpoons HCO_3^- + H^+ \tag{13.2}$$

$$HCO_3- \rightleftharpoons CO_3^{2-} + H^+ \tag{13.3}$$

The atmospheric CO_2 or biologically induced CO_2 in reaction with water content produced carbonic acid (Reactions 13.1–13.3). The natural weak acid is the primary reagent to initiate chemical weathering. However, the stability of chemical reactions depends upon the pH of the aquatic environment. The alkalinity or acidity of the aquatic medium will depend upon the molar concentration of HCO_3^-, CO_3^{2-}, and H^+ ions. The end product of Eq. 13.3 combined with Ca^{2+} released from various rocks produces $CaCO_3$ minerals. The intensity of chemical weathering over riverine ecosystems depends upon the availability of weathering agents (e.g. H^+ in aquatic medium). The availability of H^+ either from weak or strong acids is of greater interest for soil weathering. The dissolved atmospheric CO_2 from H_2CO_3 is an excellent source of H^+ in the natural environment. In addition, H^+ also contributed from anthropogenic sources (H_2SO_4 and HNO_3). The combustion product of SO_x and NO_x on oxidation produces H_2SO_4 and HNO_3, which is a source of H^+ in aquatic solution. The biologically emitted weak acid such as carboxylic acid (e.g. CH_3COOH) amplifies the rate of weathering reaction. Yadav and Rajamani (2004) have reported intense aeolian activities in and around the catchment area, which significantly affect the river sediment deposit and weathering pattern. Further, they suggest physical sorting and chemical weathering mainly that determine the chemistry of sediment in river catchment areas. Silicate weathering can be calculated on the evaluation of dissolved components in river runoff, while carbonate weathering significantly affects the chemical composition of water. A recent study suggests that agricultural and industrial activities increased the Cl^- and Ca^{2+} content in riverine water (Gao et al. 2023). Hossain et al. (2023) have observed the high and least anthropogenically affected transboundary river sediments, in Punarbhaba, Bangladesh. They reported enrichment and enhance chemical weathering due to the influence of fluvial deposition, sediment fractional deposition, and industrial disintegration. Cotton et al. (2013) have reported that the rapidly increased CO_2 level in the ambient atmosphere shifts the precipitation pattern in the twenty-first century across the world.

13.4 Human-Induced Drivers of Weathering Agents in the Riverine Environment

The exponential growth of industries and emission patterns continuously emits various weathering agents (acidic components) other than atmospheric CO_2 in the atmosphere. The emitted weathering agents include an oxide of sulfur (SO_2), nitrogen (NO_X), and nitrogen fertilizers. The emitted weathering agents in the atmosphere get deposited either dry or wet over the catchment area of the riverine system and are induced by the natural weathering processes. The acid-forming precursor gases undergo oxidation and form different acidic components. Among them, the continuous emission of SO_2 and NO_2 gets oxidized into H_2SO_4 and HNO_3 and deposited either through wet deposition or dry deposition (Liu et al. 2022). Zhonghua et al. (2022) have studied the historical (1950–1970) and current water chemistry data (2018–2021) for different rivers in the Taihu region using a mass-balance approach. They suggest that, in recent years (2018–2021), silicate and carbonate weathering increased 3.1 and 4.9 times compared to historical data, respectively. The increased rate of weathering is associated with an increased level of SO_2 in different watersheds in the Taihu region.

In addition, acidic components are continuously discharged into a nearby river or aquatic system in the form of acid mine drainage. In the downwind location from industries and mines, acid-forming precursor gases are enriched and amplify the weathering reaction.

Raymond and Hamilton (2018) have reported that the transport flux of HCO_3^- is severely affected by human activities such as mining, fossil fuel combustion, and modern agriculture practices. The increasing atmospheric load due to the continuous emission of particulate matter (PM) and precursor gases leads to contamination of the riverine catchment area. The chemically PM constituted 30–50% water-soluble inorganic ions, and it varies with variation in source, season, and meteorological condition. Among all, water-soluble inorganic ions, sulfate, and nitrate are major contributors to the total mass. Azam et al. (2018) have reported the trace element concentration in dissolved and particulate bound forms for tributaries of the Ganga River water. Further, they suggested high dissolved flux compared to particulate bound flux could be associated with human-induced activities.

13.5 Atmospheric CO_2, Carbonate, and Silicate Weathering

Atmospheric CO_2-induced chemical weathering is vital for nutrient cycling and key component of the global carbon cycle (Zhao and Wang, 2023). Therefore, the atmospheric carbon source, sink, and role in chemical weathering need to be explored in detail to understand the earth's surface development processes and nutrient dynamics. The carbon content in aquatic systems is present in DOC and DIC. The dissolved carbon content is vital for photosynthesis in an aquatic plant. The aquatic plants present in inland water (river, lake, and pond) and marine water consume the dissolved carbon. The degree of solubility of CO_2 in water depends upon the temperature, salinity, and atmospheric partial pressure. The major portion of carbon content in the carbon pool is deposited at the bottom of the aquatic system in the form of calcium carbonate and organic form and immobile in the carbon pool. The labile form of carbon content presents a dissolved form (H_2CO_3, HCO_3^-, and CO_3^{2-}).

$$CaSiO_3 + CO_2 \rightarrow CaCO_3 + SiO_2$$

$$CaCO_3 + CO_2 + H_2O \rightleftharpoons Ca^{2+} + 2HCO_3^-$$

The silicate weathering of rock controlled long-term climate change and provided the feedback mechanism on CO_2 consumption (Liu et al., 2011; Schopka et al., 2011; Raymond and Hamilton, 2018). The removal of atmospheric CO_2 through the enhanced silicate weathering technique is an emerging tool to meet the United Nations goal to reduce the global mean temperature up to 1.5–2 °C (Meysman and Montserrat, 2017). Further, fast-weathering silicate mineral was introduced over the coastal area to reduce the atmospheric CO_2 concentration. Mao et al. (2022) have discussed that silicate weathering is a key component to regulate the global climate and biogeochemical cycle. Further, based on the Na^+ depletion rate and chemical depletion factor, the weathering rate is affected by the monthly mean temperature and monthly mean water availability over the region. The atmospheric CO_2 gets utilized by the silicate rock and deposited in carbonate form. However, dissolved silica is released in aquatic systems. In contrast, deposited carbonate in a lower pH medium dissolved in water releases the bicarbonate in the aquatic system. The fast dissolution and current changing pH of marine water and wet atmospheric deposition over river, lake, ocean, and catchment area carbonate weathering are significantly affected. Wang et al.

(2022) have observed the hydrogeochemistry of the Xiaojiang River in China and suggested cations dominated with 56.7% followed by atmospheric contribution (15.2%) and silicate weathering (13.9%). Further, they suggested the high rate of carbonate and silicate weathering. The consumed CO_2 in silicate weathering over the earth's surface can be calculated by the assessment of the chemical composition of surface runoff and hydrogeochemistry of the nearby aquatic system (Gao et al. 2023). Further, each analyzed chemical species (Cl^-, Na^+, and Ca^{2+}) acts as a proxy for the particular sources over the area. Pattanaik et al. (2013) have calculated the atmospheric consumption of CO_2 using the dissolved load of various solutes of the Kaveri River and its tributaries in south India using the average annual flow and total drainage. Further, they reported silicate weathering rate is 3.8 ± 0.1 $10 \times 105\,mol/km^2/a$ while atmospheric consumption of CO_2 is $0.2 \pm 0.03 \times 105\,mol/km^2/a$ over the Kaveri River Basin and its tributaries river. The dissolved load in the river water system is derived mainly due to weathering of dolomite, atmosphere silicate, and limestone in the catchment area of the river. Zhao et al. (2023) have reported that silicate weathering increases ($0.03\,t/km^2$/year to $2.4\,t/km^2$/year) towards the downstream of river while carbonate weathering is high ($2.1\,t/km^2$/year to $32.7\,t/km^2$/year) over the upstream region of the river.

The riverine ecosystem is one of the most fertile and productive regions of the world, and it forms through long-term weathering and erosion processes. The weathering process starts with the formation of soluble and insoluble chemical species. The availability of minerals and their type largely determine the weathering over river catchment areas. The abundance of various types of minerals depends upon the rate of deposition and removal of material over the catchment area. In addition, chemical weathering over catchment area rivers depends upon the availability of carbon dioxide, moisture content, nature of minerals, temperature, and biological activities. The variations in climatic factors (temperature, pressure, and relative humidity) and edaphic factors are closely associated with geographical location. The tropical regions over the globe have quite a high temperature and humidity compared to mid-latitude and polar regions and are favored by weathering and erosion processes. Wood et al. (2023) observe that the long-term geological silicate weathering removes CO_2 due to use in chemical processes. Further, they use an engineered rock weathering approach to increase the CO_2 consumption in processes to mitigate the climate change. In addition, the column-based absorption mechanism consumes maximum CO_2.

13.6 Chemical Weathering and Its Factor Affecting

The mechanism of weathering processes in a riverine environment is regulated via lithology, average rainfall, humidity, temperature, erosion rate, and biological activity. The nature and type of rock formed during the geological time scale also regulate the rate of chemical weathering (Figure 13.2).

Schopka et al. (2011) have studied the behavior of chemicals in a tropical climate and tectonically and volcanic active region over Luzon Island, the Philippines. In addition, silicate weathering amplify bicarbonate, cation flux and CO_2 sequestration over the study area compared to other parts of the world. Hilton and West (2020) describe the mountain building and associated weathering and erosion-regulated carbon cycle on the earth.

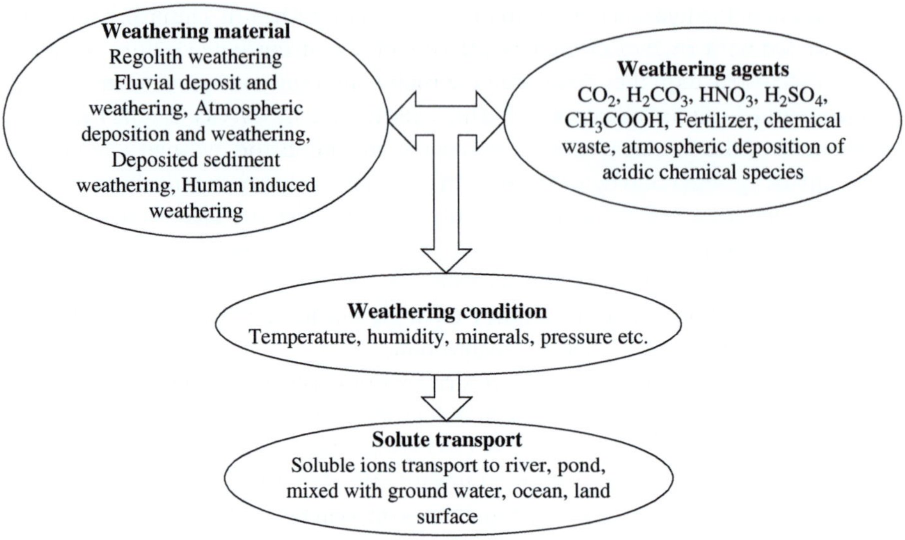

Figure 13.2 General overview of weathered material transport from the river catchment area to the ocean and its possible factor affecting.

The weathered carbon material transport toward the ocean through river transport is an excellent path for nutrient transport and carbon dynamics. Maharana et al. (2018) have observed the weathering pattern of sediment using the chemical index of alteration (CIA) values for Himalayan and peninsular rivers. Further, they suggested that the high CIA value for the peninsular river compared to the Himalayan river could be controlled by weatherable lithology, tectonically stable sediment, and alteration in climatic conditions. Human activities either increased or reduced carbon sequestration in the ocean, inland aquatic, and terrestrial systems. However, human-induced emission of weathering agents amplifies the weathering rate in a riverine environment. The carbonate chemical weathering induced by H_2SO_4 and HNO_3 up to 18.84% to total chemical weathering in the subtropical karst basin in south China (Lyu et al. 2018).

13.7 Conclusion

The captured atmospheric CO_2 in the riverine environment is the driver for chemical weathering due to the formation of carbonic acid in the natural environment. The consumed atmospheric CO_2 in biotic component indirectly participates in chemical weathering. The consumed CO_2 on the land surface in the presence of water content helps in the earth's surface transformation, improves nutrient dynamics, and maintains the global carbon cycle. In addition, vegetation density, water availability, and lithology fluvial deposits play an additional role in chemical weathering in the riverine catchment area. Enhanced silicate and carbonate weathering help in the long-term consumption of atmospheric CO_2 and helps in climate regulation and nutrient dynamics. Enhancing chemical weathering by mixing with silicate minerals helps in climate mitigation strategies. The atmospheric deposition of human-induced acidic components and changes in land use patterns enhance chemical weathering in river catchment areas. Dissolved and particulate-bound weathered materials are transported toward the ocean through the river channel during the rainy season to help in the global carbon cycle. The conservation of river catchment areas and their management help in improving terrestrial and aquatic carbon storage.

References

Achard, F., Eva, H.D., Stibig, H.J. et al. (2002). Determination of deforestation rates of the world's humid tropical forests. *Science* 297: 999–1002.

Amrish, V.N., Arun, K., Balakrishna, K. et al. (2022). Major ion chemistry and silicate weathering rate of a small Western Ghats river, Sharavati, southwestern India. *Applied Geochemistry* 136: 105182.

Azam, M.M., Kumari, M., Maharana, C. et al. (2018). Recent insights into the dissolved and particulate fluxes from the Himalayan tributaries to the Ganga River. *Environmental Earth Sciences* 77: 1–14.

Balogh-Brunstad, Z., Keller, C.K., Bormann, B.T. et al. (2008). Chemical weathering and chemical denudation dynamics through ecosystem development and disturbance. *Global Biogeochemical Cycles* 22: GB1007. https://doi.org/10.1029/2007GB002957.

Beerling, D.J., Kantzas, E.P., Lomas, M.R. et al. (2020). Potential for large-scale CO_2 removal via enhanced rock weathering with croplands. *Nature* 583: 242–248.

Brantley, S.L., Megonigal, J.P., Scatena, F.N. et al. (2011). Twelve testable hypotheses on the geobiology of weathering. *Geobiology* 9: 140–165.

Cotton, J.M., Jeffery, M.L., and Sheldon, N.D. (2013). Climate controls on soil respired CO_2 in the United States: implications for 21st century chemical weathering rates in temperate and arid ecosystems. *Chemical Geology* 358: 37–45.

Dash, C., Dhal, S.P., Pati, P. et al. (2023). Climate-induced denudation of the Eastern Ghat during the Holocene: a multi-proxy study from Chilika Lagoon (India). *Catena* 221: 106754.

Dethier, E.N., Renshaw, C.E. and Magilligan, F.J., 2022. Rapid changes to global river suspended sediment flux by humans. *Science* 376:1447–1452. https://doi.org/https://doi.org/10.1126/science.abn7980

Donnini, M., Frondini, F., Probst, J.L. et al. (2016). Chemical weathering and consumption of atmospheric carbon dioxide in the Alpine region. *Global and Planetary Change* 136: 65–81.

Dontsova, K., Balogh-Brunstad, Z., and Chorover, J. (2020). Plants as drivers of rock weathering. *Biogeochemical Cycles: Ecological Drivers and Environmental Impact* 33–58. https://doi.org/10.1002/9781119413332.ch2.

Gao, Q., Peng, H., and Liao, P. (2023). Chemical weathering and CO_2 consumption rate of montane silicate in South China: a case study of the Xizhijiang River Basin. *Environmental Earth Sciences* 82: 1–18.

Goll, D.S., Ciais, P., Amann, T. et al. (2021). Potential CO_2 removal from enhanced weathering by ecosystem responses to powdered rock. *Nature Geoscience* 14: 545–549.

Harzhauser, M., Peresson, M., Benold, C. et al. (2023). Environmental shifts in and around Lake Pannon during the Tortonian Thermal Maximum based on a multi-proxy record from the Vienna Basin (Austria, Late Miocene, Tortonian). *Palaeogeography Palaeoclimatology Palaeoecology* 610: 111332.

Helbig, M., Živkovic, T., Alekseychik, P. et al. (2022). Warming response of peatland CO_2 sink is sensitive to seasonality in warming trends. *Nature Climate Change* 12: 743–749.

Hilton, R.G. and West, A.J. (2020). Mountains, erosion and the carbon cycle. *Nature Reviews Earth and Environment.* 1: 284–299.

Hossain, S., Khan, R., Anik, A.H. et al. (2023). Natural and anthropogenic contributions to the elemental compositions and subsequent ecological consequences of a transboundary river's sediments (Punarbhaba, Bangladesh). *Environmental Research* 216: 114444.

İnal, V., Addi, H.M., Çakmak, E.E. et al. (2022). The nexus between renewable energy, CO_2 emissions, and economic growth: empirical evidence from African oil-producing countries. *Energy Reports* 8: 1634–1643.

Kump, L.R., Brantley, S.L. and Arthur, M.A., 2000. Chemical weathering, atmospheric CO_2, and climate.

Liu, J., Han, G., Zhang, Q. et al. (2022). Stable isotopes and Bayesian tracer mixing model reveal chemical weathering and CO_2 release in the Jiulongjiang River basin, Southeast China. *Water Resources Research* 58: 2021WR031738.

Liu, Z., Dreybrodt, W., and Liu, H. (2011). Atmospheric CO_2 sink: silicate weathering or carbonate weathering? *Applied Geochemistry* 26: S292–S294.

Liu, Z., Macpherson, G.L., Groves, C. et al. (2018). Large and active CO_2 uptake by coupled carbonate weathering. *Earth-Science Reviews* 182: 42–49.

Lyu, X., Tao, Z., Gao, Q. et al. (2018). Chemical weathering and Riverine carbonate system driven by human activities in a subtropical Karst Basin, South China. *Water.* 10: 1524.

Maharana, C., Srivastava, D., and Tripathi, J.K. (2018). Geochemistry of sediments of the Peninsular rivers of the Ganga basin and its implication to weathering, sedimentary processes and provenance. *Chemical Geology* 483: 1–20.

Mao, H.R., Cui, L.F., Zhang, Z.J. et al. (2022). Influence of monsoon climate on chemical weathering of granitic regoliths. *Global Biogeochemical Cycles* 2022GB007362.

Meysman, F.J. and Montserrat, F. (2017). Negative CO_2 emissions via enhanced silicate weathering in coastal environments. *Biology Letters* 13: 20160905.

Mohsin, M., Naseem, S., Sarfraz, M., and Azam, T. (2022). Assessing the effects of fuel energy consumption, foreign direct investment and GDP on CO_2 emission: new data science evidence from Europe & Central Asia. *Fuel* 314: 123098.

Moosdorf, N., Hartmann, J., Lauerwald, R., Hagedorn, B. and Kempe, S., 2011. Atmospheric CO_2 consumption by chemical weathering in North America. *Geochimica et Cosmochimica Acta*, 75(24), pp.7829–7854. https://doi.org/https://doi.org/10.1016/j.gca.2011.10.007

Pattanaik, J.K., Balakrishnan, S., Bhutani, R., and Singh, P. (2013). Estimation of weathering rates and CO_2 drawdown based on solute load: significance of granulites and gneisses dominated weathering in the Kaveri River basin, Southern India. *Geochimica et Cosmochimica Acta* 121: 611–636.

Raymond, P.A. and Hamilton, S.K. (2018). Anthropogenic influences on riverine fluxes of dissolved inorganic carbon to the oceans. *Limnology and Oceanography Letters.* 3.143-155.

Regnier, P., Resplandy, L., Najjar, R.G., and Ciais, P. (2022). The land-to-ocean loops of the global carbon cycle. *Nature* 603: 401–410.

Schopka, H.H., Derry, L.A., and Arcilla, C.A. (2011). Chemical weathering, river geochemistry and atmospheric carbon fluxes from volcanic and ultramafic regions on Luzon Island, the Philippines. *Geochimica et Cosmochimica Acta* 75: 978–1002.

Shabani, E., Hayati, B., Pishbahar, E. et al. (2022). The relationship between CO_2 emission, economic growth, energy consumption, and urbanization in the ECO member countries. *International Journal of Environmental Science and Technology* 19: 1861–1876.

Sverdrup, H. (2009). Chemical weathering of soil minerals and the role of biological processes. *Fungal Biology Reviews* 23: 94–100.

Wang, D., Han, G., Li, B. et al. (2022). Characteristics of ions composition and chemical weathering of tributary in the three gorges reservoir region: the perspective of stratified water sample from Xiaojiang River. *Water* 14: 379.

Wen, H., Sullivan, P.L., Billings, S.A. et al. (2022). From soils to streams: connecting terrestrial carbon transformation, chemical weathering, and solute export across hydrological regimes. *Water Resources Research* 58: 2022WR032314.

Wood, C., Harrison, A.L., and Power, I.M. (2023). Impacts of dissolved phosphorus and soil-mineral-fluid interactions on CO_2 removal through enhanced weathering of wollastonite in soils. *Applied Geochemistry* 148: 105511.

Yadav, S. and Rajamani, V. (2004). Geochemistry of aerosols of northwestern part of India adjoining the Thar Desert. *Geochimica et Cosmochimica Acta* 68: 1975–1988.

Zhao, Y. and Wang, W. (2023). Spatiotemporal change of carbon sink through chemical weathering of rocks over the Tibetan Plateau under climate warming. *Journal of Asian Earth Sciences* 105542.

Zhao, Y., Wijbrans, J.R., Wang, H. et al. (2023). Chemical weathering and CO_2 consumption inferred from riverine water chemistry in the Xi River drainage, South China. *International Journal of Environmental Research and Public Health* 20: 1516.

Zhonghua, C., Yan, Z., Xiaoke, Z. et al. (2022). Shift of major driver for chemical weathering from the natural control to human dominance since 1980s in the Taihu watershed, China. *Environmental Science and Pollution Research* 30: 20558–20569.

14

Geoscientific Factors Affecting Weathering and Erosion of Surface Exposure and Rock Types

Vamsi K. Kudapa[1], *Uday Bhan*[1], *Nirlipta P. Nayak*[1], *Lalit Goswami*[2], *Somenath Ganguly*[1], *and Susheel Kumar*[3]

[1] *Department of Petroleum Engineering and Earth Sciences, UPES, Dehradun, India*
[2] *Department of Chemical Engineering, Chungbuk National University, Cheongju, Korea*
[3] *Centre for Petroleum Exploration, Mizoram University, Aizawl, India*

14.1 Introduction

The Earth's crust is made up of rock that is solid and crystalline in nature, and bedrock is that rock. As an outcome of the development of weathering, bedrock is broken down into smaller particles that are referred to as sediment or soil. Salt expansion, pressure expansion, frost wedging, and root wedging are the constituent parts of the mechanical weathering process. The great majority of the processes that occur on the surface of the Earth, such as landslides, river notch, chemical weathering, and the carbon cycle, all depend heavily on mechanical weathering's contribution (Caves Rugenstein et al. 2019; Glade et al. 2019).

The term "weathering" refers to one of the most fundamental processes in geomorphology, which is the transformation of rock surfaces that are subjected to the conditions of the atmosphere. This process is responsible for the formation of many types of landforms. Alteration is caused by processes that are physical, chemical, and biological in nature. Rock deterioration, decay, crumbling, decomposition, rotting, disintegration, disaggregation, and breakdown are all terms used to describe the changes that occur as a result of these processes, all of which weaken the rock surface and increase its susceptibility to erosion (Hall et al. 2012). However, in other instances, weathering may cause the surface layers of the rock to become more resistant to erosion, at least in the short term (Robinson and Moses 2011; Zhang et al. 2023). The surface of a rock may be protected against erosion in the short term by such hardening, but in the long term, it may contribute to the rock's degradation. A crust that has been hardened, for instance, may react substantially differently to environmental stressors than the rock that lies underneath it, which may result in the breakdown of the upper layer. This occurs relatively frequently in both natural and urban settings, and it can be found in both (Smith 2003). Cracking that is caused by mechanical weathering is the primary process that is responsible for the formation of clastic sediment

Weathering and Erosion Processes in the Natural Environment, First Edition. Edited by Virendra Bahadur Singh, Sughosh Madhav, Naresh Chandra Pant, and Ravi Shekhar.

from outcrops or at the bedrock–soil interface in the subsurface. Both the sediment size at the ground surface (such as rock fragments on a slope, glacial moraine, or alluvial fan) and the sediment size within soil strata are reduced as a result of mechanical weathering (Singh et al. 2014). To gain an understanding of mechanical weathering, it is necessary to have a firm grasp on the inner structure, or architecture, of the Earth's highest crust, which includes the "Critical Zone," as well as the rate at which mobile sediment is generated from larger, more stable sedimentary clasts or bedrock (Sklar et al. 2017; Moon et al. 2020; Moravec et al. 2020).

The mechanisms responsible for rock weathering and erosion collaborate closely with each other. Erosion removes components that have already been weathered, exposing newly exposed, unweathered rock surface to further and ongoing contact with atmospheric conditions. Robinson and Moses (2011) go into considerable length in their examination of the workings of the weathering processes as well as their interactions with erosion. The practice of measuring and monitoring the effects of weathering and erosion processes is an essential part of gaining a knowledge of how these processes interact to shape rock surfaces. The advancement of technology has made it possible to observe rock weathering from a distance and at a considerably better spatial resolution than was previously possible, which has facilitated the development of new technologies. This chapter discusses the relative contributions of mass-wasting mechanisms and glaciers to the formation of the high valleys in the Central Himalayas. The Himalayan range is at its narrowest in this area, and the world's deepest gorge cuts across it (the Kali Gandaki valley).

14.2 Mechanical Weathering

Numerous geologic and geomorphic processes are slowed down by physical rock breakdown, also known as mechanical weathering (Shobe et al. 2017; Schwenk et al. 2022). For example, water is often recognized as a major factor in rock deterioration. Weathering, regolith generation, and erosion are all strongly correlated with moisture, according to significant empirical evidence (De Haas et al. 2015). Thus, it is generally assumed, without any thought being given to it, that mechanical weathering is accelerated by the presence of water and that moisture-dependent, stress-causing processes like freezing predominate (West et al. 2014; Bhan et al. 2023).

On a global scale, brittle failure (jointing) caused by tectonic plate motions is more common than mechanical weathering at or near the Earth's surface. Here, "near" refers to a depth of about 100 meters, as this is the typical depth to which people report seeing chemically and physically changed bedrock (Bazilevskaya et al. 2013). Mechanical weathering is caused by a variety of stresses, the most common of which fall into one of three categories: topographic, tectonic, or environmental. Topographic stresses are those caused by the weight of the rock or soil above it, or its absence in the case of unloading, at a certain depth (Leith et al. 2014). Tectonic stresses are the result of the constant movement of plates in the Earth's crust, which can be amplified by the topography (Martel 2006). Even while geological and topographic stresses are potentially static, the vast majority of environmental stresses have a cyclical nature and occur at regular intervals.

14.2.1 Mechanical Weathering Due to Stress-Loading

Stresses in rock, be bedrock or loose clasts, lead to the crack propagation that makes up mechanical weathering. In turn, these stresses result from external forces (stress-loading) acting on the rock (Negi et al. 2020). Rocks can be subjected to tectonic, gravitational, and environmental stresses, alone or in various combinations. The shifting of the planet's crust causes what are called "tectonic stresses" (Heidbach et al. 2016). Gravitational stresses, also known as topographic stresses, result from the weight of an overlying mass (the topography or overburden) (Molnar 2004). Rock's interactions with other Earth systems (the hydrosphere, the atmosphere, and the biosphere) give rise to environmental stresses.

The actions that take place at the fracture tip as the bond is disrupted are the second factor that promotes and restricts subcritical cracking. When rocks are stressed, the bonds between the atoms at the fracture tip weaken and stretch to the point where the molecules there start to react with water if water is available in the adjacent crack in either vapor or liquid form (Das et al. 2008; Kudapa 2022). Due to the low stresses at that time, the bonds are broken and the crack spreads subcritically. Several distinct chemo-physical mechanisms, such as fatigue, stress corrosion, and ion exchange, are hypothesized to contribute to the breakdown of molecular bonds in rock (Atkinson and Meredith 1987). Understanding the mechanisms that influence mechanical weathering as a whole depends significantly on the realization that mechanical weathering involves chemical interactions. Most importantly, the theory and experimental data on subcritical cracking tell us that all the variables that might affect the rates of chemical reactions, such as temperature, moisture availability, water chemistry, or rock mineralogy, will also affect the processes and rates of subcritical cracking. A rising body of experimental evidence (Chen et al. 2019), wherein rocks are subjected to known stress levels in climate-controlled chambers and subcritical cracking velocity is observed, amply demonstrates and supports this theory. These findings demonstrate that environmental factors, rather than stress-loading intensity, determine subcritical cracking rates.

14.2.2 Different Factors for Mechanical Weathering

Subcritical cracking requires at least three conditions to take place, regardless of the chemical reaction or the substance that is being broken: (i) an already present heterogeneity, such as a micro-fracture; (ii) moisture, either in the form of liquid water or humidity; and (iii) load-induced stress (Brantut et al. 2013), all of which are probably present in all surface or near-surface rock clasts or bedrock. Numerous experimental studies have shown that the rate of subcritical cracking is affected by a variety of environmental variables, such as temperature [24], the amount of applied stress (Atkinson 2015), the composition of the rock (Nara et al. 2011), the pore humidity (Yoshitaka Nara et al. 2010), and the pore water chemistry (Heap et al. 2009).

14.2.2.1 The Effect of Temperature and Moisture

Rising Earth-surface temperatures, liquid water, and air relative humidity increase subcritical cracking, according to experiments (Heap et al. 2009; Nara et al. 2013). Subcritical cracking occurs without liquid water. Subcritical cracking occurs in rock experiments from

30% humidity to 100% saturation (Brantut et al. 2013), even without liquid water (Nara et al. 2013). In experimental data showing the moisture dependency of subcritical cracking, the velocity curve slope is almost always constant, supporting a constant mechanism (Røyne et al. 2011).

The average relative humidity at the Earth's surface is between 70 and 80%, excluding deserts and high mountains (Dai 2006). Currently, data on the interior moisture content of rock or bedrock beneath soil are scant (McAllister et al. 2017). Good estimates of subcritical cracking rates in weathering settings, on the other hand, will require precise observations of how humidity translates into long-term rock internal moisture.

14.2.2.2 Dependence on the Water's Chemistry

Subcritical cracking rates are influenced by pore water chemistry, but the exact nature of these impacts on rock is poorly understood. Despite this, simple water vapor, as opposed to organic vapors such as methanol, has been recognized and proved to be one of the most potent subcritical cracking agents (Jeong et al. 2007). It appears that subcritical fracture development in rock is likewise pH dependent. This is notably true at low-stress intensities in quartz, where the crack velocity is much higher in basic water (Atkinson and Meredith 1981). However, just like with any other geochemical reaction, the effect of pH relies on the specific components of the solution as well as the mineral that is being reacted. For example, subcritical cracking velocities for calcite normally rise as pH increases; but, in the presence of specific bases, this trend is reversed (Dunning et al. 1994).

14.2.2.3 What Happens When Pressures Are High

Subcritical cracking rates generally decrease or become null as confining pressures rise (Brantut et al. 2014). Nonetheless, the vast majority of research has been conducted at atmospheric pressure (Brantut et al. 2013), where crack migration is most likely to occur. Topographic and tectonic stresses (Martel 2011) likely function to some extent to offset confining pressures in bedrock and may be additive to other environmental-related loading (St. Clair et al. 2015).

14.2.2.4 Stress Dependency due to Weathering

Experimental and theoretical investigations demonstrate that the rate of subcritical cracking increases with the magnitude of applied tensile stress (Amitrano and Helmstetter 2006), and this is true regardless of the kind of rock being studied. Loading not only causes mechanical tensile stresses to increase but also weakens bond strengths at crack tips, which contributes to the increased efficacy of chemical reactions associated with subcritical cracking under higher pressures. In contrast, when moisture availability is poor, for instance because of permeability, within a narrow range of stress levels between K_c and K_{th}, cracking acceleration is barely detectable (Nara et al. 2017). When stress exceeds K_c, moisture can no longer travel to fracture points at high breaking velocities, resulting in a cracking process that is more mechanical and influenced by the rocks strength (Atkinson 1984).

Furthermore, when adjacent microfractures combine to generate a greater stress concentration than a single break, a smaller total magnitude of external forces is required for subcritical cracking. The proximity of cracks has a fundamental effect on the overall stress intensity, which explains this occurrence. In these situations, crack formation is most rapid in areas with the highest fracture density or in directions parallel to a preexisting micro crack fabric (Nara and Kaneko 2006). Therefore, areas of high microfracture or fault density, such as along bedding planes, will experience faster subcritical cracking, resulting in more pronounced macro cracks.

14.3 Erosion

Sediment and soil are moved away from the site of weathering by a mechanical process called erosion, which is typically fueled by water, wind, gravity, or ice. Water in its liquid state is the primary eroding agent. Rocks and sediment are relocated by gravity and mass-wasting mechanisms. Large rock fragments and fine sediment are both transported by gravity and ice in the form of glaciers. Resistance to erosion plays a key role in the development of unique geological features. The Grand Canyon's sheer cliffs are a perfect example of this. The cliffs are formed from the rock that remains after more permeable materials have been worn away. Hoodoos are a type of rock formation found in Bryce Canyon National Park and Goblin Valley State Park in Utah. They are formed when erodible rocks of varying hardness collide.

The relationship between climate and topography, in alignment to Molnar and England (1990), who contested the substantial evidence for tectonically driven faster uplift and the projected tectonic forcing of climate in the late Cenozoic (Molnar and England 1990), also directly depends on the relative effectiveness of mechanical erosion, in addition to chemical erosion, in glaciated regions (Raymo and Ruddiman 1992). In glaciated places, the relative efficiency of mechanical erosion, in addition to chemical erosion, also plays a direct role in the relationship between climate. Molnar and England (1990) put out the hypothesis that changes in the climate may have come about initially. They hypothesized that the fast erosion of highlands, which was caused by climatic shifts during the Quaternary period, would inevitably be linked to high rebound rates as a result of increased erosional dropping of buoyant central crust. This quick rebound might be misconstrued for an active tectonic uplift if one were not paying close enough attention. The idea that the unusual features of the Quaternary, such as milder climates and the recurring presence of glaciers and ice sheets, are responsible for increased erosion is a crucial assumption that underlies (Molnar and England 1990) hypothesis. However, glaciers contribute more to erosion than rivers do. Despite the fact that this is usually accepted to be the case and is intuitively appealing, the statistics that are reviewed in this context in the literature do not demonstrate that, given an identical location without glaciers, more erosion happens in the glaciated region (Andrews et al. 1994).

14.3.1 Mass Wasting

"Mass wasting" refers to the gradual erosion and subsequent downhill movement of rock and soil. Mass wasting is commonly used as a synonym for landslide, while mass wasting is a more general phrase that encompasses any type of movement downslope. In geology, a landslide refers to any type of mass waste caused by the rapid movement of geological material. During a mass-wasting event, the loose material and soils above it are the ones that usually shift. Depending on the direction of the preponderant motion, rock topples, slides, or falls are the appropriate terms for the movement of bedrock blocks. Flows are the movement of a material that is primarily liquid. Mass waste can be a slow or quick means of transportation. Debris flows are an extreme example of the dangers associated with rapid movement. The steep topography and frequent precipitation of the California coast, the Rocky Mountains, and the Pacific Northwest make them prime targets for devastating mass-wasting disasters.

When a slope collapses, mass is wasted. When the slope is too steep and unstable for the current materials and conditions, it will collapse. There are two primary criteria that determine slope stability: the slope angle and the strength of the underlying material. The force of gravity, which contributes to mass loss, is relatively constant on Earth's surface, with some minor fluctuations due to differences in elevation and rock density. In the diagram, gravity pulls a rock block located on an incline toward the planet's core (f_g). There are two different types of gravity's pull that act on a slope: the shear, or driving, force (f_s) that accelerates the block's descent, and the normal, or resisting, force (f_n) that pushes into the slope and generates friction. Shear strength describes the ratio of shear to normal force. A block will not slide downhill if the normal force (friction) is more than the shear force. Downslope movement occurs when shear stress exceeds normal force due to a steeper slope angle or weakening ground material (Barlow et al. 2006) (Figure 14.1).

In many cases, the presence of water drastically modifies the shear strength of a slope. Sedimentary and igneous rocks both contain water within their pore spaces, which are the interstitial air spaces between the grains. Let's use the angle of repose of a dry sand pile as an example. Adding water to sand, like when constructing a sandcastle on a beach, may cause the angle of repose to rise, possibly as high as 60° or even 90°. Too much water in the sandcastle's pores, however, reduces the shear strength and angle of repose, leading to the structure's collapse (Pradhan and Siddique 2019).

Discontinuities, or "planes of weakness," in sedimentary rocks also play a role in determining their shear strength. When bedding planes are parallel to the slope, they can

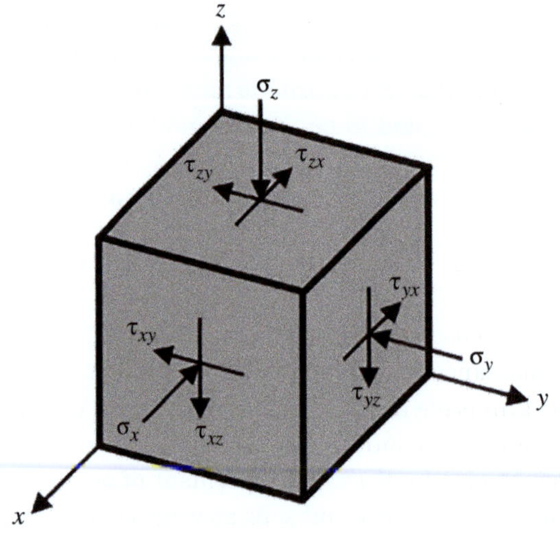

Figure 14.1 Diagrammatic representation of constant force of gravity (f_g), increased shear force, and decreased normal force with an increase in slope (*Source:* Reproduced with permission from Hungr and McDougall (2009)/ Elsevier).

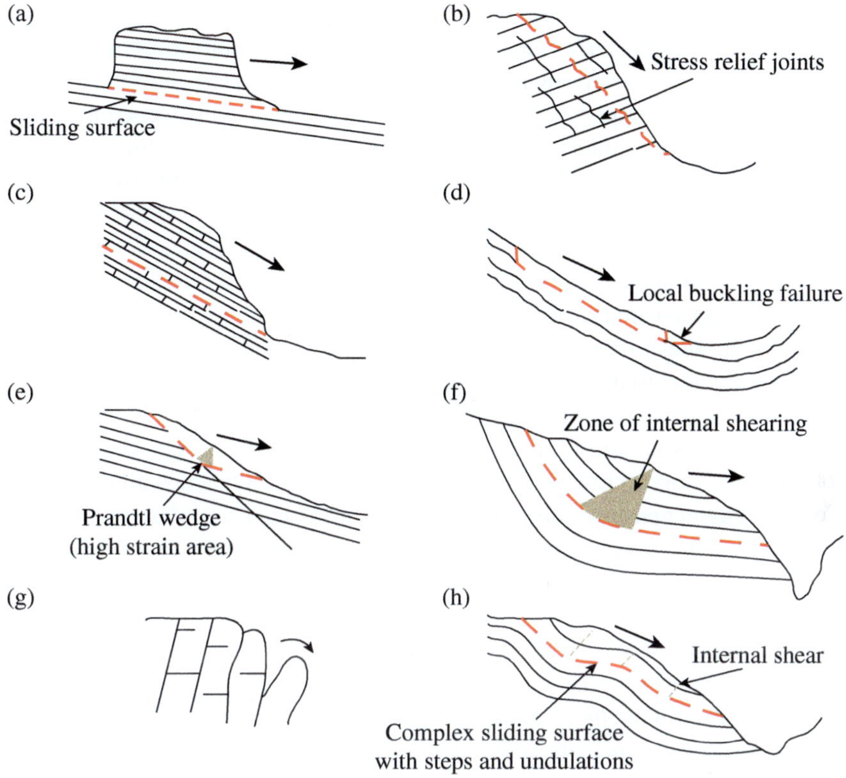

Figure 14.2 Rock stability at different sliding surfaces. (a) Large rock glide. (b) rough translational slide, (c) planar translational slide, (d) toe buckling translational slide, (e) biplanar compound slide, (f) curved compound slide, (g) toppling failure, and (h) irregular compound slide (*Source:* Stead and Wolter 2015/with permission of Elsevier).

operate as crucial planes of weakness, but when they are perpendicular to the slope, their effect is reduced (Johnson et al. 2022). From translational to complicated multimechanism failure, Glastonbury and Fell (2000) show how the shape and makeup of a rock slope and its geometry estimate the possible appliance of a landslide (Figure 14.2). The significance of structure in affecting the complication of failure mechanisms was mentioned by Stead et al. (2006).

14.3.2 Glacier Landforms

There are two types of glacial landforms: those made by erosion and those made by deposit. Landforms caused by erosion happen when material is taken away. Glacier ice melts because of the pressure and movement inside it. Glaciers slide over bedrock on a thin layer of water because of this. A lot of sediments like sand, gravel, and boulders are also found in glacial ice. Together, the movement lifts and grinds the bedrock, creating a smooth surface, a fine sediment known as rock flour, as well as other unsorted particles. The sediments that the ice leaves behind are different in shape and composition when it melts and dissipates. These sediments make up characteristic landscapes (Ganguly et al. 2015; Singh et al. 2021).

14.3.2.1 Landscapes Caused by Glacial Erosion

Alpine and continental glaciers both erode rock and shape the landscape through this process. Erosion creates a wide variety of landforms, including striations, cirques, glacial horns, arêtes, trim lines, U-shaped valleys, roches moutonnées, over deepenings, and hanging valleys. Trough valleys are so named because of their U-shaped formation when mountain glaciers erode the ground. The term "glacier" refers to large bodies of abrasive ice that move across the surface of the planet (Bhan et al. 2021; Sundriyal et al. 2021). The striations in a glacier's surface are created when the glacier's ice base scrapes against the underlying rock. Glacial polishing is the technique of bringing a hard granite or quartzite bedrock to a smooth surface using rock flour as a fine grit in the ice.

Old V-shaped stream valleys are transformed by glaciers into wide valleys with relatively flat bottoms and steep sides, giving them a recognizable "U" shape. This is due to the fact that glaciers are much larger than rivers of comparable length and, therefore, erode more at their bases than their sides (Johnson et al. 2022; Singh et al. 2022). A huge glacier occupied Little Cottonwood Canyon close to Salt Lake City, Utah, and it reached all the way to the entrance of the canyon and into Salt Lake Valley (Laabs et al. 2011; Bhan et al. 2022). The erosional landforms that may be seen in the canyon now include a *U*-shaped valley and polished and striated rock surfaces. Canyons created by rivers typically have a *V*-shaped profile when viewed from the side. The northern Big Cottonwood Canyon features a *V*-shaped bottom section, suggesting that the glacier did not reach all the way to the mouth but instead stayed in the upper reaches.

An arête is a steep ridge formed by the carving of the ridge between two nearby U-shaped valleys. Because glaciers erode a wide valley, the arêtes are separated by vertical walls. A cirque is a bowl-shaped formation found at the mouth of a glacially carved valley. It is formed when the glacier's head erodes against the mountain, pulling rock away from it. Once the glacier melts away, the tarn commonly fills the cirque's former glacial bowl. The horns we see today are the result of headward cirque erosion by three or more mountain glaciers, and they take the form of spire-like mountains with steep sides and significant cirques on three or more sides. Cols are the valleys between peaks or along the sides of arêtes (also known as mountain passes). A smaller tributary glacier erodes less as it meets a larger stem glacier. After the ice melts, the tributary valley is often left as a hanging valley complete with a waterfall. A truncated spur is a sharply triangular cliff that forms when a *V*-shaped valley is flattened and widened by a trunk glacier, which also erodes the tips of the valley's side peaks (Babu et al. 2022; Vinay et al. 2022).

14.4 Case Study – A Mighty River, Kali Gandaki in Nepal

The valley of Kali Gandaki connects the wet, hot, and low subtropical Himalayan Midlands (1000–1500 m) toward the south, between the high and dry Tibetan Plateau (6410–4825 m). A major Ganga tributary drains this valley, which passes through the Himalayan structural features. The river travels through a steep valley carved by the Thakkhola-Mustang graben deposits at the entrance of greater Himalayas.

The Mustang Thakkhola graben toward the north (at approximately 2698 m adjacent to Jomosom) and the convergence of the Kali Gandaki through the Miristi Khola (at a distance of 1295 m near Dana) to the south were studied for the effects of glacial and mass-wasting characteristics (Figure 14.3). The lithological contact between the greater Himalaya thrust sheet's gneisses and the superimposing (but equally thrusted) metasediments of the "Tibetan series" is an interesting contrast that can be utilized to pinpoint the origins of some deposits. Regional deformation styles accommodate this lithology shift. Gneissic slabs (foliation planes) dipping 25–403° north form the crustal ramp, while metasedimentary rocks are folded northwards (Fort 2000). In spite of the fact that this location is very close to some of the world's tallest mountains, such as Dhaulagiri 1 (approx. 8165 m), Nilgiri South and North Peaks which are at 6837 and 7059 m, approximately, and Annapurna 1 (8089 m), there are currently only scattered glacial remains to be found here (Figures 14.3 and 14.4). At this time, the bottom of the valley does not have any ice. In comparison, glaciers that flow down the west and north slopes of the Nilgiri range can get as low as 4200 m, whereas those that flow down the east face of Dhaulagiri do not go lower than approximately 4770 m. The latter are, in point of fact, regenerated glaciers that can be found on steep ledges that hang over the primary Kali Gandaki river. The lowest glacial tongue may be found near to 3795 m, which is hidden in the steep and gloomy valley of Pangbu Khola, which is located west of Nilgiri South Peak. In this steeply sloped terrain, avalanches provide nourishment for the glaciers, much as they do for many of the Himalayan ice masses. Because of all of these different aspects, the Kali Gandaki seems to be a significant location for distinctive glacial processes from mass-wasting ones. The glaciers and the lithotectonic processes that are occurring here have a very strong interaction with one another, which provides a large potential for slope instability. In addition, for geological setting, the recognition of well-preserved deposits needs to rely on petrographic and sediment logical evidence in addition to reconstructed morphologies (Sundriyal et al. 2018).

According to Fort et al. (Fort 2000), the majority of glacial indication is best explained as the result of considerable mass wasting with a low frequency, a process directly tied to the exact energetic geodynamic environment. The central High Himalaya has less glacial morphology than the eastern and western parts.

14.5 Erosion Rates Comparison Between Glaciated and Non-Glaciated Basins

There aren't many recent studies in the literature between erosion rates from basins with and without glaciers. Twelve basins in a very homogeneous bedrock area (fissile schist) in the Southern Alps, New Zealand, were the subject of a 1990 study by Hicks et al. (1990). The local precipitation in each basin, which ranged widely from around 1 to 10 m/y, was principally represented in the sediment yields. They only found considerable glaciation in one of the basins they analyzed. Ivory Glacier is located in this 2.2 km^2 cirque basin, but it cannot be distinguished from the others based on a link between precipitation rate and sediment yield, suggesting that the amount of glacier cover does not appear to have an impact on sediment yield. Additionally, they collated sediment outputs from basins in various places, both with and without glaciers, which revealed wide scatter and no clear relationship to the extent of glacier cover.

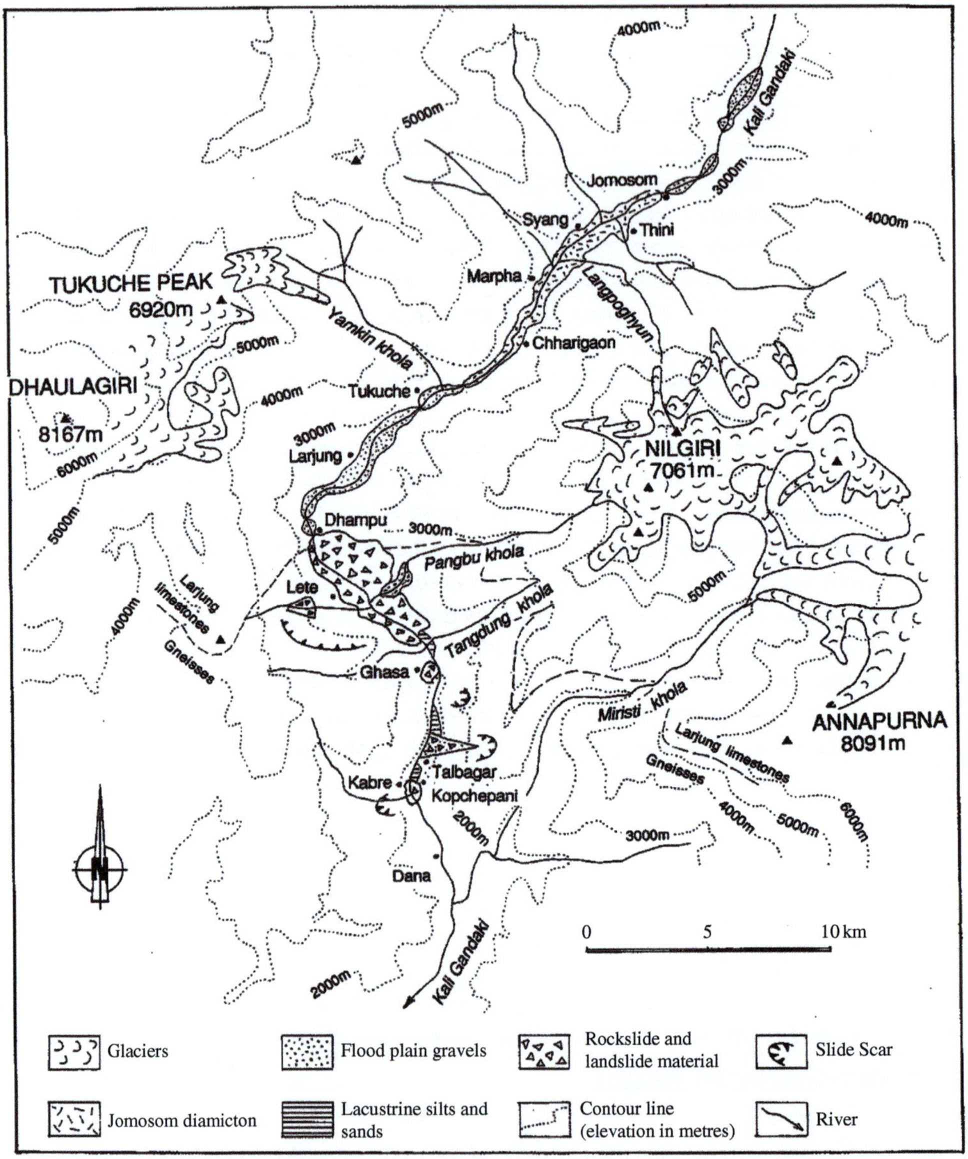

Figure 14.3 Kali Gandaki crosses the Greater Himalayas. The surrounding summits' elevation limits glaciers. Mass-wasting features caused the valley bottom's flats and narrows. The dashed line divides limestones (north) from gneisses (south) (*Source:* Fort 2000/ with permission of Elsevier.).

Other important glaciological parameters that influence sediment yields include sliding pace, glacier size, ice flow, meltwater output, and direction to the glacier terminus. In addition, because of differences in subglacial and subaerial sediment storage, it is anticipated that the connection between present-day yields from glaciated basins and sediment once produced by glacial erosion will be intricate (Harbor and Warburton 1993). Church and

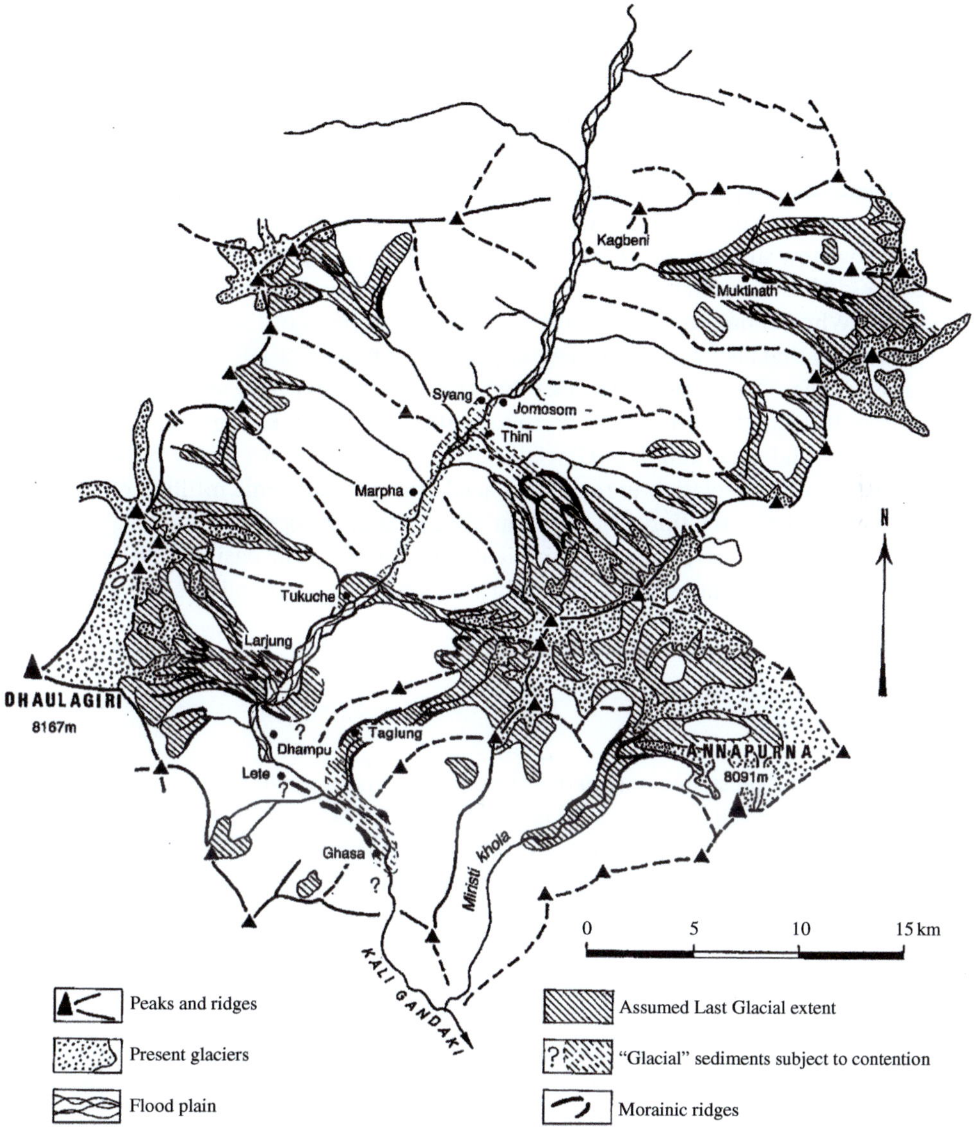

Figure 14.4 Kali Gandaki's Thakkhola glacial relics and restoration (Lower Mustang District) (*Source:* Fort 2000/ with permission of Elsevier).

Ryder (1972) argues, for instance, that the sediment outputs from Baffin Island's glacier basins are unrelated to the rates of sediment production that are now understood to be typical of so-called "paraglacial" ecosystems (Church and Ryder 1972). Instead, results demonstrate a redistribution of sediments provided during prior glacial cycles. Current glacial sediment fluxes have been created for ice-proximal marine settings like the fjords of Glacier Bay in southeast Alaska, where sediment deposition and remobilization have been studied (Hunter 1994).

14.6 Conclusion

Crack and rock morphologies are often azonal, which means that they are found in settings and climate zones of the same kind but in comparable patterns. It is highly likely that climate-dependent subcritical cracking, which is generated by environmentally regulated chemical and physical processes, is the primary driver of mechanical weathering. The following wave of laboratory tests, field observations, and numerical modeling will start to show how studying mechanical weathering might improve our comprehension of landscapes. These methods include fieldwork, laboratory experiments, and computer modeling. When compared to the eastern and western parts of the High Himalaya, the glacial activity that occurred in this center part of the range has left behind less morphological evidence. This is due to the fact that the range is relatively small, as well as the high desiccation gradient that runs north to south, which worked together to restrict the size of the glacial accumulation. Glacial erosion rates of 0.1 to 1 mm/y have been discussed extensively in relation to regional glacial erosion, glacigenic infilling of sedimentary basins, and the relationship between climate and topography. Large glaciers on Alaska's southern coast produce more yields on average than basins with or without glaciers worldwide. Smaller glaciated basin yields surpass comparable-sized glacier-flee basin yields.

References

Amitrano, D. and Helmstetter, A. (2006). Brittle creep, damage, and time to failure in rocks. *Journal of Geophysical Research: Solid Earth* 111 (B11): 1–19.

Andrews, J.T., Milliman, J.D., Jennings, A.E. et al. (1994). Sediment thicknesses and Holocene glacial marine sedimentation rates in three East Greenland fjords (ca. 68 N). *The Journal of Geology* 102 (6): 669–683.

Atkinson, B.K. (1984). Subcritical crack growth in geological materials. *Journal of Geophysical Research: Solid Earth* 89 (B6): 4077–4114.

Atkinson, B.K. (2015). *Fracture Mechanics of Rock*. Elsevier.

Atkinson, B.K. and Meredith, P.G. (1981). Stress corrosion cracking of quartz: a note on the influence of chemical environment. *Tectonophysics* 77 (1–2): T1–T11.

Atkinson, B.K. and Meredith, P.G. (1987). The theory of subcritical crack growth with applications to minerals and rocks. *Fracture Mechanics of Rock* 2: 111–166.

Babu, V., Rai, S.K., Bhan, U., and Devaraju, J. (2022). Chemical weathering of sediment (CWS): a web-based application for chemical weathering study of rocks and sediment. *Materials Today: Proceedings* 68: 979–985.

Barlow, J., Franklin, S., & Martin, Y. (2006). High spatial resolution satellite imagery, DEM derivatives, and image segmentation for the detection of mass wasting processes. *Photogrammetric Engineering \& Remote Sensing*, 72(6), 687–692.

Bazilevskaya, E., Lebedeva, M., Pavich, M. et al. (2013). Where fast weathering creates thin regolith and slow weathering creates thick regolith. *Earth Surface Processes and Landforms* 38 (8): 847–858.

Bhan, U., Boruah, A., Maurya, D.S. et al. (2022). Mineralogy, organic richness and macerated microbial studies of the Rohtasgarh shales in the Vindhyan Basin, India: implications for gas generation potential. *Journal of the Geological Society of India* 98 (4): 567–575.

Bhan, U., Kudapa, V.K., and Kumar, R. (2023). Study of the rate of CO_2 consumption with silicate and carbonate weathering in aquatic system. *Hydrogeochemistry of Aquatic Ecosystems* 199–212.

Bhan, U., Singh, D., Sharma, M. et al. (2021). A note on the Fan-Fabric Structures in the late Palaeoproterozoic Kajrahat Limestone, Katni, MP, India. *Journal of the Palaeontological Society of India* 66 (2): 315–322.

Brantut, N., Heap, M.J., Baud, P., and Meredith, P.G. (2014). Mechanisms of time-dependent deformation in porous limestone. *Journal of Geophysical Research: Solid Earth* 119 (7): 5444–5463.

Brantut, N., Heap, M.J., Meredith, P.G., and Baud, P. (2013). Time-dependent cracking and brittle creep in crustal rocks: a review. *Journal of Structural Geology* 52: 17–43.

Caves Rugenstein, J.K., Ibarra, D.E., and von Blanckenburg, F. (2019). Neogene cooling driven by land surface reactivity rather than increased weathering fluxes. *Nature* 571 (7763): 99–102.

Chen, X., Eichhubl, P., Olson, J.E., and Dewers, T.A. (2019). Effect of water on fracture mechanical properties of shales. *Journal of Geophysical Research: Solid Earth* 124 (3): 2428–2444.

Johnson, C., Affolter, M.D., Inkenbrandt, P., and Mosher, C. (2022). *An Introduction to Geology* (ed. Salt Lake Community College). OpenGeology https://geo.libretexts.org/Bookshelves/Geology/Book%3A_An_Introduction_to_Geology_(Johnson_Affolter_Inkenbrandt_and_Mosher.

Church, M. and Ryder, J.M. (1972). Paraglacial sedimentation: a consideration of fluvial processes conditioned by glaciation. *Geological Society of America Bulletin* 83 (10): 3059–3072.

Dai, A. (2006). Recent climatology, variability, and trends in global surface humidity. *Journal of Climate* 19 (15): 3589–3606.

Das, S.S., Maurya, A.S., Pandey, A.C. et al. (2008). Influence of sediment source and monsoonal variations on the late Quaternary clay mineral assemblages at ODP site 728A, northwestern Arabian Sea. *Current Science* 95: 1320–1326.

De Haas, T., Conway, S.J., and Krautblatter, M. (2015). Recent (Late Amazonian) enhanced backweathering rates on Mars: paracratering evidence from gully alcoves. *Journal of Geophysical Research, Planets* 120 (12): 2169–2189.

Dunning, J., Douglas, B., Miller, M., and McDonald, S. (1994). The role of the chemical environment in frictional deformation: stress corrosion cracking and comminution. *Pure and Applied Geophysics* 143 (1): 151–178.

Fort, M. (2000). Glaciers and mass wasting processes: their influence on the shaping of the Kali Gandaki valley (higher Himalaya of Nepal). *Quaternary International* 65: 101–119.

Ganguly, S., Tiwari, S., Bhan, U. et al. (2015). Melting of sea ice inexplicable for recent global eustatic sea level rise. *Journal of Earth Science & Climatic Change* 6 (1): 1.

Glade, R.C., Shobe, C.M., Anderson, R.S., and Tucker, G.E. (2019). Canyon shape and erosion dynamics governed by channel-hillslope feedbacks. *Geology* 47 (7): 650–654.

Glastonbury J., Fell R. (2000) Report on the Analysis of "Rapid" Natural Rock Slope Failures. University of New South Wales. School of Civil and Environ-mental Engineering Report No. R390, pp. 154.

Hall, K., Thorn, C., and Sumner, P. (2012). On the persistence of 'weathering'. *Geomorphology* 149: 1–10.

Harbor, J. and Warburton, J. (1993). Relative rates of glacial and nonglacial erosion in alpine environments. *Arctic and Alpine Research* 25 (1): 1–7.

Heap, M.J., Baud, P., and Meredith, P.G. (2009). Influence of temperature on brittle creep in sandstones. *Geophysical Research Letters* 36 (19): 1–6.

Heidbach, O., Rajabi, M., Reiter, K., & Ziegler, M. (2016). *WSM-World Stress Map Database Release*. GFZ Data Services. http://doi.org/10.5880/WSM.

Hicks, D.M., McSaveney, M.J., and Chinn, T.J.H. (1990). Sedimentation in proglacial Ivory Lake, Southern Alps, New Zealand. *Arctic and Alpine Research* 22 (1): 26–42.

Hungr, O., & McDougall, S. (2009). Two numerical models for landslide dynamic analysis. *Computers and Geosciences*, 35(5), 978–992. https://doi.org/https://doi.org/10.1016/j.cageo.2007.12.003

Hunter, L.E. (1994). *Grounding-Line Systems and Glacier Mass Balance of Modern Temperate Glaciers and Their Effect on Glacier Stability*. Northern Illinois University.

Jeong, H., Kang, S., and Obara, Y. (2007). Influence of surrounding environments and strain rates on the strength of rocks subjected to uniaxial compression. *International Journal of Rock Mechanics and Mining Sciences* 44 (3): 321–331.

Kudapa, V.K. (2022). Carbon-dioxide capture, storage and conversion techniques in different sectors – a case study. *International Journal of Coal Preparation and Utilization* 1–26. https://doi.org/10.1080/19392699.2022.2119559.

Laabs, B.J.C., Marchetti, D.W., Munroe, J.S. et al. (2011). Chronology of latest Pleistocene mountain glaciation in the western Wasatch Mountains, Utah, USA. *Quaternary Research* 76 (2): 272–284.

Leith, K., Moore, J.R., Amann, F., and Loew, S. (2014). In situ stress control on microcrack generation and macroscopic extensional fracture in exhuming bedrock. *Journal of Geophysical Research: Solid Earth* 119 (1): 594–615.

Martel, S.J. (2006). Effect of topographic curvature on near-surface stresses and application to sheeting joints. *Geophysical Research Letters* 33 (1): 1–5.

Martel, S.J. (2011). Mechanics of curved surfaces, with application to surface-parallel cracks. *Geophysical Research Letters* 38 (20): 1–6.

McAllister, D., Warke, P., and McCabe, S. (2017). Stone temperature and moisture variability under temperate environmental conditions: implications for sandstone weathering. *Geomorphology* 280: 137–152.

Molnar, P. (2004). Interactions among topographically induced elastic stress, static fatigue, and valley incision. *Journal of Geophysical Research: Earth Surface* 109 (F2): 1–9.

Molnar, P. and England, P. (1990). Late Cenozoic uplift of mountain ranges and global climate change: chicken or egg? *Nature* 346 (6279): 29–34.

Moon, S., Perron, J.T., Martel, S.J. et al. (2020). Present-day stress field influences bedrock fracture openness deep into the subsurface. *Geophysical Research Letters* 47 (23): e2020GL090581.

Moravec, B.G., White, A.M., Root, R.A. et al. (2020). Resolving deep critical zone architecture in complex volcanic terrain. *Journal of Geophysical Research: Earth Surface* 125 (1): e2019JF005189.

Nara, Y. and Kaneko, K. (2006). Sub-critical crack growth in anisotropic rock. *International Journal of Rock Mechanics and Mining Sciences* 43 (3): 437–453.

Nara, Y., Hiroyoshi, N., Yoneda, T., and Kaneko, K. (2010). Effects of relative humidity and temperature on subcritical crack growth in igneous rock. *International Journal of Rock Mechanics and Mining Sciences* 47 (4): 640–646.

Nara, Y., Kashiwaya, K., Nishida, Y., and Ii, T. (2017). Influence of surrounding environment on subcritical crack growth in marble. *Tectonophysics* 706: 116–128.

Nara, Y., Morimoto, K., Yoneda, T. et al. (2011). Effects of humidity and temperature on subcritical crack growth in sandstone. *International Journal of Solids and Structures* 48 (7–8): 1130–1140.

Nara, Y., Yamanaka, H., Oe, Y., and Kaneko, K. (2013). Influence of temperature and water on subcritical crack growth parameters and long-term strength for igneous rocks. *Geophysical Journal International* 193 (1): 47–60.

Negi, M., Rai, S.K., Ghosh, S.K. et al. (2020). Characterization of the Detrital Zircon from the Lesser Himalayan Proterozoic siliciclastics. *Himalayan Geology* 41 (1): 1–10.

Pradhan, S.P. and Siddique, T. (2019). Mass wasting: an overview. In: *Landslides: Theory, Practice and Modelling* (ed. S.P. Pradhan, V. Vishal, and T.N. Singh), 3–20. Cham: Springer.

Raymo, M.E. and Ruddiman, W.F. (1992). Tectonic forcing of late Cenozoic climate. *Nature* 359 (6391): 117–122.

Robinson, D.A. and Moses, C.A. (2011). Rock surface and weathering: process and form. In: *The SAGE Handbook of Geomorphology* (ed. K.J. Gregory and A.S. Goudie), 291–309. London: SAGE.

Røyne, A., Bisschop, J., and Dysthe, D.K. (2011). Experimental investigation of surface energy and subcritical crack growth in calcite. *Journal of Geophysical Research: Solid Earth* 116 (B4): https://doi.org/10.1029/2010JB008033.

Schwenk, M.A., Schläfli, P., Bandou, D. et al. (2022). From glacial erosion to basin overfill: a 240 m-thick overdeepening--fill sequence in Bern, Switzerland. *Scientific Drilling* 30: 17–42.

Shobe, C.M., Hancock, G.S., Eppes, M.C., and Small, E.E. (2017). Field evidence for the influence of weathering on rock erodibility and channel form in bedrock rivers. *Earth Surface Processes and Landforms* 42 (13): 1997–2012.

Singh, B., Singh, S., and Bhan, U. (2022). Oceanic anoxic events in the Earth's geological history and signature of such event in the Paleocene-Eocene Himalayan foreland basin sediment records of NW Himalaya. *India. Arabian Journal of Geosciences* 15 (3): 1–27.

Singh, D., Sharma, M., Bhan, U. et al. (2021). Carbonate Fan Fabric Structures (FFS) in time and space: a case study from the Palaeoproterozoic Kajrahat Limestone, Vindhyan Supergroup, India. *Journal of the Palaeontological Society of India* 66 (2): 290–302.

Singh, V.B., Rai, A.K., and Bhan, U. (2014). Illustration and taxonomy of late Neogene Deep Sea agglutinated benthic foraminifera from the Exmouth Plateau in the Eastern Indian Ocean. *International Journal of Advanced Earth Science and Engineering* 3: 183–200.

Sklar, L.S., Riebe, C.S., Marshall, J.A. et al. (2017). The problem of predicting the size distribution of sediment supplied by hillslopes to rivers. *Geomorphology* 277: 31–49.

Smith, B.J. (2003). Background controls on urban stone decay: lessons from natural rock weathering. *The Effects of Air Pollution on the Built Environment. Air Pollution Reviews* 2: 31–61.

St. Clair, J., Moon, S., Holbrook, W.S. et al. (2015). Geophysical imaging reveals topographic stress control of bedrock weathering. *Science* 350 (6260): 534–538.

Stead, D., Eberhardt, E., and Coggan, J.S. (2006). Develop-ments in the characterization of complex rock slope deformation and failure using numerical modelling techniques. *Engineering Geology* 83: 217–235.

Stead, D., & Wolter, A. (2015). A critical review of rock slope failure mechanisms: the importance of structural geology. *Journal of Structural Geology*, 74, 1–23. https://doi.org/ https://doi.org/10.1016/j.jsg.2015.02.002

Sundriyal, S., Bhan, U., Selvakumar, S. et al. (2021). Two decadal changes in the major ions chemistry of melt water draining from Dokriani Glacier, central Himalaya, India. *Journal of the Geological Society of India* 97 (3): 308–314.

Sundriyal, S., Shukla, T., Tripathee, L. et al. (2018). Deposition of atmospheric pollutant and their chemical characterization in snow pit profile at Dokriani Glacier, Central Himalaya. *Journal of Mountain Science* 15 (10): 2236–2246.

Vinay, B., Uday, B., Javaregowda, D., and Santosh, K.R. (2022). Estimation of chemical weather indices of sediments using R program. *Materials Today: Proceedings* 68: 1063–1086.

West, N., Kirby, E., Bierman, P., and Clarke, B.A. (2014). Aspect-dependent variations in regolith creep revealed by meteoric 10Be. *Geology* 42 (6): 507–510.

Zhang, X., Xu, B., Xie, Y. et al. (2023). Continuous mass loss acceleration of Qiangyong Glacier, southern Tibetan plateau, since the mid-1880s inferred from glaciolacustrine sediments. *Quaternary Science Reviews* 301: 107937.

15

Impacts of Climate Change on Weathering and Erosion of Rock Types Exposed on Earths Surface

Nirlipta P. Nayak[1], *Vamsi K. Kudapa*[1], *Uday Bhan*[1], *Lalit Goswami*[2], *Susheel Kumar*[3], *and Anamika Kushwaha*[2]

[1] *Department of Petroleum Engineering and Earth Sciences, UPES, Dehradun, India*
[2] *Department of Chemical Engineering, Chungbuk National University, Cheongju, South Korea*
[3] *Centre for Petroleum Exploration, Mizoram University, Aizawl, India*

15.1 Introduction

Weathering is a crucial phenomenon that has ability to both impact and get impacted by the climate of Earth with change in time. Weathering is solely responsible for quantitative control of solute and nutrients that reach ocean and are important for balancing natural processes in sea as well terrestrial environment. Also, weathering significantly impacts the migration of carbon to sedimentary rocks from ocean and atmosphere interaction (Stallard and Edmond 1983; White and Blum 1995). Climate along with other factors impacts the rate of weathering and varies from one place to another on Earth. Even weathering has been reported to alter the amount of carbon dioxide which is being circulated in our atmosphere and affects the temperature of Earth by a phenomenon known as global warming. Thus, weathering acts as a climate regulator and is important when studying the long-term change in climate (Sarin et al. 1989; Petch et al. 2000). Erosion is a geological process in which Earthen materials (i.e. soil, rocks, sediments) are worn away and transported far off by geological agents, i.e. water or wind. Weathering differs from erosion by the concept of transportation. Erosion resistance plays a major role in creating distinctive geological features. This is well-demonstrated in the cliffs of the Grand Canyon. The cliffs are made of rock left standing after less resistant materials have weathered and eroded away. Erosion is the reversal of deposition. This is a crucial phenomenon as large-scale erosion leads to the formation of several geological features, for instance, the formation of the Grand Canyon by erosion of rock by the Colorado River for a very long time. Erosion occurs in uppermost layer of surface and thus is also known as soil erosion with the removal of uppermost layer known as topsoil which is also important layer of Earth as fertility of land depends on topsoil. Erosion rate is determined by the agents that are causing erosion to occur in an area and erosion agents include water, wind, vegetation, and animals, etc. Anthropogenic

Weathering and Erosion Processes in the Natural Environment, First Edition. Edited by Virendra Bahadur Singh, Sughosh Madhav, Naresh Chandra Pant, and Ravi Shekhar.

activities can also impact the soil erosion substantially. Thus, controlling the rate of erosion by humans is another area of research that needs to be emphasized in order to maintain the fertility of land of an area.

Erosion occurs when the uppermost layer of soil is left exposed to erosion agents for a long time and these agents include wind, rain, and human activities. Human activities like agriculture where during ploughing large fields are left exposed to wind and water making them venerable to erosion. Similarly, grazing of fields by animals makes the soil loose and leads to erosion of it due to wind and other agents. Another human activity that acts as an erosion agent is deforestation. Deforestation cutting of trees leads to erosion. The roots of trees hold substantial amount of soil so, during deforestation when trees are removed the force that binds the soil particles together reduces making that area vulnerable to erosion. Also, large-scale deforestation in hilly areas can land to natural hazards like landslides along with erosion causing harm to humans and properties. Climate is also a major driver of erosion and includes phenomena like high or low rainfall, temperature variation, etc.

Erosion has a devastating impact of health of soil as it degrades the quality and quantity of uppermost layer of Earth which is used for agriculture to feed the population. Uncontrolled erosion of soil leads to reduction in biodiversity over, within, and beneath the soil surface resulting into the alteration of alters habitats and has a detrimental effect on social resilience and economies. Also soil erosion finally lands up meeting streams, rivers, and oceans along with the harmful chemicals that are used in agriculture. These chemical when get soluble in river or seawater impacts the water ecosystem severely. Algae bloom is one of the common phenomena, which is the result of soil erosion.

Two major types of erosion are water and wind erosion. The former refers to the removal of uppermost layer of soil by rain or snowmelt. Water erosion leads to the formation of several structures including sheet, rill, gully, and bank erosion. The latter process includes the movement of uppermost layer of Earth from one place to another by the activities of wind. Very strong wind is reported to move high amount of dirt soil and can cause event like dirt storm.

Thus, erosion can be defined as a variation in force that binds upper layer of Earth with Earth to the force exerted by erosion agents like wind and water. So, greater the force by erosion agent, higher will be the erosion. There are numerous ways by which erosion can be mitigated. Some of them are:

- Soil organic matter (SOM) building: A healthy and good quality soil constitutes ideal proportions of water, air, minerals, and organic matter. The force that binds soil with surface of Earth is exerted by SOM and it constitutes of animal and plants that are decaying with time. Studies suggest that increasing of SOM from 1 to 3% increases water holding capacity which leads to reduction in erosion by 20–30%.
- Reforestation: This is another method that can mitigate the erosion as the roots of trees hold the soil together making them resistive to erosion.
- Erosion control matting: A ground covering also called as erosion control blanket and is made up of biodegradable substances. This helps in supporting vegetation which is growing on ground.
- Minimal tillage utilization and rotational grazing.

Conventionally weathering is defined as alteration in chemical properties of rocks and sedimentary formations over the surface of Earth under subaerial climate (Figure 15.1),

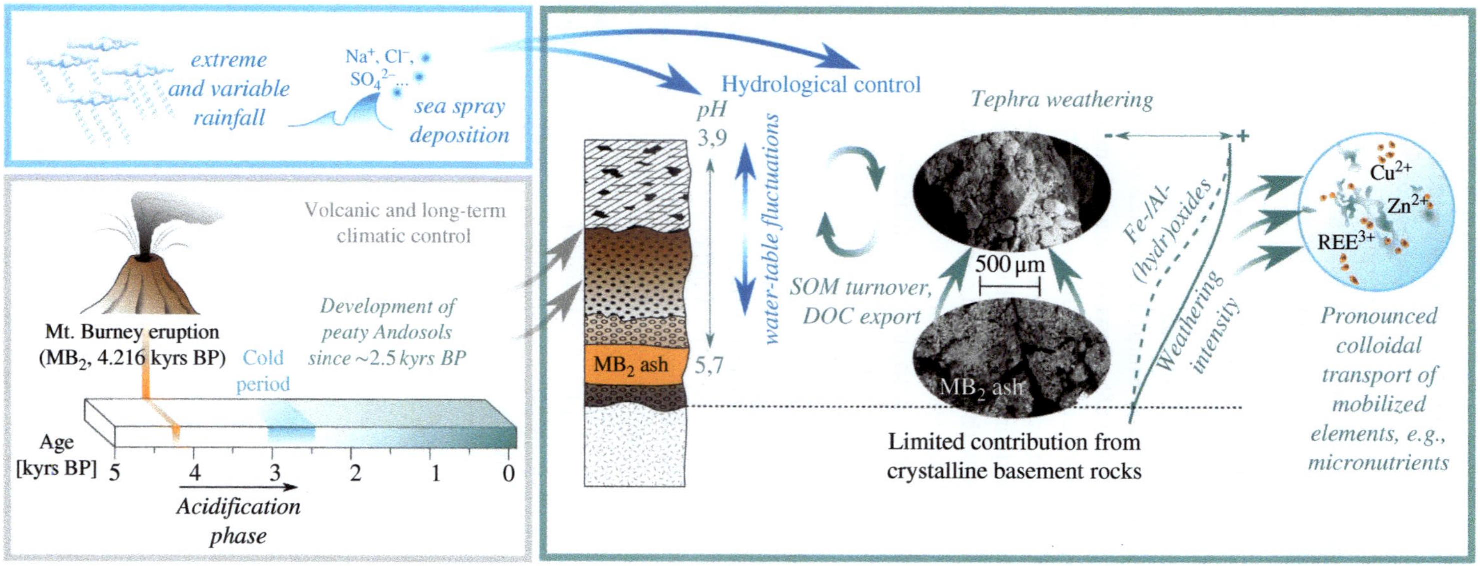

Figure 15.1 Depicts various forms of Weathering process (*Source:* Reproduced with permission Kales et al. 2022 / ELSEVIER).

although processes similar to weathering also occur at sea bottom (oceanic crust) because of low temperature. Similarly, alteration by hydrothermal activities at oceanic or continental crust also follows the same mechanism but this process does not fall under the scope of weathering. The reason for the same can be attributed to the fact that weathering is only limited to matter having silicate, carbonate organic matter nature and weathering parameters like reactions, reaction products and climate vary between these three (Lasaga et al. 1994).

Physical rock breakdown or mechanical weathering comprises all in situ rock fracturing at and near Earth's surface. It represents a critical rate-limiting factor for a broad range of geologic and geomorphic processes. Therefore, it is mostly accepted that the presence of water accelerates mechanical weathering. Mechanical weathering commonly occurs at and near Earth's surface in excess of regional-scale brittle failure (jointing) attributable to tectonic plate motions. The sources of stress that lead to mechanical weathering can generally be divided into three broad – and likely interacting – categories: topographic, tectonic, and environmental. Here we define topographic stresses as those that arise from gravitational forces acting in concert with the weight of overlying rock or soil – or lack thereof for the case of unloading – at some depth.

Tectonic stresses arise due to ongoing plate motions in Earth's crust that can also interact with topography to enhance stress. Finally, we define environmental stresses as those that comprise all climate- and environment-related stresses, including freezing, salt precipitation mineral hydration biologic activity, and thermal cycling. Both tectonic and topographic stresses might be considered static, but virtually all environmental stresses are cyclic at some periodicity (e.g. diurnal, seasonal, and annual) (Eppes and Keanini 2017)

Acid–base reaction can describe both the weathering of silicates and carbonates matter where acids that occur naturally are produced emissions from volcanic gases, weathering of sulfide minerals, CO_2, and its dissolution in water and organic matter and its decomposition interacts with minerals to imbibe acidity to release cations and leading to increase in basicity(Jenny 1941). The major difference between weathering of silicate and carbonate minerals is that the former produces clay a type of secondary mineral matter by reaction while the latter dissolves completely. Although in both processes, weathering acts as an agent for acidification, the only thing that varies is the scale of acidification which depends on the phases of minerals involved in reaction (Galy and France-Lanord 1999; Jacobson and Blum 2003).

Sulfuric and carbonic acids are amongst the most important natural acids. Carbonic acid is a weak acid displaying incomplete dissolution of proton as donor) while the generation of sulfuric acid can be attributed to sulfide minerals and their oxidation, gases escaping from volcanoes during eruption of magma in the form of SO_2, carbonyl and dimethyl sulfide combinedly known as biogenic volatile compounds (France-Lanord and Derry 1997). Biogenic volatile compounds are crucial for formation of atmospheric aerosol but do not significantly impact the acidity during weathering and the formation of carbonic acids is initiated by dissolution of CO_2 by water (Drever and Clow 1995):

$$CO_2 + H_2O \rightarrow H_2CO_3 \tag{15.1}$$

$$H_2CO_3 + H^+ \rightarrow HCO_3 \tag{15.2}$$

In reaction 15.1, hydration of carbon dioxide leads to the formation of weak carbonic acids while in reaction 15.2 dissolute partially to generate bicarbonates and acidity. Limestones and dolomites (important carbonate rocks) are significant rocks and play a very crucial role in maintaining flux of weathering at Earth surface due to reason that they are highly soluble in water as compared to other types of rock under similar conditions with their ability to act as a strong toward acidity generated either by natural or anthropogenic sources (Dessert et al. 2003). Natural acids tend to interact with carbonate rocks very instantly leading to neutralization of acidity and generation of basicity by producing bicarbonates and carbonates ions. The reaction depicting the same is (Berner et al. 1983).

In reaction 15.3, carbon dioxide either from atmosphere or through organic matter decomposition interacts with water and calcite resulting into generation of bicarbonates ion and calcium in dissolved form. Considering parameters similar to open system and under present atmospheric level of pCO_2, the value of pH for reaction 15.3 at equilibrium will be around 8.6. Thus, the action of H^+ ions in system with optimum value of pH around 5.6 will lead to reduction of activity by three times displaying strong buffer effect which dissolution of carbonates have on aqueous solution(Arrhenius 1896). Dissolved organic carbon (DIC_S) is the dominating form of bicarbonates ions at pH range of 5.7–10.2 which includes most of the natural water occurring on Earth surface. Ph level of most river on Earth depicts the range of carbonate presence making them widely available and the assurance that water body having carbonates in it will have Ph range from neutral to slightly basic is done by strong buffer effect and has ability to significantly alter the rate of reactions aluminum-silica minerals and dissolution of numerous metals (Tripathi and Rajamani 2007).

$$CO_2 + H_2O + CaCO_{3(\text{calcite})} \rightarrow Ca^{2+} + 2HCO_3 \tag{15.3}$$

Aluminum-silica minerals display more complex weathering process as compared to those displayed by carbonate minerals. An example for the same is (Sharma and Rajamani 2001):

$$\begin{aligned} 2NaAlSi_3O_{8(\text{albite})} + 2CO_2 + 11H_2O \rightarrow\ & Al_2Si_2O_5(OH)_{4(\text{kaol})} \\ & + 4H_4SiO_{4(\text{aq})} + 2Na^+ + 2HCO_3 \end{aligned} \tag{15.4}$$

The above reaction indicates albite and carbonic acid produce kaolinite that is an accessory clay mineral showing depletion in cations. The above reactions also yield silicic acid which is in dissolved form, bicarbonates and sodium ions. This type of reaction is included in incongruent weathering and displays the formation of secondary phase and is crucial for formation of soil. pH, temperature, surface area available for reaction are important factors that determine the rate of aluminum-silica weathering and are reported to be occurring at minimal to neutral pH (Rau et al. 2007). Thus, availability of carbonates in soil might lead to decrease in dissolution rate of silicate minerals by altering the pH to a range where the dissolution is reported to be at the slowest pace. Experiments carried out at laboratories indicate that rate of reaction increases by a magnitude of three with increase in each 10° Celsius for aluminum-silica minerals but Field scale studies depict different values as they consider surface area and other parameters for calculations (Nesbitt and Young 1989; Nesbitt and Markovics 1997).

Old organic carbon (OOC) is stored in sedimentary formations in the form of kerogen and this kerogen is expected to undergo weathering in terms of oxidation reaction. Oxidative weathering reactions are not well understood as compared to reaction occurring during mineral weathering because kerogen is subjected to yield functional groups that are oxidized in nature and later, they act as a site for bacterial activities. A simple schematic for the same is represented below (Mongelli 1993):

$$CH_2O + O_2 \rightarrow CO_2 + H_2O \tag{15.5}$$

The above reaction (15.5) displays that oxidation of OOC present in sedimentary formation leads to the formation of carbon dioxide, although it is ambiguous about the reaction pathway being followed in reaction 15.5. Kerogen weathering is oxidative in nature unlike carbonates and silicates weathering which follow an acid and base reaction pathway. Organic acid production has been also reported along with partial oxidation accompanying degradation of kerogen (Kaotekwar et al. 2014; Ling et al. 2018).

$$\text{R-COOH} \rightarrow \text{RCOO-} + \text{H+} \tag{15.6}$$

Sedimentary formations that are rich in organic content, for instance, coal and shale contain pyrite and thus oxidation of FeS_2 leads to generation of sulfuric acids, which further results into decreasing of pH. Carbonic acid is weaker acid as compared to sulfuric acid and thus sulfuric acid is expected to account for a major fraction of natural acidity even when the component of pyrite available in rock is very less (Khanna et al. 2009).

Jenny (1941) outlined "state factors" – which often pertain to weathering in biological ecosystems and have an impact on formation of soil. Time, climate, topography, and vegetation are some of them. As weathering relies on rate; thus, it is very slow process and can take millions of year to display significant changes in rock on geological time scale (Heiri et al. 2001; Godděris et al. 2008).

The main objectives of this study include:

- Conceptualizing of erosion and weathering along with their types and probable impact on ecosystem of Earth including humans.
- Role of erosion and weathering in driving the climate change and vice versa.

The next section describes different types of weathering mechanism.

15.2 Type of Weathering Mechanism

Weathering as discussed earlier is a normal cycle associated with disintegration of rocks. Rocks are formed when volcanic eruption occurs. These formed rocks at surface of Earth get weathered by different agents like atmosphere, biosphere, hydrosphere, etc. leading to change in mineralogy of rock via dissolution and finally turns to soil. This soil gets eroded and through agents like river finally meets the sea and turns to sediments. This sediment under continuous burial turns to rock again which with time comes in contact with magma and gets melts. This melt again erupts from volcanoes and thus completing cycle (Eggleton et al. 1987; Fedo et al. 1995). Thus, with the help of weathering and erosion, rock cycle takes place at Earth. Also, weathering and erosion are responsible for the formation of several geological features including mountains. So, it becomes crucial to understand different mechanisms that are involved in process of weathering (Babechuk et al. 2014). This section discusses about the same.

15.2.1 Physical Weathering

Physical weathering is also known as mechanical weathering that leads to disintegration of rocks over surface of earth without altering the chemical composition of rocks. There are six ways by which physical weathering can occur and includes sheeting, thermal expansion, abrasion, frost wedging, biological activities/root wedging, and salt crystal growth. Abrasion is defined as a process by which clasts of rocks are broken by collisions of other clasts. Gravity is one of the factors that help in abrasion as clast travel down through mountain cliffs. Abrasion also occurs at seabed where clast at seabed collides with clast present in sea current (Merrill 1921; Ollier 1963).

Similarly, frost wedging occurs at location having optimum temperature for melting and freezing of water. The process initiates with water percolating cracks in rocks at day time. During the night, the same water freezes leading to widening if crack as the volume of solid ice increases by 9% as compared to volume of liquid water at same amount. The very next day when temperature increases the ice melts again changing to water leading to more percolation of water in rock. This simultaneous freezing and melting of water break the rock apart leading to weathering of rock. Biological activities are also important agents that carry out weathering. For instance, Earthworms through soft rocks make several vesicles which with times get larger and lead to breaking of rock. The process is known as bioturbation (Griggs 1936; Goldich 1938). In a similar manner, roots of plants percolate rocks leading to fractures in rock which finally helps in disintegration of rock – at coastlines where seawater having very high salt content percolates the surrounding rocks, after sometimes when water gets evaporated leaving the salt behind. This salt crystallizes with increase of volume leading to disintegration of rock and the process is known as salt crystal growth. Rocks also get weathered by sheeting process also known as exfoliation where upper layer of rock gets eroded away due to pressure release resulting into exfoliation (Blackwelder 1933; Chapman and Greenfield 1949).

15.2.2 Chemical Weathering

Chemical weathering refers to alteration in chemical composition by series of reactions and through several agents like water and acid. The alterations occurring in chemical composition of rock rely heavily on environment and type of minerals available in rocks. For instance, minerals like quartz are negligibly affected by chemical weathering whereas minerals like feldspar are easily weatherable by chemical weathering. Chemical weathering proceeds mainly by dissolution, hydrolysis, oxidation, and hydration reaction (Keller 1955).

Dissolution reaction involves splitting of ion and is reversible reaction if solvent is removed. For instance, a rock containing halite when reacts with acidified water (weak carbonic acid, formed due to interaction between atmospheric and rainwater) results into splitting of halite into Na^+ and Cl^- ion and once the water in evaporated then the ions accumulate again to form halite. Similarly, Ggypsum and anhydrite also follow the same reaction with acidified water. The reactions involved are as follows (Williams 1949; Trendall 1962): 2 CO

$$CO_2 + H_2O \leftrightarrow H_2CO_3 \leftrightarrow HCO_3^- + H^+ \tag{15.7}$$

$$CaCO_3 + H_2CO_3 \leftrightarrow 2HCO_3^- + Ca^{2+} \tag{15.8}$$

Rainwater when combines with atmospheric CO_2 can only form weak carbonic acid while soil through biological activities can produce much higher concentration of carbon dioxide. Thus, water when percolates through soil can be much more acidic as compared to acid rain.

Similar to dissolution, hydrolysis also proceeds with only difference being absence of CO_2 in reaction and formation of new mineral after reaction. When a mineral of rock reacts with water, it forms another mineral along with ions. For instance, potassium feldspar yields kaolinite a secondary mineral clay along with quartz and potassium cation and hydroxyl anion. The reaction involved is (Haldar 2017; Marghany 2022):

$$2KAlSi_3O_8 + 3H_2O \rightarrow Al_2Si_2O_5(OH)_4 + 4SiO_2 + 2K^+ + 2OH \tag{15.9}$$

Similarly, silicate minerals are also anticipated to undergo hydrolysis with different end products. For example, hydrolysis of pyroxene can result into formation of smectite or chlorite and in a same manner olivine can yield serpentine mineral when subjected to hydrolysis weathering. Hydration is another reaction mechanism through which weathering occurs in natural system and is similar to hydrolysis with difference that the reaction does not yield a new product and water is added to chemical structure of mineral (Zhang et al. 2020). A possible outcome of the reaction is that the reacting mineral holds greater volume after the completion of reaction. One such example is transformation of anhydrite into gypsum. In a similar way, oxidation reaction occurs where free oxygen (oxygen which is not combined with any other elements) reacts with minerals. This is an important reaction as it talks about conversion of non-oxygenated Earth to oxygenated one. The reactions involved in oxidation of olivine are as follows (Al-Harthi et al. 1999):

$$Fe_2SiO_4 + 4H_2CO_3 \rightarrow 2Fe^{2+} + 4HCO_3^- + H_4SiO_4 \tag{15.10}$$

$$2Fe^{2+} + \tfrac{1}{2}O_2 + 2H_2O + 4HCO_3^- \rightarrow Fe_2O_3 + 4H_2CO_3 \tag{15.11}$$

15.2.3 Biological Weathering

Biological weathering is a complex group of processes, including actions of a wide range of organisms. Biological weathering describes rock disintegration arising from biochemical and/or biophysical activities. Conventionally, very limited attention has been paid to the role of biological weathering in arid regions. This is due to the perceived absence of plants in those regions. The above statement is supported by an assumption that less complex forms such as algae and lichen would play no significant role in rock weathering and breakdown (Warke 2013).

The activities of organisms initiate and enhance biological weathering. The various processes involved dissolution of rocks, generation of humic acids followed by bioerosion. The changes are mostly caused due to the growth of roots and penetration in organisms of shells, lichens, cyanobacteria, algae in rocks on which they nurture. Cyanobacteria is responsible for the bioerosion of carbonate rocks i.e. limestone and dolomite. Along with cyanobacteria, lichens, and fungi has also significant role in bioerosion. The large areas where these organisms live and thrive during the geological period damage and destroy significant portion of rock producing massive quantity of fine-grained carbonate detritus of limestone sludge (Haldar 2020).

The type of organisms that colonize rock surfaces reflects its "bioreceptivity." Bioreceptivity is dominantly governed by mineralogical characteristics along with rock surface roughness and porosity. The chemistry of the mineralogical constituents of rock has profound implications on rate of colonization of species. Species adapted to alkaline conditions, thrive on limestone, whereas for acidic condition, preferred rock type is sandstone. In arid climatic condition, due to harsh environmental condition, species survive few millimeters beneath the rock surface or in microfractures and cracks primarily for protection from the effects of prolonged desiccation (Hutchens 2009).

Biochemical activities are effective in humid environmental condition rather than arid condition. Biochemical weathering accomplished with the secretion of organic acids during metabolic functions. The metabolic activity of lichens produces oxalic acid, a chelating agent, and reacts with various minerals to form a range of different oxalates, with some such as calcium oxalates being relatively insoluble so that once formed it can play a bioprotective role on rock surfaces.

15.2.4 Nuclear Weathering

The Earth and other objects which are available in space are getting continuously bombarded with different types of radiations from solar rays that consist of solar and galactic waves which are capable of altering the atomic and mineralogical composition of rocks. This phenomenon is known as space weathering and is frequent in planets having no atmosphere as nonavailability of atmosphere makes the planet exposed to solar radiation leading to space weathering but in case of Earth the thick blanket of air prevents penetration of these solar radiations. With continuous emission of carbon dioxide, the atmosphere of Earth has been depleted leading to creation of a hole known as ozone hole. Due to this depletion of ozone hole, solar radiation is penetrating Earth atmosphere and leading to weathering of rocks. This is known as nuclear weathering and the scale of this type of weathering is almost negligible and needs further research to determine the context of the process. In a similar way to solar rays' radioactive material in ground and material that are artificially made by humans who are capable of emitting radiation can also cause this type of weathering to occur and sometimes is referred to as artificial nuclear weathering (Berner 1991; Srivastava et al. 2018).

Climate is also a crucial agent that helps in disintegration of rock and converts it into soil and sediments, the process is termed as weathering. For instance, rocks which are present in equatorial climate tend to get weathered faster as compared to rocks which are available in cold areas. Thus, climate is an integral part that determines the rate of weathering. The next section discusses the same.

15.3 Impact of Climate on Weathering and Erosion

As discussed in previous sections, weathering can occur via physical/mechanical breakdown, chemical breakdown, and biological breakdown. Physical breakdown includes processes like freezing, sheeting, melting, etc. Chemical disintegration includes processes like hydrolysis, oxidation, hydration, and dissolution. Similarly, biological weathering includes roots percolation, etc. These all processes differ with respect to climate, especially the rate of reaction in chemical weathering is temperature dependent (Walker et al. 1981). Thus, chemical weathering is expected to occur at faster rate in regions having high temperature and rainfall which gives the trend that hot and humid regions are expected to show high rate of chemical weathering. In the same manner, physical weathering occurs at very fast rate in cold regions due to the fact that minerals which constitute the rock expand and contract simultaneously due to rapid variation in temperature from day to night. This repetition of freezing and melting temperature makes the rock to get fractured. For instance, both cold and hot deserts display high chances of physical weathering as both regions experience sudden change in temperature in a day (Mora et al. 1996; Pearson and Palmer 2000). Biological weathering occurs in a region having high intensity of flora. Floral life dominates in warm and humid climates. Also, region like tropical rainforest can be seen as a region having biological weathering.

Temperature, precipitation, wind pattern, and solar pattern over a long course of time determines the climate of a region and weathering pattern varies accordingly with climate change. In a general trend, some rocks are more vulnerable to weathering in hot and humid climate while dry climate of a region makes rocks more vulnerable toward mechanical weathering (Edmond 1992; Elderfield and Schultz 1996). For example, limestone present in humid climate gets weathered faster as rain water mixes with carbon dioxide and makes limestone to get dissolve leading into the formation of valleys but sandstone on contrast gets weathered in dry climate as the quartz which is the main constituent of it does not get weathered from chemical weathering but can be disintegrated by physical weathering due to freezing and melting of water in crack leading to fractures. Similarly, rocks rich in olivine can easily get broken in humid regions (Chamberlin 1899; Derry and France-Lanord 1996). As weathering is affected by climate, the vice versa also holds true. Currently, researchers across the globe are shifting their focus toward establishing trend that can help in displaying that how weathering can affect the climate. The next section discusses the same aspect.

Erosion is assumed to be impacting climate in numerous ways and similarly change in climate also leads to variation in rate of erosion. Climate change is expected to impact the erosion in following ways:

- Frequent and high-intensity rain: This leads to large-scale erosion with degrading soil as well as rainwater quality. Areas using rainwater as resources will be impacted most from it. Also, this will lead to wash away of more soil which will be submerged into river or streams leading harmful effect on the water ecosystem.

- Hotter areas: Areas experiencing high temperature will not only display drought-like condition but will also make the soil highly vulnerable to erosion because high temperature leads to decrease in precipitation of the area leading to drying out of soil.
- Larger wildfires: As even more wildfires ravage the environment, soil erosion is predicted to worsen and regions with reduced groundcover and flora are more vulnerable to wind and water erosion.

15.4 Impact of Weathering on Climate

Chemical weathering tends to affect the climate most as it involves utilization of carbon dioxide available in atmosphere. In a general trend, CO_2 from atmosphere mixes with rainwater or water vapor present in atmosphere forming weak carbonic acid. This weak carbonic acid when interacts with carbonate rock leads to dissolution of them and which further gets eroded and leads to transportation of these sediments into oceans. Once these sediments reach sea, they get trapped for long time on geological time scale (Taylor and McLennan 1985; Bluth and Kump 1994). Due to this trapping the carbon dioxide available in atmosphere is now stored in ocean for a long time. This entrapment can lead to decrease in temperature of the region. According to a study conducted by Laboratoire Géosciences Environnement Toulouse, in collaboration with the Laboratoire des Sciences du Climat et de l'Environnement and the University of Bergen (Norway), it has been reported that oceans and river can store up to 0.3 billion tons of CO_2 from atmosphere, although the amount is negligible as compared to emission made by humans which makes up at around 8 billion tons a year. Thus, it can be concluded that regions having availability of carbonate rocks in abundance and having humid climate can entrap CO_2 from atmosphere and can decrease the temperature of region at a scale relatively negligible (Roy 2006; Jin et al. 2017; Parmar and Sharma 2020).

15.5 Conclusion

Weathering is an important biogeochemical process that is both influenced by and influences climate over the course of Earth history. Weathering processes control the flux of solutes and many nutrients to the oceans and the marine and terrestrial biospheres, and the transfer of carbon from the ocean–atmosphere system to sedimentary rocks. Weathering rates are dependent on climate, among several factors, and vary widely across the Earth's surface. In turn, weathering processes can alter the atmospheric concentrations of the important greenhouse gas, carbon dioxide. Weathering processes thus have the potential to act as climate feedback. Weathering and erosion are important phenomena that help in sustaining life at Earth. Weathering refers to disintegration of rocks via physical/mechanical, chemical, and biological pathways while erosion is the transportation of these sediments or broken rocks from their parent rock to finally at sea. This transportation of sediments also helps in movement of nutrients and minerals, which helps in maintaining life system in ocean environment. Physical weathering refers to breakdown of rock without altering their chemical composition. Some processes involved in physical weathering are salt crystal growth, freezing and thawing, sheeting, etc. A special type of physical weathering known as biological weathering helps down in breaking down rocks via insertion of roots of trees or through living organisms like Earthworms. Chemical weathering refers to weathering of rocks via change in their chemical composition and proceeds mainly via hydration, dissolution, oxidation, and hydrolysis reaction. Chemical and mechanical weathering work hand-in-hand using surface-area-to-volume ratio concept. Chemical weathering attacks the rock surfaces as water and reactants cannot penetrate solid rock. However, mechanical weathering penetrates bedrock, converting large pieces into smaller ones and thereby creating new rock surfaces. The formation of carbonic acid plays the main role in chemical weathering. Not only it helps in chemical weathering but also assists in hydrolysis and dissolution. Hydrolysis breaks down silicate rock and creates clay minerals. Another type of weathering mechanism is nuclear weathering which is insignificant on Earth due to the presence of atmosphere. Although researches are being carried out in order to establish the scale of weathering caused by radioactive material present in Earth naturally or from man-made material. Weathering is a phenomenon, i.e. both impact and get impacted by climate. Hot and humid regions are expected to display more chemical weathering while dry climates display physical weathering mechanism. Tropical rainforest areas show biological weathering. Also, in a same manner as climate affects weathering, the vice versa holds true too. Carbonate abundance regions with hot and humid climate can help in decreasing the temperature of region.

References

al Sabti, B., Samayamanthula, D.R., Dashti, F.M., and Sabarathinam, C. (2023). Fluoride in groundwater. In: *Hydrogeochemistry of Aquatic Ecosystems*, 1–31. Wiley https://doi.org/10.1002/9781119870562.ch1.

Al-Harthi, A.A., Al-Amri, R.M., and Shehata, W.M. (1999). The porosity and engineering properties of vesicular basalt in Saudi Arabia. *Engineering Geology* 54 (3–4): 313–320. https://doi.org/10.1016/S0013-7952(99)00050-2.

Arrhenius, S. (1896). On the influence of carbonic acid in the air on the temperature on the ground. *Philos. Mag. J. Sci.* 5: 237–276.

Babechuk, M.G., Widdowson, M., and Kamber, B.S. (2014). Quantifying chemical weathering intensity and trace element release from two contrasting basalt profiles, Deccan Traps, India. *Chemical Geology* 363: https://doi.org/10.1016/j.chemgeo.2013.10.027.

Berner, R.A. (1991). Atmospheric CO_2 levels over Phanerozoic time. *Science* 249: https://doi.org/10.1126/science.249.4975.1382.

Berner, R.A., Lasaga, A.C., and Garrels, R.M. (1983). The carbonate-silicate geochemical cycle and its effect on atmospheric carbon dioxide over the past 100 million years. *American Journal of Science* 283: 641–683.

Blackwelder, E. (1933). The insolation hypothesis of rock weathering. *American Journal of Science* s5-26: 97–113. https://doi.org/10.2475/ajs.s5-26.152.97.

Bluth, G.J.S. and Kump, L.R. (1994). Lithologic and climatologic controls of river chemistry. *Geochimica et Cosmochimica Acta* 58: https://doi.org/10.1016/0016-7037(94)90015-9.

Chamberlin, T.C. (1899). An attempt to frame a working hypothesis of the cause of glacial periods on an atmospheric basis. *Journal of Geology* 7: 545–584.

Chapman, R.W. and Greenfield, M.A. (1949). Spheroidal weathering of igneous rocks. *American Journal of Science* 247: https://doi.org/10.2475/ajs.247.6.407.

Derry, L.A. and France-Lanord, C. (1996). Neogene Himalayan weathering history and river 87Sr/86Sr: impact on the marine Sr record. *Earth and Planetary Science Letters* 142: https://doi.org/10.1016/0012-821X(96)00091-X.

Dessert, C., Dupre, B., Gaillardet, J. et al. (2003). Basalt weathering laws and the impact of basalt weathering on the global carbon cycle. *Chemical Geology* 202: https://doi.org/10.1016/j.chemgeo.2002.10.001.

Drever, J.I. and Clow, D.W. (1995). Weathering rates in catchments. In: *Chemical Weathering Rates of Silicate Minerals (Reviews in Mineralogy, Vol. 31)* (ed. A.F. White and S.L. Brantley), 1594–1597. Mineralogical Society of America.

Edmond, J.M. (1992). Himalayan tectonics, weathering processes, and the strontium isotope record in marine limestones. *Science* 258: https://doi.org/10.1126/science.258.5088.1594.

Eggleton, R.A., Foudoulis, C., and Varkevisser, D. (1987). Weathering of basalt: changes in rock chemistry and mineralogy. *Clays and Clay Minerals* 35: https://doi.org/10.1346/CCMN.1987.0350301.

Elderfield, H. and Schultz, A. (1996). Mid-ocean ridge hydrothermal fluxes and the chemical composition of the ocean. *Annual Review of Earth and Planetary Sciences* 24: https://doi.org/10.1146/annurev.earth.24.1.191.

Eppes, M.C. and Keanini, R. (2017). Mechanical weathering and rock erosion by climate-dependent subcritical cracking. *Reviews of Geophysics* 55 (2): 470–508. https://doi.org/10.1002/2017RG000557.

Fedo, C.M., Wayne Nesbitt, H., and Young, G.M. (1995). Unravelling the effects of potassium metasomatism in sedimentary rocks and palaeosols, with implications for paleo weathering conditions and provenance. *Geology* 23: 921–924. .

France-Lanord, C. and Derry, L.A. (1997). Organic carbon burial forcing of the carbon cycle from Himalayan erosion. *Nature* 390: https://doi.org/10.1038/36324.

Galy, A. and France-Lanord, C. (1999). Weathering processes in the Ganges-Brahmaputra basin and the riverine alkalinity budget. *Chemical Geology* 159: https://doi.org/10.1016/S0009-2541(99)00033-9.

Goddéris, Y., Donnadieu, Y., Tombozafy, M., and Dessert, C. (2008). Shield effect on continental weathering: implication for climatic evolution of the Earth at the geological timescale. *Geoderma* 145: https://doi.org/10.1016/j.geoderma.2008.01.020.

Goldich, S.S. (1938). A study of rock weathering. *Journal of Geology* 46: https://doi.org/10.1086/624619.

Griggs, D.T. (1936). The factor of fatigue in rock exfoliation. *Journal of Geology* 44: 783–786.

Haldar, S.K. (2017). Geology and geochemistry. In: *Platinum-Nickel-Chromium Deposits*, 37–61. https://doi.org/10.1016/B978-0-12-802041-8.00002-X.

Haldar, S.K. (2020). Sedimentary rocks. In: *Introduction to Mineralogy and Petrology*, 187–268. Elsevier https://doi.org/10.1016/B978-0-12-820585-3.00006-5.

Heiri, O., Lotter, F.A., and Lemcke, G. (2001). Loss on ignition as a method for estimating organic and carbonate content in sediments: reproducibility and comparability of results. *Journal of Paleolimnology* 25: https://doi.org/10.1023/A:1008119611481.

Hutchens, E. (2009). Microbial selectivity on mineral surfaces: possible implications for weathering processes. *Fungal Biology Reviews* 23 (4): 115–121. https://doi.org/10.1016/j.fbr.2009.10.002.

Jacobson, A.D. and Blum, J.D. (2003). Relationship between mechanical erosion and atmospheric CO_2 consumption in the New Zealand Southern Alps. *Geology* 31: https://doi.org/10.1130/G19662.1.

Jenny, H. (1941). *Factors of Soil Formation*. McGraw-Hill.

Jin, L., Ma, L., Dere, A. et al. (2017). REE mobility and fractionation during shale weathering along a climate gradient. *Chemical Geology* 466: https://doi.org/10.1016/j.chemgeo.2017.06.024.

Kaotekwar, A.B., Meshram, R.R., Sathyanarayanan, M. et al. (2014). Structures, petrography and geochemistry of Deccan basalts at Anantagiri hills, Andhra Pradesh. *Journal Geological Society of India* 84: https://doi.org/10.1007/s12594-014-0178-z.

Keller, W.D. (1955). *Principles of Chemical Weathering*. Lucas Bros.

Khanna, P.P., Saini, N.K., Mukherjee, P.K., and Purohit, K.K. (2009). An appraisal of ICP-MS technique for determination of REEs: long term QC assessment of silicate rock analysis. *Himalayan Geology* 30: 2361–2386.

Lasaga, A.C., Soler, J.M., Ganor, J. et al. (1994). Chemical weathering rate laws and global geochemical cycles. *Geochimica et Cosmochimica Acta* 58: https://doi.org/10.1016/0016-7037(94)90016-7.

Ling, S., Wu, X., Zhao, S., and Liao, X. (2018). Evolution of porosity and clay mineralogy associated with chemical weathering of black shale: a case study of Lower Cambrian black shale in Chongqing, China. *Journal of Geochemical Exploration* 188: https://doi.org/10.1016/j.gexplo.2018.02.002.

Marghany, M. (2022). An introduction to minerals, rocks, and mineral deposits. *Advanced Algorithms for Mineral and Hydrocarbon Exploration Using Synthetic Aperture Radar* 1–30. https://doi.org/10.1016/B978-0-12-821796-2.00014-8.

Merrill, G.P. (1921). *Rocks*. Macmillan Co: Rock-weathering and Soils.

Mongelli, G. (1993). REE and other trace elements in a granitic weathering profile from "Serre", southern Italy. *Chemical Geology* 103: https://doi.org/10.1016/0009-2541(93)90288-T.

Mora, C.I., Driese, S.G., and Colarusso, L.A. (1996). Middle to Late Paleozoic atmospheric CO_2 levels from soil carbonate and organic matter. *Science* 271: https://doi.org/10.1126/science.271.5252.1105.

Nesbitt, H.W. and Markovics, G. (1997). Weathering of granodioritic crust, long-term storage of elements in weathering profiles, and petrogenesis of siliciclastic sediments. *Geochimica et Cosmochimica Acta* 61: https://doi.org/10.1016/S0016-7037(97)00031-8.

Nesbitt, H.W. and Young, G.M. (1989). Formation and diagenesis of weathering profiles. *Journal of Geology* 97: https://doi.org/10.1086/629290.

Ollier, C.D. (1963). Insolation weathering: examples from central Australia. *American Journal of Science* 261: https://doi.org/10.2475/ajs.261.4.376.

Parmar, S. and Sharma, S.K. (2020). Estimation of soil loss and soil erodibility for different crops, nutrient managements and soil series. *Indian Jour. Pure App. Biosci.* 8: https://doi.org/10.18782/2582-2845.7984.

Pearson, P.N. and Palmer, M.R. (2000). Atmospheric carbon dioxide concentrations over the past 60 million years. *Nature* 406: https://doi.org/10.1038/35021000.

Petch, S.T., Berner, R.A., and Eglinton, T.I. (2000). A field study of the chemical weathering of ancient sedimentary organic matter. *Organic Geochemistry* 31: https://doi.org/10.1016/S0146-6380(00)00014-0.

Rau, G.H., Knauss, K.G., Langer, W.H., and Caldeira, K. (2007). Reducing energy-related CO_2 emissions using accelerated weathering of limestone. *Energy* 32: https://doi.org/10.1016/j.energy.2006.10.011.

Roy, S. (2006). Sedimentary manganese metallogenesis in response to the evolution of the Earth system. *Earth-Science Reviews* 77: https://doi.org/10.1016/j.earscirev.2006.03.004.

Sarin, M.M., Krishnaswami, S., Dilli, K. et al. (1989). Major ion chemistry of the Ganga-Brahmaputra river system: weathering processes and fluxes to the Bay of Bengal. *Geochimica et Cosmochimica Acta* 53: https://doi.org/10.1016/0016-7037(89)90205-6.

Sharma, A. and Rajamani, V. (2001). Weathering of charnockites and sediment production in the catchment area of the Cauvery River, southern India. *Sedimentary Geology* 143: https://doi.org/10.1016/S0037-0738(01)00102-6.

Srivastava, P., Siddaiah, N.S., Sangode, S.J., and Meshram, D.C. (2018). Mineralogy and geochemistry of various colored boles from the Deccan volcanic province: implications for paleoweathering and paleoenvironmental conditions. *Catena* 167: https://doi.org/10.1016/j.catena.2018.04.024.

Stallard, R.F. and Edmond, J.M. (1983). Geochemistry of the Amazon: 2. The influence of geology and weathering environment on the dissolved load. *Journal of Geophysical Research* 88: https://doi.org/10.1029/JC088iC14p09671.

Taylor, S.R. and McLennan, S.M. (1985). *The Continental Crust: Its Composition and Evolution.* Blackwell Scientific Publications.

Trendall, A.F. (1962). The formation of 'apparent peneplains' by a process of combined lateritisation and surface wash. *Zeitschrift für Geomorphologie* 6: 1228–1232.

Tripathi, J.K. and Rajamani, V. (2007). Geochemistry and origin of ferruginous nodules in weathered granodioritic gneisses, Mysore Plateau, Southern India. *Geochimica et Cosmochimica Acta* 71: https://doi.org/10.1016/j.gca.2007.01.001.

Walker, J.C.G., Hays, P.B., and Kasting, J.F. (1981). A negative feed back mechanism for the long-term stabilization of Earth's surface temperature. *Journal of Geophysical Research* 86: https://doi.org/10.1029/JC086iC10p09776.

Warke, P.A. (2013). 4.12 Weathering in arid regions. In: *Treatise on Geomorphology*, 197–227. Elsevier https://doi.org/10.1016/B978-0-12-374739-6.00060-9.

White, A.F. and Blum, A.E. (1995). Effects of climate on chemical weathering in watersheds. *Geochim.t Cosmochim. Acta* 59: https://doi.org/10.1016/0016-7037(95)00078-E.

Williams, J.E. (1949). Chemical weathering at low temperatures. *Geographical Review* 39: https://doi.org/10.2307/211162.

Zhang, G., Konietzky, H., Song, Z., and Zhang, M. (2020). Study of Cerchar abrasive parameters and their relations to intrinsic properties of rocks for construction. *Construction and Building Materials* 244: 118327. https://doi.org/10.1016/j.conbuildmat.2020.118327.

Index

Weathering and Erosion Processes in the Natural Environment, First Edition. Edited by Virendra Bahadur Singh, Sughosh Madhav, Naresh Chandra Pant, and Ravi Shekhar.

h

i

n

o

y

z